HUMAN DETECTION AND DIAGNOSIS OF SYSTEM FAILURES

NATO CONFERENCE SERIES

I Ecology
II Systems Science
III Human Factors
IV Marine Sciences
V Air–Sea Interactions
VI Materials Science

III HUMAN FACTORS

Volume 8 Mental Workload: Its Theory and Measurement
Edited by Neville Moray

Volume 9 Human Evoked Potentials: Applications and Problems
Edited by Dietrich Lehmann and Enoch Callaway

Volume 10 Human Consequences of Crowding
Edited by Mehmet R. Gürkaynak and W. Ayhan LeCompte

Volume 11 The Analysis of Social Skill
Edited by W. T. Singleton, P. Spurgeon, and R. B. Stammers

Volume 12 Coping and Health
Edited by Seymour Levine and Holger Ursin

Volume 13 Processing of Visible Language 2
Edited by Paul A. Kolers, Merald E. Wrolstad, and Herman Bouma

Volume 14 Intelligence and Learning
Edited by Morton P. Friedman, J. P. Das, and Neil O'Connor

Volume 15 Human Detection and Diagnosis of System Failures
Edited by Jens Rasmussen and William B. Rouse

HUMAN DETECTION AND DIAGNOSIS OF SYSTEM FAILURES

Edited by
Jens Rasmussen
Risø National Laboratory
Roskilde, Denmark

and
William B. Rouse
University of Illinois
Urbana, Illinois

Published in cooperation with NATO Scientific Affairs Division

PLENUM PRESS · NEW YORK AND LONDON

Library of Congress Cataloging in Publication Data

NATO Symposium on Human Detection and Diagnosis of System Failures (1980: Roskilde, Denmark) Human detection and diagnosis of system failures.

(NATO conference series. III, Human factors; v. 15)
Bibliography: p.
Includes index.
1. System failures (Engineering)—Congresses. 2. Man—machine systems—Congresses. 3. Psychology, Industrial—Congresses. I. Rouse, William B. II. Rasmussen, Jens, 1926- . III. North Atlantic Treaty Organization. Division of Scientific Affairs. IV. Title. V. Series.
TA169.5.N37 1980 620.7'2 81-8699
ISBN 0-306-40744-2 AACR2

Proceedings of a NATO Symposium on Human Detection and Diagnosis of System Failures, held August 4–8, 1980, in Roskilde, Denmark

A Division of Plenum Publishing Corporation
233 Spring Street, New York, N.Y. 10013

Printed in the United States of America

PREFACE

This book includes all of the papers presented at the NATO Symposium on Human Detection and Diagnosis of System Failures held at Roskilde, Denmark on August 4-8, 1980. The Symposium was sponsored by the Scientific Affairs Division of NATO and the Risø National Laboratory of Denmark.

The goal of the Symposium was to continue the tradition initiated by the NATO Symposium on Monitoring Behavior and Supervisory Control held in Berchtesgaden, F.R. Germany in 1976 and the NATO Symposium on Theory and Measurement of Mental Workload held in Mati, Greece in 1977. To this end, a group of 85 psychologists and engineers coming from industry, government, and academia convened to discuss, and to generate a "state-of-the-art" consensus of the problems and solutions associated with the human's ability to cope with the increasing scale of consequences of failures within complex technical systems. The Introduction of this volume reviews their findings.

The Symposium was organized to include brief formal presentations of papers sent to participants about two months in advance of the meeting, and considerable discussion both during plenary sessions and within more specialized workshops. Summaries of the discussions and workshop reports appear in this volume.

We are indebted to many individuals for any success achieved by the Symposium and this book. First of all, we gratefully acknowledge the authors and participants who agreed to follow our guidelines regarding reading the preprints before the meeting, making very brief formal presentations, and, most of all, meeting the various deadlines necessary for this state-of-the-art volume to be quickly published. The Symposium Secretary, L.P. Goodstein, deserves considerable gratitude for orchestrating a multitude of administrative and logistical functions necessary to holding the Symposium and producing this volume. The efforts of the Session Chairmen, Session Secretaries, and Workshop Chairmen were also of prime importance. Among the staff members of Risø National

Laboratory, Bitten Svendsen, Anne Marie Eichen and Lene Ekelund were of invaluable assistance in retyping the manuscripts for the book, and together with Bodil Aarup and Lene Wittrock served admirably in a host of secretarial and hospitality-oriented functions during the conference. Furthermore we very much appreciate the topical indexing of this volume by S.H. Rouse. Finally, of course, the support of NATO and Risø National Laboratory must be acknowledged as essential to our efforts.

Jens Rasmussen
William B. Rouse

CONTENTS

INTRODUCTION

As transportation systems, production processes, power plants, and communications networks have become more integrated and centralized, they have become increasingly complex. In an effort to cope with complexity, computers have been introduced for performing many monitoring and control functions. From the perspective of normal operations, this trend of integration, centralization, and automation appears highly desirable. However, in abnormal and emergency situations, the increasing complexity and the automation designed for dealing with this complexity can present considerable difficulty.

One of the more obvious difficulties is the scale of the consequence of system failures. Large scale complex systems tend to produce far-reaching and often unpredictable effects as the result of a failure. Typically, system designers handle this problem by introducing automatic protective functions while keeping the human in the system for monitoring and supervisory control - also in emergency situations where the human is expected to act as a backup in the event of unforeseen failures. Unfortunately, this approach is not always successful because increased complexity tends to cause humans to produce "human errors", not only during emergency control but also when monitoring the automatic system. Also, errors are not infrequently caused by inadequate system design.

As might be expected, an increasingly popular idea is to design systems that preclude human errors either by eliminating the human with more automation or, by designing computers to monitor the human. While this approach is likely to achieve some success, it is fundamentally limited by the fact that all systems must ultimately rely on humans. Increased automation simply shifts human responsibilities from active control to an intermittent intervening in terms of detection, diagnosis, and correction of system failures. Future automation may possibly result in humans mainly serving as system managers and maintenance personnel. Nevertheless, humans will remain very much involved, coping with the complexity which the computers fail to handle. While such a future may sound exciting, it also leads one to wonder what type of support will be necessary to provide humans with the abilities to successfully fill such roles.

The goal of this book, and of the symposium on which it is based, is to provide a state-of-the-art review of at least partial answers to this question. This includes both theoretical and practical perspectives as well as an assessment of trends and future issues relating to human detection and diagnosis of system failures.

A SET OF ISSUES

In this introduction, we will briefly summarize a wide range of issues that are discussed in the papers and that also arose throughout formal and informal discussions of these papers. However, we will make no attempt to relate these issues to individual papers. The reader should consult the overview papers written by each session chairman for a more detailed linking of particular issues to specific authors.

A continuing topic of discussion was design criteria. Some individuals stressed availability in terms of the probability that the system is available for use. Others, particularly those in the aircraft and maritime industries, were also concerned with efficiency because energy costs have risen so drastically. As might be expected, safety was also an important criterion. However, except for those advocating a simple system reliability metric, a definition of safety was somewhat elusive.

A few participants were keenly aware of the distinction between operability and maintainability. It was suggested that many of the aids envisioned for the operator would result in a very substantial increase in the complexity of the maintainer's tasks which would in turn have important impacts on availability if not safety. Finally, very little of the discussion of design criteria focused on the human's overall role in the system. While this may reflect the nature of this particular conference rather than the general state of the art, it nevertheless needs more attention as our later discussions will indicate.

Considerable attention was devoted to various aspects of modelling. Discussions of models of the machine or process (i.e., the non-human portion of the overall system) focused on our relatively poor abilities to model processes during abnormal situations. This is especially true when multiple abnormal events occur and when these events are both rare and unforeseen in the design of the process. Because consideration of multiple events leads to a combinatorial explosion of alternative abnormal situations, the modelling problem could appear insurmountable if correctness and completeness are required. However, this difficulty can be considerably lessened by employing high levels of representation of system function (e.g., mass and energy flows) and recognizing the fact that the operator often does not need as much information about abnormalities since the goal is to eliminate rather than fine-tune such situations. Thus, for example, a model of the dynamic nature of abnormality propagation may not be necessary. A non-dynamic model may be quite sufficient for diagnostic purposes.

Also of interest within the issue of process models was the distinction between hardware and software. Models of software complexity and verification procedures for software are beginning to emerge. However, it is quite reasonable to conjecture that the problem of diagnosing software failures will become increasingly important and a topic of many future discussions.

Another model-oriented issue concerns behavioral models. Most participants appeared to agree that we should not expect to develop a single quantitative model of human behaviour in detection and diagnosis tasks. Instead, we need qualitative frameworks which categorize human behaviour and systems in a manner that serves to relate various specific quantitative models which are useful within particular categories. From this perspective, methods such task analysis can be useful for developing frameworks and delineating the boundaries of the task categories. Within this volume, several qualitative frameworks are discussed but, as yet, no single framework has been validated to the extent of achieving the status of a basic paradigm.

A great deal of discussion was devoted to the human's internal or mental model. While we have finally reached the point that this rather obvious notion is well accepted, researchers' models of internal models range from sets of differential equations to functional block diagrams to snapshots of physical form. The current problem is not really one of finding the "correct" internal model, but instead trying to find where each alternative applies and is useful, while also convincing the research community that such ambiguity is completely tenable.

As might be expected in a discussion of problem solving in fault diagnosis tasks, search strategies received some attention. The usefulness of different strategies has clear implications for display design and training. Individual differences in terms of level of experience, for example, are also important here. It was perhaps surprising that search strategies did not emerge as one of the central issues of the discussion. However, it is possible that lingering remnants of the idea of complete proceduralization (see discussion on training) were sufficient to cloud this issue.

In terms of aiding the operator, there was clear agreement that many conventional alarm systems leave much to be desired with the operator often becoming innundated with hundreds if not thousands of alarms when a high-consequence abnormality or emergency occurs. It is somewhat paradoxical that individual alarm requirements are accepted unconditionally from component and subsystem suppliers while one later invests considerable effort into trying to eliminate many alarms - at no small cost and with some uneasiness about the resulting validity of the overall alarm system. Methods of alarm analysis such as fault trees and cause-consequence diagrams may prove useful for sorting out this plethora of signals, although the value of these methods for unforeseen events is open to debate and, surprisingly, is not often discussed. One approach to dealing with the unforeseen event problem may be to adapt the level of presentation to be compatible with the human's task. Thus, unforeseen events may be dealt with most effectively by providing the operator with a high level transformation of the problem that lets the operator attempt to solve the problem rather than trying to solve it for him.

It was, of course, inevitable that discussion should eventually focus on human-computer interaction. The ubiquitous notion of "transparency" again emerged with the perhaps surprising result that many participants questioned the value of a transparent system in failure situations. In fact, from a perspective of dealing with failures, the group was not even able to agree on a definition of transparency. Discussions of human-computer interaction also dealt with the idea of embedding models of human behaviour within the computer. Finally, some discussion was devoted to the problem of convincing the operator to accept the suggestions produced by computer-based decision aiding systems.

As one of the central themes of the conference, training was an important issue. While a continued lowering of the basic abilities of trainees in many domains has led to many advocates of complete proceduralization, it was generally agreed that proceduralization could not be the total answer - unforeseen events have to be handled by somebody. However, there were many skeptics with regard to the practicality of more knowledge-based training. This is reflected in the fact that the notion of the human's internal model was seldom mentioned in conjunction with training. (Actually, proceduralization eliminates the need for an internal model!). While a few of the papers in this volume are at least partially directed at this issue, there is a real danger that continued proceduralization will make jobs increasingly boring and lead to poor retention as well as difficulties in finding job entrants with reasonable levels of ability.

Training simulators were also discussed at length. Several issues surrounding fidelity were considered. However, surprisingly little consensus was reached. Seldom is one type of simulator compared to another in terms of transfer to real equipment. Instead, most effort is devoted to development and demonstration of particular simulator concepts. Perhaps this state of affairs parallels that of the research community's inability to resolve the issues surrounding the need for motion in aircraft simulators, or the emergence of full-scale, physically identical control room simulators that avoid the whole issue of fidelity. A related issue concerns the problem of specifying clear and unambiguous performance criteria for trainees although, obviously, this problem is not unique to the theme of this conference.

At various points throughout the discussions, organizational considerations were emphasized. The social structures of problem solving teams were stressed and the relatively mediocre problem solving performance of committees noted, once with a plea that team training methods be developed. Several participants noted that the accountability of the individual or group involved in problem solving can have a tremendous impact on performance. It was further emphasized that the organizational structure and procedures can significantly affect motivation and consequently performance.

Another recurring theme was human error. A distinction was made between inadvertent "mental lapses" or "slips of the mind" and intentional erroneous acts based on misconceptions. The group was continually amazed how many control panels seem to have been designed in complete ignorance of basic human factors principles. Beyond that realization, however, definitions and classification of human error received only limited attention. Although, one particularly interesting notion did emerge. Namely, the human should be trained so that he can recognize when the situation exceeds his abilities such that the probability of human error is greatly increased unless assistance is sought. It is worth noting that any computer-based aiding systems should also be able to recognize when such a situation occurs.

The concept of complexity was repeatedly mentioned. Two papers discuss specific measures of complexity of fault diagnosis tasks and other papers refer to the implications of complexity. These implications include not only those faced by the operator but also those faced by the researcher who is trying to understand and aid the operator. The choice of models to fit in the various categories of the previously mentioned qualitative frameworks will, to an extent, be dictated by the impact of complexity.

Several methodological controversies also managed to emerge. These mainly took the form of at least two behavioral scientists complaining that engineers and their mathematical tools were not relevant to understanding human behaviour. While these polemics may have been partially based on a mixture of misunderstanding due to a lack of communication and misrepresentation of some engineering-oriented research, the central issue is nevertheless valid. Succinctly, the two sides of the issue appear to be reasonably stated by noting that engineers are so pragmatic that they sometimes risk solving the wrong problem while behavioral scientists are purists to the extent that they do not solve any problems at all. Clearly, a hybrid of these two approaches is needed. A prerequisite to such a development is a continuation of the dialogue and healthy debate that emerges in these papers and continued in the discussions at the conference.

THE STATE OF THE ART

In view of the above set of issues, what can be said about the state of the art? It is certainly quite safe to conclude that understanding and aiding the human to cope with system failures will continue to increase in importance. Further, there appears to be agreement that detection and diagnosis of system failures is not just another task like scanning or tracking - instead, a wide range of functions on a variety of cognitive levels are needed. Thus, not only are systems becoming increasingly complex; the tasks of humans are correspondingly increasing in complexity and our abilities to understand those tasks are being challenged.

To deal with these two types of complexity, there is a great need to develop widely accepted frameworks or qualitative models within which problems can be viewed, researched, and resolved. It appears that such models cannot be mere extensions of more traditional theories and approaches. Instead a conceptual leap is necessary to allow us to view systems and humans from a somewhat broader perspective. Several of the papers in this volume appear to offer the seeds of such a transition. Others papers provide methods, models, and theories that will inevitably fit within a new framework. And, some papers review indefatigable efforts to keep the old framework. However, as these traditional approaches whittle away at the problem, what remains is increasingly complex. A transformation of the problem is fundamentally necessary if we are to avoid the consequences of a piece-meal solution. Thus, it is certainly a time of transition for both the human operator and the research community.

Risø, August 1980

Jens Rasmussen
William B. Rouse

REAL LIFE PERSPECTIVES

CHAIRMAN: D. L. PARKS

SECRETARY: A. WINTER

REAL LIFE PERSPECTIVES

Donald L. Parks

Boeing Commercial Airplane Company
Seattle, Washington 98124
U.S.A.

CHAIRMAN'S REVIEW

This introductory section demonstrates that techniques to prevent, detect and diagnose system failures are needed and used throughout the system life cycle, from early design through field trouble shooting. Papers emphasize the breadth and depth of diagnostic methods that are used in system development, operation and maintenance for human detection and diagnosis of system failures. It is easy to infer from these papers that almost any complex system one might imagine is a candidate user of such methodology, as will be shown by the progression of applications from nuclear power plants to accident appraisals, to software development, and on day-to-day operation and maintenance.

All papers in this applied section show use of a modeling concept or some relatively formal analytic method in solving practical problems in complex systems, in order to collate, organize and interpret the widely varied information and data on operations needs versus human detection and diagnostic capabilities. It will become clear that some such organizing technique is essential for such complex systems if one is, first, to select relevant questions from a universe of candidates, and then to set appropriate study priorities within the constraints of time and budget. While the modeling approach still varies somewhat from person to person, the approaches are becoming more consistent and the problem solving potential of such approaches is becoming more apparent, and in many respects correlate with earlier overviews (e.g. Parks 1979, and "Final Report" 1979).

Discussion subjects for this section are highlighted in Figure 1 and summarized briefly in the following introductory remarks.

REAL-LIFE PERSPECTIVES

Methods, Models, and Diagnostic Techniques

Author	Discussion vehicle	Subject area	Tools/information presented
Sheridan	Nuclear power	Design and operability	Methods Control models Workload
Thompson	Caution warning systems	Operations questions and perceptual-motor criteria	Functional methodology Task analysis Logic structures Perceptual constraints in operation
Gardenier	Ship operations	Accident factors, analysis, and prevention	Scenario Task analysis Fault tree analysis Modeling alternatives/ indulgences
Bond	Computer maintenance	Online competitive maintenance operations	User orientation and profile Utility of higher structural models and methods
Brooke	Computer programming	Software development, formulation, and debugging	Methodologies Sources of error Diagnostic techniques
Christensen	Operational readiness	Maintenance impact and improvement methods	Observed constraints Improvement areas and concepts Trouble shooting strategies Checklists

Fig. 1

The first paper by Sheridan provides a profile of a nuclear power plant operation and the operator role. This work re-emphasizes the impact on diagnostic behavior from limitations in the man-machine interface and presents a series of thoughts for improving man-machine interface technology. For example, it is noteworthy that, while there are standard and practiced accident routines in nuclear operations, there is a heavy emphasis on adherence to prescribed procedures in all control room activities. Accordingly, the paper presents the question whether the system effectively prepares operators for unrehearsed events. Several attributes of diagnostic behavior and error potential are summarized, and a model is offered for use in problem solving for this type of operation.

Next, Thompson presents a discussion of two of the most rapidly progressing display technology areas of today, i.e., the electronic display technology and advanced caution-warning system technological developments. Thus, while there are now over 20 years experience in experimental and military flight use of Cathode Ray Tube (CRT) integrated symbolic displays, committment for commercial airplane uses are recent. We might add that color CRT displays also have only recently been committed for these purposes, on the B-767, B-757 and A-310 aircraft. The new airplane programs will develop and use more current, correlated state-of--the-art solutions; at the more basic level, a major research program is now under way to develop standards for future electronic caution-warning systems. These programs will answer the many practical questions that must be resolved in order to assure suitable operability for late 1981 and early 1982. Of course, refinements can be expected to continue for many years, as occurred with the electro-mechanical displays. In passing, we might comment that this area is presently progressing so fast that state-of-the-art information on proven capabilities can become dated or obsolete in a year or two.

Gardenier's paper demonstrates sea-going variations on three familiar modeling aids for analysis of ship operations - an accident scenario, one type of task analysis and one type of fault tree analysis. These techniques have been useful in many other settings, and demonstrate the breadth of their utility once more in this paper. This paper presents a number of pertinent philosophical questions, such as how far to carry development of man-machine interface provisions for a system that "normally" is adequate, and it also presents a question regarding the merits versus the ease of model proliferation for any given question. Indirectly, the author presents us with a challenge to make methodology more efficient and to better control model proliferation. The author points out that, as an alternative and in view of the typically expanding demands of so many research approaches (and associated growth in costs), many system developers would prefer to move directly to known and "reliable" methods. Sometimes this is because they do not perceive the magnitude of the problem and sometimes it is to minimize or avoid repetition of a bad experience.

Bond's paper highlights the merits of standardized methods, in its description of development and use of a more efficient diagnostic method for maintenance. Computer trouble shooting and maintenance has become a highly competitive industry, which in turn has led to improved and very efficient techniques and procedures for fault detection and diagnosis. Basically, this paper demonstrates that proceduralized diagnostic techniques to organize and resolve issues can be quite helpful in routinely performing complicated problem solving activities - at least for known, though complicated, systems. The process described shows the merits of emphasizing and applying so far as possible a structured approach to problem solving, even to incorporating a system of priorities and criteria for seeking higher level advice. One cannot help but ask the question whether such methodology or problem solving techniques might have a more generalized application, to become an aid to solving more complex problems. After all, information from Rouse (personal discussions) suggests that problem solving technique is a skill and may be trainable. Perhaps greater emphasis in training on problem solving technique and priority schemes is advisable in order to improve and expedite solutions.

In the next presentation Brooke presents a contrast to Bond's success story - in an overview of software development tools for debugging software programs from "error" prevention through seeking the "hidden flaw". Software programs in particular demonstrate that flaws will creep into the most painstaking efforts and both error avoidance techniques and debugging efforts will be required. This paper starts from a major point for today's systems - complicated software may be as expensive as the hardware it is supporting - and provides a summary and discussion of a number of diagnostic techniques for software development and trouble shooting. It gives deserved emphasis on closed loop logic, modular construction, top down - bottom up development and other techniques for helping the programmer to assure self-discipline and then confirm consistency of his approach. From experience, we might add in passing that computer graphics have also proven useful - skeptics citing unnecessary costs are becoming believers as moving graphics illustrating the operation of a given program demonstrate flaws in less time than one might expect from other techniques (an added benefit of such techniques is in the improved communication of software operations and operability with the non-programmer reviewer). Additionally, added experience has demonstrated that software development should be treated much like a hardware design-development program with a planning and development procedure similar to that used in hardware programs, from requirements definition, specifications and design definition through actual design, development and confirmation (including a systematic design review cycle).

Finally, Christensen's information in the last paper of this applied section reinforces the position that designing a system for ease of maintenance is a major factor, first in facilitating detection and diagnosis of system failure and second in fully fixing the failure in order to avoid subsequent and/or related malfunctions. This summary underlines the extent of continued need for improved man-machine interface provisions, applying the wealth of known information on how to minimize or prevent human initiated failures. Also included is a convenient checklist of design guide information for minimizing human error potential in maintenance. It is interesting that some 40 or so years after the first "Human Engineering" efforts, these problems still exist as continuing evidence of human factor limitations and high-cost maintainability items that could be significantly improved by design. Indirectly, this paper reminds us that a lot of basic human engineering information for man-machine interface design and trade-offs is no longer so easily accessible in reference guides and handbooks nor is it part of academic training to the extent it was 15 or 20 years ago.

In overview, this section on real life perspectives presents a reasonable cross section of real experiences to which later sections can be related. Each author shares some meaningful experiences, information and methodology.

DISCUSSION OVERVIEW

Distinctions between operator, designer and maintainer interests and goals were a prime discussion topic. While normal for the varied interests areas represented, the controversial nature of the debate sometimes masked the fact that the real need is for an integral systems approach that incorporates all such interests for developing complex systems. Less formal approaches most typically provide inadequate visibility of all interface relationships for complicated systems such as nuclear power plants.

This symposium reconfirmed that even full automation requires an operator monitor backup and maintainer role. No single participating technology could offer 100% reliable performance and we can hope to approach such goals only by combining the best from each of several technologies. For example:

- One discussion brought out that one concept of instrumentation and control models for a complicated nuclear plant problem reduced ambiguity to a choice between two possible decisions. However, this choice could not be made from existing technology (more time, effort and state-of-art progress would be needed). Discussion did not expand on alternative supportive concepts, although another discussant offered that his similar dilemmas were typically resolved by a "display"; in one particular case he had experienced, the choice could be made by the operator on the basis of odor or other distinctive characteristics.

- Training experts could not assure that 100% effectiveness would be produced by the most ideal combination of training and/or technical manuals and/or procedures.

- 100% operator reliability is also unlikely; reported experience in overall discussions indicated that adding operators does not improve operator performance or reliability beyond that of the "best" operator. In fact, performance may be worsened by adding operators; discussions described research results showing that one operator's assumptions, expectancies and desires about probable performance of another operator will influence the first operator's interpretations and performance. (The concept of "Team Training" is one approach being explored as a method of improving such assumptions.)

In overview, then, "system" performance depends on the degree to which correct information will be available, interpreted and used in operation and maintenance. Methods and procedures to enhance applications are evolving. Additionally, continued improvement in information display was discussed with proposals to avoid operator exposure to unnecessary levels of detail; to improve data utility; and to re-examine display philosophy in context with new electronic display-control systems (for example, nuclear displays might benefit from (a) the aircraft concept of status, predictor and command displays, and (b) use of programmable electronic displays to access, integrate and present information in more meaningful and productive ways).

The following summaries of specific discussions are identified by the author whose paper was being discussed:

<u>Sheridan</u>

Nuclear power plant discussion produced three relevant additional topical objectives for future progress.

First, functional cause-effect models are needed in order to better understand dependent relationships. It is apparent in such events as occurred at Three Mile Island that one error frequently leads to another; there is need to better understand the relationships and associated circumstances. However, the way things are organized for nuclear operations, there is no ready taxonomy or classification method for operation and maintenance errors that have been observed.

Secondly, Sheridan presented the notion of a "disaggregation" concept to produce information through the smallest level of system detail necessary for effective operation and maintenance. While Sheridan's discussions appear to draw a dichotomy between the notion of aggregation (i.e., his supervisory control model) and disaggregation, his concepts appear correlated with Rasmussen's functional scheme presented elsewhere in this symposium and Parks' functional breakdown structure (Parks 1979) to provide successive layers of detail. Rasmussen's and Parks' positions would argue for a synthesis to that minimum level of detail to permit meaningful and understanding response by the operator/maintainer, and representation of abnormal as well as normal interactions.

Third, in discussing the incident at Three Mile Island, Sheridan identified conditions he identified as "cognitive narrowing" and "cognitive lock up"; that is, a decreasing span of attention and concentration during problem solving. In some circumstances, the operator becomes so involved in resolving a given problem that he loses the overall context of operations and misses the deterioration of other parameters. Possibilities for reducing this phenomena (called "fascination" in other contexts) include assured operator awareness of the real possibility of such phenomena, by emphasizing discipline, emphasis on procedures, specific training, and improved display features.

Such situations describe some of the events that led to the Aerospace Industry and FAA caution-warning system study covered by Thompson's paper. Expected results of the caution-warning system are to make the system more adaptive, improve the scheme to establish priorities for the caution-warning effort, and improve the display of priority activities.

Thompson

Discussions extended the thesis of Thompson's paper, to include long term plans for using advanced electronic display-control systems, such as cathode ray tubes (CRT's), light emitting diodes (LED's), multifunction switching and touch sensitive displays.

State-of-art progress in electronic display control systems demonstrates sufficient utility in that (a) fighter aircraft have been using such displays extensively, (b) NASA has been flying "commercial" versions for some time in a specially equipped 737, and (c) the new generation of commercial aircraft will be using them. However, the concepts were new to non-aerospace participants in this symposium. Totally new concepts from the traditional electro-mechanical display-control systems are now feasible via electronic display techniques. For example, the new displays can present either the integrated or the detailed information (or both) associated with a given situation and could actually "lead" the operator/maintainer through the necessary procedures by appropriate alpha-numeric verbal procedures and diagrams. Discussions demonstrated "intelligent" displays of integrated, computer processed data that offer far superior information presentation to assist the crew in system operation, failure detection and diagnosis.

Gardenier (not present)

Gardenier's paper on ship navigational failures demonstrated that "systems" techniques developed for aircraft are generic and are not necessarily limited to aircraft (see also Parks 1979); they should also transfer to the nuclear operations. A major point of this paper described a ship accident that paralleled Sheridan's description of nuclear incidents reinforcing the rarity of a "single cause" incident or accident - Gardenier reports a scenario of mildly deteriorating events under otherwise "ideal" conditions which led to irreversible problems creating the accident. In turn, as also pointed out by Sheridan, the loosely connected series of events created severe problems for analysis in order to determine methods for improving failure detection and diagnosis.

Discussions supported Gardenier's contention that related problems are being reduced and human factors limitations on systems are also being reduced as a function of increased consistency in use of man-machine interface principles. Gardenier's philosophy, that fascination with models sometimes precludes recognition of simpler solutions in the real world, also was considered to deserve added thought.

Bond

Trouble shooting in computer repair demonstrates that reducing levels of abstraction for the user-maintainer should be a major system development goal for maintenance. This permits a broader population to use diagnostic methods more meaningfully and intelligently and thus better than the more typical rote-procedural approaches. The level of maintainer understanding relates directly to effectiveness of maintenance; however Bond points out that this does not mean that the maintenance personnel should comprehend abstract engineering concepts.

Detection and diagnostic methods that work are apparently common in commercial operations that are financially dependant on efficient and effective computer trouble shooting operations. Several success stories were reported in addition to Bond's. Most notably, the Bell telephone system was reported to be "down" only 8 minutes per year, and even then the only effect of being "down" is a slower completion of a given operation.

Brooke

Structured programming in particular was emphasized as "basic", and more effort is needed on software reliability and validity. Improved technology transfer is necessary and desirable. Discussions only touched the surface of added techniques that could be used to improve software debugging, such as: an improved software design development process; more efficient path checking techniques; broader distribution and use of specific languages and checking techniques; software semantic content; clearer, more transparent illustrations of the software path; and clearly defined program control and maintenance procedures.

Interestingly, no discussion was dedicated to human error tendencies and methods to avoid them for routine tasks, e.g., entering and checking, or in tracing concepts through the tabular format of a typical listing. Other technologies have recognized this inherent human problem for years - it is a main reason for complex checks and balances in accounting practices. However, software technology has paid little attention to the interaction of traceability errors and of errors that derive from the human interface with tabular formats.

Christensen and Howard (not present)

Reported distribution of downtime activity was supported from other military experience, i.e., 65-75% diagnosis, 15-25% remedy and 5-15% verification with an unexplained residual. Numbers for the amount of downtime were considered to be somewhat inflated since more use is made of "higher level" military maintenance facilities for "routine" maintenance and there is more rigid adherence to maintenance standards than might be the case for civilian, non-certificated equipment. Additionally, some military systems are going through a life cycle extension (e.g., the B-52) so part of the 30% DOD maintenance costs are related to "rebuilding" activities. However, the fact remains that a fair proportion of systems are unavailable for a significant part of the time.

Christensen's ideas for using "controlled English" (limited vocabulary to 800 words of high school English) stimulated interest as did his maintainability checklist. More extended feedback can be expected in the future.

In conclusion, this section on "real life perspectives" presents some of the problems encountered in applications. They represent typical conditions that must be resolved in order to meet operational and maintenance objectives. They also represent challenges to more basically oriented researchers to develop techniques to improve efficiency and scope of such efforts.

REFERENCES

"Final Report of Applications Group", 1979, in Mental Workload - Its Theory and Measurement, Neville Moray, ed., Plenum Press.

Parks, D.L., 1979, "Current Workload Methods and Emerging Challenges", in Mental Workload - Its Theory and Measurement, Neville Moray, ed., Plenum Press.

UNDERSTANDING HUMAN ERROR AND AIDING HUMAN DIAGNOSTIC BEHAVIOUR IN NUCLEAR POWER PLANTS

Thomas B. Sheridan

Massachusetts Institute of Technology
Cambridge, MA 02139

THE CURRENT OPERATING TASK

Human operators in nuclear power plants have a frustrating job, fraught with dilemmas. On the one hand they take on a terrible responsibility to protect the public from risk of exposure to radiation; on the other hand they are paid to keep the turbines rolling and to avoid outage, so as to make money for their employer. On the one hand they are expected to follow explicit rules and regulations laid out by both their employer and the government regulatory body; on the other hand they are expected to be resourceful human beings, ready to add their intelligence and common sense where purely automatic operation fails. On the one hand they are judged by their peers according to their speed of decision and action; on the other hand they know that they might best leave their hands in their pockets and count to 100 if a real emergency were to occur. They are aware that under the stress of such emergencies, their judgement may not be reliable; yet there is the danger that they will err on the side of assuming observed abnormalities are not serious, and <u>not</u> take the appropriate emergency action. They know the critical reason they are there is to take over in rapidly changing situations; but they know they may not be up to it. On the one hand, they know that as highly skilled technicians but not professional engineers they are not expected to have sophisticated theoretical understanding of the plant; on the other hand they know that their employer might have a hard time convincing engineers to put up with a job that is dull and boring ninety-nine percent of the time.

In the U.S.A. the Nuclear Regulatory Commission requires three licensed operators to be in or near the control room of a nuclear power plant 24 hours per day (at least one must be at the controls). One of these, the shift supervisor, must have passed a higher level of certification than the others. In addition several unlicensed "auxiliary operators" perform various tasks out in the plant, usually in telephone communication with the control room. The NRC and industry are now, following the Three Mile Island accident (Report of the President's Commission on the Accident at Three Mile Island, 1979; TMI-2 Lessons-Learned Task Force, Final Report NUREG 0585, 1979) debating whether an additional person with a "regular engineering degree or equivalent" should always be available.

Plant operations are learned and practiced as specific procedures. Creative, inventive, trial-and-error behaviour is frowned upon. Operators, typically secondary school graduates, learn system nomenclature, function, and procedures by military--style classroom training, simulator training and on-the-job training. Some accident situations are encountered in the simulator, but even these are mostly "standard accidents". Bizarre, completely unexpected situations are not normally encountered in training. After Three Mile Island this is likely to change somewhat.

Operators usually work eight hour shifts. This requires the utility company to employ three working shifts, plus one shift in training and the equivalent of one shift to allow fill-in for sickness, etc. Shifts rotate by eight hours once per week. Some physiologists claim that this is the worst possible schedule: operators'circadian rhythms barely become adjusted when it is time to rotate.

The human engineering of nuclear plant control panels themselves has come in for lots of criticism since Three Mile Island. Numerous violations of commonly accepted design principles have been cited (Human Factors Review of Nuclear Power Plant Control Room Design, Electric Power Research Institute, report NP-309, 1977) including: the left-hand of a pair of displays driven by the right-hand of a pair of controls; panel meters which cannot be read from more than several feet away; controls for these panel meters located thirty feet away; critical displays located on the back side of a panel while non-important displays occupy centralized front panel space; two identical scales side by side one of which the operator must remember is actually a factor of 10 different from the other because it is not so marked; controls which jut out so that the operator may inadvertantly operate them with his body when he walks too close; ambiguous labels; nomemclature on alarm annunciators which differs from the supposedly corresponding nomenclature in the procedures, etc.

As if to be further insulted in their task, operators often must move back and forth between control boards (in control rooms of two-unit plants) which are laid out as mirror images of one another. Or they must receive their refresher training in emergency procedures on a control board on which displays and controls are laid out differently from the one they will actually operate.

Many of these design flaws can and will be corrected by simple control panel modifications (Human Factors Methods for Nuclear Control Room Design, Electric Power Research Institute, 1979). Others may have to wait for entire new control rooms - while others, because of the costs of outage and potential requirement for relicensing, are not likely to be corrected at all.

HUMAN ERRORS AND HOW THEY DIFFER FROM MACHINE ERRORS

In the nuclear industry reliability analysis is now extensive (Reactor Safety Study: An Assessment of Accident Risks in U.S. Commercial Nuclear Power Plants, U.S. Nuclear Regulatory Commission, 1975). Reliability analysts, in setting up their event trees (which characterize with what probabilities different major events follow other events) and fault trees (which characterize how boolean "and" and "or" logic determines the various ways that major systems failures might occur, the "top events" of which are transferred to the event trees) would like to treat machine error and human error just the same. But this is ill advised, for people err or fail in ways different from machines, and objectivity is more difficult with people than with machines.

Below are listed ten categories of difference and of special concern in understanding human error (Sheridan, 1980).

1. Definition of Human Error

To define error operationally means to make an objective distinction between satisfactory and unsatisfactory behaviour. But sometimes doing this with machines seems easier and less arbitrary than doing it with people. For example, is allowing a certain variable to exceed a nominal fixed limit twice for short periods twice as bad as allowing it to exceed the limit once for a long period? Sometimes literal interpretation of company or government regulations means that some minor discrepancy in behaviour from some norm, for example the inversion of two procedural steps, constitutes an error. Yet operators, using their common sense, know that some such rules are not important. Further, there is no satisfactory objective means to measure whether an operator "monitored", "checked" or "observed" a display - even eye movement photography would not record what was actually seen. Measurement is clearly even more difficult in the case of incorrect plans or judgements.

2. Cause of Error

Most machine failures are explainable by the laws of physics. But there are no such well codified laws which are generally accepted as explanations of human failures. There are, of course, various theories. For example there is the behaviorist theory that errors occur because the physical or social environment does not provide the person sufficient reinforcement. Another theory assumes that each person carries in his head a model of how the environment will respond to each of his actions, and he plans his future actions based on this model. When error occurs, that is because his internal model went out of calibration with the real world.

3. Classifying Errors

One may classify errors in many different ways, (Sheridan, 1978). One is according to whether they are errors of omission or commission. Another is whether the errors are associated with human sensing, information processing, memory, muscle action, and so on. And there are other schemes. The problem is that presently there is no agreement on any one taxonomy, and consequently objectivity in the collection of error data has suffered.

4. Common Mode Error

A single machine failure may set off subsequent machine errors. A human error may also set off subsequent machine errors, OR it may embarass, fluster, frighten or confuse the person so that he is more likely to make additional errors himself. The waiter who spills food in the customer's lap is more likely to drop his tray in returning to the kitchen; the driver who commits a stupid error in rush-hour traffic is more likely to commit another. Thus any given human error is conditional upon other errors which that person just made.

5. The Opportunity for Error

Neither machines nor people make errors unless given the opportunity. Then the absolute probability of error is the probability of error, given the opportunity (and previous errors which may have contributed to common-mode) times the probability of the opportunity arising (Swain, 1964). The latter is clear-cut for machine errors, namely the number of times a relay or pump was operated, or the total time a pipe or vessel was pressurized, etc.

With people it is not so clear-cut to stipulate "opportunity events" because people may look at a display or operate a control at any time, may operate out of sequence, etc.

6. Correction of Error

People tend to correct their own errors. Machines do not. Thus, the absolute probability of an error must be multiplied by the probability that the error is not caught and corrected before it is too late - which finally gives a net probability of error (Swain, 1964).

7. Social Interaction in Relation to Error

Since the two or three operators in the control room are constantly communicating verbally, it is far from clear how one operator's words or actions affect another's tendency to err. One might hope that two or more heads are better than one. But that is not necessarily so, for it is easy for each operator to assume the other knows best, and, in being a good fellow, to reinforce the other's misperceptions. Thus error probabilities for people working together may be better or worse than error probabilities for individuals working independently and not communicating.

8. Staged Responsibility

Errors may occur not only due to human actions and interactions in the control room, but throughout the plant, and these may accrue in fragments and develop with time. Marginal errors may occur at any one of the stages of design, construction, installation, calibration, operation, maintenance, or overall management. These may be propagated from one stage to another by tacit acceptance at each new stage, without the realization that an error is being passed along which will aggravate conditions at the next stage.

9. Criteria for Human Takeover from Automatic Systems

It is far from clear when to let automatic systems have their way and when the operator(s) should take control themselves. It is not even clear whether the human operator should always (or ever) have the option to take control. Interference with the automatic functions of the safety system was one contributive factor in the Three Mile Island accident. Machines are programmed with explicit criteria for when to take over and when not. Should that be done with operators?

10. Error Data Base

The data base for most types of machine error is large and

robust. Life tests are done automatically. Machine components are stressed to failure. Unfortunately, no trustworthy and useful data base exists for human error. Data are hard to get, and when gotten are anecdotal, subjective and in some cases falsified. Reported errors are only those which are not corrected before it is too late. Reliability testing on people is not done - to do it in a fashion corresponding to machine testing (i.e., "life tests") would be against the law. "Licensee event reports" provide some such data, but a better prospect is probably the use of training simulators suitably instrumented and programmed to record operator data (Performance Measurement System for Training Simulators, 1978).

11. Attributes of Human Error and Tradeoffs

Human errors are woven into the fabric of human behaviour, in that, while not intending to make any errors, people make implicit or explicit decisions, based on what they have been taught and what they have experienced, which then determine error tendencies. In a nuclear plant one might consider that any human error has three kinds of consequences - (1) safety consequences, or relative risk to the public and to plant employees; (2) economic consequences, due to outage or to damage to equipment; and (3) personal consequences, effect on that person's job and relationship to other people. One may then develop a three-attribute scale by which any error may be rated - the error would have some value on each of the three component scales. The attributes on such a scale are not independent, but need not be very closely correlated either. Different people, when faced with a similar situation, may tend to err in different directions, i.e., one may be prone to keep the plant running, but at a greater risk to the public; another may err in the other direction; a third may be more prone to irritate his peers and superiors by undue concern for certain kinds of errors, etc. Such a scale proposed by the author is currently being evaluated by General Physics Corp. for the Electric Power Research Institute.

DETERMINERS OF OPERATOR DIAGNOSIS BEHAVIOUR

To study human behaviour of detection and diagnosis of failures in nuclear plants one must start by admitting to great ignorance in the psychology sub-category called cognition, or thinking. Limiting consideration specifically to a nuclear power plant specializes the problem only slightly, namely to monitoring and supervisory control of a highly complex and nonlinear dynamic system by two or three persons. We are probably far from having a generally useful and unifying quantitative model. Instead we have multiple disconnected qualitative theoretical pieces which may help in describing our empirical observations and intuition. In a subsequent section I suggest a framework for organizing the pieces. The proposed framework is sufficiently general that it can be used for human supervisory control of any dynamic system. It embeds the fault diagnosis problem in the overall task. For this reason it is important to consider first some factors which concern the transition from the normal monitoring task to the fault diagnosis task.

1. The Nature of Failure Detection from the Operator's Viewpoint

Mostly the detection of system failure and alerting of the operator is done by machine. That is, when certain system variables reach predefined limits, audible alarms and visual annunciator displays are triggered.

To silence the audible alarm an operator must actuate the "acknowledge" button on the appropriate panel (where the annunciator light came on). Sometimes, however, there is a common "acknowledge" button for the whole control room so that the operator may have no idea what annuciator triggered a particular alarm. It is easy for the operator, in order to silence an unpleasant sound, to hit the button and then look for the annunciator light which triggered it. This is not an adviseable practice, especially in the case of triggering events such as loss-of-coolant accidents where, in one typical plant, over 500 annunciators or control status lights changed state (went on or off) the first minute, and over 800 the second minute (Kozinski). Thus even when detection of system failure is automatic, the sheer mass of display activity in the first few minutes of a major event can leave the operators bewildered.

2. Stress, Mental Workload and "Cognitive Lock-up"

With the detection and annunciation of system failure we associate the onset of stress in the operator. Some evidence suggests that when people are under great stress their reliability is approximately zero (their probability of error approaches unity), then increases exponentially with the passage of time. This generalization formed the basis for Swain's section on Human Reliability in the NRC Reactor Safety Study (WASH-1400) (Reactor Safety Study, 1975) and the new NRC human reliability handbook (draft handbook on human reliability analysis being prepared by A. Swain for the U.S. Nuclear Regulatory Commission). One problem with this generalization is, and this was born out in the Three Mile Island accident, that the operators may not realise that they are coping with a crisis until some time after they have made a series of inappropriate responses. In this case the monotonic exponential reliability curve cannot be correct.

Whether or not the operators experience high emotional stress - which is certainly one form of "mental workload" (Moray, 1979) - they may be overloaded mentally in terms of sheer "busy-ness" (fraction of time occupied, as measured, for example, by a secondary task) or in terms of cognitive difficulty or complexity of the problems they are solving. Judges can discriminate between these types of contributive factors to mental workload on a three attribute scale (three dimensions are emotion, busy-ness, problem difficulty) which is now being tested in an aviation context (Sheridan and R. Simpson). It could also apply to the nuclear operations context. It is not known how the different types of mental workload differentially affect diagnostic capability.

One particular phenomenon which apparently results from stress and mental workload may by termed "cognitive lockup" or "cognitive narrowing". This is the tendency, observed in aircraft, nuclear and other settings, for operators to "focus in" on one hypothesis of what is wrong and what to do about it, and narrow (sometimes literally) their attention to one or two displays and controls. The result may be that they keep trying to operate the same control to make a particular display change in an expected way. The longer they observe that the response is not as expected the harder they try. It is like the foreign (typically American) visitor who, when the natives do not understand his language, simply keeps repeating himself louder.

Some recent evidence on mental workload associated with multiple task monitoring and attention sharing suggests (Daryanian) that of various contributive factors - such as the rate of new independent tasks to be serviced, the inefficiency with which they can be serviced, and the urgency of getting them serviced - it is the rate of appearance of new tasks which most contributes to the mental workload. If lots of tasks queue up and the operator cannot "keep ahead of things", the subjective sense of mental workload seems to saturate.

3. Operator Effort Allocation Tradeoffs

The human operator(s) of a complex system such as a nuclear power plant must be seen as managers who time-share or allocate their sensory, motor and cognitive efforts among a great many different and seemingly independent demands. Happily the dynamic time constants of the key state variables (at least those upon which some control is feasible) are long; with rare exceptions the sampling rate need not be greater than several per minute.

It is more or less accepted by psychologists that people can only attend to one independent activity at a time. Reportedly, Julius Caesar could read, write and speak three different texts simultaneously, but I am skeptical. Clearly, when sensory and motor functions are tightly coupled, as in a sport, people can and do integrate these. In a nuclear plant, however, many operator tasks are quite disparate in space and function. Thus operating skill requires appropriate effort tradeoffs of at least five types:

1) processing internal mental constructs vs. observing or manipulating things in the environment,
2) observing or manipulating at one location vs. doing it at another,
3) taking more sensory data to improve confidence vs. commiting to control action before it is too late,
4) focussing narrowly and carefully in an ordered way on a few displays and controls at a time vs. scanning broadly and continuously over the whole control board,
5) letting automatic systems operate vs. intervening to adjust or override them.

A FRAMEWORK FOR ANALYZING HUMAN SUPERVISORY CONTROL AND DIAGNOSTIC BEHAVIOUR

Figure 1 schematizes a proposed framework for analyzing human supervisory control and diagnostic behaviour in nuclear plants and similar systems. It is not called a model in that it is not developed to the point of having predictive capability. It has benefited from discussions with and papers by many colleagues who, like myself, have been contemplating the human operator in complex systems for many years (Sheridan and Johannsen, 1976; Krishna-Rao, Ephrath, and Kleinman, 1979; Rasmussen, 1978; Johannsen & Rouse, 1979). I believe that this particular conceptual structure is a useful one for the nuclear plant context, particularly in light of the error and determinants of diagnostic behaviour of the human operator discussed above. The capitalized M and S variables are explicitly those separately identifiable manipulative and sensory behavioral components which can be measured. The lower-case m and s variables are hypothesized interactions with an internal model in the operator's head, which can be measured only imprecisely by subjective assessment or "verbal protocol". m' and s' refer to corresponding interactions with an explicit computer based model, which could be measured if such a model were available.

The principal features of the "framework" are as follows:

1) A human supervisory controller allocates his attention among a number of sensory and motor subtasks.
2) The human supervisory controller interacts with printed procedures, with controls on the control panel, with the plant system dynamics, and with consequences of what the plant does. He also interacts with his own internal

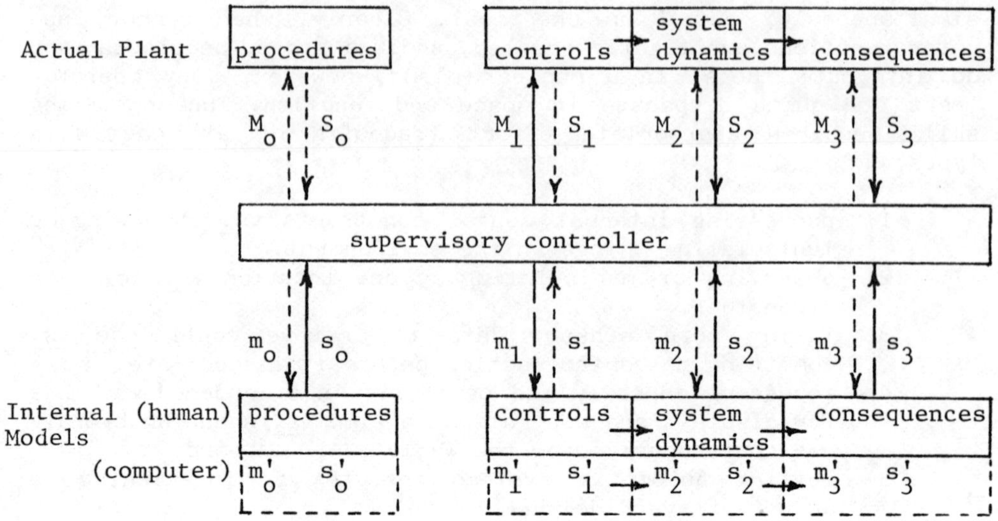

1. Procedures
   - M_o motion to consult a procedure (or to modify it)
   - S_o what is read in procedure
   - m_o effort to remember procedure (or to modify memory)
   - s_o what is remembered in procedure
2. Controls
   - M_1 motion to activate a specific control
   - S_1 what is observed as to which and how control is activated
   - m_1 making assumption of move to activate specific control
   - s_1 decision of what would be observed if m_1
3. System State
   - M_2 motion to observe system state display
   - S_2 what is observed from system state display
   - m_2 alteration of $m_1 \rightarrow s_2$ model to conform to $M_1 \rightarrow S_2$ experience
   - s_2 decision of how system state would change if m_1
3. Consequences
   - M_3 motion to observe alarms, consult with other persons
   - S_3 what is observed from alarms, other persons
   - m_3 alteration of $m_1 \rightarrow s_3$ model to conform to $M_1 \rightarrow S_3$ experience
   - s_3 decision of what consequences would result if m_1

Fig. 1 A framework for modeling human diagnostic behavior (M and m are manipulated variables for real and model plants respectively. S and s are corresponding sensory variables. m' and s' relate to a computer-based model.

representation or mental model of these.

3) Normally he cannot deal with more than one sensory signal (S or s) and/or one motor signal (M or m) at a time. If a closed loop is operative it is usually one of the pairs shown.

4) The principal operation consists of: operating actual controls (M_1) based on a remembered procedure (s_o); observing actual system state (S_2) from the control panel; observing actual consequences (S_3) from the alarms (annunciators) or from communication with other operators.

5) However, this principal activity is mediated by: specific effort (m_o) to remember procedures or update that memory; consultation with actual procedures (M_o, S_o); observation (S_1) that he is operating the intended control; explicit strategy (M_2) for moving his body and eyes to monitor panel displays; strategy (M_3) to consult annunciators and other operators.

6) The internal model of controls, plant system dynamics and consequences is used in planning, controlling, and in diagnosis if an abnormality is detected by alarm. m_1, s_2 and s_3 parallel the principal variables for controlling the actual plant.

7) If s_1 deviates from S_1 that indicates incorrect operation of controls, and M_1 is corrected.

8) If S_2 and S_3 deviate from what is intended or expected, then an internal model run (thought experiment) is performed with control manipulations m_1 modified by simple negative feed-back. If s_2 and s_3 are thereby improved, then M_1 is made to conform to the best m_1. By this technique some alarms can be corrected by simple feedback controls actions which drive S_2 and S_3 into conformity with no great effort spent on diagnosis.

9) At the same time, if s_2 and s_3 deviate slightly from S_2 and S_3 respectively (for m_1 and M_1 well matched) that indicates that the model of system dynamics and/or the model of consequences are not well calibrated to reality. These discrepancies can usually be corrected by slight manipulations m_2 and m_3 respectively of the parameters of the internal model in a linearized (parameter feedback) sense, i.e., repeated trials are run with a different parameter, each time slightly altered until the model matches reality. Having to do this can mean either that the model parameters have deviated from those of a normal plant, or that the plant parameters have deviated from those of a normal model.

10) If s_2 or s_3 continue to deviate from S_2 or S_3 respectively, whether or not an alarm has lit or sounded, somehow the model and actual system have gotten well out of conformity and may be diverging. In this case, and especially if S_2 and S_3 are unexpected, it is likely that the actual system has become abnormal. Then, in order to force conformance between

actual and model systems and thereby identify the new parameters, small adjustments to the old parameters of the human operator's internal model will not suffice. Other means will be required, such as computer aiding. This is discussed in the next section.

AIDING DIAGNOSIS BY COMPUTER

Computers are used in most nuclear power plants today for buffering information about selected plant variables, which finally is printed on paper with a time index. Microelectronics pervade instrumentation systems, including, for example, processing and multiplexing of signals and alarm logic. In some plants computers take signals from the alarm/signal recorder and generate colorgraphic CRT displays. The very newest plants have so-called "advanced control rooms" which are entirely computer-based, with general purpose CRT and small push-button panels replacing the large dedicated control boards.

However, the most pressing problems remain to be solved. How can the computer and the operator best work together, in both routine and emergency situations? What allocations of function should be assigned to each?

It is generally assumed that the computer is better at performing numerous precise and rapid calculations, while the human operator is best at making associative and inferential judgements. Consistent with this assumption, the computer is more able to predict the plant response to various kinds of forcing functions, given a good model of the plant. Thus the nuclear plant computer should incorporate a dynamic plant model similar to the operator's internal model discussed in the previous section, and the operator should utilize the computer's "internal model" in addition to his own - see Figure 1 at the bottom. This means the one-independent-thing-at-a-time human attention allocator has more things to allocate among. It also means that unless the computer model is a good model it may be more of a distraction to the human operator than an aid. But these are some of the risks in using computers for anything. However, the concern that computer hardware is not reliable is being allayed by new developments in "fault-tolerant" computer architectures.

As suggested above the on-line dynamic computer model is not necessary if (1) the operator is alerted to the abnormal variable by an alarm, or otherwise, and (2) the operator through direct feedback can adjust the proper controls to make that variable normal (step 8). However, neither of these conditions will be true necessarily.

Sometimes alerting or detection (1 above) is deficient and it is not enough to measure whether a variable has gone beyond a stipulated high or low "limit". It has been suggested that the computer provide, for a selected set of key plant variables X_i, a weighted "urgency function" of the form

$$\text{"urgency"} \qquad U_i = a|e_i| + b \frac{e_i \dot{e}_i}{|e_i|}$$

where $e_i = X_i - X_{i\ normal}$.

Whether the present methods of fixed upper and lower alarm limits are used, or whether the urgency function relative to an $X_{i\ normal}$ is used, the decision as to what is normal X_i is not simple. The normal value of X_i changes with the plant situation, i.e., whether it is in the process of start-up or shut-down, fractional power or full power, etc. Various investigators are working to have the computer determine which levels of which variables should set off alarms.

If identification (2 above) is deficient there are various computer aiding techniques now being explored. One method (Deyst) is to find sets of variables such that simple transformations may establish equivalencies:

$$G_a(X_i X_j X_k) = G_b(X_l X_m X_n) = G_c(X_q X_r X_s)$$

For example, in a steam generator, the time integral of the difference between inflow and outflow should be equivalent to the water level and should also be equivalent to the pressure head measured at the bottom of the tank. If there is disagreement among all three, or if two of these measures agree, but the third one does not, the operator is so informed. Then there are various possible explanations that can be summoned automatically from the computer's memory for operator consideration.

A second method, (Hasan), is to continually compare, at each of a large number of test points, the actual plant with a computer model running in real time. Any discrepancies which exceed "noise threshold" are called to the operator's attention. Further, any covariations of pairs of variables which are radically different from normal (e.g., normally X_i and S_j rise and fall together, but now they are going in reverse directions) are also noted. Trends can be recorded and displayed at the operator's convenience, etc.

Further, a good computer model can be run in fast-time. Thus, when there are major discrepancies between actual plant and model, identification can be accomplished by parameter tracking - a series of fast-time runs with various parametric settings to determine what settings provide the best fit to recorded data from the actual plant. Because of the complexity and non-linearity of nuclear plants the choice of parameter variations for such fast-time identification experiments must be guided by hints from other sources as to which parameters may have gone awry. There are far too many parameters to vary all of them in any systematic way.

Figure 2 illustrates two ways the computer model can be implemented. In the upper part (a) the "plant" (represented here by three interconnected matrix-dynamic elements) is replicated as closely as is practical by a fully connected or aggregated model. Thus a supra-threshold discrepancy between an actual plant variable (output of A, B or C) and its model counterpart (corresponding output of a, b or c) reveals a failure, provided that the model is a close approximation to the plant. However, a major difficulty with the aggregated model is that a discrepancy at a given location does not necessarily mean that the cause of that discrepancy, i.e., a failure, has occurred just upstream. Any component might have failed.

The lower part (b) shows a disaggregated model, where the inputs to a, b and c respectively are the same signals as enter the corresponding A, B and C components. Now a discrepancy between the outputs of corresponding components in plant and model reveals not only that there is a failure but exactly where that failure has occurred.

That the disaggregated model is superior in failure isolation does not mean that the aggregated version is not useful. For estimating future states of non-directly measurable variables (as in a Kalman estimator), the aggregated model is essential - run in fast-time with initialization on any computation cycle from the actual plant.

The question naturally arises as to how good a replica the model must be under all circumstances. There are really two questions: (1) how good in normal operating range; (2) how good in abnormal range. For detecting and isolating initial deviations from normality (1) is of concern. For detecting and isolating secondary failures produced within the plant by the stresses of variables in abnormal states (2) is of interest.

It should also be noted that the disaggregated model technique is closely related to techniques used in "fault tolerant" computers and instruments where two to four components are fed the same signals at each functional location in a system. The outputs are compared and a median or mean is used as a functional output sent on to the next (set of) components. If there is enough discrepancy in outputs at any functional location, an emergency is declared.

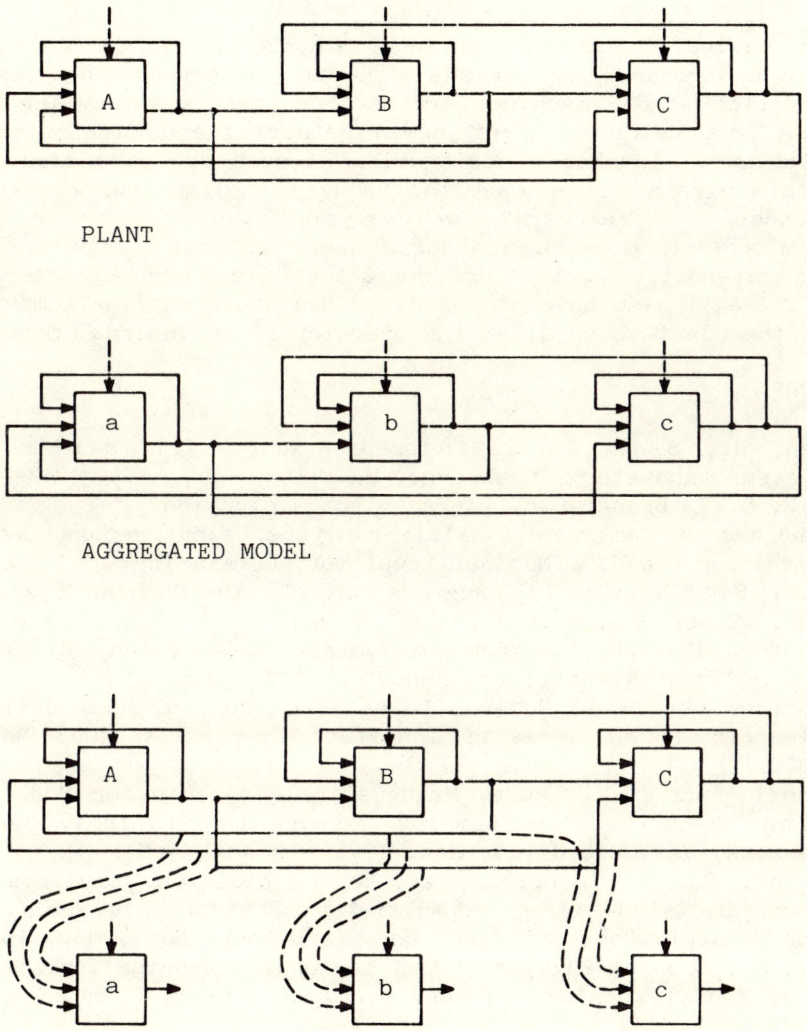

Figure 2. Aggregated and disaggregated models. In the upper part (a) the external forcing functions are meant to be the same for the model (a,b,c) as the plant (A,B, C). Then for a normal plant, the output of any model component should agree with the corresponding plant component. A discrepancy indicates a failure somewhere. In the lower part (b) a discrepancy isolates the failure to the corresponding component.

CONCLUSION

The nuclear plant operator faces a number of dilemmas.We are coming to understand the various ways he can err and how human errors differ from machine errors. We are also coming to understand the determiners of human detection and diagnosis of system failure. A number of categories of each are discussed. As an aid to analyzing human operator behaviour in nuclear plants I have proposed a "framework" for analysis, characterized by the idea of attention allocation among a number of motor and sensory tasks in interaction with both the actual plant and an internal model of same. I also have proposed the use of an on-line computer model of the plant to parallel the operator's own internal model.

REFERENCES

Daryanian, B., S. M. Thesis in preparation, M.I.T. Man-Machine Systems Laboratory, Cambridge, MA.

Deyst, J., C. S. Draper Lab., personal communication.

Draft handbook on human reliability analysis being prepared by A. Swain for the U.S. Nuclear Regulatory Commission.

Hasan, M., S.M. Thesis in progress, M.I.T. Man-Machine Systems Lab., Cambridge, MA.

Human Factors Methods for Nuclear Control Room Design, Electric Power Research Institute, report NP-1118, June 1979.

Human Factors Review of Nuclear Power Plant Control Room Design, Electric Power Research Institute, report NP-309, March 1977.

Johannsen G. and W. B. Rouse, Mathematical Concepts for Modeling Human Behaviour in Complex Man-Machine Systems, Human Factors, Vol. 21, No. 6, Dec. 1979. pp. 733-748.

Krishna-Rao, P., A. R. Eprath, and D. L. Kleinman, Analysis of Human Decision-Making in Multi-Task Environments, Univ. of Connecticut, Dept. of Elec. Eng., and Comp. Sci., Nov. 1979.

Kozinski, E., General Physics Corp., personal communication.

Moray, N., Mental Workload: Theory and Measurement, Plenum, N.Y., 1979.

Performance Measurement System for Training Simulators, Electric Power Research Institute, report NP-783, May 1978.

Rasmussen, J., Notes on Diagnostic Strategies in Process Plant Environment, Risø National Laboratory, rep. M-1983: also Notes on Human Error Analysis and Prediction, Risø National Laboratory, rep. M-2139, Nov. 1978.

Reactor Safety Study: An Assessment of Accident Risks in U.S. Commercial Nuclear Power Plants, U.S. Nuclear Regulatory Commission, report WASH-1400, Oct. 1975.

Report of the President's Commission on the Accident at Three Mile Island, Oct. 1979, Washington, D.C.

Sheridan, T.B., Taxonomy/Checklist for Human Errors in Nuclear Power Plant Operation, Unpublished memo, 1978.

Sheridan, T.B., Human Errors in Nuclear Power Plants, Technology Review,Feb. 1980, pp. 22-33.

Sheridan, T.B. and G. Johannsen (eds.), Monitoring Behaviour and Supervisory Control, Plenum Press, N.Y., 1976.

Sheridan, T.B. and R. Simpson, draft of report on mental workload scaling techniques being prepared for the U.S. Federal Aviation Administration.

Swain, A.D., THERP, Report SC-R-64-1338, Sandia Laboratory, Albuquerque, N.M., Aug. 1964.

TMI-2 Lessons-Learned Task Force, Final Report, NUREG 0585, Oct. 1979, U.S. Nuclear Regulatory Commission, Washington, D.C.

COMMERCIAL AIR CREW DETECTION OF SYSTEM FAILURES: STATE OF THE ART AND FUTURE TRENDS

David A. Thompson, PhD

Professor of Industrial Engineering
Stanford University
Stanford, California 94305

INTRODUCTION

The generally safe and dependable commercial aviation industry has never had properly designed Caution and Warning Systems (CAWS) to alert the aircrew to operational or system malfunctions or emergency situations. When flight systems were simpler, relatively crude CAWS were manageable. Today, however, the complexity and size of modern avionics systems makes it crucial to have optimal systems to alert the crew to problems, and to assist them in handling them.

To the extent that CAWS does not assist the aircrew in identifying the underlying cause of an alerting stimulus, its relative importance at that point in the flight, alternative actions available to the flight crew to handle the indicated problem, and the best way to implement the optimal alternative action, then their information processing capacity and ability are loaded with having to perform these actions themselves. During high workload periods (e.g., landings and takeoffs), this is a possible contributor to the buildup of dangerous situations.

STATE OF THE ART

A 1976 Boeing Aircraft Company study (Boucek & Veitengruber, 1976) analyzed the significant and growing proliferation of Caution and Warning system lights, bells, buzzers, gongs, chimes,

---

Research supported by the NASA Ames Research Center, Mountain View, California, Contract NCA2-DR 745-710.

etc. used to alert the pilot to a present or potential systems problem. It found that the number of alerts has increased 61% from the Boeing 707 to the 747 (the L-1011 now has 886 separate warning lights, instruments, and sounds) and that there is no consistent philosophy governing the optimal design of CAWS anywhere in the aircraft industry. Each aircraft manufacturer designs the CAWS for his next aircraft without coordination with other manufacturers - in fact, uses secrecy as a competitive advantage at times, (see Arnoldy, 1971). Nor are CAWS designs prescribed and evaluated by a central agency, such as FAA, CAB, or NASA. The consequence is the present state of non-standardization of non-optimal CAWS in various commercial aircraft designs, and proliferation of CAWS displays.

More recently, a broader survey was made (Cooper, 1977) of all major foreign and domestic commercial aircraft manufacturers, military aircraft manufacturers, general aviation manufacturers, and government (civilian and military) aviation R&D labs. The principal results of this study, finding general acceptance from this group, were that:

a) There are too many warnings, particularly requiring immediate attention.
b) Both audio and visual warnings are generally required.
c) Reliability of warnings needs to be improved to reduce false alarms.
d) Appropriate, broadly administered standards and guidelines are badly needed.
e) Warnings should be prioritized, with some inhibited during certain flight phases.
f) New warning systems should be completed, thoroughly evaluated - with Full Mission simulation - before installation.
g) Central vision warning systems needed.
h) Lighting intensity and contrast generally poor.
i) Darkened cockpit preferable.
j) No more than 4-5 audio warning systems.
k) Voice warnings desirable.
l) More automatic systems to handle system problems, avoiding the CAWS demand on the aircrew.

The survey was an excellent one, and much of its findings about future system design preferences are included in later sections of this report.

What these two reports point out, as well as other studies (Cane, 1971; Graham, 1974; FAA, 1967; Miller, 1969; Vreuls, 1968), is that the relatively good safety record of the airlines notwithstanding, the state of the art in CAWS is one of nonstandardization, inadequate evaluation of new displays, excessive numbers of non-prioritized alerting stimuli, and lack of intelligent automaticity in CAWS design. Although excellent design improvements have been suggested (Beddow, 1975; Munns, 1971; Paananen, 1969; Palmer, 1979; Thompson, 1978), and appropriate analysis techniques are available (Barnhard et al., 1975), much work obviously needs to be done.

STUDIES PRESENTLY UNDERWAY

A Federal Aeronautics Administration (FAA) study is presently underway, headed up by Boeing Commercial Airplane Co., with the participation of Lockheed California Co. and Douglas Aircraft Co. as subcontractors to Boeing. The 1976 Boeing study pointed up large gaps in the Human Factors literature related to display design in general and CAWS design in particular. The present study has focused on defining prototype alerting concepts, and on design of a test facility for evaluating prototype system hardware. Testing was scheduled to begin in April, 1980, with results available in December, 1980 (Elson, 1979).

The testing will involve stimulus detection tests of aircrews while performing simulated flight-related tasks, at high, medium, and low workloads. A typical experimental design would then present visual stimuli at four different brightness levels using two different styles, over a widely varying ambient light level, at each of the three workload levels. Other tests will evaluate the interaction of brightness, location, distracting signal number, and ambient light level; false signals, their frequency of occurrence, distracting signals; verbal and non-verbal auditory signals, and confined visual/auditory signals. For further description, see Veitengruber and Boucek (1976).

Cooperating with the FAA in monitoring of the project is NASA's Life Sciences group at Ames Research Center.

The Society of Automotive Engineers is presently drafting a standard for aircraft alerting systems. Their studies are under the sponsorship of SAE's S-7 committee, chaired by J.R. Gannett at Boeing.

FUTURE THRUST OF CAWS DISPLAYS

What is indicated by the future cockpit display designs in general are displays that are:

Dark and Quiet. A cockpit as dark and quiet as reasonably possible. No red/yellow/blue/green/white status lights unless crew attention is necessary or unless requested by the crew.

Soft. (CRT's etc.) rather than Hard (lights and electromechanical displays). Figure 1 illustrates a typical avionics information display concept.

Information rich (words, sentences, graphics) rather than simple (lights, buzzers, single variable). This is true for both visual displays (specific problem identification/recommendation rather than just an indicator light) and auditory displays (spoken sentences rather than just buzzers or tones). The information content of each message may be as high as the crewmember's information reception capacity, given his likely workload at the time, so as to optimize the man-machine communication linkage (Siegel et al., 1975; Van Gigch, 1970).

Integrated (all related variables pictured or included in a gestalt description of the situation) rather than separated (individual variables being displayed on separate dials or indicators). This is now done to some extent on the flight director, but does not include all flight-related variables such as speed, climb/sink rate, etc. It should be done on all systems of related variables (e.g., individual and comparative jet engine performance).

Assistive rather than non-assistive. Systems will help with procedural tasks by displaying check lists (on CRT's) and acknowledging when various items have been taken care of. Supplementary information, such as navigation way-points or airport layouts and procedures will be displayed upon command. When emergencies arise, the appropriate check list may be automatically displayed rather than one's having to access hard copy stored in a flight bag. Newell and Simon (1972) found the limited capacity of short-term memory (7 ± 2 chunks) to be a major deterrent to effective problem solving. Streufert (1970) argues that in seeking and selecting data to evaluate an on-going situation, men tend, on one hand, to gather information indiscriminately, resulting in more information that can be used effectively in problem solving, and on the other hand, to restrict search to only a limited subset of the alternatives relevant to the problem at hand. In addition, since emergencies follow patterns (Edwards, 1968, 1978), the display could show the typical pattern that the emergency tends to follow, or ask for the necessary information to determine the particular pattern that would probably result. As Edwards points out, pilot's training consists largely of learning to recognize patterns of relationships that represent emergencies. However, assisting this process (e.g., prompting, outlining) could be an important supplement.

Figure 1. British Aerospace's advanced flight deck design simulator is built around seven CRT's portraying flight, engine and systems information (North, 1979).

Intelligent rather than stupid. This is an even more assistive feature that upgrades the system to that of a very competent, experienced colleague that can instantly diagnose all problems that arise (run fault-logic trees to find root causes), display solutions, recommendations, and consequences; and even implement the selected solution if so ordered. Moreover, the technology and methodology exists (from automated medical diagnosis research - MYCIN - at Stanford University) (see Feigenbaum, 1978; Shortliffe, 1976) to imbed into the system the pooled intellectual skills of the best available pilots to instantly recommend, or command (in the case of FAA or company regulation), the optimal solution to most problems that may arise. In addition, if the pilot wishes to ask "What if?" questions of the system ("How would fuel consumption change if altitude or way points were changed?") he would not only get immediate answers but also get explanations to all of his questions about the system's reasoning (logic, data, and assumptions). This would not make the pilot's job trivial, but would offer him an intelligent support system of data and recommendations when and if he needed them, rather than simply flashing or buzzing when a symptom of some deeper problem was detected.

Alerting display technology will of course be imbedded in this general display philosophy. CAWS displays will consequently tend toward being:

Soft (displayed on CRT's) rather than Hard (lights and buzzers).

Information rich (descriptive sentences written across the top of a CRT or the below-glide slope portion of a graphic display shown in red) rather than simple (buzzer, indicator lights).

Integrated (all related alarms displayed together in their logical relationship or in their sequence of importance) rather than separated (alarms indicated anywhere in the cockpit with no displayed relationship).

Assistive (triggered alarms call up check lists to insure correct, complete response) rather than non-assistive (merely display a symptom of some undefined problem).

Intelligent (diagnose all systems symptoms before displaying them, determine the underlying root-cause, and display this together with alternative and recommended solutions) rather than stupid (triggered by symptom sensors). Noncritical alerts would be inhibited during high workload periods like takeoff and landing.

Consequently, the CAWS visual displays of the future would be complete, concise explanations and recommendations/commands, probably on the top few (dedicated) lines of one or more CRT's. Alarms would appear color coded as is now planned for conventional displays (Red for Warning, Yellow for Caution, Blue for Advisory, etc.) and would probably flash rapidly for Warning alarms until acknowledged.

The three categories of alerting functions presently under active consideration are:

Warnings: emergency operational or aircraft system conditions requiring immediate corrective or compensatory action by the crew.

Cautions: abnormal conditions that require immediate crew awareness and require corrective or compensatory crew action.

Advisories: conditions that require crew awareness and may require crew action.

The CAWS auditory displays of the future would be concise problem statements and recommendations/commands communicated by synthesized pilot's speech. Each priority category would probably be proceeded by a unique alerting tone.

New categories of alarms may also emerge with the expanding flight control technology of the 1980's. The ground controller's audio communication with the flight crew may be supplemented by a digital link, so that course/speed/waypoint changes may be entered by the controller into a numeric keyboard supplemented by selected function buttons. The controller's commands would then be transmitted to the aircraft to be displayed on the pilot's navigation CRT as well as heard by him (increasing accuracy and reducing confirmation delays). In the event that the pilot was told to come to 090° (or to reduce speed by 50 knots) and he failed to do so within a reasonable time, he would be automatically alerted to this command.

In addition, the ground controller could identify, either numerically (9-o'clock, 9-miles) or graphically (lightpen on his scope) other traffic in the area which could also be transmitted via digital link up to the flight crew and displayed on their routing plan CRT's as red flashing boxes or " 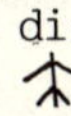 ". This enriched information exchange, verbal, alphanumeric, and graphical, between the pilot and the ground controller will insure an absolute minimum of missed communications and better job performance by both.

FUTURE IMPACT ON AIRCREW

An issue that must be addressed is whether the systematic information flow changes on the flight deck resulting from the above hardware and flight procedural changes will result in safer flight. The indications are that they will, as illustrated by the examples given in the preceding sections. But the most optimistic indication that these suggested changes will result in substantive reductions in human error is provided by Norman (1979) in his pioneering work in "slips" of the mind. He departs from "Freudian slips" and their therapeutic significance, and instead categorizes slips as simply consequences of the way the mind functions and its capacity for storing and outputting the proper information in the correct time frame.

A complete description of Norman's analysis is not appropriate here, but a brief breakdown of his taxonomy of slips will assist the reader. He classifies "human error" into (1) mistakes, or actions based on incorrect or incomplete information, and (2) slips, or actions with mistaken intention. Slips (of the mind, resulting in slips of the tongue and of the hand) may be further broken down into:

a) <u>Description errors</u> - acting at too high a level of abstraction to be meaningful for the present situation, such as confusing one control for another.
b) <u>Actuation or Triggering errors</u> - failing to interrupt a primary task (tracking a glide slope) to perform a secondary task (lowering the flaps), or activating it at the right time, but in reverse.
c) <u>Capture errors</u> - having new, unlearned behavior "captured" by habituated patterns, such as a new Boeing 737 pilot turning around to talk to the flight engineer, when there is none.

These error types, or slips, are not arbitrary categories of statistical crash reports, but directly follow from the behavior involved. One cannot read his analyses without personally identifying with his taxonomy of slips.

What is most encouraging is that the new types of hardware and procedural changes discussed earlier seem to fit Norman's error analysis very well. That is, the type of hardware that is evolving, information-rich, integrated, assistive, intelligent controls and displays, is most appropriate to reduce substantially human errors related to the (a) Description of events, (b) Activation and Triggering of actions, and (c) Overriding habituated "capture" errors. It would be most appropriate if, prior to evaluation and testing these new, enriched, assistive, intelligent (diagnostic) systems, and in-depth analysis was made of the flight crews' behaviors that will change from a slip-reduction, error-reduction standpoint.

The experiments being run at Boeing will certainly produce individual pieces of the final solution (e.g., appropriate brightness/contrast relationships for day vs. night displays), and will relate indirectly with soft CAWS displays. Studies must be made of CRT visibility (see Smith, 1979), information formatting, color, and flash rate (is a flashing message harder to read?); and of CRT location and orientation for optimal individual or shared viewing, including constraints imposed by over-the-shoulder sunlight in the cockpit. The problem diagnosis/recommendation algorithms must be acceptable to aircrews - logical, consistent, able to be queried and altered if appropriate (and younger pilots may respond differently to these than older pilots). Task assignment among the aircrew members may change as a function of these new information sources and abilities being available; the flight engineer may become the flight information manager, inputting operational information and decisions, commands, and exploring "What if?" questions when evaluating alternatives.

Once the individual system elements (CRT display paradigms, verbal system warnings and responses, etc.) are evaluated, then full missions simulation evaluation needs to be made with as much fidelity as feasible to evaluate systems integration. Complete integration of CAWS into the basic avionics display system may raise some interesting issues, such as the feasibility of predictive CAWS that adaptively recognize problem patterns in system variables prior to their becoming critical, and calling this to the attention of the crew in time for them to respond in something other than a "panic" mode (e.g., engine exhaust gas temperature displayed first in yellow, then orange, then red prior to triggering the ENGINE FIRE alert) and all as part of normal power plant displays.

There may also be some sociological advantages to having more information available for the pilot. At present, the pilot has the ultimate, final responsibility for the safety of his aircraft, but has to depend on a ground controller for much of the information he needs (in fact, is commanded by the controller) for flight decisions. With enriched displays of his own aircraft's flight progress and that of the aircraft around him, and enhanced emergency decision-making, he is reclaiming some of the self-sufficiency pilots enjoyed in the earlier days of aviation.

It has been true for some time that a pilot is no longer just a flyer, but rather manages a complex vehicle in order to implement a variable flight plan. The above hardware developments will continue the flight crew member's evolution toward that of a flight information manager.

The transition to this new flight paradigm, however, will probably not be easy. Not only will the classic Transfer-of-Training problems arise, but there will undoubtedly be schisms between the younger pilots trained on CRT displays in the Military, and the older, senior pilots who, in fact, specify and approve final cockpit designs for airlines. The overall impact may be as great on pilots and airlines as was the transition from flying propellor planes to flying jets.

And, of course, all soft displays are not necessarily good displays. Much good Human Factors analysis and testing will be necessary in the evolving cockpit of the late 1980's and 1990's. But the opportunity exists to do it properly, rather than piecemeal as before, so as to be uniformly consistent and appropriate over aircraft types and airlines.

REFERENCES

Arnoldy, C., "Cockpit Display: Users vs. Makers" Information Display, July/August, 1971, pp. 27-36.

Barnhard, W. et al., A Method for the Study of Human Factors in Aircraft Operations, NASA TMX-62, Ames Research Center, Sept., 1975, 42 pp.

Beddow,S., "A New Centralized Warning Display in Aircraft", 8th International Aerospace Instrumentation Symposium Proceedings, Cranfield, England, March, 1975, A75-28785.

Boucek,G.P. and J. E. Veitengruber, Human Factors Guidelines for Caution and Warning Systems (Draft Report for FAA Contract FA73WA-3233-Mod 2), Boeing Commercial Airplane Co., Document No. D6-44200, November, 1976.

Cane, A. P. W., "Displays in Flight Management-Evolution and Revolution", British Air Line Pilots' Association Technical Symposium, London, November, 1971, A72-13423.

Cooper, George, E., A Summary of the Status and Philosophies Relating to Cockpit Warning Systems, NASA Contractor Report NAS2-9117, June, 1977, 45 pp.

Elson, B.W., "Cockpit Alert Standardization Urged", AW&ST, May 21, 1979, p. 99.

Edwards, W., "Controller Decisions in Space Flight", Applications of Research on Human Decision Making, NASA, STID, Washington, D.C., 1968, pp. 93-106.

Edwards, W., "Should Pilots Need to Make Emergency Decisions?", Air crew Emergency Decision Training: A Conference Report, Perceptronics, Woodland Hills, CA, November 28-30, 1978.

FAA Symposium on Pilot Warning Instruments Proceedings, Washington, D.C., December, 1967, AD666 122.

Feigenbaum, E. A., "The Art of Artificial Intelligence - Themes and Case Studies of Knowledge Engineering" AFIPS Conference Proceedings, NCC, Vol. 47, 1978, pp. 227-240.

Graham, W., and Mangulis, V., Results of the Visual Detection Simulation Experiment for the Evaluation of Aircraft Pilot Warning Instruments (APWI), Vol. 2, Final Report, December 1974, AD-A017023: FAA-RD-75-59 Vol. 2: N76-20103.

Miller, B., Optimum Response to Alerting Signals and Warning - Messages, System Development Corp., Santa Monica, N69--26158, 14 March, 1969.

Munns, M., "Ways to Alarm Pilots", Aerospace Medicine, Vol. 42, July 1971. A72-11291.

Newell, A., and H. A. Simon, Human Problem Solving, Prentice-Hall, 1972.

Norman, D., Slips of the Mind and an Outline for a Theory of Action, CHIP88, Center for Human Information Processing, U.C. San Diego, La Jolla, CA., 92093, November 1979.

North, D. M., "Advance Flight Decks Gain Acceptance", Aviation Week & Space Technology, April 23, 1979, pp. 60-65.

Paananen, R., et al.,Compilation of Data from Related Technologies in the Development of an Optical Pilot Warning - Indication System, NASA, Washington, D.C., May 1969, NASA TN D-5174, N 69-27267.

Palmer, E., "Experiments on Interrupted Monitoring of a Second--Order Process", Proceedings of the 15th Annual Conference on Manual Control, Wright State U., Dayton, Ohio, March 20-22, 1979, 22 pp.

Rowland, G. E., and C. T. Reichwein, Functional Analysis of Pilot Warning Instrument Characteristics, FAA-NA-71-40, Rowland & Co., Haddonfield, N. J., September, 1971, NTIS N71-35216.

Siegel, A.I., M.A. Fischl and D. Macpherson, "The Analytic Profile System (APS) for Evaluating Visual Displays", Human Factors, 17(3), 1975, 278-288.

Shortliffe, E., Computer-Based Medical Consultations: MYCIN,New York, Elsevier, 1976.

Smith, W., "A Review of Literature Relating to Visual Fatigue", 23rd Annual Human Factors Society Conference Proceedings, Boston, Mass., Oct. 29-31, 1979, pp. 362-365.

Streufert, S., "Complex Military Decision-Making", Naval Research Reviews, 23/9, 1970, p. 12.

Thompson, D. A., Human Factors Issues in Proposed Aircraft Caution and Warning System Experimentation, Technical Report for Contract NCA2-OR745-710, Nasa Ames Research Center, 1978, 22 pp.

Van Gigch, J. P., "A Model Used for Measuring the Information Processing Rates and Mental Load of Complex Activities", J. Can. Operational Res. Soc. 8, 116, 1970.

Veitengruber, J. E., and G. P. Boucek, Collation and Analysis of Alerting Systems Data, Summary Program Plan, FAA Contract DOT-FA73WA-3233 (Mod 11), Document D644199, Boeing Commercial Airplane Company, Seattle, Washington, October, 1976.

Vreuls, D., et al. "All Weather Landing Flight Director and Fault Warning Display Simulator Studies", Human Factors Society 5th Annual Symposium Proceedings, Los Angeles, CA, June, 1968, (Western Periodicals, Co.), pp. 79-93.

SHIP NAVIGATIONAL FAILURE DETECTION AND DIAGNOSIS

John S. Gardenier, D.B.A.[x)]

U.S. Coast Guard (G-DMT-1/54)
Washington, D.C. 20593, U.S.A.

INTRODUCTION

The navigation of ships through the world's waters presents a continuing problem of system failure detection and diagnosis. In open, unrestricted waters with little or no traffic, the ship navigation system is fairly tolerant of errors and other failures. As ships approach narrow, restricted channels and increasing traffic densities, the systems failure tolerance decreases.

Increasing hazards of large oil and chemical shipments create demands for systematic analysis of ways in which to minimize the risk of failures while still getting the needed cargoes moved economically. Such analysis points to the human controller as the key system element. The human being, when functioning as an adaptive controller, is anti-systematic, integrating all variables, time frames, and information bits in ways that defy precise identification, flow definition, or prediction of decisions or behaviours.

In problems of this type, the systems analyst must either deal with artifical control processes, which he can model but not observe in realistic operations; or else he must deal with very unstructured, empirical observations of real behaviours, which do

---

x) The author is strictly responsible for this paper; it does not necessarily represent U.S. Coast Guard official positions or policy.

not lend themselves to modelling. Achieving a sound balance of these two types of research is a central issue to all forms of system failure detection and diagnosis.

The slow, relative simplicity of ship navigational control relative to complex process controls of nuclear power plants and high speeds of aerospace systems offers us a chance to focus on the core problem.

SHIP NAVIGATION AND PILOTAGE

I must down to the seas again,
to the lonely sea and the sky.
And all I ask is a tall ship
and a star to steer her by.

Masefield - Sea Fever

Ships are high technology systems, but they have only a few rather simple navigational sensors, controls, and indicators. With a few exceptions, these are sufficient to allow knowledgeable, skilled, and alert crews and pilots to navigate ships safely wherever they need to go in the world's waters. Despite this, ships are lost worldwide at the rate of about one a day, and less disastrous casualties occur more frequently in United States waters alone (U.S. Coast Guard, annual).

In the vast majority of accidents - at least eighty percent - there is no apparent failure of any critical control mechanism nor is there any overwhelming environmental force involved. The critical factors in the failures are mostly human factors. Several studies oriented to the human role in ship accidents have found that in many vessel accidents the responsible personnel appeared to be experienced, competent, adequately rested, and free of the influence of drugs, medications, alcohol, emotional stress, or environmental stress (Paramore et al., 1979, Gray, 1978).

We also know that there is seldom a problem in control actuation. Once the decision is made for a propulsion or steering change, the change can usually be made simply, reliably, and subject only to well-understood time delays. Interactions of the ship with variable winds and currents and bottoms in response to the controls are somewhat more problematical.

In air or highway vehicles, except when automatic systems are in use, there is a steady and demanding workload of making minor adjustments to maintain speed and direction, correcting for routine environmental or control system variations. The ship steering function is in the hands of a helmsman. The ship's master, pilot, or watch officer, being free of routine hands-on control tasks, is dedicated to the functions of system failure detection and diagnosis in his role as a navigator. Of course, all such personnel have other duties aside from navigation.

Scientifically, we begin to capture this or any human-based system failure detection and diagnosis problem in the traditional behavioral science way - through task analysis. Figure 1 presents one type of task analysis description for a portion of the ship pilotage function which employs the ship's radar and fathometer (Smith et. al., 1976).

There have been several other ship task analysis studies, including a simulator validation study in which the watch officers' functions were observed at sea, recorded in minute detail relative to the external and internal ship situation, and the observed data were computerized for workload analysis (Hammell, 1979).

In a real and important sense, the ship navigational failure detection and diagnosis task is mental and invisible, which constrains our scientific ability to observe, record, and experiment with it. If it were a purely intellectual endeavor, then we might capture it through functional analysis or verbal protocols, as has been attempted (Huffner, 1978). Some of restricted waters shiphandling and pilotage seems to be intuitive. An athlete, artist, or scientist can only explain the general theory of those functions and provide a few examples. One can seldom say precisely why one moves left in one case and right in another; why one attacks (proceeds boldly) in one instance, but waits cautiously in another, apparently identical situation. Yet the pragmatic achievements resulting in part from those intuitive mini-decisions mark the difference between the seasoned professional and the novice or layman. Consistency in correct mini-decisions of this type helps to separate the occasional great individual in any field from the majority of competent, seasoned professionals. This is as true of the mariner as of the athlete or artist. This fact should not inhibit us in documenting systematically as much of a function as we can. We seek light and clarity as researchers; we do not glorify mystery. Conversely, we must not overestimate our ability to lay out all of human performance on the laboratory table.

TASK CODE: III.C.3

WORKER FUNCTION LEVEL AND ORIENTATION						WORKER INSTRUCTIONS	GENERAL EDUCATIONAL DEVELOPMENT		
DATA	%	PEOPLE	%	THINGS	%		REASONING	MATH	LANGUAGE
2	55	1A	5	3A	40	3	3	3	1

TASK CODE: III.C.3 **GOAL:** Navigate through (maneuver in) restricted waters as required in order to reach destination safely and expeditiously.

OBJECTIVE: Identify and respond to potentially hazardous conditions in order to avoid collisions, rammings, and groundings while simultaneously maintaining position within the limitations of the restricted waterway when some emergency arises.

TASK: Operates the radar and fathometer in order to detect and identify navigational hazards and aids to navigation.

PERFORMANCE STANDARDS

Descriptive:

- Selects the optimum combination of range scales, sector search, intensity, etc., for the most accurate and prompt detection of navigational hazards and aids to navigation.
- Accurately detects various aids to navigation and navigational hazards on radar.
- Accurately detects any navigational hazards on fathometer.

Numerical:

- In 100% of the cases, all necessary navigational aids and all navigational hazards are detected.

TRAINING CONTENT

Functional:

- How to manipulate radar unit, i.e., vary range scales, sector search selector, intensity, range and bearing circles and lines, true or relative motion mode, etc.
- How to manipulate fathometer unit, i.e., vary depth scale, intensity, etc.
- How to detect navigational hazards and aids to navigation on radar and fathometer.
- How to identify navigational hazards and aids to navigation on radar and fathometer.

Specific:

- Knowledge of navigational aids along track, or man-made and geophysical characteristics which present good radar targets.
- Knowledge of special hazards known along route which present radar targets.
- Knowledge of individual ship's particular radar unit.
- Knowledge of individual ship's particular fathometer unit.

Fig. 1. Radar/fathometer lookout task.

SHIP NAVIGATIONAL FAILURE DETECTION AND DIAGNOSIS

An unstable pilot steers a leaking ship, and
the blind is leading the blind straight to the pit.

Holy Bible, Matthew 15:14

The failures of interest in this paper are not failures of mechanical functions. Ships do break up at sea as a result of age, poor maintenance, design faults, and heavy weather. Detection and diagnosis of potential structural failure is basically a matter for diligent inspection and sound application of non-destructive testing methods. Also, engines do fail, leaving a ship helpless in the seas, but that is not the subject at hand. As mechanical failures occur on ships, they are normally detected, diagnosed, and repaired or bypassed safely. That is the engineering process control problem and is the subject of many other papers.

Ship Collisions and Groundings

The failures - accident situations - addressed by this paper are collisions and groundings of ships. These almost invariably occur close to land or in inshore waters (Card, Ponce and Snider, 1975). This is the part of navigation called pilotage, navigation by visual or sensor reference to landmarks. When these accident cases are sorted into many categories (boxes), they tend to group with one or two cases in a box (Faragher et al., 1979). The ship's master is normally on the bridge. There often is a local pilot on board. Such accidents most often occur in clear day or clear night situations, with no mechanical failure involved, and with alert, experienced, highly qualified personnel in charge. They rarely result from a single cause. Most often, several failures coincide. As a result, each accident may appear as a freak accident. The biblical reference in Matthew, then, better describes ordinary temptation and sin than common ship hazards.

One must admit, of course, that collisions and groundings do occur in fog or storms and to novice or drunken mariners and on improperly maintained ships. I suspect that such factors are not the core problem. These factors may simply worsen error tendencies that are present even under the best of conditions.

A Ship Collision

Given the great variety of errors that occur, no one case is really "typical". Nonetheless, let us take one case for illustration. The Navy submarine tender, USS L.Y. SPEAR, with a local pilot aboard, was upbound in the lower Mississippi River in the early morning of February 28, 1978. The Commanding Officer was on the bridge with the Navy helmsman and numerous other personnel. The local pilot was to testify later that the large number (seventeen on duty plus some unspecified number of others off-duty) of naval personnel on the bridge hampered his view of indicators and the external scene.

The weather was clear; there were no apparent unusual currents, winds, or traffic. The L.Y. SPEAR pilot decided to overtake the tanker, ZEPHYROS, which was properly holding the right bank, allowing ample way for a normal overtaking. Whistle and radio signals were exchanged achieving mutual agreement. Shortly after she would pass ZEPHYROS, the L.Y. SPEAR would be approaching a leftward bend in the river. Slightly above that bend was known to be a downbound tow of gas barges.

The steering was under control of the pilot and helmsman. The Commanding Officer retained ultimate responsibility and supervised the passage. He would normally defer to the pilot's specialized local knowledge unless he detected some reason to assume direct personal command. During the passage upriver, the pilot would order minor course corrections to follow the channel or to position the ship for a later maneuver. When the course order was "steady" (rudder amidships), the helmsman would steer as necessary to keep the ship on a straight path.

The pilot ordered 10° right rudder for a minor course correction as the L.Y. SPEAR began to pass the ZEPHYROS at a lateral distance of about 150 to 180 meters. The rudder remained at that angle until the bow of the L.Y. SPEAR was noticed to be turning toward the middle of ZEPHYROS. The pilot of ZEPHYROS and the Commanding Officer of L.Y. SPEAR simultaneously alerted the L.Y. SPEAR pilot, who immediately ordered left full rudder. In now-critical seconds, the following ensued:

1. The helmsman quickly turned the rudder as ordered.

2. There was a normal mechanical time delay for the ship's rudder to move to the ordered position.

3. There was a further normal delay for the hydrodynamic force of the water flowing over the rudder to begin to turn the ship, after which delay the leftward turn was rapid.

4. The pilot of ZEPHYROS stopped, then backed his engines, trying to reduce speed to help the faster L.Y. SPEAR get past him. (There was insufficient time for this maneuver to be fully effective).

5. On the L.Y. SPEAR, the Commanding Officer (C.O.) also ordered his engines full astern. This was an error because slowing

could not conceivably be accomplished in time. Also, the attempt to slow would disrupt the water flow over the rudder, lessening the effectiveness of the turn, which was the only means of avoiding the collision. The pilot immediately reordered engines full ahead, correcting that error.

6. The C.O. ordered the collision alarm sounded and word to be passed to the eleven hundred personnel of the L.Y. SPEAR that collision was imminent. These actions probably avoided many injuries.

7. The C.O. observed that his ship was responding rapidly to the left rudder command. As the ship's head swung forward and away from ZEPHYROS, the stern was swinging right and toward her. The C.O. told the pilot to shift the rudder. The pilot ordered right full rudder, which did slow the stern swing, but not enough to prevent L.Y. SPEAR's stern from glancing off ZEPHYROS' side.

8. After L.Y. SPEAR passed ZEPHYROS, the pilot gave several more full left and full right rudder orders, then took the helm himself and finally steadied the ship.

9. Radio contact confirmed that neither ship was badly damaged; both proceeded to their destinations.

Errors Involved in the Collision

Let us recoup the errors: the L.Y. SPEAR pilot let the helm remain on 10° right rudder too long at an inappropriate time; he may have forgotten that order was operative. He claimed later that a shear current pulled the ship to the right. The rudder setting, however, seems sufficient to explain the rightward movement of the ship especially at the 18-20 knot speed the pilot had ordered. Even if a shear current were experienced, as is plausible, he should have allowed and compensated for such currents. The pilot also failed to compensate for stern swing until reminded to do so by the C.O.

The Commanding Officer seems to have let too many people on the bridge. He was slow in alerting the pilot to the collision risk. He should not have reversed engines. Basically, he appeared to rely excessively on the pilot until collision was imminent. As the avoidance of the collision became hopeless, his performance improved; he helped to minimize both ship damage and personnel injuries.

The pilot of ZEPHYROS also allowed the situation to become critical before alerting the L.Y. SPEAR pilot (National Transportation Safety Board, "Marine Accident Report",1978).

What we see in such a case are a number of mental lapses, all of a momentary nature. We are led to believe that similar lapses occur vastly more frequently than accidents. Usually the difference between "harmless errors" and accidents can be expected to be the combination of errors in a close tolerance situation. Here the errors were not independent of each other; in many other cases, the errors are independent.

Collision and grounding cases are frequently detection and diagnosis problems. People fail to see a detectable aid to navigation, a fixed object, or an oncoming vessel. Sometimes they see a ship or object and initially evaluate it properly as not being a threat. Subsequently the situation changes, but the individual has failed to monitor the ship or object, perservering in the original decision that it is a non-threat until too late. Sometimes a hazard such as severe current shear is not in fact detectable until it is adversely affecting the ship. The possibility of such an event is often foreseeable, but required precautions are not employed. Frequently, communications fail in collisions. People fail to use their radios, or the channels are saturated with chatter, or they misidentify the vessel they are talking to, or each becomes certain that both have agreed on a course of action, but each has agreed to an action opposite to the other's (Paramore et al., 1979).

VIEWS OF MARITIME HUMAN ERROR

The trouble is he's lazy! The trouble is he drinks!
The trouble is he's crazy! The trouble is he stinks!
The trouble is he's growing! The trouble is he's grown!
Krupke, we've got ... troubles of our own!

Stephen Sondheim, West Side Story

The above quote is sung in the musical comedy by young men who are in frequent trouble with the law. They satirize the various well-intended psychologists, policemen, and social workers who try to deal with them by manipulation, punishment, and sympathetic support in alternating cycles.

As we engineers, psychologists, and operations analysts begin to examine human behaviour in various system failure detection and diagnosis areas, similar dangerous tendencies may arise.

1. We may look at the full job setting - both normal operations and emergencies/failures, but we concentrate on the failures.

2. Not being mariners (or reactor operators, airline pilots, or whatever), we tend to think of them as "them", who are therefore different from "us". They may see the eggheads and academics as very odd people with questionable (at best) qualifications to evaluate them.

3. We try to produce fast and definite answers to problems. Once we form a cherished hypothesis, then individual differences, the variety of unrelated studies bearing on these problems, and the richness of human experience allow us to select those facts and observations which reinforce our opinion and to dismiss contradictory indicators. If several of us do this, independently, then we will produce a variety of contradictory characterizations of human performance and human error within and among the specific subject areas we study.

4. If, on the other hand, we carefully weigh all the uncertainties of human error and throw up our hands, then new dangers arise. There is a classic rumor, at least, of a steam engine accident in which three automatic safety devices were wired open. Rather than deal with the "messy" human factors involved, the investigators recommended adding a fourth automatic safety device. How many of you have seen similar nonsense in your own field?

Diversity of Marine Safety Studies

Studies of human factors in maritime safety have addressed subjects as diverse as: the nature of the work, design of specific equipment, controls, and displays, layout and equipment lists for the work station (ship's bridge), seafarer psychology (especially authoritarianism and personality rigidity), sociology of shipboard life, shipboard work organization, inter-company and inter-ship differences in amenities, working atmosphere, and specification of required navigational procedures, literacy, physical health, and the relative roles of shipboard and shore personnel, such as pilots and longshoremen (Mara et al., 1968, Moreby, 1975, Anderson (ed.), 1977).

None of these studies has linked these factors to ship accident experience. Lacking the discipline of such linkage, one finds safety studies reaching broad and multi-faceted conclusions. For example, a Panel on Human Error in Merchant Marine Safety for the National Academy of Sciences concluded that ship accidents are caused by: 1. inattention, 2. the ambiguous master-pilot relationship, 3. inefficient bridge design, 4. poor operational procedures, 5. poor physical fitness, 6. poor eyesight, 7. excessive fatigue, 8. excessive alcohol use, 9. excessive personnel turnover, 10. high calculated risk acceptance, 11. inadequate lights and markers, 12. misuse of radar, 13. uncertain use of sound signals, 14. inadequacies of the rules of the road, 15. inadequacies of the casualty data base (National Academy of Sciences, 1976).

Recommendations that result from such studies cover everything from requiring expensive new equipment to changes in nearly every aspect of law, regulation, staffing, training, design, and operation of ships. Generally, none of these recommendations is suitably evaluated for range of applicability, expected benefit, social and economic cost, and locus or form of implementation. Many of the recommended actions have much less value upon close study than at first glance.

An Example of a Mechanical Safety Improvement

One new maritime safety measure is a type of device which automatically plots radar contacts. Now called simply Automated Radar Plotting Aids (ARPA), these devices have been widely promoted by their manufacturers as "Collision Avoidance (Radar) Systems". Who, after all, could object to putting even an expensive "collision avoidance system" on a ship? Seriously, such devices are based on plausible needs:

1. On oceangoing vessels which employ local pilots in inland waters, only brief periods of crossing coastal waters call for pilotage by the ships' crews. Even then, if the weather is clear, the primary mode of detecting and avoiding hazards is by visual lookout. Thus, the crews tend not to be thoroughly proficient at radar plotting. Computer-aided plotting is relatively quick and effective. It reduces rote mechanical chores, giving the watchstander more time to do what a machine cannot - think and decide.

2. Personnel may fail to detect an approaching vessel either due to multiple duties at different positions on the bridge or due to mental lapses or negligence. An alert which sounds whenever a target approaches (or is predicted to approach) within some preset distance can alleviate some forms of "inattention".

3. In multi-ship encounters, particularly with high closing speeds, the ARPA can provide a situation display far more rapidly than could a human being. The mechanics of operating such displays can be taught quickly and easily.

4. Various simulator studies have compared ARPA and unaided radar in open waters and moderately restricted coastal waters. ARPA sometimes improved miss distance or maneuver range. When scenario differences were tested, it was found that ARPA did not improve performance in cases where collision situations or fixed hazards were clear-cut. It did help in ambiguous or subtle situations (CAORF Research Staff, 1979).

Hoping to reduce the probability of collisions, many shipowners have installed ARPAs voluntarily. Planned U.S. and international actions will require these devices on many more ships. Yet, based on U.S. ship casualty experience, I would not expect to find any statistical decline whatever in ship collisions due to so-called "collision avoidance systems".

1. The vast majority of collisions occur in inland waters in clear weather with a local pilot on board. Often the radar is not even turned on. If it were on, the alerting function would almost certainly be turned off or ignored because of the large number of objects that routinely approach within collision threat detection range.

2. Even in lower visibility, with the radar and ARPA turned on, the pilot is unlikely to use it. Many pilots do not seek or accept information from the ships' crews or onboard systems.

3. ARPAs depend on the ships' radars and share their limitations. Radars are defeated by intervening land masses, are subject to target masking by a nearer object with a strong radar return, and are degraded by precipitation. Inherent angular error of the radar may easily cover the full ship channel width.

4. ARPAs use linear extrapolation to plot other vessel tracks and to compute the amount of time to closest contact and the projected passing distance. If either own ship or the target maneuvers, the projections become inaccurate until both have steadied onto straight tracks again. But it is these nonlinear tracks and unexpected maneuvers which are most commonly involved in accidents. Proponents counter that ARPA can aid in the rapid detection of unexpected maneuvers by other vessels. It seems to me that this is generally true only in congested coastal waters, such as the English Channel, where there is very heavy irregular traffic, no local pilot, a lack of fixed obstructions, and sufficient angular discrimination to favor radar-ARPA characteristics. The U.S. ship casualty data base contains very few such accidents.

5. In a study of U.S. coastal and inland collisions, the potential benefit of computer-aided radar systems was evaluated by carefully drawn criteria. For example, if the threat detection and evaluation function was properly performed and timely without ARPA, so that no ARPA could have done more, or if the target was obscured by an intervening land mass until too late for collision to be avoided, the possibility of benefit was eliminated. All doubts (such as whether the radar would have been turned on and/or utilized) were resolved in favor of the system. The maximum possibility of ARPA utility was found to be only 10 to 13 percent of the cases (Stoehr et al., 1976, Gardenier, 1976).

6. Reducing accidents by the full 10-13 percent would require perfect reliability, perfect use and interpretation by operators, perfectly timely and effective communication of the information, and perfect use of the information for decisions. Failure in any of these functions would preclude preventing some accidents.

7. Some well-maintained ARPA sets have been found operable as much as 80 to 85 percent of the time desired; others have far lower availability. Radars are not used during many clear weather, restricted water passages. High rates of crew turnover and limited chance to work with the system mitigate against assurance that a given mate is qualified to use and interpret the device. Mate-master and mate-pilot relationships inhibit volunteering information. Different languages and cultures between crew and pilot may further degrade communication.

For all of these reasons, the inherent capability of ARPA to reduce collision frequency is well below the theoretical maximum 10 to 13 percent, even before deducting for the ability of ARPA to contribute to collisions.

Concern has been expressed that reliance on ARPA might further degrade the level of radar plotting skills in the fleet, such that failures of the computer system would leave the ships more vulnerable to misdirection than they are now. Mates initially trained on one system who are standing watch on a ship with a different analog display may make serious mistakes. Use of ARPA to increase voluntary crew communications to pilots could easily do more damage by distracting or confusing the pilot than good by providing assistance.

On the positive side, an ARPA can be used very well as a navigational aid in making landfall, as by distinguishing stationary buoys from slow-moving small craft. It can be valuable in promptly resolving rare instances of complex multiship encounters. Some systems may alert an inattentive mate. Accidents preventable by ARPA could involve major environmental damage. Conversely, very few accidents will be prevented, and some may be caused by ARPA.

An Example of a Human Factors Improvement

Let us look at a very different solution concept, relying on human factors, which is being widely adopted in the United Kingdom. It is called "bridge team training", which was given considerable impetus by Shell following an intensive study of ship accidents to their vessels. The concept was summarized for me by a Shell official as follows:

"As can be seen, most navigational accidents occurring to Group vessels have been attributed to human error experienced by competent Officers on well-found vessels. Investigation determined that the Officers did not use all the information available to them, and often made decisions based on only one type of aid or technique. It was further found that Officers did not communicate well with each other. While passage planning was accomplished, the plan was seldom explained to Junior Officers or Pilots. This led to situations where Junior Officers would not question actions of their superiors, even though they suspected that a grave mistake was occurring. Finally, it was found that mariners were not alert to the possibility of human error. They did not expect error ... and therefore did not check to see if it was present".

"After over two years of sponsoring bridge team courses, we believe them to be the single most effective means of improving navigational safety. By going beyond the realm of traditional skills and focusing on the total problem of navigation management, we believe that such courses can do much towards eliminating human error" (Barth, 1979).

Let us examine the evidence concerning this assertion. The first difficulty is that there is no source of information that will reliably tell us the navigational management practices utilized on various vessels now. Hardly any master will respond to a question or series of questions by stating or implying that his current navigational planning or teamwork has serious controllable deficiencies. If he believed that, he would change them.

Accident data tell us that collisions in the United States occur generally in pilotage waters; yet the integration of local pilots with the ship's bridge team is not nearly as controllable as is the crew navigational practice. The perception of the value of team training varies widely from the United Kingdom, where it has become rather widely popular, to the United States, where its use is primarily on military vessels. It is also true that accident data reveal numerous instances where, in the absence of mental lapses, vessels do not perform as intended due to wind or current forces and/or local waterway geography or traffic and/or controllability peculiarities of individual ships. Finally, team training being a human factors solution, the effort required to achieve the potential benefits extends well beyond the mere provision of, say, a one week training course. The required implementation includes significant continuing efforts at motivation, enforcement, and perserverance in indoctrination of crews which experience more rapid turnover than is common on oceangoing tankers such as Shell's.

That is the bad news on team training. Now let us look at the good news. Lacking a sound data base for empirical evaluation of the concept, we will rely on normative analytic methods. Figure 2 is the summary level of a collision, ramming, and grounding fault tree (Graham, 1979). It was prepared in conjunction with an offshore deepwater port (DWP) hazard analysis, but applies to other scenarios as well. This particular tree (which is certainly not uniquely descriptive of the general problem) emphasizes successively constrained time frames from bottom to top and a wide variety of factors to be monitored. As suggested earlier in Figure 1, some factors can be monitored by several different sensors concurrently.

This sort of analysis suggests that no one individual (pilot or master) could reliably monitor all cues with all independently useful sensors concurrently under all normally encountered conditions of pilotage. Without effective teamwork, the full sensor and information processing capability which is inherent in even a poorly designed ship's bridge must not be utilized. It is easy to see that practitioners might readily tolerate such a situation if the sensor capability which they normally use is more than adequate for the majority of situations encountered, and sufficient to have somehow gotten them through the worst situations they have so far encountered without serious incidents. The value of third or fourth level redundancy is only apparent, we should recall, where multiple errors combine with a close tolerance situation. Furthermore, it is easier to dismiss each accident as a "freak" occurrence than it is to recognize the value of expending considerable continuing effort to preclude an accident that might otherwise have only a thirty percent probability of occurrence in a mariner's full career.

It appears that even where wind and current effects in particular transits are appreciable hazards, the support potentially available to a master or pilot from a well-organized bridge team is valuable. Utilization of that support would appear to require only three achievable prerequisites:

1. Pilots must explain their intentions in detail in advance of the transit, or as early in the transit as it is safe to do so;

2. Watch personnel must have specific reporting requirements, mainly factual in nature; and

3. The master must have the confidence and sensitivity to use "double check" information or questions in a way that reinforces the pilot's shiphandling without questionning his competence.

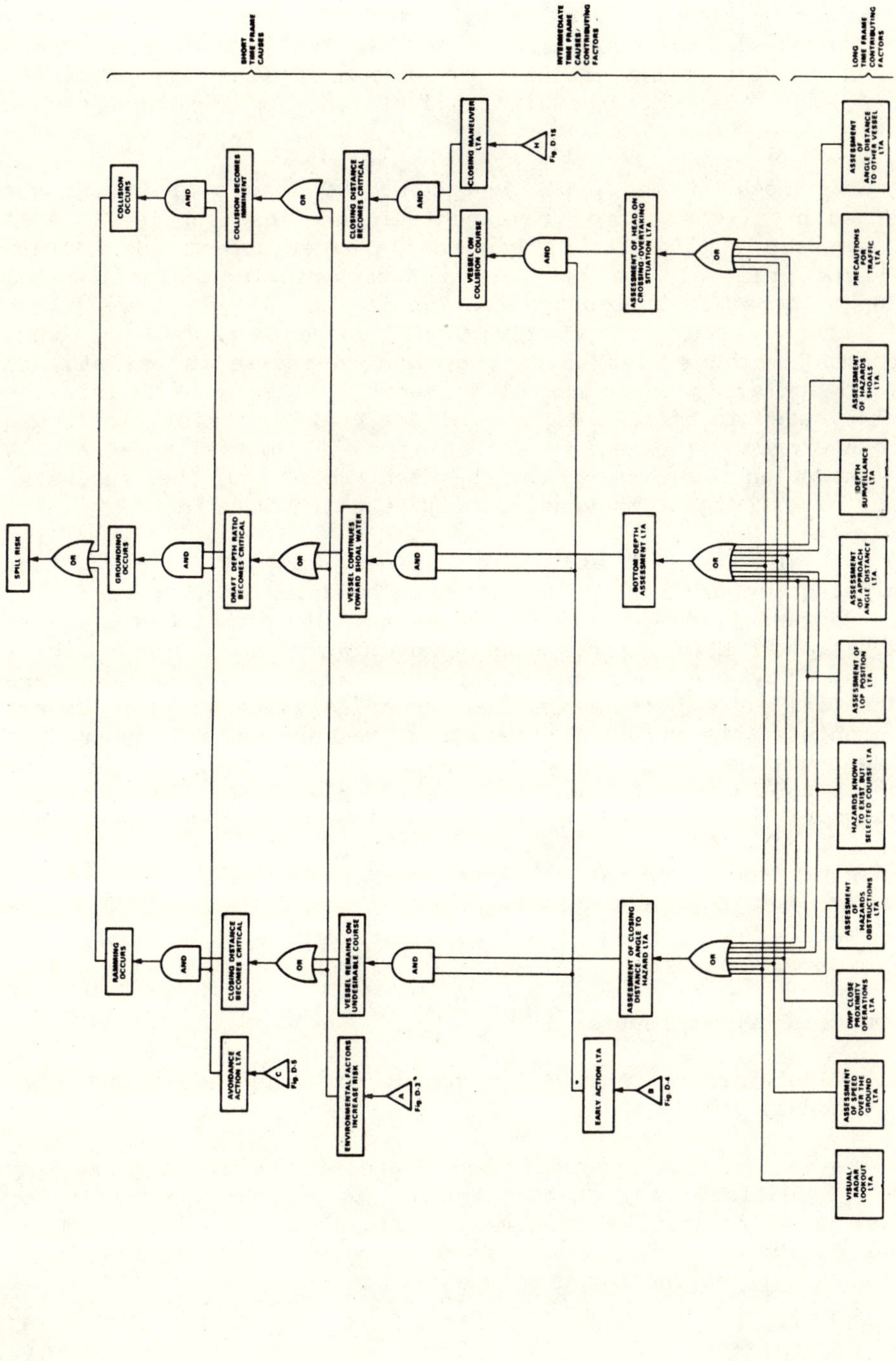

Fig. 2. Summary fault tree.

Finally, using another type of normative process, engineering analysis, we find further support for the team training concept. The principle of redundancy in engineering design extends readily to operational watchstanders. The "buddy" principle in swimming, co-pilots in airliners, and shift supervisors in many process control applications, all embody the principle of human redundancy. With teams of three or more professionally qualified personnel, sophisticated combinations of functional redundancy are feasible.

This look at the pro's and con's of team training is not supported by the amount of research evidence that applies to ARPA devices. Even Shell has not had enough experience to be certain that this principle is actually reducing their accident rate. Looking to more widespread application of the principle, we do not know enough about the variety of current practices, what enforcement methods might apply, or how to advise the pilots and crews to achieve greater useful teamwork. Pilots have to maintain a healthy scepticism that efforts of the crew to assist them could be counterproductive and distracting until mutually agreeable guidelines can be worked out between them and the extensive variety of clients of many nationalities which they face.

All of the evaluation of devices like ARPA or of team training, or of other efforts to reduce human error (Henry et al., 1968, Roggema, 1977, Inter-governmental Maritime Consultative Organization, 1978) face the characteristics of the above examples. They are complex, and partly because of that they are subject only to persuasive, not conclusive argument. Expert assessments of the value of safety measures are not merely subject to disagreement; they are certain of disagreement. Is there no criterion or model by which safety values can be reliably assessed?

MODELS AND METHODS FOR SAFETY IMPROVEMENT EVALUATION

> "Nature has some sort of arithmetic-geometrical coordinate system, because nature has all kinds of models. What we experience in nature is in models, and all of nature's models are so beautiful".
>
> R. Buckminster Fuller (inventor of the geodesic dome) (Thomas, 1966).

System failure detection and diagnosis is subject to many forms of modeling. Many of these relate to evaluation of risks or to the value of safety measures. Certainly, to one analyst or another, virtually every form of model is "beautiful". Models do very well in improving our understanding in many scientific areas, from simple mechanical models such as the inclined plane to complex, living system models like the structure of the DNA molecule.

In a Platonic sense, it is tempting to believe that there exists an ideal model for evaluating the system failure detection and diagnosis process. Our various efforts at modeling (such as fault trees, adaptive controller simulations, and others), we may hope, are approximations to that ideal model - some better than others. Once some critical mass exists of enough empirical data, enough imperfect models, and enough dedication, then the "true" model will be revealed as a robust synthesis and simplification of all the best in the earlier efforts. Much of science has, in fact, progressed in this manner (Margenau, 1950).

Until that golden day arrives in this field, however, we have numerous problems with the individual characteristics of models. Let me illustrate in the ship navigation area, although I feel certain that the same principles apply to parallel applications:

1. Enough is known by naval architects to model the movement of ships through water at many levels of complexity, using mathematics alone or physical scale models as well. Modern ship design would be impossible without such models. In safety, they help us to find parametric boundaries of encounter situations where ships are barely unable to avoid each other or channel boundaries. They do not tell us much about why such situations arise or how to prevent them.

2. Statistical models of accident rates or track deviations give us various plausible distributions of increasing deviation from intended track with decreasing frequency of occurrence. They are least potent, however, in comparing relative accident rates of the more thoroughly described (hence smaller) subcategories of accident characteristics and in describing the most hazardous (least frequent) deviations. The first problem is the fact that sample sizes become intractably small as the extent of identification of the sample becomes more complete. The second problem is the fact that statistical models are generally more reliable and powerful in characterizing the means of distributions than they are in characterizing the tails of those distributions.

3. The relatively new concept of "catastrophe theory" offers the possibility of describing the radical departures from normal operation which characterize many accidents. Its application appears to require, however, that we must first have a descriptive understanding of the causal elements and the relevant metrics of human error in order to apply the model elements. Those requirements, of course, are our most prominant deficiency.

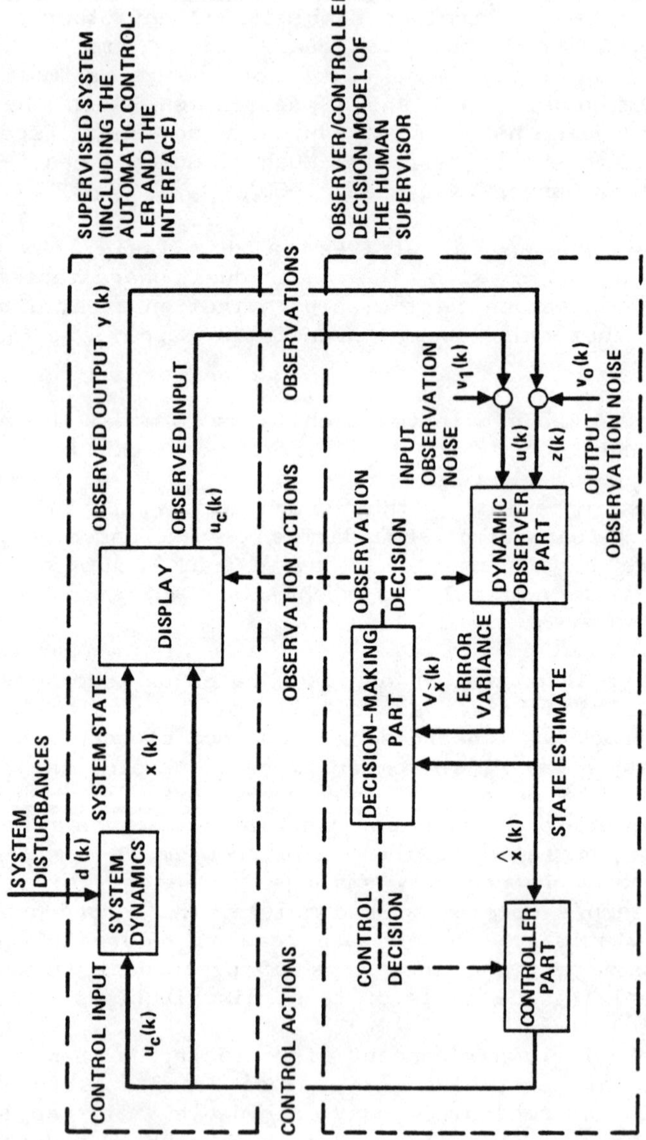

Figure 3. Structure of the observer/controller/decision model

4. Various forms of system feedback, or controller, models also look tempting. A generalized form of a controller model which we are currently exploring is displayed in Figure 3. An interesting feature of this form of model is the explicit separation of the actual interactions (of system disturbances, system dynamics, and displays) from the psychological-perceptual nature of input and output estimation and decision. The basic difficulty here is that the model forces us to postulate explicit observation and decision functions, but allows us to "fudge" the results with several "noise" models. It would appear likely, therefore, that numerous alternative functional process algorithms could be made to fit fairly well with observed performance. How, then, would we select among alternative and incompatible hypotheses? As in correlation analysis, the best fit does not guarantee the most truth; the best description does not necessarily support the best prediction.

A less demanding approach to evaluation criteria for system failure detection and diagnosis is through "open system" methods instead of "closed system" models. We have explored the use of quasi-experimental methods, as described by Campbell and Stanley in the 1960's (1966). The basic principle is to select and/or reinterpret on-hand data in a manner which approximates the output of a controlled experiment. One may then apply analysis of variance, or time series, or regression methods to explore for statistically significant interactions of accidents and hazards or safety measures. We have had narrative accident reports re-read and re-coded successfully so as to explore accident causes and the probable usefulness of potential or recent changes (Dunn and Tullier, 1974). Some of the problems with this approach, however, are:

1. It is very difficult to devise "controls" for the large variety of potentially confounding factors.

2. It is laborious and expensive to get large enough samples to allow exploration of realistically complex variable interactions; given a large data base it is hard to find efficient discrete multivariable analysis routines to process it.

3. It is hard to get practical managers or engineers to use the method scientifically. The temptation is to dispense with all controls, read a sample of cases, and make a simple bivariate judgment in each case: "Yes, showing motivational films to watchstanders monthly would have helped prevent this accident, but not that one".

Provided that one can afford them, controlled behavioural experiments are very nice. Ship bridge simulators are relatively new, but they are used to explore behavioural impacts of waterway design, ship design, training effects, and external assistance (aids to navigation or vessel traffic services) (MARSIM 78; Third CAORF Symposium 1979). Apart from standard problems of sound experimental design, the major difficulty lies in the great expense of adding variables. This leads us to gamble with low degrees of freedom, to forsake desirable controls, and/or to use screening designs and nonparametric correlations rather than more robust experimental designs, such as full factorial analysis of variance.

Also within the general area of fairly rigorous quantitative models and methods are the wide variety of mechanical and electrical engineering design and evaluation methods. As valuable as these have shown themselves to be over many years, they lack any standard acceptable means of predicting the quality of the behavioural interface with operating and maintenance personnel.

IN FAVOUR OF MULTI-FACETED RESEARCH

> We should not pretend to understand the world only by the intellect; we apprehend it just as much by feeling. Therefore, the judgment of the intellect ... must, if it be honest, also come to an understanding of its inadequacy.
>
> Carl Jung, Psychological Types

As we move from mathematical models to more empirical forms of investigation, we lose intellectual rigor, but gain the value of the feelings operators have for their jobs. The rigorous unreality of the model fades into the often paradoxical credibility of the "sea story".

Direct collection of the wit, wisdom, and diverse insights of operators does not have to be messy and intractable to scientists. The limits of ways to organize data collection are the limits of imagination. We may, for example, use some variation of the critical incident technique, wherein operators describe hazardous situations they have experienced (Zeitlin, 1975). We may do a process control trace, where we use mock-ups or diagrams on paper to ask operators to describe to us their functions and decisions in a logical sequence (Smith et al., 1976).

There is an American Indian saying, "Do not judge a man until you have walked a mile in his moccasins". Coast Guard researchers have often found it invaluable to observe and discuss maritime operations directly as they are performed, provided that we can do so safely (Faragher et al., 1979; Huffner, 1976).

We can use checklists, structured interviews, open form interviews, workshops, games, role-playing, and symposia as tools of behavioural research in system failure detection and diagnosis. Each reader can add substantially to my little list.

The quote from Jung above foreshadows a major recent discovery in psychological science - that different types of associative processing tend to be done by the left and right hemispheres of the brain. Generally, language use and other discrete, sequential types of processing are controlled by the left side while orientation in space, pattern recognition, and other forms of nonverbal, holistic, analog processing are more frequently controlled by the right side, at least for right-handed people. Concurrently, the left side tends to process vision from the right visual field and muscular control of the right side of the body while the right hemisphere tends to process information from the left visual field and controls the left side of the body (Glass, Holyoak, and Santa, 1979). Much more needs to be learned than is already known about this distinction, which is not the same in all people.

Dr. Myriam Smith (1980) has suggested that much of ship pilotage may be based on right hemisphere processing. Much of engineered information processing involves reduction of a complex, patterned scene to a few discrete, sequenced symbols. Whether engineers attempt to regroup the processed data into an analog (graphic) display or leave it to an operator to recreate a quasi-visual impression based on the coded data, it is likely that much of the information of value in the visual scene is lost in engineered processing involving data extraction, compression, coding, and sequencing. This concept, combined with the relationship between visual field and brain hemisphere may have significant implications for design and arrangement of control consoles and work stations.

Similarly, it may be that classroom-style training and licensing examinations are not as relevant as they could be to some practical skills like shiphandling, due in part to the reliance of the former on left-side types of processing whereas the practical performance may rely heavily on right-side types of processing. The fact that individuals vary in their skills in right-side versus left-side types of mental processing (due both to genetic and environmental influences) could imply that those personnel who are best selected for pilotage functions may be somewhat less compatible with the training and licensing techniques to which they are commonly exposed. We need to explore the implications of this concept for human detection and diagnosis of system failures in many applications.

As a practical matter, we can never find enough money, time, or manpower to do all of the research that should, in some sense, be done. The mere effort to do substantial amounts of research seems to be unpopular with operational managers, commanders, regulators, taxpayers, and others who are less interested than scientists in the deductive aesthetics of models and the rich variety of empirical inference. One would almost suspect that they would actually forego any research whatsoever if there were some way to move directly to a 99.99 percent reliable method of system failure detection and diagnosis.

There are established rules of thumb for human engineering design, work organization and scheduling, operational quality-control, and continuing training which tend to be ignored even in areas of crucial practice. For example, U.S. nuclear plants are said typically to use three eight hour shifts, rotating each shift by eight hours each week. Physiologically, this pattern is terrible because operators barely adjust their circadian rhythms before they are disrupted again (Sheridan, 1980).

A few common sense rules of human engineering design which are routinely ignored are:

1. Test staff-machine interfaces in the design phase, just as you would test machine functions or machine-machine interfaces.

2. Try to minimize the safety criticality of inter-personal communications where possible (as between piping maintenance crews and control room watchstanders?) because missed communications are not only common, but also difficult to detect (Henry et al., 1968).

3. Recognize that everyone has occasional mental lapses, In situations where these could have serious consequences, provide for human functional redundancy through as carefully organized a planning process as you would use to provide for mechanical redundancy.

4. Hold fire drills. Hold drills and provide training for other emergencies as well.

A POINT OF VIEW

System failure detection and diagnosis involving staff-machine interfaces is not well understood even in fairly straightforward, slow systems, such as ships. Complexity and/or speed of operation may worsen the basic problem. Alcohol, drugs or fatigue do make it worse.

Before one can realistically hope to understand the problem in an established operating environment, one must compile and examine carefully detailed records of normal operations and of things that go wrong.

Because of the rich complexity of socio-technical systems, it is easy to say that fourteen or fifteen categories of problems should be attacked immediately and, by implication, with unlimited resources. Careful examination of most "solutions" will reveal numerous favourable and unfavourable points regarding each proposal. In general, I have come to distrust every claim that a machine or an automated system would solve a human factors problem. Claims of effectiveness for proposed improvements in system failure detection and diagnosis of staff-machine systems are increasingly credible as they meet more rather than fewer of the following criteria:

1. They are shown to be consistent with established principles of human engineering, as well as systems engineering.

2. They relate well to organized behavioural science knowledge of normal operations and maintenance, as well as failure experience.

3. They are evaluated by a systematic logical process which is uniformly applied to two or more serious, but radically different, alternative approaches.

4. They are subjected to a demanding experimental or pilot process for rigorous evaluation prior to full implementation.

5. They are credible to highly professional, experienced operators, preferably as a result of an iterative process wherein such operators have contributed to the design and evaluation of the improvement.

6. They are implemented in a manner designed to remove any implicit threats from the change to all existing employees and to motivate their cooperation with the improvement.

Considering these factors, some of the most hopeful developments emerging currently in the ship navigation field are clarification of the pilot-crew interface, bridge team training, improved bridge design, and use of ship simulators for training and, potentially, for licensing (Gardenier, 1979).

Despite the understandable sense of urgency that managements have about safety of critical systems, it is an inescapable fact that behavioural and operational science in this field lag behind the physical sciences. Much research is required in applying routine operational studies to additional environments, in accident investigation and analysis, in systems modeling, and in empirical/experimental studies of systems failure detection and diagnosis behaviours.

Conversely, scientists must not, in their fascination with advancing the state of the art, ignore principles of improvement that can be made quickly and cheaply.

REFERENCES

Anderson, D. (ed.), 1977, "Human Factors in the Design and Operation of Ships", Proceedings of the First International Conference on the title subject, Gothenberg, Sweden, February 1977.

Barth, R., 1979, Letter to J.S. Gardenier re: "Bridge Team Training", London, Shell International Marine Limited, 9 November.

Campbell, D.T. and Stanley, J.L., 1966, "Experimental and Quasi-Experimental Design for Research", Chicago, Rand--McNally.

CAORF Research Staff, 1979, "A Compendium of Collision Avoidance Performance Using Various Shipboard Electronic Aids", Computer-Aided Operations Research Facility Report 13-7901--01, Kings Point, New York, National Maritime Research Center, April 1979.

Card, J.C., Ponce, Paul V., and Snider, W.D., 1975, "Tankship Accidents and Resulting Oil Outflows, 1969-1973", 1975 Conference on Prevention and Control of Oil Pollution. Washington, D.C., American Petroleum Institute.

Dunn, W.A. and Tullier, P.M., 1974, "Spill Risk Analysis Program, Phase II: Methodology Development and Demonstration", U.S. Coast Guard Report CG-D-15-75, Springfield, Virginia, NTIS AD A785026, August 1974.

Faragher, W.E. et al., 1979, "Deepwater Ports Approach/Exit Hazard and Risk Assessment", esp. Appendix G, U.S. Coast Guard Report CG-D-6-79, Springfield, Virginia, NTIS AD A074529, February 1979).

Faragher, W.E. et al., 1979, "Deepwater Port Hazards and Risk Assessment", esp. Appendix G, U.S. Coast Guard Report CG-D-6-79, Springfield, Virginia, NTIS AD A074529, February 1979.

Gardenier, J.S., 1976, "Toward a Science of Merchant Marine Safety", Marine Traffic Systems, Rotterdam, Netherlands Maritime Institute (1976) Major extracts reprinted in Schiff und Hafen 7:613-616.

Gardenier, J.S., 1979, "Where Are We Headed With Vessel Bridge Simulators?", Proceedings of the Third CAORF Symposium: Marine Simulation Research, October 15-16, 1979" Kings Point, New York, National Maritime Research Center.

Glass, A.L., Holyoak, K.J., and Santa, J.L., 1979, Cognition, Reading, Mass., Addison-Wesley.

Graham, W.C., 1979, "Fault Tree Analysis", Appendixes D and E of Deepwater Port Hazard and Risk Analysis, U.S. Coast Guard Report CG-D-6-79, Springfield, Virginia, NTIS AD A074529, February.

Gray, W.O., 1978, "Human Factors", Oil Companies International Marine Forum Safe Navigation Symposium, Washington, D.C., 17-18 January 1978.

Hammell, T.J., 1979, "Validation of Mate Behaviour on CAORF", Kings Point, New York, Computer-Aided Operations Research Facility, February 1979.

Henry, W.O. et al., 1968, "Human Engineering Guidelines Applicable to Merchant Marine Bridge Design", Vol. III of Human Factors in Ship Control, Connecticut, General Dynamics Corporation, Springfield, Virginia, NTIS AD PB 179357.

Huffner, J.R., 1976, "Pilotage in Confined Waterways of the United States: A Preliminary Study of Pilot Decision-Making", U.S. Coast Guard Report CG-D-96-76, Springfield, Virginia, NTIS AD A029715, July 1976.

Huffner, J.R., 1978, "Pilotage in the Port of New York", U.S. Coast Guard Report CG-D-81-78, Springfield, Virginia, NTIS, July 1978.

Inter-Governmental Maritime Consultative Organization, 1978, "Final Act of the International Conference on Training and Certification of Seafarers, London, IMCO (STW/CONF/12, 5 July 1978.

Mara, T. et al., 1968, "Human Factors in Ship Control", Vols. I-III, Groton, Connecticut, General Dynamics Corporation.

Margenau, H., 1950, "The Nature of Physical Reality: A Philosophy of Modern Physics", New York, McGraw-Hill.

"MARSIM 78: First International Conference on Marine Simulation, September 5-8, 1978", London, The Nautical Institute.

Moreby, D.H., 1975, "The Human Element in Shipping", Colchester, England, Seatrade Publications, Ltd.

National Academy of Sciences, 1976, "Human Error in Merchant Marine Safety", Washington, D.C., NAS Commission on Sociotechnical Systems, Maritime Transportation Research Board, June 1976.

National Transportation Safety Board, 1978, "Marine Accident Report: Collision of USS L.Y. SPEAR (AS-36) and Motor Tankship ZEPHYROS (Liberian), Lower Mississippi River, February 22, 1978", Washington, D.C. N.B. The fact and error summaries herein were extracted and interpreted by the author from this one report, with no supplementary source of data. As of this writing, the official U.S. Coast Guard findings on the same accident were not available to the author or to the public.

Paramore, B. et al., "Task Performance Problems in Reports of Collisions, Rammings, and Groundings in Harbors and Entrances", U.S. Coast Guard Report CG-D-28-79, Springfield, Virginia, National Technical Information Service NTIS AD A071658, March 1979.

Paramore, B. et al., 1979, "Human and Physical Factors Affecting Collisions, Rammings, and Groundings on Western Rivers and Gulf Intracoastal Waterways", U.S. Coast Guard Report CG-D-80-78, Springfield, Virginia, NTIS AD A074290, January 1979.

"Proceedings of the Third CAORF Symposium: Marine Simulation Research, October 15-16, 1979", Kings Point, New York, National Maritime Research Center.

Roggema, J., 1977, "The Design of Shipboard Organization: Some Experiences with a Matrix-Type of Organization in Norway", Maritime Policy Management 4:265-276.

Sheridan, T.R., 1980, "Human Error in Nuclear Power Plants", Technology Review 24-33.

Smith, J. et al., 1976, "Task Analysis Report Relative to Vessel Collisions, Rammings, and Groundings", Vol. II. U.S. Coast Guard Report CG-D-1-77, Springfield, Virginia, NTIS AD A037317 (Associated Volumes are I, AD A037316 and III, AD A037442.

Smith, Myriam, 1980, Personal communication.

Stoehr, L.A. et al., 1976, "Spill Risk Analysis: Methodology Development and Demonstration", U.S. Coast Guard Report CG-D-21-77, Springfield, Virginia, NTIS AD A043054.

Thomas, C., 1966, "In the Outlaw Area", New Yorker.

U.S. Coast Guard, Annual, "Marine Safety Statistical Review", Commandant Instruction M16700.2, Washington, D.C.

Zeitlin, L.R., 1975, "Human Causal Factors in Maritime Casualty and Near Casualty in the United States Merchant Marine", Kings Point, New York, National Maritime Research Center.

TROUBLESHOOTING IN THE COMMERCIAL COMPUTER INDUSTRY: A SUCCESS STORY

Nicholas A. Bond, Jr.

California State University
Sacramento, California 95618, USA

INTRODUCTION

There are certainly extremes of "success" and "failure" in the troubleshooting domain. Some of the most spectacular failures occur in the military setting. A good example was the Mayaguez affair in 1975, which followed the Vietnam war, and was America's most recent combat activity. Some months after that incident, Secretary of Defense Schlesinger admitted that:

> "... The thirty-one-year-old carrier Hancock ... operating without one of its four shafts ... never reached the scene. The helicopter carrier Okinawa, ... with part of its boiler plant off the line ... also never arrived at the scene. The escort vessel Holt, the first ship at the scene, had power-supply problems, and consequently its main battery was down the night before the engagement."

During the presidential campaign of 1980, other military maintenance horrors were cited: one candidate, when commenting on the possibility of U. S. Naval action in the Iranian crisis, remarked that only half of Navy ships could reach a designated area on schedule, and that only half of the planes aboard the carriers would be able to fly in a combat state.

But many success stories can be cited too. There are several areas wherein complicated equipments are well maintained, and are quickly returned to service when they fail. To name just five of these areas, consider broadcasting companies, the TV repair shops, NASA space instrumentation program, physical and chemical laboratory operations, and large commercial computer centers. In all these cases, the diagnostic problem appears to be rather well solved. For purposes of this volume, it seemed worthwhile to examine the commercial industry. Prime computers and their associated hardware are among the finest technical achievements of the era; they are exceedingly complex; they usually demand logically-directed search when they malfunction; and yet they are maintained by ordinary non-professional people. Perhaps the general success of fault-locating efforts in the computer industry can serve as a model for other domains with maintenance problems. Also, since minicomputers now appear in schools, homes and small businesses, the commercial experience should indicate what maintenance picture we can expect when more powerful digital systems are very widely distributed.

A few months ago, I was present when a mainframe went down in a university computing center. A technician arrived within an hour, and after a little discussion with the operators he began to load some special test routines and consult his manuals. The search seemed to go fairly smoothly, and at least in the early stages the work appeared to be a matter of loading software packages, and comparing their printouts with desired outputs in the manuals. To an outsider, the manuals seemed completely unintelligible, but the technician referred to them and seemed to "branch" his behaviour accordingly. At no time did the technician exhibit "inspiration" or sudden insight; nor did he attempt to replace whole segments of the equipment. After a while the technician used an extender probe to take a few voltages, and then he replaced two modules. In this case the replacements restored the machine to operation, and the troubleshooting episode was over. As it happened, the technician made three more service calls that day, all of them on input-output equipment such as readers, printers, and disc drives. He said that the day was fairly typical.

Obviously, many factors contributed to this success story. The technician appeared well trained and confident, and he seemed to enjoy what he was doing. The reference materials he used were clearly essential in this particular problem, and the observer had the strong feeling that they would generally be useful in bracketing a fault. From all appearances, the mainframe itself was designed for relatively easy maintenance; because of the high parts density, the voltage checks took some careful application of extender probes, but the little pins and boards were legibly labeled; the voltage probe "clicked in" with a solid feel. At a superficial level, this success story is "explained" by good people, good equipment design, and good manuals. However, just how were all these good items actually obtained and utilized? To get some preliminary answers to this question, observations and interviews were carried out in two contract maintenance centers in California.

Both of the companies operating these centers were large and successful; most of the equipment they maintained was manufactured by other divisions of the company. One center had 18 full time technicians; the other had nearly seventy.

ADMINISTRATIVE CONSIDERATIONS

Anybody who spends even a little time at a contract maintenance center will be impressed with the major operating axiom: the equipment must be kept on the air; long down-times simply cannot be tolerated. From all indications, a company which services its own products is highly motivated to make them serviceable. The maintenance requirement extends well into the design divisions as well, so the field service division can insist on good "design for maintainability" in its major equipment items. And it can require that the manual writers, parts suppliers, and other support people do an effective job. There is rapid feedback from the customer on how well the whole maintenance enterprise is going along. Users quickly complain if a major system is down frequently; they can even cancel equipment leases, and everybody knows this.

Both centers I visited have time limit policies for correcting faults in the field. An ordinary service call is expected to be accomplished within an hour or two after the technician gets to the site; if it takes longer than that, the supervisor will be notified. If the case runs four hours or longer, an "alert" procedure is pursued. During alert, a specialist may visit the site, or consult with the technician on the site via phone. Remote "support centers", perhaps back at the factory, can also be interrogated 24 hours a day to help. If the malfunctioning item is rather small, such a disk drive, a whole replacement unit may be sent out, though this is not often done. On rare occasions, a large mainframe computer may have a fault that is due to unusual circumstances prevailing between a very complex software system and the operating situation; and so both software and hardware people will be sent out to the scene.

A contract maintenance center approaches a true meritocracy. People get ahead strictly on their merits; you cannot "bull" your way into a soft job. Each technician is expected to produce, and to be self motivating. Competence is recognized and highly regarded; there are mistakes, to be sure, but I never saw a slipshod or really inadequate performance.

SELECTION AND TRAINING

Most computer technicians come from two "pools"; one pool is made up of the junior college or technical school people. Here a typical applicant would have two years or so of technical training, with perhaps an Associate in Arts degree, and some twenty college hours of electricity, electronics or shop courses. The second major pool consists of military dischargees who have had at least six months of electronics or avionics, with another few years of field technician experience. All applicants take a small battery of commercial tests; verbal, quantitative and basic electronics scores are considered to be predictors of future success, but validity data were not available. A field supervisor may interview a batch of candidates, and send his ratings in to the personnel office. Of those who survive initial screening, less than one-third are hired. This suggests a fairly low selection ratio, and thus a high effective validity. In the two companies visited, layoffs and hiring freezes were rare over the past decade; the industry has enjoyed steady growth, though the rate of increase in manpower is less than that of company sales.

Nearly all technicians start at the bottom. There are four to six job classification levels for technicians, with 1979 salaries in the $ 9,000 to $ 20,000 range. Supervisors are chosen from within the company, and their salary expectations are about twice as high as the technician's.

Practically all technician training is given by the company itself; this is due to the heavy practical emphasis on "this company's products". Recruits are often taught to service input-output devices first, in courses lasting 9 to 12 weeks. There are two reasons for this: about 50 percent of service calls are due to these devices, and so a person trained on them can be sent out to do useful work a couple of months after being hired; also, mainframe computer casualities are becoming relatively rare. Classes are very practical and intensive, with small numbers of students; there are rarely more than 25 in a class. Many practice problems are solved on the actual hardware, so that at the end of a course, the trainee is expected to service the item. Every technician, and every supervisor too, can expect to be in class several weeks a year.

An unusual kind of training is the assignment of technicians back to the factory, for work in the production departments of the company. A man may work in the department producing a new high-speed printer, for example. Several benefits can accrue from this production experience: the trainee may gain better understanding of what is really in all those little circuits, the production testing of modules and systems may transfer to the later service work, the factory-trained man may elect to specialize in that item, and there can be useful interaction among the field people and production people with respect to manuals and test routines.

Since the training is so intensive and practically oriented to one set of products, there are questions about how much basic electronics knowledge the technicians have. As one crude measure of this, I asked a grab sample of technicians to "explain" to me three bipolar IC technologies which I had hand-copied from the Gasperini (1979) textbook. Actually, I had studied these circuits, and knew the answers; my purpose was to see if the technicians understood them. One circuit was a standard and-gate, using diodes and two transistors. All the technicians I queried had a very thorough understanding of this circuit, and the truth table expressed by it. The other two circuits were a Schottky-clamped level shifter and an emitter-coupled or "current mode logic" circuit. On these items, respondents were less sure of operational details. For instance, few seemed to appreciate that in emitter-coupled switching, load on the power supply does not change; but most of them knew that the circuit was faster because the transistors were not driven hard into saturation. On the basis of this fragment of data, I would guess that computer technicians have a very good functional appreciation of the logical operations performed by IC modules, but that they could not carry out a quantitative analysis of things like excess charges accumulating between base and collector. Perhaps such details are perceived by the technicians as matters for designers to handle.

The computer training people that I consulted were not psychologically trained. Their instructional approach, though, clearly contains elements that were suggested by Gagné (1967) and by Miller (1961). Gagné's general hypothesis is that certain kinds of learning are necessary prerequisites to other kinds. Prior training on subordinate skills, such as verbalization of "rules" in electrical circuit interactions, will increase efficiency of more complex skills such as troubleshooting. Gagné's model of training design then consists of identifying the component tasks of the final performance, insuring that these tasks are fully achieved, and arranging a sequence so that mediational effects among components will be realized. Principles such as reinforcement and distribution of practice will be far less important than will the task component analysis. Miller's treatment says that trouble-shooting can be taught in two rather contrasting ways. Model I would focus on <u>principles of operation</u>; with enough of this kind of learning, the trainee can deduce a reasonable search strategy by himself, regardless of the symptom situation. Model II would specify a <u>set of procedures</u> which can reliably isolate most troubles; the technician is primarily a skilled follower of the procedures. Model I should be more flexible, and Model II should be more reliable, for the set of problems it can solve. Present computer maintenance training leans more toward Miller's model II, and also follows Gagné in the sharp designation of prerequisite subskills. Most of the trainers, incidentally, are from the maintenance division, not the design or manufacturing departments.

FINDING TROUBLES

All troubleshooting proceeds by an eliminative logic. When there are many "possibles", as in a computer or complex tape drive, some scheme for carrying out the eliminations will be necessary. "Bracketing" attempts to restrict the range of the trouble, perhaps by inserting a known good signal into a chain of components, or by seeing where it is "lost". If a radio signal is put into an RF circuit and there is no sound in the speaker, you do not know much about where the trouble is. However, you can "walk the signal" through the IF and audio stages, and if a speaker output results, then the trouble should be between the antenna and the "good signal" point. In digital computer systems, bracketing proceeds by monitoring logical "1's" and "0's" at strategic test points, and there is much attention to "what should be there" under certain input conditions.

Various proposals have been made to systematize testing of serial chains. The "half-split" method recommends that the next check be made at the midpoint of the unchecked components. The half-split concept is really a special case of the information-theory model of troubleshooting (Guiasu, 1977), which calculates the entropy of each possible test, and advises the searcher to examine the test characteristic with the highest entropy; this strategy supplies the largest amount of information, given present knowledge. Tests can be utility or time weighted to reflect difficulties or costs, and so a theoretically optimal sequence can be calculated.

Experimental studies have shown that troubleshooters do not strictly follow rules such as half-split or maximum uncertainty-reduction (Rigney, 1968; Rouse, 1978). Rigney's (1968) subjects operated at about one-third of maximum efficiency, and in another experiment only about half of the electronics tests "made sense", to experts who scanned records of the behaviour (Bryan et al, 1956). This relative inefficiency should not be attributed to laziness or stupidity; rather, the technician's information base about normal-abnormal readings may be incomplete or incorrect, and so much redundancy will be seen, as he tries to "overdetermine" the location of the trouble. Another factor leading to inefficiency might be the tendency to "do something"; Rasmussen and Jensen (1974) found that technicians seemed to be little concerned with the engineering elegance of their search logic, but highly interested in correcting this particular trouble. Such an outlook should encourage many checks, and these would appear on later analysis to be redundant. My strong impression is that there is less redundancy in computer trouble search than in, say, military or automotive troubleshooting, but this is only a conjecture. The matter deserves investigation.

As Miller's Model II training model suggested, search logic can be well defined, in advance, for some computer symptoms. For these occasions the maintenance person can be supplied with a logic-tree diagram or list of successive checks to make. The results of each check then branch down eventually to the fault. Psychologically, the technician is unburdened by having both his information-base and search logic encapsulated into the tree. There is no doubt that prescribed sequences can be quickly learned (Potter and Thomas, 1976); perhaps on the order of 90% of the troubles in real equipment can be found by briefly-trained men (Foley, 1978). The difficulties with the canned-sequence approach are obvious, however; not all troubles can be anticipated and programmed; the readings actually made may be ambiguous; the guide may contain mistakes; unless the sequence is successful, the searcher does not know what to do. Still, the proceduralized guide has been adopted by the computer industry, and it is carefully debugged and revised.

There are many specific techniques that can be observed in contract computer troubleshooting. For mainframe troubles, automatic test routines can be of immense assistance. The most advanced diagnostics actually tell the technicians which modulus to check or replace; such programs are termed "card callers", because of the explicit suggestions they make. Some routines even "thank" the technician for replacing that bad 7442 decoder with a good one. Few diagnostics are so elaborate; most merely print outputs of each subsection of the mainframe under certain "stopped" logical conditions; the technician's job then is to scan the expected string of logical 1's and 0's in the manual, and compare them with the test output. After many practice runs, this sequencing seems to be rather routine.

A direct technique for testing integrated circuits (IC's) employs the logic comparator. The searcher can connect the inputs of a suspected IC (say, a NAND gate) to another NAND gate. the comparator will display any difference in outputs. Logic clips show the 1-0 state of all the pins on a standard module and thus facilitate monitoring of the device. Simple heat indications may be useful. If an IC circuit gets too warm, it will malfunction. A short blast of freeze spray may cause it to operate again. Then the troubleshooter can apply a "heat gun", and watch for the malfunction to recur.

Piggybacking entails the introduction of a new IC, in parallel to the suspected IC. This is done one IC at a time, and may be handy for intermittent faults which are caused by unpredictable heat effects. Selective shotgunning amounts to replacing several modules, once the trouble is isolated to a small area. If the fault is isolated to two or three units, it may be cheaper to replace all the units, rather than spend more diagnostic time.

When a computer technician smokes it out, he forces more current into a set of his suspect IC's; one of the units will get hot, so a touch may indicate which one is pulling the line voltage down. The current tracer permits the troubleshooter to follow current levels along the lines on a circuit board. The operator places the probe at a point when (good) current is known to be flowing, adjusts the instrument until the light just barely turns on, and then follows the trace along the board until the light goes out; it is at this place that the circuit current is sinking.

For service work in counters and shifters, a logic analyzer may be the key instrument. Essentially, an oscilloscope with many channels, the analyzer can be set to show logic levels on each channel, in frozen time. If a decade counter is suspected, then a 1 Hz signal can be fed in, and a line-by-line display shows the way that the counter is incrementing.

For checking out standard chips such as operational amplifiers, a series of idealized circuit assumptions may serve as the logical basis for troubleshooting. The input impedance may be assumed infinite, the output impedance zero, the voltage gain infinite, and the output voltages constrained to +V and -V voltage "rails". These approximations are not really true (infinite input impedance would demand that the amplifier need no input current), but they are accurate enough to permit the determination of whether the unit is operating satisfactorily, which is all the technician has to know.

All of the above techniques start from a "what should be there" logic base. If "what should be there" is not there, then the next bracketing step is more or less evident. A main skill of the technician, then, is to set up the instruments and analyses so that the normal-abnormal decision at each test point becomes a rather simple discrimination.

JOB DESIGN AND MOTIVATION

The job design movement says that characteristics of jobs can enhance the intrinsic motivation of workers. Though the list of characteristics varies somewhat from one author to another, there are about half a dozen key structural factors: flexibility and variety of the tasks performed, work challenge, goal-setting by the workers, access to all relevant job information, loose discipline and employer-employee trust, a feeling of personal responsibility for the work, intrinsic meaningfulness of tasks, feedback about the quality of performance, and opportunity for personal growth. Jobs that are high in these characteristics should produce high work satisfaction (Davis and Taylor, 1979).

To my knowledge, job satisfaction data on an adequate sample of computer technicians have not been published. Certainly computer maintenance people display informal signs of high job satisfaction and intrinsic motivation. It is not unusual to see the service people describing troubleshooting histories to each other, and going over the analysis. The work itself may be near the optimal level for challenging the capabilities of the workers; all those manuals and banks of IC's are tough, but they eventually can be mastered, and with the mastery comes a real sense of competence. There may be just enough "surprise" value in meeting new systems and new situations to produce curiosity and "play" satisfactions (Csikszentmihalyi, 1975).

A second factor was a genuine allowance for a personal growth. In any occupation, much work is routine, and many operators are satisfied with their moderately challenging routine work. But if you need more challenge, the companies will support you; they will give you more advanced training and equipment diversity, and often they will rotate you around through manufacturing and sales divisions, if that seems to fit your needs. If you express management ambitions, those can be realized too, though again in a meritocratic context. Technical specialization, and eventual status as an expert, can be planned with some real likelihood of achievement. The companies do not seek prima donnas or superstars, and I did not see any at the locations I visited; but individual differences in aptitude and knowledge are recognized.

RECAPITULATION

Contract computer maintenance is successful, then, not because of any technical or personnel management secrets. The things that contract maintenance agencies do are simply done better. Starting from a rigid determination to "keep it on the air" and a simple "what should be there" aiding model to guide search behaviour, the companies have evolved an effective way of integrating equipment design, personnel training, aiding, and administration. Since the service agencies usually maintain their own products, there is rapid feedback regarding the maintenance adequacy of design, support, and personnel factors; this feedback has permitted adaptive adjustments which lead to high performance. Of course, a very large investment, in both time and dollars, was needed to realize this performance, and to make all the adjustments over the years.

Behavioral scientists should ponder the fact that the computer maintenance movement has benefited very little from academic psychology and related disciplines. The industry has avoided some of the usual people problems, by applying technology and by intelligent administrative decisions. If it was hard for mechanics to visualize the electrical actions of the breakdown diodes in a high threshold logic board, then psychological aids to effective imagery were not developed; instead, the board was reengineered and cascaded so that technicians did not have to visualize the details. When human memory could not encompass a big system, the computer industry turned to better manuals, aids, and software diagnostics; these materials were assiduously adjusted and refined, until they worked. When graduate electrical engineers became impatient with maintenance duties, as they did in the 1950's, then the companies began to recruit their people from junior colleges and the one-term military population. The excellent motivational and career aspects seem not to be derived from the job design theorists, but rather from a heads-up view of what highly-trained-but-non-professional people want. The computer maintenance domain represents an admirable balancing job by mangement, and could well be emulated by other groups who have to operate and repair complex equipment items.

This brief review of contract computer maintenance has explored only a small corner of the domain; yet it leads to a more optimistic view of maintainability. The computer companies have shown us that complex systems can be designed and built so as to furnish "clear" troubleshooting cues. If no automobile manufacturer has done this yet, then a great opportunity to revolutionize the repair shop is being missed. Proceduralized logic diagrams and aids can unburden a searcher, and can turn an ordinary person with only a few weeks of training into an effective troubleshooter; so perhaps the military personnel system should realize and exploit this fact. Finally, the computer industry experience has demonstrated how near-optimal levels of job and career satisfaction can lead to a remarkably productive work force.

REFERENCES

Bryan, G. L., Bond, N. A., Jr., LaPorte, H. R., Jr., and Hoffman, L., 1956, "Electronics Troubleshooting: A Behavioral Analysis, "University of Southern California, Los Angeles, California.

Csikszentmihalyi, M., 1975, "Beyond Boredom and Anxiety", Jossey-Bass, San Francisco, California.

Davis, L. E., and Taylor, J. C., 1979, "Design of Jobs", Goodyear, Santa Monica, California.

Foley, J. P., 1978, "Executive Summary Concerning the Impact of Advanced Maintenance Data and Task-oriented Training Technologies in Maintenance, Personnel, and Training Systems", Wright-Patterson Air Force Base, Dayton, Ohio.

Gagné, R. M., 1967, "Learning and Individual Differences", Charles Merrill, Columbus, Ohio.

Gasperini, R. E., 1975, "Digital Troubleshooting", Movonics, Los Altos, California.

Guiasu, S., 1977, "Information Theory with Applications", McGraw-Hill, New York.

Miller, R. B., 1961, Comments, in: "The Training of Astronauts", National Academy of Sciences, Washington.

Potter, N. R., and Thomas, D. L., 1976, "Evaluation of Three Types of Technical Data for Troubleshooting Results and Project Summary", Human Resources Laboratory, Brooks Air Force Base, Texas.

Rasmussen, J., and Jensen, A., 1974, Mental procedures in real-life tasks: A case study of electronic trouble-shooting, Ergonomics, 17:293.

Rigney, J. W., 1968, "An Analysis of Structure and Errors in Corrective Maintenance Work", University of Southern California, Los Angeles, California.

Rouse, W. B., 1978, Human problem solving performance in a fault diagnosis task, IEEE Trans. on Systems, Man, and Cybernetics, SMC-8:258.

TOOLS FOR DEBUGGING COMPUTER PROGRAMS - HOW MUCH DO THEY HELP?

J.B. Brooke

University of Wales Institute of Science and Technology
Department of Applied Psychology
Penylan, Cardiff CF3 7UX

INTRODUCTION

The computer is firmly embedded in the fabric of our industrial and commercial life, and it seems likely that our dependence on information processing technology will increase in the new decade. However, there is a widening gap both in cost and in sophistication between the hardware and the software of the computer systems. We find only too often that the provision of hardware by suppliers is delayed because of hitches in software development; alternatively, the hardware may be supplied with inadequate software. Anyone who has had any part in the maintenance of a computer system will know of the chore of continually patching the manufacturer's latest correction into the operating system or language processors, and of writing his own software to surmount the shortcomings of that supplied.

Furthermore, we find that whilst we are using hardware that has been developed in the late 1970's, we are still, in general, programming in languages developed and standardised during the 1950's and 60's, such as FORTRAN, BASIC, ALGOL and COBOL. Even the "new" languages (such as PASCAL, CORAL, ADA and RTL/2) were first conceived (and their standards frozen) in the early 1970's at the latest. User programming in assembler code is still necessary, in many instances, to enhance these "high level" languages to the level required to perform the functions we now expect computers to perform (e.g. real time operation). It might be argued that this lag is a consequence of having languages eminently suitable for the expression of solutions to our computing problems. The cost, in man-hours, of developing a piece of applications software indicates that this not the case. The increasing ratio of software to hardware costs is making it imperative that we should find ways of improving the performance of programmers in writing error-free programs that perform the function they intend.

An important factor in the development of any piece of software is the location and correction of errors. This activity is generally referred to as debugging. Gould and Drongowski (1974), reviewing actuarial studies of debugging found that it constitutes a major component of the time and money spent in large software projects. They found estimates that programmers spend three times as long debugging programs as initially coding them; in terms of cost, debugging can represent from 25-50% of the total.

TYPES OF ERROR

Errors in computer programs can take many forms, and there have been numerous (largely unsatisfactory) attempts to classify these errors. For instance, a distinction is often made between syntactic and non-syntactic errors (e.g. Gould and Drongowski, 1974). A syntactic error is defined as any deviation from the defined syntax of the language; in practical terms, this is based on whether or not a program compiles or is interpreted correctly. (Unfortunately, not all compilers are a complete mapping of the syntactic rules of a language; it is often possible to discover some construction in a program that is syntactically incorrect but which is rejected by the compiler for the wrong reasons). A non--syntactic error is anything else that causes a program not to achieve its desired goal.

However, in terms of the actual incidence of these two types of error, there is little point in so simple a dichotomy. Boies and Gould (1974) found that only about one sixth of all programs (written either in high level or assembler languages) contained syntactic errors on first compilation. Miller and Thomas (1977), reviewing studies of syntactic errors, concluded that investment in more comprehensive syntax checking facilities in language compilers and interpreters may be unnecessary, and found evidence that many syntac checking facilities already available are unused. This simple classification of errors thus pays scant attention to the real distribution of errors and needs refinement.

Shneiderman and McKay (1976) propose a classification which mirrors the error distribution slightly more realistically. These authors distinguish between composition errors and formulation errors. Composition errors include syntactic bugs as well as minor errors in the programming language representation of an algorithm. Formulation errors result from an improper understanding of the solution to a problem. This classification is satisfactory from the programmer's point of view, for he is capable of deciding whether a minor deviation from the correct solution to a problem is the result of his poor coding or of a logical mistake in his solution. For the external observer, however, such classification is next to impossible.

These examples demonstrate that it is unlikely that any entirely suitable classification of program errors can ever be made. In the present paper, the following distinctions are made, although those approaching debugging from another point of view may find them unsatisfactory.

a) Syntactic errors.

b) Incorrect formulation of algorithms so that program never achieves its intended goal.

c) Inadequate formulation of algorithms so that the program will work within certain limits; it will fail however, when trying to handle information beyond these limits.

DIRECT AND INDIRECT APPROACHES TO AIDING DEBUGGING

The diverse nature of errors that can occur in programs has led to differing approaches to the provision of assistance for debugging. The aids that are available can be broadly classified as being of two types: direct and indirect. Direct and indirect refer to the relationship between a job aid and a particular program. Direct aids are generally software tools that can be incorporated in or that operate on a specific piece of program code; they allow the programmer to study the operation of a particular program. Indirect aids, on the other hand, are ways of writing programs, or features of languages, or alternative ways of representing programs that are intended to reduce the initial number of errors made by programmers and to ease the error tracing process when they are made. Indirect aids do not actually give the programmer information about the particular program he is writing and testing; they only tell him how he should write, or how he should go about finding errors, in programs in general.

This distinction does not mean that we can necessarily consider the two types of aid in isolation. It will become apparent that the use of certain direct aids depends on the use of other, indirect aids, so that debugging may proceed efficiently. Additionally, it may be that the direct aids provide us with a convenient way of assessing the indirect aids.

In the following sections, various types of direct and indirect debugging aids will be examined. The main intention of this paper is to look at work that has been done on the assessment of the aids and to consider other ways in which such evaluation could be done. Thus the selection of job aids that are considered is necessarily biased towards those where some assessment has been done. Furthermore, the author is primarily a psychologist and not a computer scientist; thus the discussion, especially of the direct job aids, is conducted in general terms, since the nature of hardware and software is so volatile in the present technological climate. This may mean that descriptions of certain types of aid do not fully describe facilities that are available. However, in justification of this approach, many programmers are now working, and are likely to continue to work under the types of operating system software described, at least for the near future.

DIRECT AIDS TO DEBUGGING

a) Syntax Checking

Usually, the first aspect of a program that is checked is the syntax. This takes place when the program is compiled or interpreted by a language processor; errors are logged as they are found. Error messages can vary widely in detail, even within a single language processor; it is not uncommon to find that a single syntax error is reported several times at different levels of detail by successive passes of a compiler through the program. Some messages are vague to the point of saying "questionable syntax"; others are extremely detailed, even pointing out the location of the fault in a line of code (e.g. see Barr, Beard and Atkinson, 1976, on the Stanford BIP system).

From the psychological point of view, there has been little work done on syntax checking and the correction of syntax errors. As the process is largely automatic, at least at the checking stage, there is probably little aid that psychologists can offer. However, there are some all-too-familiar problems that psychologists could offer programmers some help with; these probably constitute indirect aids to debugging, but will be included here. One major problem with syntactic errors is that, just as in any other process, there can be referral of symptoms. A single syntactic error may give rise to a number of others, although the statements in which the other errors occur are syntactically perfect; the errors result, perhaps, from the omission of the incorrect statement by the compiler or interpreter. The location of this single syntactic error is a search task of a fairly complicated nature. The error must be located in a flow structure that derives not simply from the flow structure of the program itself, but from its interaction with the sequence of compiler operations. Psychologists have investigated structured search and symptom referral in simpler, static processes (Miller, Folley and Smith, 1953; Dale, 1958; Duncan, 1971). Their techniques could be modified and improved to deal with this more dynamic situation.

Another area in which psychologists can help is the investigation of the best representation of syntax, so that the programmer can avoid making errors. Fitter and Green (1979), for instance, report a comparative study of diagrammatic representations of syntax and of syntax represented as rewriting rules (e.g. Backus-Naur notation). Diagrammatic notation proved to be better for the location of syntactic errors; however, it had no advantage when the subject's task involved answering questions relating to the structure of the grammar.

b) Program Structure Checking

The checking of syntax is a common feature of all language processors. Less common are features of compilers or interpreters that check for gross errors in program structure, such as illegally nested loops or dubious programming practices such as entering an iterative structure at a point other than the beginning. A good example of a structural checking tool is the ERROR DOCTOR of the BASIC interpreter used in the Stanford BIP project (Barr, Beard and Atkinson, 1976). This examines the program submitted by the user, identifies gross structural errors and asks the user to correct them before execution is allowed to proceed.

c) Run-Time Debugging Tools

Once a program has compiled "correctly", the programmer enters the difficult realm of dealing with non-syntactic bugs. The simplest, and most common form of debugging tool available to help the programmer track down errors in his program are the run-time error messages which are usually provided as part of the operating system. At their lowest level, these identify only the instruction which failed and the type of error; more sophisticated error handlers will provide a "trace-back" facility. A typical "trace--back" in a language using subroutines or procedures will identify the sequence of procedure calls leading to the instruction that failed. This can be useful if the user's program consists of a number of small procedures; if the program is one self-contained unit there is no real advantage.

Another kind of trace-back is an execution log of the user's program, indicating the sequence in which program statements were executed. Such error tracing is typical of systems in which programs are run in batch mode, since it is a time consuming process. However, comparisons of batch systems with interactive systems (e.g. Sackman, 1970) have generally displayed a slight superiority for usage of interactive systems. Whether there is an advantage or not, there has been a general movement away from batch systems to interactive systems, if only because falling costs have made it feasible. Interactive systems tend not to have facilities such as execution logs of this detail.

Interactive systems can, on the other hand, provide a different type of run-time error tracing facility in addition to simple error logging. This facility is the inclusion in the program of a module which allows the program to be stopped at specified points (breakpoints) and the contents of memory locations to be inspected and modified. Until recently, such debugging tools were limited to use with programs written in assembly language and the programmer had to be able to identify which symbolic labels in his program corresponded to absolute locations in memory. High level language programmers wishing to achieve the same functions had to insert their own breakpoints using high level language statements, and had to add extra I/O statements to inspect the contents of their variables. This tended to obscure the real structure of the program. However, symbolic debugging aids for high-level programs are now becoming available so that the high-level programmer is able to examine interactively the operation of his program and to inspect and modify the contents of named variables without recourse to these methods.

These interactive symbolic debugging tools are necessary, since no matter how specific error messages are, only in very simple errors will the actual failure of a program instruction be in the same place as the error that caused the failure. A program may execute all its statements and terminate without logging a single error, yet not achieve its desired goal because of incorrect formulation of the algorithm. The interactive debugging tools allow the programmer actively to investigate the "unseen" workings of the program and to deduce what is going wrong. Error logging and tracebacks on the other hand, only allow the programmer to examine execution passively.

It must be stressed that deduction is necessary. The software tools only provide the programmer with a basic repertoire of useful operations; he must decide on a sensible and efficient way of using these operations. Unfortunately, there seems to be little, if any, experimental work on the strategies used by programmers in selecting breakpoints in programs and in deciding which variables to inspect and modify. In many ways, these operations are analogous to the processes of signal injection and signal tracing in electronic troubleshooting. In fact, many of the concepts of electronic troubleshooting can be applied to program debugging, e.g. the use of charts displaying system function, with rigid rules regarding direction of flow (Garland and Stainer, 1970); or conceptualising the equipment or process as "blocks" serving particular functions, and testing blocks rather than individual components. In the section of this paper dealing with indirect approaches to aiding debugging it will be seen that the development of techniques for the composition of programs has taken a similar path. Program debugging may have something to learn from research on hardware troubleshooting.

Despite the theoretical appeal of interactive debugging tools, we do not know to what extent they are used. Gould (1975) found that a small sample of programmers who were asked to debug FORTRAN programs did not use interactive debugging facilities when they were available. However, one suspects that this may be a result of programming techniques learned using batch systems and not a reflection of the value of the debugging tools. With interactive systems now common, a further study of the techniques used by programmers is appropriate. Furthermore, we must ask whether debugging performance can be improved by training programmers to use these tools, and not simply whether or not the tools are used.

d) Testing Packages

We have already seen that a program may be incorrectly formulated and still run through to completion without logging any errors that can be picked up by the operating system software. It is up to the programmer to decide if the result of the program is the one he desires. To this end it is usually necessary for the programmer to generate test data to be used by the program; and to know in advance what effect subjecting these data to the process supposedly described by the program will have.

In the case of an incorrect formulation of the algorithm, it will immediately be obvious that there is a mismatch between the desired and actual outcomes of the program. However, if the algorithm has been inadequately formulated, the program may achieve the desired result with certain sets of data, yet not with others. As a trivial example, a program involving a division will work adequately as long as the divisor is not zero; the program may pass its testing if the test data is non-zero but may fail in use if zero data is part of the range with which it is expected to cope.

To cope with this problem, packages are needed to generate a wide range of test data on which the program can operate. Miller and Thomas (1977) reviewing attempts to solve the problem say that testing packages exist, but that there is room for improvement. From the point of view of the casual programmer there are no direct aids available to test his program with wide-ranging data sets; it seems also that novice programmers do not employ the device of testing programs with particular sets of input data (Miller, 1974). It is vital on occasions that such testing should take place, to identify whether bugs are present, so development of these packages and, more importantly, of training of programmers to use them is important.

e) Other Direct Debugging Aids

Most of the debugging aids that have been described so far are tools that we might expect to find available on commercial and industrial systems. However, there are other interesting debugging tools available on systems designed for educational use.

The Stanford BIP system (Barr, Beard and Atkinson, 1976) has already been mentioned in other contexts. This is a tutorial system which teaches students BASIC programming by setting tasks and comparing their solutions to model solutions. Among its debugging features are error diagnostics which can be expanded on request through several level of detail (finally providing a reference in a book!). Hints on the correct way to solve problems, based on the model solutions are given, if the student requests them. A facility is also provided that interactively traces the execution of the program written by the student. This latter feature is rather different to the interactive debugging tools mentioned earlier in that a portion of the text of the program is displayed and a pointer indicates the instruction currently being executed. Up to six of the program variables can be selected and their current values displayed simultaneously with the program text. Iterative loops and conditional jumps are graphically represented by an arrow pointing from the end of the loop or the conditional statement to the destination instruction.

The clarification of flow of execution in a program is generally assumed to be a useful debugging aid. There is little empirical evidence to support this assumption, and it will be seen in the discussion of empirical studies of flowcharts that what evidence there is, is equivocal. Despite this, the debugging features of the BIP system are extremely sophisticated and deserve empirical investigation especially as they relate program operation to a representation familiar to the programmer. There are those who argue that providing a concrete representation of computer system function which need not necessarily have anything to do with computer system structure is essential in the training of programmers (DuBoulay and O'Shea, 1978). The graphic execution flow displays of BIP seems to be an excellent working example of this.

Another, rather different, approach to training programmers to debug their programs is represented by the SPADE-Ø program (Miller, 1979). Whilst it has apparently little immediate relevance to industrial and commercial computing, its interesting feature is that the monitoring program includes plans for designing and debugging the user's programs. Thus the user can avoid errors in the design stage by consulting SPADE-Ø and can ask for help when errors are made.

It may be argued that such systems are designed to help only novice programmers and that they are too inefficient in terms of their demands on machine time and resources to be of practical value to the experienced programmer. The problem of machine resources is not one that need concern us over much with present hardware trends. Despite a tendency for programs to expand to fill the space available, it is reasonable to assume that the machinery available will be able to cope with the demands of the more sophisticated operating systems. On the question of helping the experienced programmer, we can only wait and see how far these operating systems can be developed. Before much more development is done, however, it is important that empirical studies should be made of their effects on programmer efficiency.

If we consider the nature of most of the direct debugging aids mentioned above, it is obvious that, in isolation, they will not correct errors in programs by themselves. They are merely tools, and their efficient use depends on the programmer's comprehension of the program; he must be able to spot the mismatch between the way his program is executed and the way he _wishes_ it to be executed. To this end, various attempts have been made to clarify programs in a more indirect way. These attempts will be considered in the next section.

INDIRECT DEBUGGING AIDS

a) Language Design

Debugging is the location and correction of errors in programs. However, "prevention is better than cure". Considerable debate has taken place on the design criteria of a "good" language, which, among other things, dissuades the programmer from making errors. The debate has polarised opinion into two groups: an active "structured programming" camp and a more passive camp preferring established programming methods. It is extremely difficult to pin down _exactly_ what structured programming is; however, according to Wirth (1974), one of its leading advocates, it is an approach reflecting an admission of the limitations of our minds. Rather than programming through the use of tricks and trade secrets, we should tackle problems systematically using a limited number of "building blocks". By restricting the programmer's degrees of freedom, he is forced to make explicit the way in which he solves his problem. In Wirth's opinion; "The recognition of these limitations can be to our advantage if we restrict ourselves to writing programs which we can manage intellectually, where we fully understand the totality of the implications".

These attitudes have led to the development of "structured languages" such as PASCAL and ADA. These are usually contrasted with scientific and business languages such as FORTRAN and COBOL. What evidence is there that "structure" improves programmer performance?

The "structure" debate has largely taken place on a theoretical level. Probably the most relevant empirical work has been done by Sime and his associates (Green, 1977; Sime, Arblaster and Green, 1977; Sime, Green and Guest, 1973, 1977). These workers have been concerned with the relative merits of different constructions of certain features of computer languages - specifically, different types of conditional construction. The types of construction they chose to study are typical of the two types of language detailed above - the conditional jumping to a labelled statement of FORTRAN-type languages and the if-then-else construction typical of PASCAL and ALGOL type languages. Assessment criteria were the ease with which programs were constructed and the errors that were made in program writing. Generally the if-then-else constructions turned out to be superior to the branch-to-label constructions in terms of ease of writing; the number of errors that were made in program composition; the comprehensibility of programs written by other people; and the lifetimes of those errors that were made. (These advantages were to an extent lost when conditionals became heavily nested).

Sime, Green and Guest (1973) initially suggested that the reason that the "if-then-else" constructions were superior was that they involved less memory load than the branching construction. However, in a later paper (Sime, Green and Guest, 1977) they withdrew this explanation, since the programmer can always refer back to his listing. Instead, they suggested that the difference lies in the ease with which the program can be "deprogrammed" back to the original algorithm. This is an important point. One of the main planks of the structured programming school is that we should not sacrifice program clarity for the sake of reducing the size of a program. Thus, where an "unstructured" program may jump to the same point in a program from several different places, thus saving on the amount of coding necessary, a "structured" program will repeat the same sequence of code several times. Untangling the unstructured program for the purpose of error diagnosis is extremely difficult since a failure at a particular point of a program may have been reached by a number of different routes. In the structured program, on the other hand, the route should be far more explicit.

The work on evaluation of conditional structures does not answer the specific question of whether structured languages are necessarily better than unstructured ones. It is possible to write a badly-structured program in a structured language; by imposing discipline on one's programming style, it is possible to write a fairly well structured program in a language with features conducive to bad structuring. Fitter (1979) says: "There is as a rule little to be gained from a wholesale comparison of existing full-scale languages: the points of difference are too numerous ..... The target of behavioural experiments on programming, as we see it, is to help the designers of future languages and the teachers of present ones". The techniques used by Sime et al. indicate how we can evaluate particular aspects not only of programming languages per se. but also of ways of programming in particular languages with the aim of reducing error. In an ideal world, the programmer's writing style would be determined by the language in which he was writing; as this is a goal we are never likely to achieve, we need to know what effect the prescription of writing techniques can have.

b) Writing Techniques

Much of what is advocated by the structured programming school depends on discipline in the composition of programs. Writing techniques can be prescribed both in general terms and in specific terms. An example of a general approach to writing techniques is "top-down programming/bottom-up testing" or "stepwise refinement" in the composition of programs, used by authors such as Findlay and Watt (1978). Top-down programming is essentially the same process as task analysis (Duncan, 1969).

The overall problem is split into several subproblems; each of the subproblems is further defined as a sequence of subproblems until the solutions to subproblems are trivial. Combining the final solution of all the subproblems gives the overall solution to the initial problem. Thus a hierarchy of subproblems is generated, starting from the top.

Bottom-up testing means that as a solution is found for a subproblem at the lowest level of the hierarchy, the solution is tested in isolation. When the programmer is satisfied that this solution is correct, it can be combined with the solutions of other subproblems already tested in order to provide a solution to a subproblem at the next level up in the hierarchy. This solution is then tested, combined with solutions to other subproblems to form a solution to another subproblem ..... and so on. The advantage of this "top-down/bottom-up" approach is that errors are trapped as they occur. The possibility of interactions between errors is minimised, thus reducing the likelihood of untraceable errors. A similar technique is "modularisation" where the program is split into logical "chunks" although not necessarily in such a systematic way. An unpublished experiment by Kinicki and Ramsay (reported in Shneiderman and McKay, 1976) indicates that modular programs are better comprehended and are more easily modified than non-modular programs. However, Shneiderman and McKay point out that it is not easy to provide guidelines as to how a program should be modularized, and that what may be comprehensible as a module to one programmer need not necessarily be so to another.

These general rules for program writing can be applied irrespective of language, although the programmer needs to be more careful in some languages than in others. In high level languages, the programmer can define his subproblem solutions as independent procedures, deciding what information should be passed from one procedure to another. In lower level languages, on the other hand, such as assembler languages and BASIC, the programmer must take care to ensure that information used in one subproblem solution does not interfere with the solution of another subproblem e.g. through inadvertent use of the same variable names. The advantage of structured languages in particular, and high level languages in general is that they protect the programmer from himself by reducing his degrees of freedom in writing the program; e.g. he must specify his data types and what data is available for use by various parts of his program.

The notion of prescribing writing procedures inevitably means that there must be some restriction of the syntax available to the programmer. Now there are currently two approaches to providing programming systems for non-programmers. One attempts to make the computer understand "natural" language, with all the ambiguity inherent in it; the other advocates the use of a restricted syntax which the programmer must learn, but which avoids ambiguity. Gould, Lewis and Becker (1976) compared the effects of different levels of restriction of syntax on the writing of procedures and description of objects. Whilst subjects were tolerant of ambiguity both in writing and carrying out instructions, they often voluntarily employed restricted syntax notation after exposure to it, even though they were not constrained to do so. Thus arguments that restriction of syntax through prescription of writing procedures is "unnatural" may be irrelevant.

Sime, Arblaster and Green (1977) examined the prescription of writing procedures at syntactic level. They studied non-programmers writing programs under three different regimes - an automatic syntax condition where syntactic errors were impossible because programs were made up from syntactic units rather than words; a "procedural" condition where programs were written word by word, but subjects were made to follow a well-defined procedure; and a "plain" condition where subjects were informed of the syntax but were then allowed to write their programs freely. Both the automatic and procedural conditions were better than the "plain" condition in terms of the number of problems solved without error (although it was obviously impossible to err syntactically in the "automatic" condition). There appeared to be a higher tendency for "semantic" errors in the "automatic" condition than in either of the other two conditions; less syntactic errors were made in the "procedural" condition than in the "plain" condition. There were no significant differences between the three conditions in terms of error lifetimes - that is, the number of attempts made to correct each error. Thus it seems that prescription of writing procedure on a syntactic level does little to help the programmer once he has made errors, although it reduces the initial number of errors made.

In the light of these results, it would seem that the claims made for general programming techniques (e.g. stepwise refinement) need to be inspected experimentally. One suspects that any prescription of writing techniques will only affect the initial number of errors; but if there is subsequently even one bug that the programmer has difficulty in finding and fixing, he is still lumbered with a program that does not work.

c) Commenting, Variable Names and Layout

Much of the preceding discussion has been to do with ideal situations - how languages could be improved, the use of automatic program construction. However, many programmers have to work in languages which are not conducive to good programming and on small systems where sophisticated composition programs are not available. What can these programmers do to ease the debugging task?

One suggestion that is generally made is that programs should be commented, so that the programmer trying to debug his own or somebody else's program can understand what is going on. An unpublished experiment by Yasukawa, reported by Shneiderman and McKay (1976) indicates that commenting has no effect on the debugging of FORTRAN programs. Subjects using uncommented programs performed as well as those using commented programs. Many subjects, in fact, said that they did not use the comments. Shneiderman and McKay suggest that this is because programmers have already learnt that comments may obscure the error by misleading the programmer as to what the code usually does. Gould (1975) found that comments were occasionally used but does not indicate in what way.

A second possibility is to make the program code as meaningful as possible by the choice of sensible variable names. Shneiderman and McKay (1976) report experiments that indicate that the more complex a program is, the greater is the advantage in debugging gained by the use of mnemonic variable names although the effects were not significant. Again, it is possible that the selection of variable names may tend to mislead unless extreme care is taken to ensure that a particular variable only serves the function indicated by its mnemonic.

A third factor that may help in program comprehension is the physical layout of the program listing. Many languages allow underlining of keywords or indenting of particular blocks of program code, providing a perceptual coding in addition to the symbolic coding of a statement of problem solution. This coding of information in two ways is called redundant recoding (Fitter and Green, 1979). Again, the only direct empirical evidence relating to this topic is reported by Shneiderman and McKay (1976), who once again found that it had no effect on debugging performance. Fitter and Green (1979) argue, however, that experimental results reported by Sime, Green and Guest (1977) indicate an improvement in performance occasioned by the use of redundant recoding. The results in question concern two different forms of "if-then-else" conditional constructions. One form of the construction expressed the "else" by a redundant restatement of the condition in a negative form. Using this second type of conditional construction, programmers more often wrote correct programs and found their mistakes faster. Fitter and Green say that the considerable boost in performance gained through the use of this symbolic redundant recoding can only indicate that a much larger boost could be obtained if perceptual recoding were to be used as well. Unfortunately, this assertion remains untested empirically.

d) Diagrammatic Representations

There are, of course, certain types of program representation which have some of the information contained in the program recoded perceptually. These are the representations of programs in the form of flowcharts and other types of diagram. It would not be correct to call a diagrammatic representation a redundant recoding of a program, since a diagram often summarises the more detailed information contained in the program.

Diagrams to represent computer operations take many forms. Figure 1 shows examples of two different notations representing a sequence of four operations, labelled A to D. Test 1 and Test 2 represent "conditional statements" with yes-no answers. Operations B and C are performed depending on the outcome of a nested conditional. Figure 1(a) shows a conventional flowchart representation of the operation sequence; Figure 1(b) shows a flowchart based on "structured" principles developed by Nassi and Shneiderman (1973).

The merits and weaknesses of these and many other types of diagram are discussed in theoretical terms by Fitter and Green (1979). There is a trend towards the use of diagrammatic notations as the medium through which the solutions to problems are expressed, rather than using them as an adjunct to a program expressed in the usual form of a serial listing.

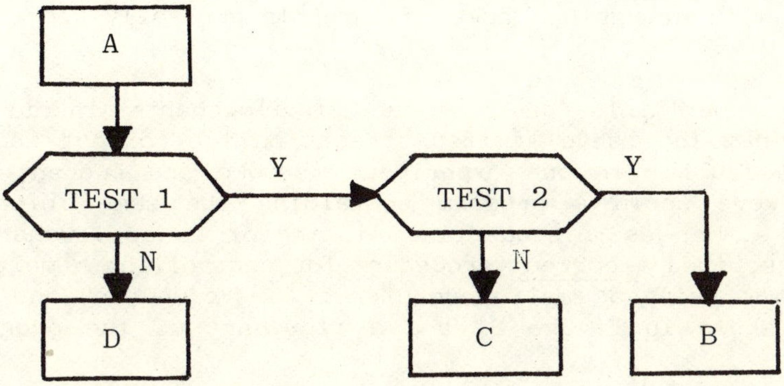

a) Conventional flowchart notation for nested conditional

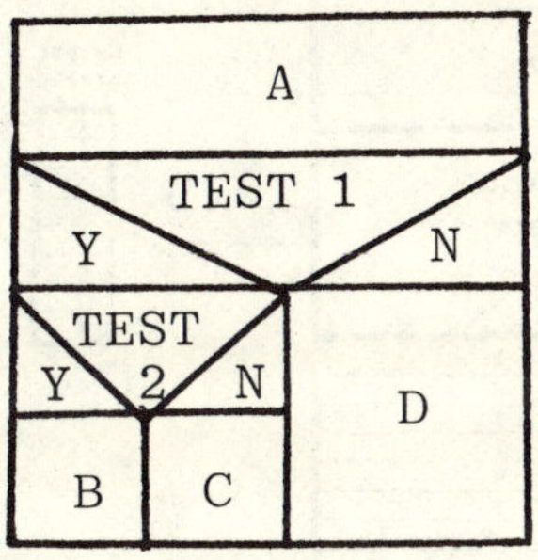

b) Nassi - Shneiderman notation for nested conditional

Figure 1

This movement, however, has a weak basis of empirical support. Studies of flowchart use are divided as to their usefulness, with some experimenters finding them to be superior to serial listings (Wright and Reed, 1973; Blaiwes, 1973, 1974; Kammann, 1975) and others finding that they make little or no difference (Shneiderman and McKay, 1976; Shneiderman, Mayer, McKay and Heller, 1977). Fitter and Green (1979) suggest that this difference results from different types of chart, saying that in those experiments where flowcharts were superior, they represented complex tree-like decision procedures, whilst in those where they showed no superiority, they included iterative loops and jumps, thus representing networks rather than trees. However, another interpretation is that in the "successful" experiments the task the subject was required to perform was to <u>follow</u> and not necessarily comprehend the procedure described by the chart; in the "unsuccessful" experiments, subjects had to correct faults that occurred within the procedure (a task requiring comprehension). Thus in debugging tasks, flowcharts generally proved to be of little use.

Our experiments on the use of flowcharts in debugging indicate that they have limited effects on the correct identification of faults. In one experiment (Brooke and Duncan, 1980), subjects were shown a program containing the same information expressed either as a sequential listing or as a flowchart. The program described a correct procedure for controlling a multigrade petrol pump, using a small "computer". The "computer" and petrol pump are shown in Figure 2, and a flowchart of the program in Figure 3.

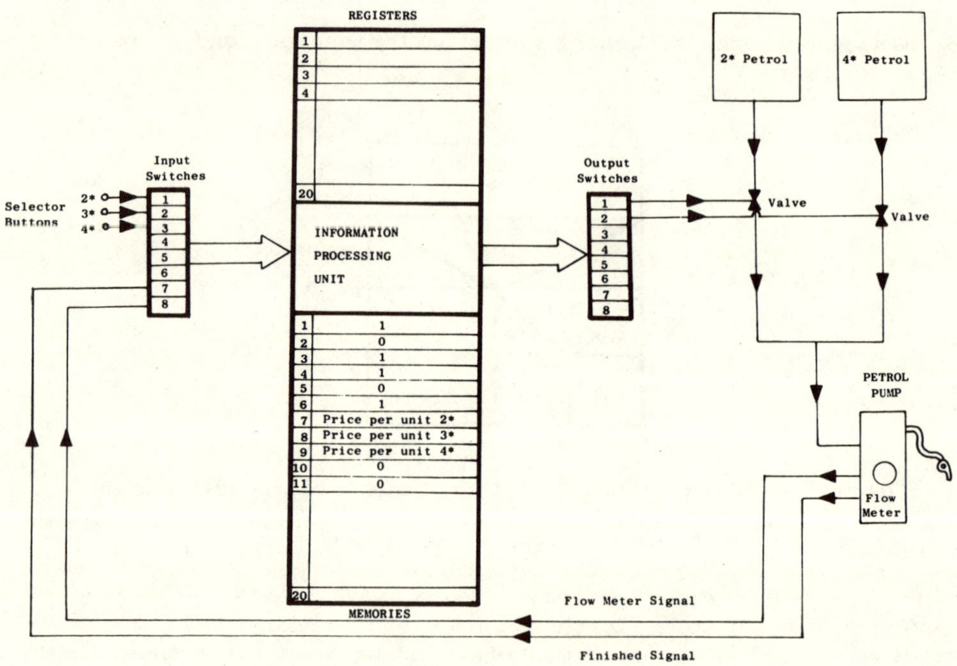

Fig. 2. Block diagram of a "computer controlled petrol pump". Different grades of petrol are produced by setting appropriate combinations of valves (3+ is a mixture of 2+ and 4+ petrol). The valve combinations for the different grades of petrol are coded in pairs in memories 1 to 6, where 1 represents an open valve, 0 a closed one. Valves can be opened or closed by the output switches; signals from the grade selector buttons, from the flow meter or from the pump indicating "end of delivery" are received by the computer through the input switches.

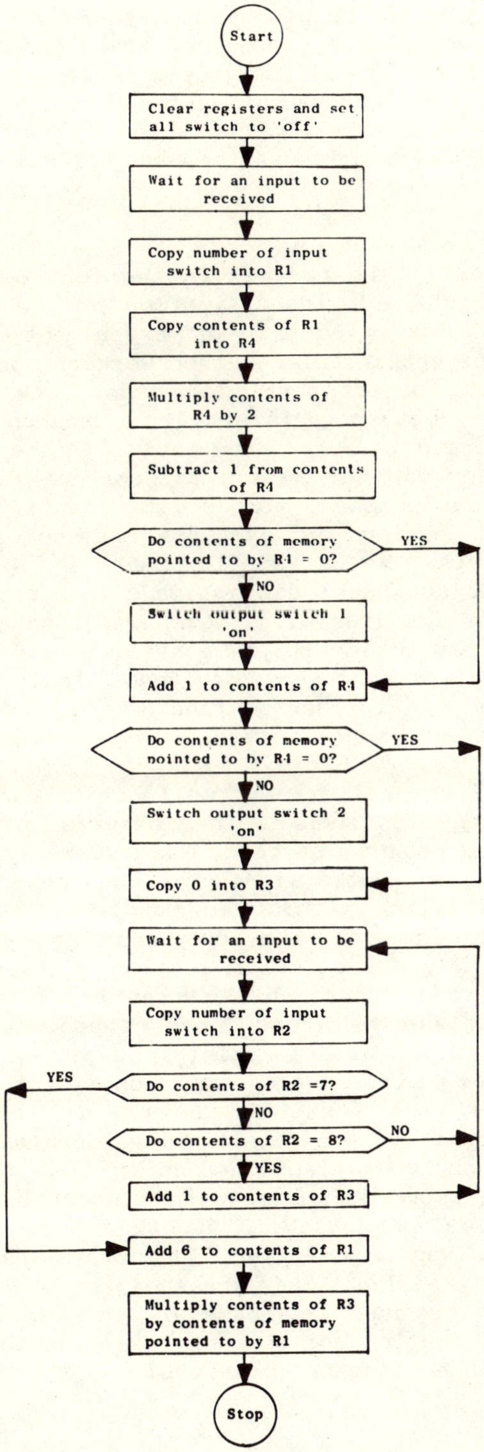

Fig. 3. Flowchart description of procedure for controlling petrol pump shown in Fig. 2.

Subjects were given descriptions of ways in which the petrol pump was malfunctioning, and were required to select one of a list of faults which, when introduced into the correct procedure, would cause these symptoms. The list contained the actual fault, another fault in the same area of the program which would give similar, though not identical, symptoms, and several faults in different areas of the program which would give rise to very different symptoms.

The presentation of the program in the form of a flowchart did not affect the correct identification of faults at all. However, the subjects taking part in the experiment came from two different sources - psychological research workers, and volunteers from the departmental subject panel (a motley crew ranging from housewives to trainee meteorologists). The research workers who used the flowchart tended to select the faults in the correct area of the program if they did not select the correct fault, whereas the subject panel volunteers tended to identify completely incorrect faults as causing the symptoms. Psychologists have a certain familiarity with the use of flowcharts and other directed graphs, which the subject panel did not, and this result seems to indicate that to those familiar with flowcharts, there may be some advantage in using them to locate the general area of malfunction in a program. There was also some evidence that the group of research workers using flowcharts tended to identify faults involving conditional statements better.

The location of general area of malfunction can be achieved using direct debugging aids such as the insertion of breakpoints into a program. If a program proceeds in a largely "invisible" manner (i.e. it is all executed within the machine with little or no I/O) and it is failing without logging error messages, the programmer must trace its execution by inserting breakpoints to determine whether or not a particular section of program code is being reached. A second experiment, currently in progress, is studying the effects that different forms of procedural description have on the selection of breakpoints. Subjects are presented with descriptions of simple programs containing conditional statements either in the form of conventional or structured flowcharts (see Figure 1) or as listings containing either "if-then-else" or "branch-to-label" conditional constructions. Using an interactive program, subjects can insert breakpoints in the programs described and establish whether a particular step succeeded, failed, or was not reached. Their task was to identify the instruction that failed. The breakpoints they select are assessed by another program in terms of whether or not they provide redundant information and whether the solutions they reach are logically impossible (diagnostic errors).

Initial results indicate that subjects using flowcharts select breakpoints more efficiently, making very few diagnostic errors; however, those subjects using serial listings with the "if-then-else" conditional construction (which contains a certain amount of redundant recoding in the form of tabulation) soon learn to insert breakpoints efficiently. Subjects using serial listings with "branch-to-label" conditional construction continue to make diagnostic errors throughout the experiment.

These preliminary results seem to indicate that flowcharts may be useful for certain aspects of debugging. Further experiments are planned which will increase the complexity of the debugging task in order to establish the level at which job aids of this type cease to provide assistance. Our second experiment described above, for instance only requires the subject to identify the failing instruction. Another experiment currently being developed will require the subject to locate both the failing instruction <u>and</u> the cause of failure.

General Considerations in the Psychological Study of Debugging

This paper has reviewed various empirical studies of the debugging of computer programs, and it is evident that many of these studies have failed to support conventional wisdom regarding aids to debugging. Debugging is a complex task, and it is probable that in many instances this complexity has obscured effects arising from the variables manipulated by experimenters. In this final section, some methodological considerations in the psychological study of debugging will be examined.

A preliminary consideration is whether the experimenter should attempts to study debugging in an observational or a comparative way. Observational paradigms have, by and large, failed to produce any clear results, probably because they are looking for <u>the</u> way in which people debug programs (Green, Sime and Fitter, 1975). Programmers are adaptable beasts, (just as humans are in most human activities), and they will adopt different programming styles in order to meet different task objectives (Weinberg and Schulman, 1974). This applies equally to debugging; Gould and Drongowski (1974) concluded that it was impractical to talk in terms of a "typical debugging strategy". Moreover, observational studies are rarely concerned with ways of <u>improving</u> programmer performance.

Comparative studies, on the other hand, allow the experimenter to assess different approaches to programming problems. As the earlier quote from Fitter (1979) indicates, the results of such studies can be used to recommend how languages and debugging aids could be developed and how programmers can be trained to take advantage of these things.

Nonetheless, there are many difficulties inherent in comparative studies of programming problems. It is impractical to compare full scale languages. They are often designed to cope with different types of problem, and thus may have trivial areas of overlap and manifold differences. Working within a single language and comparing the effects of different programming styles on debugging (e.g. Gould, 1975; Gould and Drongowski, 1974; Shneiderman and McKay, 1976; Shneiderman, Mayer, McKay and Heller, 1977) presents problems as well. It is necessary to use experienced programmers, whose own programming styles and habits may not include those being studied by the experimenter. In order to ensure comparability of the problems being solved by subjects, it is usually necessary to present programs written by other people, rather than allowing programmers to write their own. It is difficult to control the levels of experience of subjects; many programmers use only a subset of a language, and the programs presented by the experimenter may exceed this subset. Assessment of problem solutions is difficult. In defence of this approach, however, results will have at least face validity, and should be generalisable at least within the language.

The use of full scale languages can be considered as a "top-down" approach to studying debugging. At the other end of the spectrum of comparative studies, representing a "bottom-up" technique is the use of "Microlanguages" to isolate language features or to provide a simple context in which naive subjects can work under different programming regimes (e.g. Green, 1977; Sime, Arblaster and Green, 1977; Sime, Green and Guest, 1973, 1977). Green, Sime and Fitter (1975) point out that in this type of paradigm it is a great deal easier to provide statistical analysis of experimental results; that the points of difference between experimental conditions are more strictly controlled by the experimenter and experimental effects are less likely to be artefacts of other features of the language; and that subjects need have no programming experience. On the negative side, the results cannot be said to have the same face validity that experiments using "proper" languages have; problems that programmers encounter are very likely the result of interactions between a number of language features (although a "bottom-up" approach can study these interactions by gradually increasing the task complexity); and the problems which subjects are asked to solve may seem trivial or unreal. Despite these limitations, the bottom-up approach seems more likely to provide useful data.

The criteria on which we assess the performance of programmers in solving debugging problems is a crucial concern. On the level of software projects, Brooks (1975) has criticized the "man-month" as a measure of programming efficiency since it takes no account of the different ways in which programmers work. On the more microscopic level of laboratory studies, similar considerations apply. A problem may be solved quickly but inefficiently, or more slowly but in a manner unlikely to give rise to further problems. Some experimenters use measures such as error lifetimes - that is, the number of attempts made at solving a problem. However, this tells us nothing about the internal efficiency of the strategies used by the programmer, and we need some mechanism that allows us to examine this aspect of his performance.

The use of some of the debugging aids classified as "direct" in this paper seems to provide this mechanism. By getting the programmer to use such devices as breakpointing, inspection of memory locations and the insertion of data to be manipulated by the program, it is possible to classify his diagnostic moves in terms of their relevance to the problem in hand. We have attempted to do this in a small way in our breakpointing study, and it seems to provide useful objective measures of comparison between indirect debugging aids. It is reasonable to assume that as long as the programmer is making sensible diagnostic checks in his approach to a problem, he is performing efficiently. This may take more time than a brute force approach where all possible points of failure are checked, even if they are irrelevant; but in the long run, with more and more complicated programs to debug, the efficient diagnostician should come out on top in terms of time as well.

The difficulties of studying program debugging and, indeed, programming in general, has led to most development work being carried out on a theoretical level. This paper has shown that psychological studies of programming problems may have something to offer to computer scientists and to those who train programmers. Although there is much work to be done, there are encouraging signs that research in this area is gathering momentum.

ACKNOWLEDGEMENTS

The work reported in this paper was supported by grant no. HR6045 from the Social Science Research Council (United Kingdom).

REFERENCES

Barr, A., Beard, M. and Atkinson, R.C., 1976, "The Computer as a Tutorial Laboratory - The Stanford BIP System", International Journal of Man-Machine Studies 8, 567-596.

Blaiwes, A.S., 1973, "Some Training Factors Related to Procedural Performance", Journal of Applied Psychology 58, 214-218.

Blaiwes, A.S., 1974, "Formats for Presenting Procedural Instructions", Journal of Applied Psychology 59, 683-686.

Boies, S.J. and Gould, J.D., 1974, "Syntactic Errors in Computer Programming", Human Factors 16, 253-257.

Brooke, J.B. and Duncan, K.D., 1980, "An Experimental Study of Flowcharts as an Aid to Identification of Procedural Faults", Ergonomics (in press).

Brooks, F.P., 1975, The Mythical Man-Month London: Addison-Wesley.

Dale, H.C.A., 1958, "Fault Finding in Electronic Equipment", Ergonomics, 1, 356.

DuBoulay. J.b.H. and O'Shea, T., "Seeing the Works: A Strategy for Teaching Programming", Proceedings of Workshop on Computing Skills and Adaptive Systems, University of Liverpool.

Duncan, K.D., 1969, "Task Analysis Evaluated", in: F. Bresson and M. de Montmollin (eds): La Recherche en Enseignement Programme: Tendances Actuelles, Paris: Dunod.

Duncan, K.D., 1971, "Long Term Retention and Transfer of an Industrial Search Skill", British Journal of Psychology 62, 439-448.

Findlay, W. and Watt D.A., 1978, "PASCAL - An Introduction to Methodical Programming", London: Pitman.

Fitter, M., 1979, "Towards More "Natural" Interactive Systems", International Journal of Man-Machine Studies 11, 339-350.

Fitter, M. and Green, T.R.G., 1979, "When Do Diagrams Make Good Computer Languages?", International Journal of Man-Machine Studies 11, 235-261.

Garland, D.J. and Stainer, F.W., 1970, "Modern Electronic Maintenance Principles", London: Pergamon.

Gould, J.D., 1975, "Some Psychological Evidence on How People Debug Computer Programs", International Journal of Man-Machine Studies, 7, 151-182.

Gould, J.D. and Drongowski, 1974, "An Explanatory Study of Computer Program Debugging", Human Factors 16, 258-277.

Gould, J.D., Lewis, C. and Becker, C.A., 1976, "Writing and Following Procedural, Descriptive and Restricted Syntax Language Instructions", IBM Research Report 5943, IBM Thomas J. Watson Research Centre, Yorktown Heights, N.Y.

Green, T.R.G., 1977, "Conditional Program Statements and Their Comprehensibility to Professional Programmers", Journal of Occupational Psychology 50, 93-109.

Green, T.R.G., Sime, M.E. and Fitter, M., 1975, "Behavioural Experiments on Programming Languages: Some Methodological Considerations", Memo No. 66, MRC Social and Applied Psychology Unit, Sheffield.

Kammann, R., 1975, "The Comprehensibility of Printed Instructions and the Flowchart Alternative", Human Factors 17, 183-191.

Miller, L.A., 1974, "Programming by Non-Programmers", International Journal of Man-Machine Studies 6, 237.

Miller, L.A. and Thomas, J.C., 1977, "Behavioural Issues in the Use of Interactive Systems", International Journal of Man-Machine Studies 9, 509-536.

Miller, M.L., 1979, "A Structured Planning and Debugging Environment for Elementary Programming", International Journal of Man-Machine Studies 11, 79-95.

Miller, R.B., Folley, J.D. and Smith, P.R., 1953, "Systematic Trouble-Shooting and the Half-Split Technique", Lackland AFB, Human Resources Research Centre, July 1953 Tech. Report.

Nassi, I. and Shneiderman, B., 1973, "Flowchart Techniques for Structured Programming", SIGPLAN Notices 8, 12-26.

Sackman, H., 1970, "Experimental Analysis of Man-Computer Problem Solving", Human Factors 12, 187-201.

Shneiderman, B. and McKay, D., 1976, "Experimental Investigation of Computer Program Debugging", 6th International Congress of the International Ergonomics Association, College Park, MD.

Shneiderman, B., Mayer, R., McKay, D., and Heller, P., 1977, "Experimental Investigations of the Utility of Detailed Flowcharts in Programming", Communications of the Association for Computing Machinery 20, 373-381.

Sime, M.E., Arblaster, A.T. and Green, T.R.G., 1977, "Reducing Programming Errors in Nested Conditionals by Prescribing a Writing Procedure", International Journal of Man-Machine Studies 9, 119-126.

Sime, M.E., Green, T.R.G. and Guest, D.J., 1973, "Psychological Evaluation of Two Conditional Constructions Used in Computer Languages", International Journal of Man-Machine Studies 5, 105-113.

Sime, M.E., Green, T.R.G. and Guest, 1977, "Scope Marking in Computer Conditionals - A Psychological Evaluation", International Journal of Man-Machine Studies 9, 107-118.

Weinberg, G.M. and Schulman, E.L., 1974, "Goals and Performance in Computer Programming", Human Factors 16, 70-77.

Wirth, N., 1974, "On the Composition of Well-Structured Programs", Computing Surveys 6, 247-259.

Wright, P. and Reed, I., 1973, "Written Information: Some Alternatives to Prose for Expressing the Outcomes of Complex Contingencies", Journal of Applied Psychology 57, 160-166.

FIELD EXPERIENCE IN MAINTENANCE

Julien M. Christensen, Ph.D and John M. Howard, B.S.

Stevens, Scheidler, Stevens, Vossler, Inc.
Dayton, Ohio, U. S. A.

INTRODUCTION

After reviewing 31 different descriptions of maintainability, Rigby and his associates developed the following definition:

"Maintainability is a quality of the combined features and characteristics of equipment design, job aids, and job supports which facilitate the rapidity, economy, ease, and accuracy with which maintenance operations can be performed, and the system thus kept in or returned to operating condition, by average ... personnel, under the environmental conditions in which the system will be maintained." (Rigby, et al, 1961, as quoted in Crawford and Altman, 1972).

As Crawford and Altman point out, the essential features of this definition include equipment design, job aids and job supports. Criteria of maintainability include the rapidity, economy, ease and accuracy with which the equipment can be kept in operating condition (preventive maintenance) or restored to operating condition (corrective maintenance). Personnel capabilities and environmental conditions round out the definition. It is

Notice:
This paper will also appear as a chapter in a forthcoming book entitled : What Every Engineer Should Know About Human Factors Engineering, to be published by Marcel Dekker, Inc. New York and Basel. Permission of the authors and Marcel Dekker, Inc. to include this material in this volume is appreciated.

clear that maintainability is designed into equipment as certainly as is operability. Poor maintainability degrades system effectiveness just as surely as does poor operability.

With the increasing initial investments in equipment and systems, the increasing associated life-cycle costs and the unpropituous effects of down-time, maintainability, or "design for ease of maintenance", is assuming even greater importance. (Maintenance costs, for example, currently absorb approximately thirty percent of the budget for the Department of Defense). No longer can maintainability be subverted to operability. Instead, a balance that will maximize equipment and systems effectiveness must be established between the two.

Besides operability, maintainability is also related to availability (probability that the equipment/system will operate satisfactorily at any given time) and to reliability (probability that the equipment/system will operate in accordance with specified operational requirements of function, time and environment).

The interrelated nature of these factors, the so-called "ilities", is obvious. Equally obvious is the substantial responsibility that maintainability places on the design engineer--attention to interrelationships with operability, availability and reliability; attention to personnel skill requirements and availability; attention to software requirements; attention to the field environment (heat, vibration, etc.) in which the equipment/system must operate; and, finally, attention to the ever-present constraints of time and money.

The design trade-offs are multi-dimensional and incredibly complex. The resultant design will determine life-cycle cost (LCC) and thus whether or not the equipment/system will survive in the market place. Assuming the initial and operating costs are known or can be reliably estimated, comparative costs and alternative designs can be assessed by use of the formula:

$$LCC = IC + \sum_{i=1}^{L} \frac{OC}{(1+d)^{i}}$$

where,

IC = initial cost
OC = operating cost
L = life in years
d = discount rate

(This is borrowed from the economists and, in their profession, is used to compute "present discounted value").

The "OC" term causes the designer considerable difficulty. Even where there is substantial experience with similar equipment, designers seem to be eternally and unrealistically optimistic when it comes to estimating the resources that will be required to keep their products operating. (The energy crunch has finally caused Americans to begin to look at LCC and most can hardly believe that, on the average, it costs approximately 16 cents to drive a compact car one mile. The energy shortage may serve as a vehicle for introduction of, or increased emphasis on, LCC for everyone--designers, industrialists, and consumers).

It is the thesis of this paper that the ultimate criterion is how well the equipment and systems can be maintained <u>in the field</u>. It is felt that attention to field experiences provides a rich source of information for the design engineer who is responsible for design for ease of maintenance. Some of our evidence is reasonably objective; some is based on controlled interviews; some is little more than anecdotal. Even the latter, however, can sometimes serve as a source of hypotheses for more objective study.

A WORD ABOUT CRITERIA

Historically, time, specifically down-time, has been a preferred measure of maintainability. While this criterion has enjoyed considerable popularity and is relatively easy to measure, it is of limited value to the design engineer. It's sensitivity to specific maintenance design variables is generally unknown; it does not give the designer specific clues as to what to do to improve maintainability. It requires further analysis to be useful.

Errors, if properly defined and accurately recorded, constitute a somewhat better criterion than time from the designer's point of view. However, what the designer usually finds more helpful is information directly related to variables such as those shown in the Maintainability Design Checklist (Appendix I), evidence of lack of skill requirements, evidence of mal-diagnosis, unnecessary removal of parts, etc.

<u>How Does the Maintenance Man Spend His Time?</u>

Since time is such a popular criterion, it is important to know how it is spent.

Most studies disclosed that the majority of the maintenance man's time is spent on fault-diagnosis. One of the writers summa-

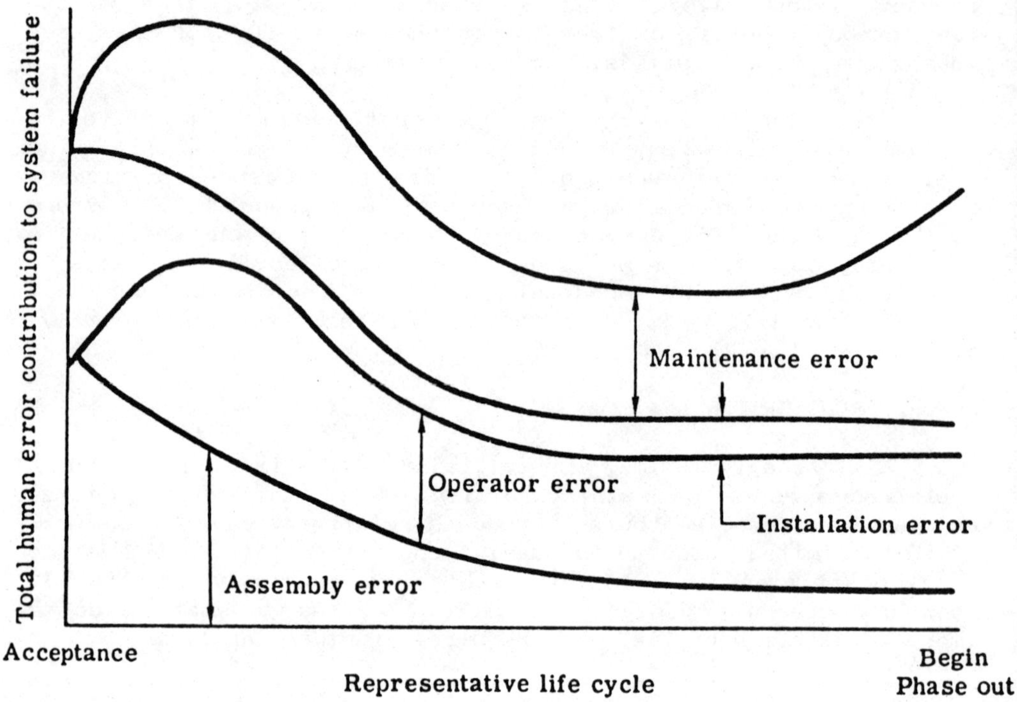

Figure 1 Proportional Contribution of the Different Species of Human Error to System Failure.

Source: Rigby (1967)

rized a number of these sometime ago (before extensive use of automatic test equipment - ATE), and found that the maintenance man's time was spent approximately as follows: diagnosis, 65-75%; remedial actions, 15-25%; verification, 5-15%. The equipment selected was primarily electronic equipment. It would appear that time spent designing equipment for ease of fault-diagnosis would contribute significantly to the effectiveness of many maintenance programs. The realibility of ATE and built-in test equipment (BITE) has improved to the point where they can be of considerable aid to the maintenance man. McDonnell Douglas reports the Hewlett-Packard Automatic Test System (a mini-computer system with a semi-conductor memory data and a disc drive) has reduced their annual testing costs for numerical control (NC) equipment from $ 400,000 to $ 7,000 (exclusive of software costs). Anon, (1979) Post (1979) reports on the greatly increased efficiency with which welding operations can be monitored with the help of micro-processor control. For example, the micro-computer checks thermal protection, balances power input, etc.

The writers came across a piece of ATE equipment designed for use with computer-aided manufacturing systems. This equipment is capable of simulating the operating conditions of over 200 different printed circuit boards used by various NC pieces of equipment. It used to require an average of 35 minutes to locate and repair a single fault in a single board; now it requires only 80 seconds, which includes a print-out of the results - a savings in time of 96 percent.

However, ATE does not do away with the maintenance man; it first requires him to spend his time differently and modifies his skill requirements. He now needs more skill in electronics, more skill in analyzing check-out results, and must be prepared emotionally and skill-wise to handle breakdowns that, when they do occur, can frequently be quite devastating.

HUMAN INITIATED FAILURES

Incidence and Incidents

Rigby has charted the approximate contribution of various types of human error to system failure. The substantial contribution of maintenance error to the total is shown in Figure 1 (Rigby 1967).

Feineman (1978) reports the following alarming statistics, derived from a series of field studies:

A. Seventy percent of U.S. Army electronic equipment was inoperative.

B. Even when on maneuvers, 30 percent of U.S. Navy equipment was inoperative.

C. Over a five-year period, the maintenance costs of electronic equipment for the air arm of the U.S. Army was over ten times the original cost of the equipment.

D. Naval aviation equipment costs twice its purchase price simply to maintain it the first year.

A summary of several studies by the first author revealed that human-initiated failures accounted for 50-70 percent of failures in electronic equipment, 20-53 percent of failures in missile systems and 60-70 percent in aircraft. Life cycle costs (LCC) can vary from 10 to 100 times the original cost of the system[x)]. The "causes" of most of these can be ascribed to four categories: improper design and manufacture, improper installation, improper maintenance actions and improper operation. (Christensen, 1979). A summary of specific errors, based on maintenance errors in missile operations, revealed the following:

Cause	% of Total
Loose nuts/fittings	14
Incorrect installation	28
Dials and controls (mis-read, mis-set)	38
Inaccessibility	3
Misc.	17
	100

The DC-10 accident of May, 1979 at O'Hare Field in Chicago in which 272 persons perished was attributed by the U.S. National Transportation Safety Board to improper maintenance procedures. We would only observe that in maintenance, as in other pursuits, procedures that are unusually difficult or time-consuming are subject to short-cuts and the possibility of less than adequate treatment. To quote from the report of the Board of Inquiry, "... vulnerability to critical damage during maintenance apparently was not considered by either the manufacturer's design personnel or the FAA's certification review team". (King, 1979).

Calculations from figures gathered by Goldstein and Rosenfield (1977) reveal that maintenance constitutes varying proportions of LCC for home appliances - 35 percent for television sets, for example, versus only six percent for refrigerators.

x) Obviously, these astounding figures of 10X to 100X must be interpreted with caution. If one leaves a system in operation longer, then, of course, its overall LCC will be greater. (Consider the B-52!). Clearly, it is mean LCC figures that must be compared.

Maintenance for a standard-sized automobile costs approximately 4.2 cents per mile of driving. (Liston and Aiken, 1977).

Handbooks, Manuals, Tech. Orders

As mentioned previously, some of the factors that should receive attention during design are shown in Appendix I. In this sense, handbooks and manuals may be considered job performance aids. Many maintenance handbooks have several common flaws. Firstly, they are usually "over-written"; that is, the level of reader ability required is often that of a college sophomore whereas studies by the U.S. Army and the U.S. Air Force show that maintenance men read at the seventh to ninth grade level. Many materials seem to have been written by engineers for engineers. Secondly, illustrations are often non-existent or inadequate and poorly integrated with the textual material. Thirdly, context goes from cause to effect when in reality the maintenance man works from effect to cause. Fourthly, the general layout of tables, graphs and illustrations often do not follow well-established rules of exposition. Finally, their cost is far from trivial. Informal inquiry by the first author disclosed cost figures for one page of technical manuals to be between 80 dollars and 1000 dollars with a mean of approximately 400 dollars. Manuals are big business! (Christensen, 1979).

On the other hand, care must be taken not to insult the user. A recent experience in the U.S. Army is a case in point. Manuals should not take on the air of comic books; maintenance men resent such treatment as an insult to their intelligence. (Chaikin - personal communication). It should be possible to design clear, easily understood written materials and diagrams without resorting to comics - such materials are not particularly amusing the second time through or when one is in the midst of trying to solve a very serious problem.

A recurrent problem in the field is that of availability not only of parts but also of manuals. Further, revisions in procedures often do not reach the field until six to eight months after the new parts or components have arrived.

Mention should be made of the work going on at Caterpillar Tractor and other places regarding the development of "controlled English" as an aid to the promotion of understanding across national language boundaries. (Kirkman, 1978). We feel that some of the findings would be helpful within language groups as well as between language groups. For example,

1. The variety of words is strictly limited (ILSAM - International Language for Service and Maintenance - contains less than 800 words).

2. Each word must have one and only one meaning.
3. A word used as a noun will not also be used as a verb or as an adjective.
4. Synonyms are to be avoided.
5. Words with the widest recognition are used in preference to less commonly used words.
6. Number of verbs is kept to a minimum.
7. Statements are short, direct and positive.
8. Repetition replaces reference to another part of the source material.
9. Sentence structure is simple.
10. Each sentence contains only one idea, command or instruction.

English written with the above restrictions in mind was used effectively after only 30 to 60 hours of instruction, even though not completely understood, by people who had no previous knowledge of English.

To repeat: adherence to the ten rules cited above might provide helpful guidance to writers of materials within language groups as well as between language groups. Proper use does not seem to lead to dull stultified writing. "... When Caterpillar published the first service literature in its restricted language, the difference was not detected by native English readers!" (Ostwald and O'Toole, 1978).

Errors of Diagnosis

Much of the time devoted to diagnosis is apparently not well spent. Before his untimely death, and while with McDonnell Douglas, Alan Burrows conducted a study of maintenance operations among commercial airlines. He divided the aircraft system into subsystems (auto-pilot communications, etc.) and found that between 40 and 50 percent of the time the elements removed for repair were not defective! (Personal communication -- before my friend's death). The members of this symposium need not be reminded of the effects on reliability of needless removal and replacement of electronics components -- "removal-induced failure", we might term it. Burrows also found what would appear to be identical components would have a MTBF of 10:1 on the best airline as compared to the worst airline. Detailed inquiry into the reasons for such disparities might be very revealing.

The Burrows figures agree very well with those given to the undersigned by a maintenance officer in the ground forces who states that 40 percent of the tank engines could have been repaired locally rather than sent back to a depot.

Installation Errors

The number of instances of equipment components being installed upside down or backwards is legion. One of the latest incidents is that of the blowout-preventer (assembly of valves) at the Ekofisk oil field in the North Sea. The device was inadvertently installed upside down. After one week of uncontrolled eruption, the well finally was brought under control by Red Adair and his team from Texas. Cost: $ 50,000,000 dollars and a lot of nervous Norwegians who, had there not been a wind shift, probably would have had their fishing grounds wiped out for a generation. It should not be too difficult to design item so they can only be installed one way -- the correct way.

In the early days of missile operations, it was not uncommon to find gyro cases installed upside down.

Environmental Factors

The working environment and climatic environment at the maintenance site obviously have an effect on effectiveness. For years, the first author has urged that the systems engineer of a system and one or two of his engineers should be paid to follow their systems into the field so that they could participate directly in the maintenance process. Numerous interviews and informal conversations with maintenance officers and men in U.S. military forces suggest that the following conditions are not uncommon in the field.

1. Fatigue. Because of irregular hours, enemy action, apprehension, etc. people in combat zones are frequently near physical exhaustion.
2 Weather. Conditions that vary from sub-zero to cool to hot and from desert dryness to swamp-like humidity.
3. Incomplete Tool Sets. Tools are frequently misplaced or borrowed in the field. Special tools are an anathema. Their inclusion should be approved only when there is no other way to get the job done.
4. Inappropriate/Inadequate Training. The effect often is disclosed most vividly in erroneous diagnosis. As suggested elsewhere in this paper, quick and more accurate diagnosis is sorely needed in field maintenance. Lack of familiarity with model changes is often a problem.
5. Manuals, tech. orders, etc. Often these are not current. Often they are not available at all for some components.
6. Personal problems. The trend toward increased use of narcotics and/or drugs has not by-passed field maintenance operations.

7. Inventory control. Errors in inventory caused by such things as erroneous transcription of a federal stock number (usually an 11-digit number) to a requisition form and/or erroneous punching of the same number in an IBM card are common. More use of error-correction codes seems to be indicated. Further, changes must be disseminated immediately to all users.
8. Computer fixation. Computer printouts seem to have a face validity that apparently is accepted even where there is compelling evidence of error. (One officer told the first author of a supply clerk literally having to be ordered to a warehouse to convince him that he had over 10,000 of a certain item -- simply because the computer said there were none!).

Physical Anthropology Considerations

Improper handling of equipment is another common cause of premature failure. Components that are too heavy, awkward to handle, not supplied with handles, etc. will be treated improperly.

While a wealth of information exists, it needs verification under field conditions -- loss of sleep, extreme fatigue, etc. Information on females is now becoming available; it also needs validation under field conditions.

Training and Experience

The argument between the advocates of specific versus generalized training and practical versus theoretical training goes on and on ... and on. Determinants are highly situation - specific (e.g. amount of available competence, etc.); further discussion here would contribute nothing to the resolution of the argument.

A study by Sauer, et al (1976) based on ratings of the performance of weapons mechanics/maintenance technicians and nuclear weapons specialists/technicians suggests that those ranked high have the following characteristics:

a. more experience
b. higher aptitude
c. greater emotional stability
d. fewer reports of fatigue
e. greater satisfaction with the work group
f. higher morale

A correlational analysis revealed significant positive correlations between task performance and:

a. years of experience
b. time in career field
c. ability to handle responsibility
d. morale

Significant negative correlations were found between task performance and:

a. anxiety level
b. fatigue symptoms

It is probably worth mentioning that over a wide variety of tasks (align, adjust, remove, etc.) Sauer and his associated found a human reliability mean of .9871. Thus, as a rule of thumb, we might expect errors by maintenance persons on the order of 13 times in 1000 attempts at the task element level.

The characteristics of the trainee population for maintenance jobs are important to both training and design specialists. In the Department of Defense, one can figure that only about three out of four trainees will have finished high school; they will read at about the level of a seventh or eighth grader; they will have little, if any, mathematical ability; their ability to integrate information from a variety of sources will be severely limited, and they will lack either the ability or the patience to do a very thorough job of filling out forms. (This is not meant to be an indictment of the trainee; it is definitely meant to be an indictment of our educational system).

Maintenance Records

Maintainability can be facilitated by proper treatment of the equipment by the operators and accurate description by them of malfunctions. Studies have shown that operator's reports suffer from several deficiencies. Many operators have their "favorite" malfunctions. (Leuba, 1967).

The Air Forces Maintenance Data Collection System (MDCS) and the Navy's 3M system have come under severe attack, yet Stranhagen (1978) avers that the systems are "highly reliable". Stranhagen suggests that in the coded data fields, less than one-half of the forms would have one error. (Stranhagen, 1978). Interestingly (and to some surprising), Stranhagen's research suggest that error is almost entirely a function of the type of character used (alpha characters have the highest error rate, numbers have the lowest and alphanumerics are in between) and is independent of knowledge of the equipment, data field size, frequency of use of a character, or position of a character in the data field.

The U.S. Air Force Acquisition Logistics Division collects and maintains a corporate data bank of "lessons learned" to provide improvement of Air Force acquisition programs. There are two categories of lessons: hardware or technical lessons and management or not-technical lessons. An informal review of a sample of these by the writers revealed maintainability design features to be a substantial cause of operational problems.

Troubleshooting Strategies

As mentioned previously, diagnosis consumes an inordinate amount of time of the maintenance man. Numerous trouble-shooting strategies have been devised. Unfortunately, none seems to be overwhelmingly preferred by maintenance men nor to show particularly significant advantages when evaluated in terms of available criteria. Common trouble-shooting strategies are shown in table 1.

Table 1. Trouble-Shooting Strategies

Strategy	Explanation and/or example
Sensory checks	Loose connections, odors, etc.
Historical information	Maintenance records and summaries
Reliability	Check most likely failure first
Conditional probability	If "A", then "B" most probable
Syndrome	Patterns of symptoms
Signal Tracing	Enter known signal and trace
Split half	Often preferred if no probability information available
Bracket	Narrow bracket from general to specific
Least effort	Make easiest checks first
Information uncertainty	Check first that which will eliminate the greatest number of failure causes
Computer	Maintenance man enters symptoms
Miscellaneous	Fuzzy sets, heuristics, etc.

Signal tracing, split-half, bracket and least effort are probably the easiest to learn. Other things being equal, split--half and bracket are probably, on the average, a bit more efficient than the others.

However, as stated previously, human response characteristics must also be considered. Leuba showed that operators have strong propensities to use certain trouble-shooting techniques and to report certain malfunctions. In this study, over fifty percent of the uncertainty was removed simply by knowing which operator made the complaint! Leuba recommends allowing the maintenance man to use his favorite troubleshooting technique because, to date, it has not been shown to make that much difference (Leuba, 1967).

Maintenance men have been known also occasionally to verify complaints that do not even exist, simply to improve their batting averages. (Leuba, 1967).

Operator reports of malfunctions often suffer from a number of deficiencies. Frequently, their reports are glib, lacking in critical data, information being withheld so as not to implicate themselves, equipment being blamed for operator errors, etc. (Leuba, 1967). Much could be done to improve the quality of operator reports of malfunctions.

Research such as that of Rouse and his associates is very much needed. (Rouse, 1978 (2), 1979 (3)). Our experience in the field suggests, however, that employment of any trouble-shooting technique will have to overcome rather well-established habits, intuitive feelings resulting from personal experience, lack of meaningful experience, and so on.

Problems of Special Importance in Computer Aided Manufacturing (CAM)

The successful introduction of CAM to a plant requires a sensitivity to its special maintenance requirements from top management down. Significant changes in attitude may be required. While the points that follow are not particularly new, they are some that have been found to be of special importance in CAM programs. (See also Blandow, 1979 and Cameron, 1978).

First, the maintenance people themselves should have a prominent voice in deciding which equipment should be purchased. Hopefully, they will have been in touch with others who have used the system that is proposed for purchase and will have estimated of the resources required to install and maintain the equipment being considered. Proper installation will give consideration to such problems as sound-proofing, stability of foundations, etc.

Second, the plant may often be expected to run 24 hours a day. This puts a special premium on efficient maintenance, whether it be preventive or corrective. It also makes in-house availability of the necessary skills even more important. Efficient inventory systems that allow quick withdrawal and availability without over-stocking assume even greater importance. This should be supported by detailed records of down-time, parts failure and machine utilization.

The advantages of design standardization for CAM equipment are several. The development of standard modules and/or sub-assemblies is to be encouraged, different purposes being met by different configurations of the standard modules. Such a program will reduce parts inventories, simplify training programs and reduce skill requirements. The design of special machines and special tools should be discouraged.

Design standardization will also result in fewer manuals, making it easier to keep them current. The manuals themselves should be written at about the eighth grade level, making liberal use of diagnosis, clear codes and schematics, etc.

Selection criteria need reconsideration; there will be a need for increased numbers of skilled electronic techniques. A whole new technology of mini - and microcomputers must be mastered. Job requirements may be tougher.

Service industries are springing up that offer comprehensive service/packages. International Data Corporation estimated that this will be a billion dollar industry by 1982. Whether or not they are indicated in a specific plant depends on such factors as size, skills of present maintenance force, availability of needed skills, etc.

SUMMARY

The general conclusion we draw from our field experiences and the field experiences of others is that, while extensive improvements can and should be made in design for ease of maintenance, development of improved trouble-shooting models, etc., immediate and extremely significant contributions can be made by improvements in supply (parts, manuals, etc.), inventory control, working conditions and attention to the personal needs of maintenance men. Their importance to the overall systems effort must not be overlooked. It must receive attention at the highest levels. To quote Geise and Haller (1965), "The causes, i.e. failures), are attributable to systems engineering management, for by absorption in technical state-of-the-art advancements, sight has been lost of the importance of the everyday operations of people in making a system function".

The introduction of more automatic equipment into both production and operational systems will cause significant changes in maintenance operations -- electricians must be replaced by electronic technicians, the maintenance man must understand the latest developments in software (where almost certainly the greatest breakthroughs will occur in the next decade or two), he must understand digital circuitry and micro-processors.

Concurrent with the above, maintainability must receive increased attention throughout the design, development and procurement phases of systems evolution. Appendix II represents an attempt to record the nature of the contributions that human factors can make during these phases (Christensen, 1979). Timely attention to these requirements should improve the maintainability of future systems. Insufficient information exists to meet many of the responsibilities listed in Appendix II; such points might serve as sources of fruitful hypothesis for additional investigations.

REFERENCES

Anon, "One Way to Minimize NC Downtime", Iron Age, 222, May, 1979.

Blandow, R.W., "A Maintenance Overview of CAM Technology", Manufacturing Engineering, July 1979.

Christensen, J.M., "Human Factors Considerations in Design for Reliability and Maintainability", In Pew, R.W., "Human Factors Engineering", Course No. 7936, Engineering Summer conferences, The University of Michigan, 1979.

Crawford, B.M. and Altman, J.W., "Designing for Maintainability" in VanCott, H.P. and Kinkade, R.G. (eds.), Human Engineering Guide to Equipment Design, Washington, D.C.: U.S. Government Printing Office, 1972.

Feineman, G., "How To Live with Reliability Engineers", Spectrum, Spring, 1978.

Geise, J. and Haller, W.W. (eds.), Maintainability Engineering, Martin-Marietta Corporation and Duke University, 1965.

Goldstein, D.B. and Rosenfeld, A.T., Energy Conservation in Home Applicances Through Comparison Shopping: Facts and Fact Sheets, LBL-5236, Energy Extension Services, State of California, 1977.

King, J.B. (Chm.), Safety Recommendations A-79-98 through 105, National Transportation Safety Board, Washington, D.C., December 21, 1979.

Kirkman, J., "Controlled English Avoids Multi-Translations", Industrial Engineering, February, 1978.

Leuba, H.R., "Maintainability Prediction - The Next Generation", Proceedings of the Spring Session or Reliability, Maintainability, etc., IIIE, Boston, 1967.

Liston, L.L. and Aiken, C.A., Cost of Owning and Operating an Automobile, U.S. Department of Transportation, Federal Highway Administration, Washington, D.C., 1977.

Ostwald, P.F. and O'Toole, P.I., "I.E.'s and Cost Estimating", Industrial Engineering, February, 1978.

Post, C.T., "Microprocessor Control Diagnoses Welder Faults", Iron Age, 222, 41, November 1979.

Rigby, L.V., Cooper, J.I. and Spickard, W.P., Guide to Integrated System Design for Maintainability, ASD Tech. Report 61-424, WPAFB, Ohio, October, 1961.

Rigby, L.V., The Saudia Human Error Rate Bank (SHERB). SC-R-67--1150, Saudia Labs. Albuquerque, New Mexico, 1967.

Rouse, W.B., "Human Problem Solving Performance in a Fault Diagnosis Task", IEEE Transactions on Systems, Man, and Cybernetics, Vol. SMC-8, No. 4, 1978.

Rouse, W.B., "A Model of Human Decision Making in a Fault Diagnosis Task", IEEE Transactions on Systems, Man and Cybernetics, Vol. SMC-8, No. 5, 1978.

Rouse, W.B., "A Model of Human Decision Making in Fault Diagnosis Tasks That Include Feedback and Redundancy", IEEE Transactions on Systems, Man and Cybernetics, Vol. SMC-9, No. 4, 1979.

Rouse, W.B., "Problem Solving Performance of Maintenance Trainees in a Fault Diagnosis Task", Human Factors, Vol. 21, No. 2, 1979.

Rouse, W.B., "Problem Solving Performance of First Semester Maintenance Trainees in Two Fault Diagnosis Tasks", Human Factors, Vol. 21, No. 4, 1979.

Sauer, D., Campbell, W.B., Potter, N.R. and Askren, W.B., Relationships Between Human Resource Factors and Performance on Nuclear Missile Handling Tasks. AFHRL-TR-76-85/AFWL-TR-76--301, Air Force Human Resources Laboratory/Air Force Weapons Laboratory, 1976.

Stranhagen, J.F. Jr., How good are Maintenance Data? Spectrum, Spring, 1978.

APPENDIX I: MAINTAINABILITY DESIGN CHECKLIST

1. Accessibility (general)

_____ 1.1. Physically accessible (removal and replacement convenient)
_____ 1.2. Visually accessible
_____ 1.3. Room to use tools and aids
_____ 1.4. Anthropometric considerations (height, reach, posture, etc.)
_____ 1.5. Protective clothing (gloves, hard hats, diving suits, pressure suits, etc.)
_____ 1.6. "Plumbing" on outside, where possible
_____ 1.7. Prober balance between maintenance and operations
_____ 1.8. Relationship to other items to be maintained
_____ 1.9. Safety interlocks (electrical power)
_____ 1.10 Rounded corners, no sharp edges

2. Accessibility (specific)

2.1. Access openings

_____ 2.1.1. Sufficient number
_____ 2.1.2. Sufficient size (one large versus two small)
_____ 2.1.3. Not hidden by other components
_____ 2.1.4. Same plane as related controls and displays
_____ 2.1.5. Convenient height, reach, posture, etc. requirements
_____ 2.1.6. When required, safe, study, convenient stands,-ladders, etc.
_____ 2.1.7. Convenient removal (drawers, hinged units, etc.); power assists for removal or units over 100 pounds (or availability to two or more persons).

2.2. Access covers

_____ 2.2.1. Hinged or tongue and slot
_____ 2.2.2. Easily opened
_____ 2.2.3. Easily held open
_____ 2.2.4. Positive indication of improper securing

2.3. Access cover and component fasteners

_____ 2.3.1. minimum number
_____ 2.3.2. captive
_____ 2.3.3. hand-operable
_____ 2.3.4. Same type on all components (interchangable; no special tools)
_____ 2.3.5. Screw heads for wrench or screw driver
_____ 2.3.6. Quick opening
_____ 2.3.7. Durable (sturdy, corrosion resistant)

_____ 2.3.8. Easily replaced, if damaged

2.4. <u>Access and component labeling</u>

_____ 2.4.1. Opening instructions (if not obvious)
_____ 2.4.2. Legible, permanent, unambiguous wording
_____ 2.4.3. Items in enclosure
_____ 2.4.4. Warnings, if appropriate

3. <u>Packaging</u>

_____ 3.1. Clear identification (number codes, color codes, etc.)
_____ 3.2. Convenient size, shape, weight, etc.
_____ 3.3. Convenient number of units/package
_____ 3.4. Handles for units over 10 pounds
_____ 3.5. Logical, consistent flow, if troubleshooting important
_____ 3.6. Component grouping, if mass replacement used
_____ 3.7. Protection from vibration, heat, cold, dust, etc.
_____ 3.8. Easily removed and replaced (plug-in, rollers, roll-out drawers, etc.)
_____ 3.9. Error-free replacement (guides, keys, alignment pins, etc.)
_____ 3.10 Easy inspection
_____ 3.11 Easy servicing (tightening, lubricating, etc.)

4. <u>Connectors</u>

_____ 4.1. Labelled
_____ 4.2. Physically accessible
_____ 4.3. Visually accessible
_____ 4.4. Hand-operated (at least no special tools)
_____ 4.5. Quick disconnent
_____ 4.6. Screw Terminals (rather than solder)
_____ 4.7. U-lugs rather than O-logs
_____ 4.8. Alignment aids (keys, colored strips, asymmetry etc.)
_____ 4.9. Unique (prevent mismating by differences in color, size, number of pins, pattern of pins, etc.)
_____ 4.10 Receptacle "hot", plugs "cold"
_____ 4.11 Self-locking

5. <u>Conductors</u>

_____ 5.1. Labelled or coded
_____ 5.2. Avoided sharp edges, etc. in routing
_____ 5.3. Automatic rewind, if appropriate
_____ 5.4. Out of the way (clamps, tie-downs, etc.)
_____ 5.5. Ample length without stretching

6. Displays

_____ 6.1. Use of characteristic odors, sounds, etc.
_____ 6.2. Consider maintenance display requirements independently of, and co-equally with, operator display requirements
_____ 6.3. See also MIL-STD1472B

7. Controls

_____ 7.1. Consider maintenance control requirements independently of, and co-equally with, operator control requirements
_____ 7.2. Non-shared controls under quick access covers
_____ 7.3. No special tools
_____ 7.4. See also MIL-STD1472B

8. Maintenance Support Equipment

8.1. Stands

_____ 8.1.1. Adjustable height
_____ 8.1.2. Part of primary equipment, if possible
_____ 8.1.3. Stable on incline up to 15°
_____ 8.1.4. Stable over entire surface
_____ 8.1.5. Label showing capacity in pounds
_____ 8.1.6. Easy to position and adjust

8.2. Work platforms and walkways

_____ 8.2.1. Sufficient space (six ft.^{2}/person)
_____ 8.2.2. Work with both hands free
_____ 8.2.3. If wheels, wheel locks
_____ 8.2.4. Access stairs at angle - 35°
_____ 8.2.5. Non-slip surfaces
_____ 8.2.6. Adequate handrail/guardrails
_____ 8.2.7. Walkways minimum of 12 inches wide
_____ 8.2.8. Label showing capacity in pounds
_____ 8.2.9. Shelter from elements

8.3. Tools

_____ 8.3.1. Minimum of different kinds of tools
_____ 8.3.2. Few (preferably none) special tools
_____ 8.3.3. Adequate grips
_____ 8.3.4. Insulated handles
_____ 8.3.5. Easily positioned

8.4. Test equipment

_____ 8.4.1. Same principles as basic equipment for controls displays, connectors, etc.
_____ 8.4.2. Maintenance-free
_____ 8.4.3. Easy to check calibration

_____ 8.4.4. Simple to learn to operate
_____ 8.4.5. Obvious or built-in labels and operating instructions.
_____ 8.4.6. Circuit protection (both primary and test equipment).
_____ 8.4.7. Convenient to set up and store

8.5. Bench mockups

_____ 8.5.1. Same principles as basic equipment
_____ 8.5.2. Quick-disconnects on all cables
_____ 8.5.3. Extra length cables
_____ 8.5.4. Extra heavy insulation on cables
_____ 8.5.5. Test points for checking signal flow
_____ 8.5.6. Correct signal values and tolerances prominently displayed
_____ 8.5.7. Accessibility for each unit
_____ 8.5.8. Test point for checking signal flow

9. Maintenance Procedures

_____ 9.1. All materials consistent with knowledge and skills of users
_____ 9.2. Written supplemented with diagrams, schematics, etc.
_____ 9.3. Brief but clear
_____ 9.4. Procedures that give unambigious results
_____ 9.5. Cross-checks where possible
_____ 9.6. Realistic tolerances

10. Job Performance Aids (JPA)

_____ 10.1. Content more important than mode of presentation
_____ 10.2. List all relevant to JPA
_____ 10.3. Develop step x step procedures for each task
_____ 10.4. For each step determine (a) information to be learned and (b) information to be put in JPA
_____ 10.5. Best method of presenting (b) above (schematics, functional diagrams, block diagrams, tables etc.).
_____ 10.6. Include only needed information
_____ 10.7. Consider characteristics of users (verbal fluency, experience, etc.)
_____ 10.8. No inconsistencies between JPA and primary equipment
_____ 10.9. Systematic procedure for maintaining currency in JPA's

11. Design of Protective Gear for Maintenance Personnel

_____ 11.1. Minimum reduction in performance capability
_____ 11.2. Detailed description of operational environment (temperature, rainfall, etc.)
_____ 11.3. Physical characteristics of maintenance population
_____ 11.4. Special equipment design features for compatibility with protective gear (e.g., larger handles on screw drivers for users with pressure gloves)
_____ 11.5. Easy on - easy off
_____ 11.6. Optimum sizing tariffs
_____ 11.7. Possible substitution of shelter for personnel equipment
_____ 11.8. Pleasant appearance
_____ 11.9. Easy to clean, maintain, etc.

12. Design of Training Equipment and Equipment for Maintenance of Proficiency

_____ 12.1. Usual good human engineering practices
_____ 12.1. Identification of critical tasks
_____ 12.3. Compatibility with "training system"

SOURCES: Crawford and Altman (1972)
Pope (Personal Communication) (1964)
Miscellaneous sources.

APPENDIX II: HUMAN FACTORS CONTRIBUTIONS TO MAINTAINABILITY DURING SYSTEMS DEVELOPMENT (Christensen, 1979)

(The implementation and application of human factors techniques, data, and principles must begin with the establishment of system requirements and continue through all phases of systems development, production and deployment).

A. Requirements Phase

1. Establishment of realistic system objectives
2. Definition of the operational environment
3. Experience from previous systems
4. Estimates of future maintainability capabilities
5. Criteria of design for maintainability
6. Coordination with customer

B. Concept Formulation Phase

1. Evaluation of alternative concepts in terms of impact on maintainability
2. Development of a maintenance concept (personnel requirements, replacement vs. repair, job aids, etc.)
3. Coordination with customer

C. Systems Definition Phase

1. Development of a Maintenance Plan (elaboration of the maintenance concept for the alternative chosen for further development)
2. Maintainability functions and task analysis (preliminary)
3. Maintainability prediction
4. Definition of maintenance population requirements
5. Software requirements
6. Requirements for job performance aids and simulators
7. Coordination with customer

D. Engineering Design Phase

1. Final maintainability functions and task analysis
2. Advice on capabilities of maintenance personnel
3. Refinement of maintainability predictions
4. Integration and resolution of operability and maintainability design features
5. Packaging and installation recommendations
6. Job performance aids
7. Simulators
8. Software (manuals, computer programs, etc.)

9. Coordination with personnel selection and training experts
10. Production specification
11. Development of Test Plan
12. Development of troubleshooting techniques
13. Coordination with customer

E. Design Verification

1. Implementation of Test Plan
   a. Method of failure inducement
   b. Number and types of failures to be induced
   c. Use of representative personnel under representative field conditions
   d. Sufficient sample size for statistical purposes
   e. Careful record of functions/tasks and comparison with predictions
   f. Records of cost, down-time, accidents, lack of tools, manuals, etc.
   g. Check on maintainability predictions

2. Preparation of test report, including recommendations for redesign, personnel changes, etc.
3. Manual and procedures revision
4. Develop Installation Plan
5. Customer participation

F. Production and Installation

1. Final recommendations to production engineers
2. Carry out Installation Plan
3. Check on Installation Plan with actual operational field trial
4. Coordination with maintainability engineers
5. Field training for maintenance men
6. Coordination with customer

G. Operations

1. Accurate and effictive maintenance records
2. Measurement of activities of field maintenance men
3. Feedback on field experience to design engineers, training specialists, systems engineers and system managers
4. Studies of impact of modification programs on reliability and maintainability
5. Coordination with customer

THEORIES AND MODELS

CHAIRMAN: T. B. SHERIDAN

SECRETARY: A. MICHAELS

THEORIES AND MODELS

Thomas B. Sheridan

Massachusetts Institute of Technology
77 Massachusetts Ave.
Cambridge, Massachusetts, U.S.A.

CHAIRMAN'S REVIEW AND DISCUSSIONS OVERVIEW

The twelve papers in this session constitute an excellent cross section of research directed toward theorizing and modeling human failure detection and diagnosis behavior. Several papers struggle with theoretical issues by reporting experimental data. Others discuss models directly. Others try to relate the two.

The paper by Ephrath and Young takes up an issue that has worried man-machine system designers for a number of years, namely whether the human operator can do a better job of detecting and diagnosing system failures as an active in-the-loop controller or as a monitor of an automatic system. The authors have found that in their own data they have conflicting evidence on the point - in some experiments the controller is significantly better, while in others the monitor is better. They resolve the problem by showing that "mental workload" is the differentiating factor, and that when the workload of being a controller would be sufficiently high it is better to be a monitor. With lower workloads better detection is achieved by being a controller.

Wickens and Kessel deal with essentially the same experimental question, but in the context of a somewhat different experiment. They found that involvement as an active controller was always better for detecting failures - in their case a change from first to second order undamped process. But, recognizing that often there are compelling reasons why control loops must be automated, they suggest that operator understanding would be enhanced by having each operator control each loop he is to operate in a training session. They further suggest that the differences between their results and those of Ephrath and Young may be explained by the differences in dynamic processes used as control tasks.

Discussion of these two papers included questioning of how quality of controlling affected the error size, and hence detectability, whether stress might have affected the results, how training affected detection performance, and what form of evidence there might be that operators employed "internal models" (their ability to report the latter was called "poor"). Kessel reasserted that plant managers should encourage active manual control. One person noted data that showed that aircraft which are "difficult to land" actually have fewer accidents than "easy" ones.

The paper by Curry provides a theoretical index of how useful data are in discriminating abnormal from normal conditions. Starting from a log-likelihood function, based on a given set of "symptoms", he develops an attention-surprisal function which outwardly looks like an information measure. Actually it is a sum of logarithms of symptom probabilities in the null or no-failure situation, each weighted by the attention allocated to it. Curry shows by example how it may be used.

Most of the discussion of Curry's paper dealt with the applicability and robustness of the index. It has not been applied externally as yet but appears to have considerable promise.

The next paper, by Moray, reviews a number of relevant current issues such as sampling, speed-accuracy tradeoff, and memory. He prevents some new data on forgetting (which can be very fast) and makes some engineering suggestions. The paper also contains an interesting list of features of real systems.

Discussion focussed on the relations between speed, stress, memory and uncertainty of data. A particularly salient comment noted "lapses" in human behavior which are very serious in monitoring automatic systems. In some cases it may simply be a way of efficient sampling, but in other cases it may be a way to disaster.

Rouse reviews a number of experiments his group has performed using a clever failure detection/isolation experimental paradigm. The experiments explore such factors as problem size, computer aiding, type of personnel, pacing, redundancy, type of training and transfer of training. One important result is the inability of people to make good use of "non-failure information". In generalizing his results he discusses "fuzzy" models and rule-based models.

Discussion of Rouse's paper brought out questions of whether training in context-free rules is useful and whether such rules work (Rouse affirms both points) and the efficacy of computer-aided training (it did not always work in these experiments).

The next paper, by Wohl, is the only one to deal explicitly with repair time data in a quantitative way - in this case in the context of field maintenance of sophisticated electronic equipment. Earlier modeling experiments experienced difficulty - because of an unusually long tail to the density function which had to be plotted on a Weibull distribution as a separate straight line from the first part of the density function. However, the introduction of a new theory about the influence of number of wires per component and number of components per junction led to a reformulation in a three-parameter exponential model (a special case of the Weibull) which fit the data much better.

In the discussion Wohl opined that a "mental model" was not necessary for his type of diagnosis problem. He felt his new model to be robust, possibly even extendable to computer software debugging.

Brehmer switches to the other end of a spectrum - from well-defined circuit diagnosis to much more complex forms of diagnosis, namely psychiatry and social psychology. His experimental technique involves the use of a computer-aided "judgement heuristic" technique based on linear regression theory, whereby overall judgements are made for given sets of symptoms. The computer then feeds back to the judge estimates of the relative weightings he used for each of the symptoms and the forms of the weighting functions as related to the amount of each symptom. Brehmer's results showed considerable intersubject differences and inconsistencies in using this technique, but he nevertheless felt it was a good idea for individuals to be conscious of how their diagnoses are being weighted with respect to symptoms.

The discussion underscored the "noise" in results, even in judgements among experts, in this context as compared to electronic circuits. Brehmer remarked that the data show marked differences between psychiatrists, psychologists and social workers in their diagnostic styles.

Rasmussen's paper treats the reader to some very interesting and insightful diagrams and classifications regarding cognitive processes for failure detection and diagnosis which have emanated from his research on error data and verbal protocols. He discusses at some length the distinction between what he calls "symptomatic search" - a fitting of present symptoms to a library of symptoms - and "topographic search" - the comparison of present to normal behavior using a more complex topographic map or cognitive structure.

His paper stimulated considerable discussion regarding the relation of "data processes" to "value formation". For example, success or failure in data processing may lead one to modify his values. There was also discussion regarding the levels of behavior - from rote skills, through rule-based to knowledge-based behavior - and how all three come into play in one person at various times.

Bainbridge, in her paper, discusses how present "engineering models" of the human operator, while quite satisfactory for fast responding systems, are less than satisfactory for slowly responding ones such as are most often encountered in process plants. She reviews various studies by Beishon, Crossman, Umbers and herself, performed within the process control context. In doing so she emphasizes the open-loop nature of much control, the importance of operator planning activities, prevalence of conflicting goals, and the irregularity of sampling. She describes how ordinal, "fuzzy" mental models may be obtained from verbal protocols, which she herself has pioneered.

In the discussion which followed her presentation there were questions as to whether she intended to indict all mathematical models (no, she only wanted to set boundary conditions) and questions concerning the importance of human memory in process control (sometimes operators have significant lapses of memory or commit significant omissions). There was mentioned the importance of giving process operators useful information at the appropriate level, and perhaps with sufficient manual involvement, at least in training.

Leplat's paper emphasizes the definition of operator tasks. He also provides as a definition of "optimal" task description for the operator - just enough precision that the operator can fulfill the task. He distinguishes the "task assigned" from the "task undertaken" - an important distinction. He concludes by giving some cautions about making field investigations.

The Kantowitz and Hanson paper follows in somewhat the same rein as Bainbridge and Leplat - with cautions about the "engineering approach" to modeling human detection and diagnostic behavior. But their language is considerably stronger. They assert that optimal models are inappropriate to describe real human behaviour, that they model tasks but not humans, that they are too complex. A plea is made for simpler paradigms.

Actually, since authors of neither of the last two papers were present there was no formal discussion on them. But the Kantowitz paper especially created some informal discussion and some defense. One was the contention that Kantowitz and Hanson miss the point of optimal models, that any precise theory carries with it a normative model of how people should behave (if they conform to that theoretical norm) and experimental comparisons naturally show some discrepancy from that precisely specified theory. (This is in contrast to the descriptive model, of course, wherein the best model is the simplest yet reasonable summary of the data). The optimal control model is not quite simple, but it certainly is robust.

The final paper, by de Kleer and Brown, gives readers the flavor of the new computer-science-based or "artificial intelligence" approach to the study of cognition. The authors develop an approach to "qualitative modeling" which is rather different from all the others, and in particular, predicate their approach on two principles: (1) that every event is caused by another event, and (2) that component input-output behavior is not affected by how the components are configured. They develop their ideas in the context of an example - modeling the behavior of an electro-mechanical buzzer.

There was no formal discussion of this paper either because the authors were not present, but had they been, there would have been some question as to why this approach is not more arbitrary and more complex than conventional engineering models, and why the "no function in structure" principle is so different from what systems engineers usually assume.

In the closing general discussion for the whole session it was pointed out that what is important is not what operators can do but what they will do - that often it is a little human error like remembering to replace an oil cap which grows to a bigger fault.

There was some further discussion about whether manual control models could be extended to most of what process operators do - beyond the obvious continuous feedback control tasks. S. Baron, champion of the optimal control model, affirmed that they can.

Finally there was discussion of ubiquitous individual differences - that they will not go away, that detection and diagnostic aids must account for them, and that multiple operators' mental models might have to be brought into mutual conformance.

In closing this summary of the papers and discussion on theories and models the writer makes the following general observations:

1) Speculating about thinking has become popular, and engineers as well as psychologists do it with enthusiasm. Neither is inhibited by the old behaviorist dicta that modeling can be based only upon operationally defineable and measureable variables.

2) The "internal" or "mental model", emanating from ancient philosophy, implemented in specific form in Kalman estimators by modern control theorists, is an accepted construct. But there is still great confusion as to how to put it to use to model detection and diagnosis.

3) There is now no single theoretical approach which seems preemenant. Control theory is struggling to prove it can apply to process control. Information and decision theoretical constructs keep popping up, but the classical theories do not go the distance. Artificial intelligence models have not even reached the point of looking relevant in most people's minds, mostly, perhaps, due to lack of experience with them. There is still plenty of respect for simple predictive models which fit empirical data for relatively specific contexts (example, see Wohl's paper). At the same time it is clear that in very complex diagnoses (example, see Brehmer's paper) we are far from any predictive model. And whether context-free diagnostic mechanisms are viable is still being questioned.

4) One senses an increased tolerance for complexity in models, even though it is well known that model complexity grows with the number of interactions between real world elements, not simply in proportion to the number of elements but geometrically (until the number of real world elements approaches infinity, at which time simpler thermodynamic type paradigms apply). Computers now allow us to "run" these more complex models. But perhaps we are losing the generality of the simpler classical models.

5) As the capability of computers and automation grows man's role becomes primarily that of planner and programmer, plus monitor, detector and diagnoser of that system behavior which deviates from those plans. This set of papers uncovers a manifold of problems in modeling such human activity. But there is clearly some way to go to have satisfactory answers available.

MONITORING VS. MAN-IN-THE-LOOP DETECTION OF AIRCRAFT CONTROL FAILURES

Arye R. Ephrath and Laurence R. Young

Department of Aeronautics and Astronautics
Man-Vehicle Laboratory, M.I.T.
Cambridge, Mass. 02139, U.S.A.

INTRODUCTION

The rapid technological advancements of the past decade, and the availability of higher levels of automation which resulted, have aroused interest in the role of man in complex systems. Should the human be an active element in the control loop, operating manual manipulators in response to signals presented by various instruments and displays? Or should the signals be coupled directly into an automatic controller, delegating the human to the monitoring role of a supervisor of the system's operation?

From a purely technological point of view, both approaches are feasible in most systems. The current state of the art makes possible, for instance, the construction of a passenger jet aircraft which would climb, cruise, navigate to any point on the globe, descend, approach and land with nary a human intervention in the process. The automatic control of industrial manufacturing facilities, of spacecraft, of urban rapid transit systems and of nuclear reactors are examples of other cases in point.

Indeed, the very availability of the applicable technology has spurred a number of investigators to suggest intriguing scenarios for the future. Draper et al. (1964) raised seriously this question of the role of the human. Wempe (1974) offered the view that, as control and navigation requirements of future aircraft grow more complex, they may exceed human capabilities and make at least partial automation mandatory. This, in turn, raises the problem of dynamic allocation of tasks between man and machine (Walden and Rouse, 1978).

Given an option of either manual or automatic control, the selection of the actual control mode for any system under consideration may be governed by a number of factors, ranging from economics to psychological motivation of the human operator. One of the most important of these factors, however, especially in the design of systems with potentially-catastrophic failure modes, is the question of safety; to wit, the ability of the human to detect a malfunction and to provide redundancy in case of a failure.

It is axiomatic that the operator should be capable of detecting and identifying failures in the system accurately, reliably and with minimum time delay. It is not at all clear, however, which control mode provides a better path to this desirable end.

An argument can be made favouring automatic control, with the human monitoring the system; a malfunction would be detected rapidly, the theory goes, since the human is not burdened by mundane (and attention-consuming) manual control tasks. The human can thus devote his entire effort and all of his attention to the task of monitoring the system, diligently scanning the displays, searching for the telltale deviations and discrepancies which indicate trouble.

Not so, may say those who favour keeping the human in the control loop. Systems with potentially-catastrophic failure modes are normally designed with a high degree of reliability; during that extremely long mean-time-between-failures the human, with nothing to do but monitor a perfectly-normal system, may become so bored and complacent that he might miss the rare failure when it does occur. Furthermore, even in the event that a failure is correctly detected and identified, the operator needs to switch from monitoring to manual control to assume the role of the malfunctioning automatic device. Such mode-shifts are rarely instantaneous, especially after extremely long periods of monitoring, Humans need time to adapt to the role of an active control element; in an emergency, that time may not be available.

In an attempt to provide an answer to this question, we have investigated, via experiments, the effects of the mode of the operator's participation in the control task on his failure detection performance. Two experiments were carried out; one was based on a single-loop control task, the other - on a complex, multi-loop task simulating an aircraft instrument landing.

SINGLE LOOP CONTROL

This first experiment utilized the human operator in his most "mechanical" role to close the manual control loop of a single axis compensatory system with random continuous input. At unknown times, the controlled element would suddenly change to a new element with different sign or number of integrations (Young, 1969). Three types of human operators were used: The "active" controller performed conventional compensatory tracking. The "inactive" manual controller also tracked but, without his knowledge, the error he observed was that of the active controller. The inactive controller's commands were not reflected in the error he observed. Finally, a "passive monitor" simply observed the active controller's error. Each participant was instructed to indicate any controlled element change if it were detected. For this simple task, with no side loading, the subjects with greater involvement in the control loop demonstrated shortest detection times. The active controller, who could compare his commands accurately with the subsequent error changes, detected failures in about 1 second. The inactive controller, who was involved in the tracking loop but less capable of performing the above comparison, took about 50 percent longer. The passive monitor, on the other hand, with only error patterns to analyze, required from 3 to 5 times longer to make the detection. These results support the internal model concept of human adaptive manual control, in which actual system performance is continuously compared with the performance of an internal reference model.

Subsequent extensive experiments of a similar nature (Young and Tada, 1980) lend further support to the model. In these experiments, three subjects used a displacement control stick to track a sum-of-sines pseudo-random signal via a compensatory display. The controlled element was a first-order integrator with a gain, 2/s. Between 20 and 60 seconds after the start of a run the controlled element dynamics were changed to -2/s, 8/s, -8/s or $5/s^2$ in such a manner that the displayed error and its rate remained smooth, and the subjects indicated a detection of this change by pressing a push-button.

Three sets of experiments were conducted. Each set consisted of 63 runs per subject, and one change in the controlled element dynamics occurred in each run in a randomized order. The same three subjects participated in all 3 sets.

In the first set, the subjects tracked the pseudo-random signal and their tracking error, control inputs and detection times were recorded. In the second set, the subjects started each run by tracking in the conventional manner; the tracking error and error-rate were continuously compared, however, to those recorded for that subject during the same run in the first set. Approximately 2 seconds before the change in dynamics were to occur, and when the differences of e and $\dot{e}$ between the two compared signals were below a pre-set threshold, the display was switched to the recorded error signal. The subject thus became an "inactive" or "shadow" controller, excluded (without his knowledge) from the control loop and observing his own error signal as recorded during the first set of experiments. It was verified in post-experiment debriefings that none of the subjects was aware of this fact.

In the third set of experiments the subjects acted as pure monitors, observing their own recorded error signals and indicating via the push-button when they detected a change in the dynamics.

The results of these experiments reveal quite dramatically the superiority of the active manual controller as a detector, over both the inactive controller and the monitor. The active controller always detected the change in the controlled element dynamics; in approximately 70% of the runs the change was detected within 1 second. By contrast, both the "inactive" ("shadow") controllers and the monitors took a good deal longer to make the detection, requiring 3-5 seconds to report the change in 70% of the runs. Furthermore, in quite a few instances the event had not been detected at all by the time the run ended, 12 seconds after the dynamics changed.

Since the "active" controllers, the "inactive" controllers and the monitors all observed identical error signals, the superiority of the "active" controller as a detector must be attributed to the availability to him of proprioceptive feedback of hand position; this feedback was denied both the monitor and the "inactive" controller. This, therefore, tends to support the hypothesis that knowledge of hand position in conjunction with system error is very important for rapid correct identification of single loop system failures.

MULTI-LOOP CONTROL

This experiment was carried out in a static cockpit simulator and utilized fifteen professional airline pilots as subjects. The simulator was a mock-up of the captain's station in a Boeing transport aircraft, and it was programmed to duplicate the dynamics of a large transport aircraft in the landing-approach flight envelope.

In addition to failure-detection performance, we were interested in measuring the subjects' workload under various control-participation modes and simulated wind disturbance conditions. To this end, a warning-light-type subsidiary task had been installed in the cockpit. It consisted of two small red lights mounted above each other outside the subject's peripheral field of vision. At random times during the run either light, with equal probability, was lit. A correct response by the subject consisted of extinguishing the light via a control-yoke-mounted, three-position thumb switch. In the absence of a correct response, the light remained lit for two seconds; then it was turned off and a "miss" was recorded. A random time-delay, uniformly distributed between 0.5 and 5.0 seconds, separated successive lights.

Our workload index was based on the ratio of the number of "misses" to the total number of light stimuli presented. It is grounded in the hypothesis that, as the primary task of flying the simulated aircraft becomes more demanding, more of the secondary task lights will be missed by the subject. This type of workload measure had been used by a number of other investigators as well (e.g., Spyker et al., 1971, and Brown, 1964).

Each subject flew a number of simulated instrument approaches from a point approximately 12 miles out to touchdown. The experiment involved four levels of control participation:

a) Monitoring, with autopilot coupling in all axes.

b) Manual in the lateral axis, with autopilot coupling in the pitch axis.

c) Manual in the pitch axis, with autopilot coupling in the roll axis.

d) Fully manual.

There were three levels of wind disturbance:

a) No wind.

b) 45° tailwind of 5 knots gusting to 15 knots.

c) 45° tailwind of 10 knots gusting to 30 knots.

Three failure conditions were used:

a) No failure.

b) A failure occurs in the lateral axis. In this condition the autopilot, if coupled, or the flight director steered the airplane away from the localizer course. The deviation was such that the horizontal situation indicator reached a one-dot angular error (1.25^{o}) approximately 100 seconds after the initiation of the failure. The effect was quite slow and subtle, providing a good measure of the limits of the subjects' failure-detection capability.

c) A failure occurs in the pitch axis, identical in type to the lateral failures and resulting in a one-dot deviation (0.35^{o} of

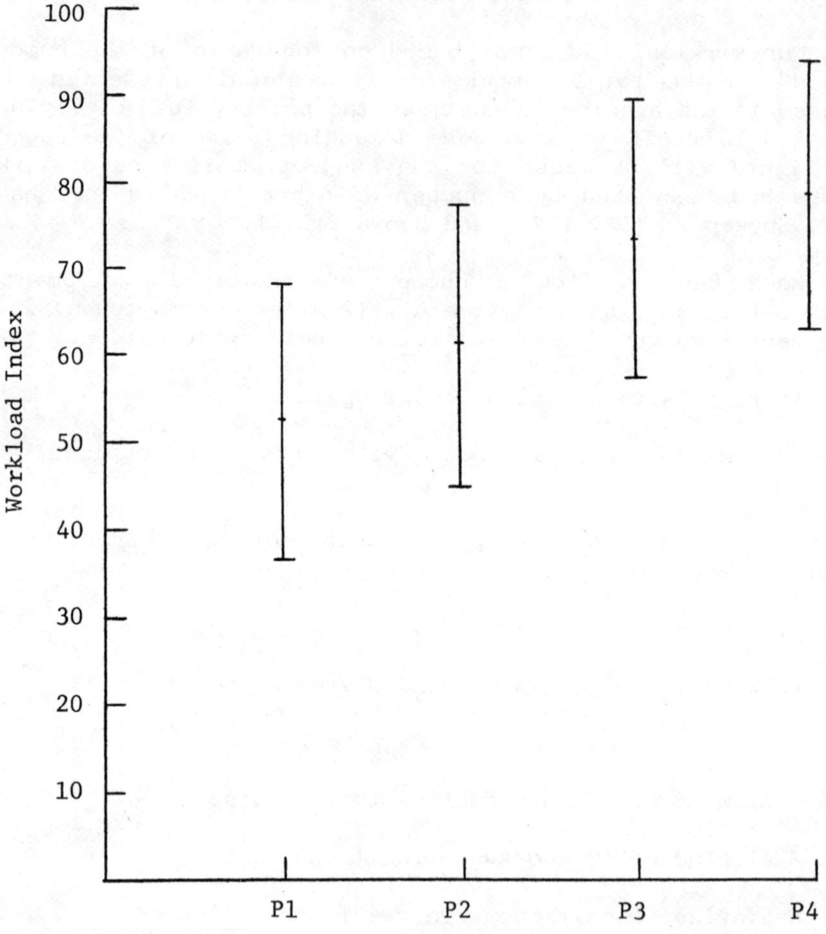

Figure 1: Workload Index at Four Participation Modes ($\mu \pm \sigma$)

P1 - Fully Automatic

P2 - Split Axis, Yaw Manual

P3 - Split Axis, Pitch Manual

P4 - Manual

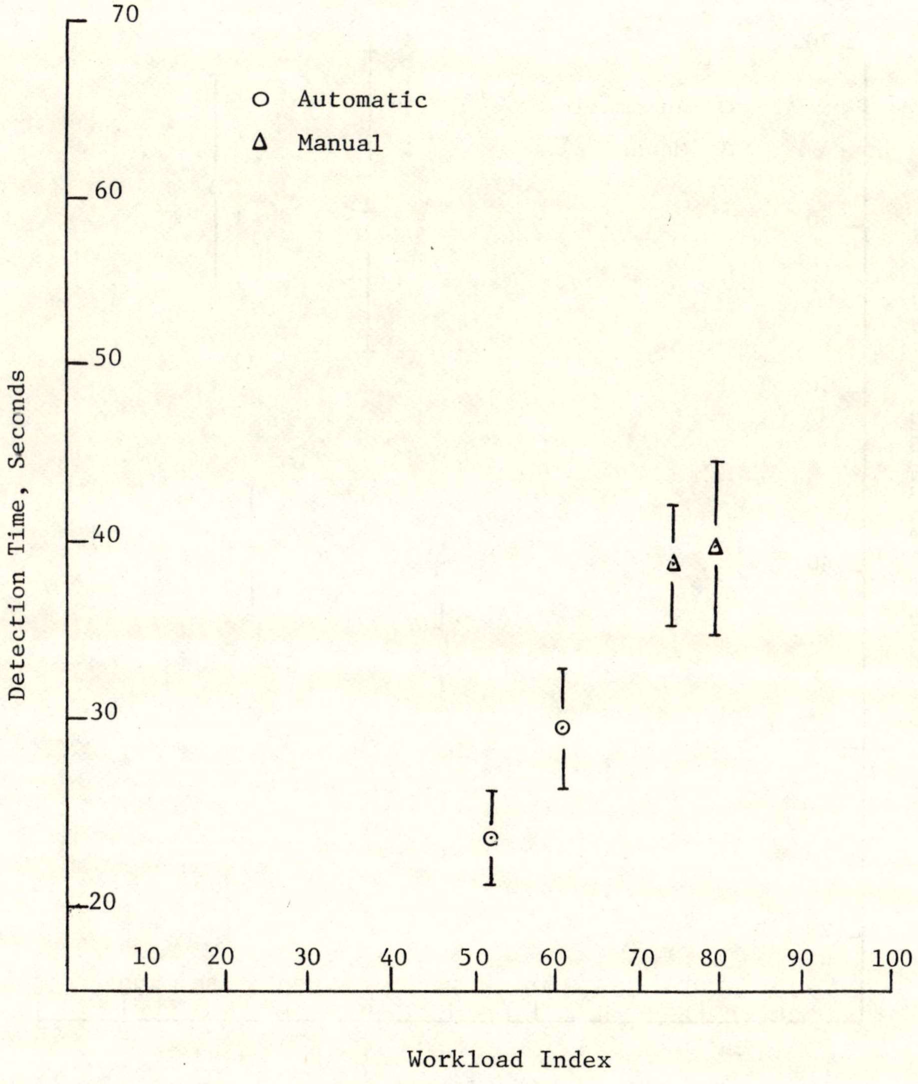

Figure 2: Detection Times of Longitudinal (Pitch) Failures

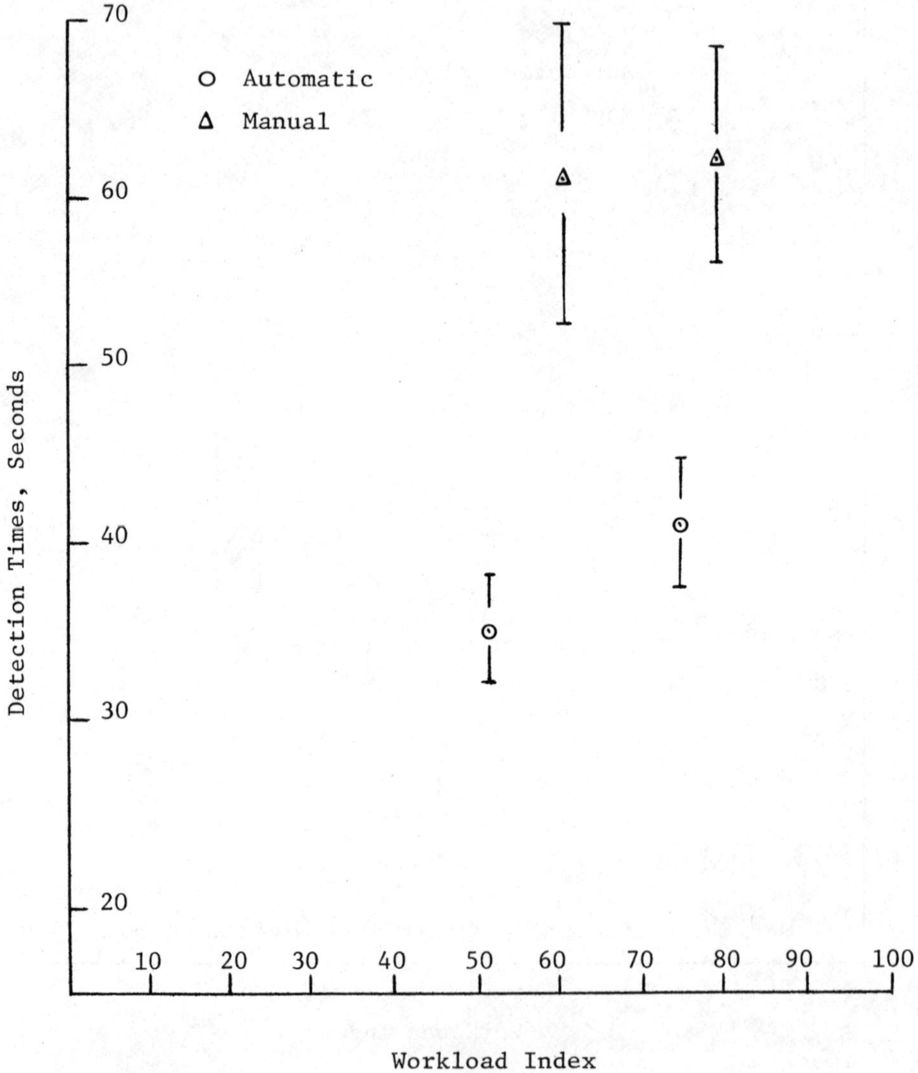

Figure 3: Detection Times Lateral (Yaw) Failures

angular error) approximately 30 seconds after the occurrence of the failure.

To avoid possible contamination of the failure-detection data by the presence of a concomitant subsidiary task, two separate experiments were carried out: Workload levels were calibrated in the first experiment via the lights' side-task, without any failures being presented. Failure detection performance was investigated in a separate experiment, in which the subsidiary light-cancelling task did not operate. The second, failure-detection experiment consisted of 90 runs for each failure condition for 270 runs in all, in a randomized order.

Results. The main effect of the control-participation mode on workload is plotted in Figure 1.

Failure detection performance was analyzed in terms of the time needed to detect a failure. As Figure 2 reveals, the time required to detect a longitudinal failure seems to increase with increasing workload. Figure 3 shows, however, that detection time is not monotonic with workload. In fact, it shows a large increase in detection time of lateral failures with little increase in workload between pure monitoring and failures in the manually-controlled axis.

This trend toward poorer detection performance under manual control becomes even more evident when one analyzes the performance in terms of detection accuracy. We measured accuracy by the fraction of failures that were missed altogether. In all, 90 approaches were flown in which a longitudinal failure occurred; of these, eight went unreported. Of the 90 lateral failures presented, nine were missed. Tables I and II show the percentages of missed failures, broken down by experimental condition. It is quite obvious that failures were missed only when the failed axis was being controlled manually. No failures were missed in an automatically-controlled axis in this experiment.

DISCUSSIONS AND CONCLUSIONS

At first, the results of our two experiments seem contradictory: Detection performance by the active controllers was superior to that of the passive monitors in the first experiment, while the opposite was true in the case of the instrument-landing simulation. Are we, then, back where we started?

The trends indicated by the results of each experiment are rather conclusive, with a high degree of statistical reliability ($p < 0.01$). Any differences between the results of the two experiments must be attributed, therefore, to the difference in experimental conditions.

Table I. Fraction of Missed Longitudinal Failures in Percent of All Longitudinal Failures

Participation Mode	Gust Level 1	2	3	Overall
Monitor	0.	0.	0.	0.
Manual Yaw	0.	0.	0.	0.
Manual Pitch	12.5	14.3	12.5	13.0
Manual Control	12.5	14.3	37.5	21.7

Table II. Fraction of Missed Lateral Failures in Percent of All Lateral Failures

Participation Mode	Gust Level 1	2	3	Overall
Monitor	0.	0.	0.	0.
Manual Yaw	37.5	14.3	37.5	30.4
Manual Pitch	0.	0.	0.	0.
Manual Control	14.3	0.	14.3	9.1

The subjects' workload level was not measured explicitly during the first experiment; it was low, however, under all experimental conditions. In the landing-simulation experiment, on the other hand, the workload index was over 50% even in the pure-monitoring mode, due, at least in part, to the larger number of displays to be monitored. The pilots, who used the raw-data situation instruments as their primary displays when in automatic control, shifted their attention to the flight director when in a manual mode. It is our conjecture that the higher overall levels of workload associated with the second experiment resulted in a lack of attention to the situation displays under manual control which, in turn, effected poorer failure detection performance.

The operator's workload has been shown (Wickens and Kessel, 1979) to affect detection performance adversely. Wickens and Kessel's study is interesting in that it employed a paradigm whose complexity lay between that of our single-loop and multi-loop experiments. It is perhaps not surprising that their results fall somewhere between the extremes of our results, as well. By utilizing a dual-axis pursuit tracking task of moderate difficulty, Wickens and Kessel showed that detection accuracy was better under automatic conditions, while detection speed was superior when the subjects tracked manually. This last result they, too, attribute to the availability of proprioceptive feedback in the manual control mode.

These results seem to lead to the conclusion that the level of workload associated with a situation may dictate the preferred operator-participation mode. In tasks involving low workload levels, such as single-loop compensatory tracking, with a single display, failure detection performance is superior when the human operator is kept in the control loop. On the other hand, when the dynamics of the plant are complex, the displays are many and, consequently, the workload level is high, then the additional increase in the operator's workload when shifting from monitoring to manual control more than offsets the advantages of being in the control loop. The result is a net deterioration in the failure detection performance.

Within the limitations of this study, our results seem to suggest that, in poor meteorological conditions, a coupled, fully-automatic approach monitored by the crew via cockpit displays is the preferred participation mode from the point of view of failure detection. Performance monitors and fault annunciators may alleviate the problem somewhat; it is not known at this time, however, whether or not they will change the preference ordering of participation modes.

REFERENCES

Brown, I.D., 1964, "The Measurement of Preceptual Load and Reserve Capacity", Trans. Assoc. Ind. Med. Off.

Draper, C.S., Whitaker, H.P. and Young, L.R., 1964, "The Roles of Man and Instruments in Control and Guidance Systems for Aircraft", Proc. of 15th Intl. Astro. Congress, Warsaw, Poland.

Ephrath, A.R. and Curry, R.E., 1977, "Detection by Pilots of System Failures During Instrument Landings", IEEE Trans. on SMC, 7:841.

Spyker, D.A., Stackhouse, S.P., Khalafalla, A.S. and McLane, R.C., 1971, "Development of techniques for Measuring Pilot Workload", NASA CR-1888.

Walden, R.S. and Rouse, W.B., 1978, "A Queueing Model of Pilot Decision-Making in a Multi-Task Flight-Management Situation", IEEE Trans. on SMC, 8:867.

Wempe, T.E., 1974, "Flight Management - Pilot Procedures and System Interfaces for the 1980-1990's", Proc. AIAA Conf. on Life Sciences and Systems, AIAA-74-1297.

Wickens, D.C. and C. Kessel, 1979, "The Effects of Participatory Mode and Task Workload on the Detection of Dynamic System Failures", IEEE Trans. on SMC, 9:24.

Young, L.R., 1969, "On Adaptive Manual Control", Ergonomics, 12:635.

Young, L.R. and Tada, A., 1980, "The Role of Hand Movement Information in Subjective Detection of Controlled Element Dynamics Change in Adaptive Manual Control" (in preparation).

FAILURE DETECTION IN DYNAMIC SYSTEMS

Christopher D. Wickens and Colin Kessel

University of Illinois
Department of Psychology

Israeli Air Force
Human Factors Division

INTRODUCTION

With the increased automation of a wide range of man-machine systems, a characteristic of skilled human performance that gains increasing importance relates to the operator's ability to monitor a system under automated control, to ensure that any departures from normal functioning are efficiently and promptly detected. Systems that humans must monitor vary widely both in their complexity (e.g., the number of variables that must be employed to describe their state) and also in terms of the salience or directness by which the occurrence of failures is indicated to the operator. In some systems, the existence of a malfunctioning component may be indicated simply by the enunciation of a visual or auditory indicator. However, with other systems, often those involving automated control or regulation of a dynamic time-varying process, the existence of a malfunction must sometimes be induced by the human operator from subtle changes in the relation between environmental inputs and system response.

We report below the summary findings of a series of investigations of the failure detection process, carried out at the University of Illinois over the past three years. In the general paradigm, which will be employed throughout the five experiments reported below with some variations, subjects either control (track) or monitor under automatic autopilot control, a first order dynamic linear system (e.g., a system whose output is the time integral of its input). These two modes of participation required of the operator are referred to as the Manual (MA) and the Autopilot (AU) mode respectively. At unpredictable times during the 2½ minute tracking or monitoring session the system increases in order in a step fashion to a level which is approximately second order. Subjects normally indicate such changes, if detected, by means of a discrete button press response. If not detected, after a period of six seconds the system dynamics make a four-second ramp return to the original, pre-failure first order (Figure 1).

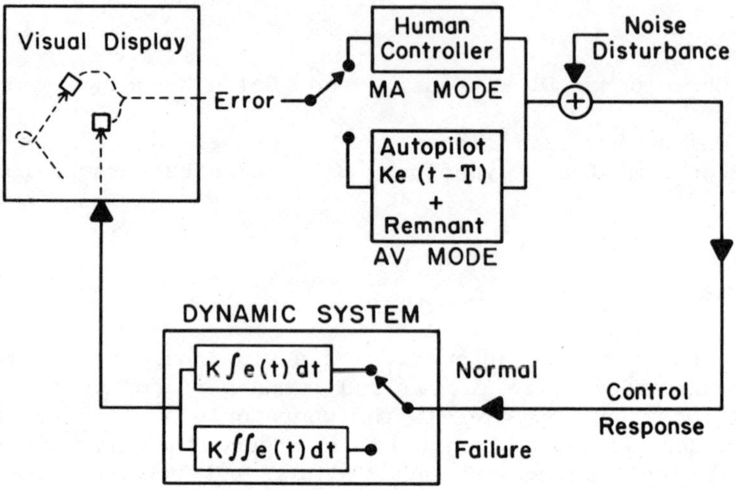

Figure 1. Schematic representation of failure detection paradigm.

Our research paradigm then focusses upon the subject's "internal model" of the dynamics of the first order system. Conceptually our model of the human operator borrows heavily from applications of statistical decision theory to reaction time (e.g., Lappin & Disch, 1972) and to failure detection (Curry & Gai, 1976). We assume that the internal model of the system mentally maintained by the operator consists of a set of expected system outputs to known system inputs, given that the plant is operating normally. Effective monitoring is accomplished by constantly comparing the observed outputs with the expected outputs to the observed inputs. If a discrepancy between the observed and expected output is noted, beyond the margin of error of the latter, it is stored, and such discrepancies are then accumulated over time. If this accumulation of differences then exceeds some internal criterion within a given interval of time, a failure is detected.

In the above conception, the latency of detection is dictated by a number of factors, four of which are: (1) The setting of the criterion. If the criterion is low, detections will be prompt, but false alarms will be more in evidence. Therefore variation in criterion setting induces a speed accuracy tradeoff in detection, dictating that both aspects of performance must be examined to assess detection efficiency. (2) The fidelity of the internal model. Here we refer to the range of possible expected outputs to a given observed system input. Clearly, deviations of observed from expected outputs will be accumulated more rapidly if this range is reduced with the higher fidelity model. (3) The number of channels of input available from the observed system display vs. the internal estimate of the current state of the system. As more channels are present, conveying either different information or the same information perturbed by independent external or internal noise sources, detection will be facilitated. (4) The processing resources or attention allocated to the accumulation and detection process.

This conception then represents the framework underlying our formulation and interpretation of the following experiments.

EXPERIMENT I: EXPECTANCY AND FAILURE DETECTION

Failures almost by definition are events that occur infrequently. As a consequence they are normally unexpected by the human supervisor of the failed system. This fact presents somewhat of a paradox to the scientific investigator of failure detection performance. While the essential criteria for a statistically reliable description of any phenomenon is repeated measurement, such repeated measurement in the laboratory, by its very nature, tends to create the higher levels of expectancy atypical of real world failure detection. Only a few investigators (e.g., Sadoff, 1962; Ruffle-Smith, 1979) have attempted to impose failures in controlled investigations under truly unexpected conditions.

The purpose of Experiment I, the Master's Thesis of Robert Earing (1977), was to assess manual mode failure detection performance under both expected and unexpected conditions within the same experiment. The objective was to determine the extent to which these are comparable, and therefore the extent to which the results of high expectancy detection studies in the laboratory may be generalized to, or at least extrapolated to, real world scenarios.

Thirty subjects received extensive training, tracking both first order (pre-failure) and second order (post-failure) dynamics, on separate trials. Following this training subjects began a trial of first order tracking, but unpredictably the dynamics changed in step fashion to an increased order. A small order increase (subtle failure) was followed on a subsequent trial by a large order increase (salient failure). These failures were clearly unexpected, because nothing in the prior briefing or instructions had led the subjects to believe that failure detection was under investigation. These trials were then followed by an identical pair, prior to which subjects were explained the nature of the experiment and were warned that the failures were forthcoming. Subjects were requested to try to regain stable control as soon after failure occurrence as possible.

In the paradigm employed, failure detection latency could not be assessed directly since to impose an overt response requirement might bias subjects towards expectancy. Instead, detection latency (along with the latency of other phases of control adaptation (Weir and Phatac, 1966)) was inferred from statistical analysis of the first change in control strategy, following the failure. While it is not clear precisely the manner in which our inferred times would map into conscious detection had this been required, in a subsequent phase of the study we attempted to relate the two times by analysis of trials in which overt, button press responses were called for.

The results suggested that detection in terms of the initial adaptive response (or change in response characteristics) to the failure was uninfluenced by the expectancy manipulation, for either small or large failures. The latency of the first phases of the adaptive response was statistically equivalent between the two conditions. However, the later stages of control adaptation were influenced by expectancy for the small failures. Again, however, for large failures, the later phases like the earlier ones did not show an "expectancy" effect.

The lesson that these results provide to research on failure detection and adaptation to dynamic system changes should be apparent. If the transitions to be investigated are salient, as the full changes in system order employed in the large failure conditions, the operator's response to transitions seems to be roughly equivalent between expected and unexpected conditions. In other words, it is appropriate to make inferences to abrupt "inflight" control failures from data obtained in the laboratory when the failure is salient. This is presumably because the alerting function of the initial increase in error, or adaptation to that error, is great enough to trigger conscious detection and resulting adaptation, independent of the operator's prior expectancy biases.

When, however, relatively subtle changes in system dynamics are investigated, such as the small failure condition employed, or the slow degradation of system states characterized by "non-catastrophic" failures, expectancy seems to be a relevant variable in the later phases of response. Experimental subjects may behave quite differently if they are told that a failure is forthcoming (or even that they are participating in an experiment on failure detection) than would operators under low failure-expectancy "in-flight" conditions.

EXPERIMENT II: THE ISSUE OF MANUAL VS. MONITORING SUPERIORITY

Experiment II (Wickens and Kessel, 1979) was designed to address the question of whether the operator's ability to detect system failures was superior when he is in the control loop (MA mode) or is removed from the loop, monitoring automatic control (AU mode). The practical implications of this question are directly evident. When a system designer is faced with the choice of whether or not to automate a particular control function, and other considerations are equivocal (e.g., control precision is not much better with AU than MA control), then considerations of the extent to which the state of the system is currently and accurately perceived by the operator should be of considerable importance. A major instigating source for our research was a specific conflict in the experimental literature concerning which mode of participation was superior. In previous investigations in which the two modes had been compared, Young (1969) provided evidence for better MA detection, while Ephrath and Curry (1977) obtained results indicating the opposite conclusions.

In contrasting analytically the two participatory modes, it is possible to identify characteristics of each that might enhance failure detection over the other. We have listed and described these characteristics in detail in Wickens and Kessel (1979) and in Kessel and Wickens (1978); however the most salient of these will be briefly restated here. It is certainly plausible to assert that detection of system failures might be superior while that system is actively under manual control. The operator in the MA mode is constantly interacting with the system; he receives both visual input concerning system state and proprioceptive input concerning the control commands that he has delivered to the system, the latter unavailable to the AU monitor. Furthermore, unlike the AU monitor, he has the option of introducing "test" signals into a system suspected to be malfunctioning, and observe the subsequent response. Finally the MA controller may have constructed a better "internal model" of the system by virtue of his greater degree of active participation. Thereby he should have more reliable expectations of system outputs to known inputs under normal operating conditions and, as a consequence, a greater ability to detect departures from normality.

While these factors all favour MA detection, this superiority may be diminished or even eliminated altogether by differences in workload favouring AU detection. The MA controller must perform two tasks concurrently, controlling and detecting, and the workload imposed by the controlling function may be sufficient to interfere with the detection/decision making task. The AU monitor naturally has only the latter task to perform, and this difference in concurrent task load could enhance AU detection. A second source of potential AU superiority relates to operator adaptation. To the extent that the MA controller adapts his control response to preserve normal tracking performance after a failure, and yet is unaware of this adaptation (as McDonnel, 1966, and others have noted may occur, and was observed with small failures in Experiment I), there will be less visual evidence of a failure from the display and thus a reduced likelihood of detection. A non-adapting autopilot on the other hand will continue to produce salient visual evidence of a changed system response following the failure. Naturally in the MA case, there will exist a change in response characteristics - a proprioceptive cue indicative of the failure. However, information from this cue may be suppressed by the visual signal, as a manifestation of the "visual dominance" phenomenon (Posner, Nissen and Ogden, 1976).

In the specific paradigm chosen, operators detected step increases in the order of a system that was tracking in two dimensions on a CRT display. The system was either controlled by the operator himself via a 2 dimensional joystick (MA mode) or by a computer autopilot that simulated as closely as possible the human operator's control describing function (McRuer and Jex, 1967). Autopilot parameters were further adjusted in value so that AU tracking "performance" (RMS error) was equivalent to MA performance. There was both a low frequency command input and a high frequency disturbance input. Failures, which occurred at an average frequency of five per two minute trial, were detected with a trigger press.

Five subjects, well practised in the detection task, performed in the AU and MA mode on alternate trials. Analysis of detection performance measures as a joint function of response latency and accuracy indicated that the MA mode was reliably superior. Latency was considerably shorter while there was minimal difference between modes in terms of response accuracy. Fine grained analysis techniques were then performed on the detection and tracking data in an effort to identify what characteristics of the operator and/or the two modes were responsible for the obtained MA superiority. The composite evidence derived from these analyses indicated that MA detection benefited from the presence of qualitatively different information available to the decision--maker in the first second or two after the failure. We concluded that this information consisted of proprioceptive cues generated by the operator's initial adaptive response (change in control behavior) to the changed dynamics.

EXPERIMENT III: TRAINING AND TRANSFER OF DETECTION ABILITIES

One potential source of difference between the two modes, whose effect we were unable to examine in Experiment II, related to possible differences in the internal model between a monitored and controlled system. Since all subjects in that experiment received training under both MA and AU conditions, it is reasonable to assume that a uniform internal model was in force in both conditions. A major goal of Experiment III (Kessel and Wickens, 1978, the PhD dissertation of the first author) was to ensure the presence of a different internal model between AU and MA detection. This was accomplished by adopting a between-subjects design. If, as hypothesized, MA training allows for a more stable model to develop, then MA superiority should again be demonstrated and in fact this superiority should be enhanced relative to the within subjects design of Experiment II in which AU detection could benefit from a model developed in part under MA training.

Employing experimental procedures similar to those of Experiment II, the results obtained by Kessel and Wickens supported this prediction, as MA superiority was again demonstrated. Moreover in the between-subjects design, MA detection was not only of shorter latency but also of considerably greater accuracy than AU detection. In Experiment II, the difference was only evident in detection latency. In fact, the overall degree of MA superiority assessed in terms of a combined speed-accuracy performance index, was five times greater than in the first study, thereby clearly demonstrating the enhanced differences in learning and internal model development between the two participatory modes.

In order to further validate these differences, a second phase of Experiment II included a transfer condition. If the overall MA superiority was in fact related to what was learned (internal model consistency) as well as to the other performance--related diffences (e.g., the added proprioceptive information channel), then some benefit in detection should be provided to subjects detecting failures in the AU mode, if they had previously received MA detection training (MA-AU) when compared to a corresponding AU-AU control group. AU detection of the MA-AU group should benefit from better model development during the prior period of MA training. To create these conditions, following three sessions of training each training group (AU and MA) transferred to receive 3 further days of failure detection in the AU mode. The results substantiated this prediction, since positive transfer in the MA-AU transfer group was observed. Information acquired while tracking clearly benefited detection performance while monitoring. Finally in an additional transfer group that was investigated (AU-MA), no positive transfer was observed from AU training to MA transfer: the internal model acquired from monitoring appeared to benefit neither failure detection nor tracking performance itself, in a later session of MA performance.

The fine grained analyses performed on the detection and control data of Wickens and Kessel (1979) were repeated on the training and transfer data, in order to determine what characteristics of the task were transferred positively from the MA training to the AU detection. Somewhat surprising here was our observation that, in terms of these indices of control and detection performance, the data of the AU transfer group appeared to show much greater similarity to the data of all of the MA groups (from both experiments) than to those of any of the other AU conditions. As stated earlier, we had previously attributed the differences between the MA and AU groups to the availability of proprioceptive evidence in the MA condition. However, since the AU transfer group showing these same characteristics clearly had no proprioceptive information available, it appeared that our proprioceptive argument was insufficient. The tentative conclusion offered in light of the data from the third experiment is that MA training served to focus attention on particular kinds of displayed visual information, particularly that related to the perception of higher derivatives of the error and cursor signals. This information - acceleration and change in acceleration - which must be perceived to achieve effective manual control of the system in its failed state, also can serve as a relevant cue indicating the initiation of a system failure. Thus the essence of the transferred information from MA to AU performance (and one probable source of MA superiority) appears to be perceptual, and attributable to the requirements that effective manual control imposes on the operator to extract higher derivatives of visual information from the display.

In resolving the apparent discrepancy between the recurrent findings of MA superiority in our studies and of AU superiority by Ephrath and Curry (1977), it should be emphasized that the particular superiority relation observed is the resultant of a number of factors, and that the kind of failures employed were qualitatively different in the two investigations. In our investigations, a change in the system transfer function was imposed, inducing in the MA condition a corresponding change in required response behavior. This manual adaptation fundamentally altered the frequency and velocity with which control was exerted, inducing changed proprioceptive information relayed back to the central decision center. The presence of this second channel of information was argued to favor MA detection.

The failures employed by Ephrath and Curry, on the other hand, were continuous linear deviations, induced into the flight path of the monitored/controlled aircraft. These would not entail any fundamental adaptation in the higher derivatives of control response (e.g., a change in mean control velocity) but only a gradual lateral bias in the mean position of the control. Without the added benefit of the proprioceptive cue, it is plausible in Ephrath and Curry's study that the cost of added workload imposed in MA detection outweighed any benefits of this mode, to the relative advantage of AU detection.

EXPERIMENT IV: THE PROCESSING RESOURCE DEMANDS OF FAILURE DETECTION

Our investigation has concerned not only the mechanisms underlying the detection process *per se*, but also the mental workload imposed by the monitoring of dynamic systems. That is, the demand that such a task places upon the human's limited resources for processing information. At issue are both the qualitative and quantitative nature of these demands.

In considering the workload imposed upon the operator by the failure detection process, characteristics of both the task and of the human operator must be considered. Such considerations can easily resolve discrepancies concerning the magnitude of demand, whether substantial (e.g., Levison and Tanner, 1972) or negligible (e.g., Keele, 1973). First, it is undoubtedly relevant here to consider whether the events to be monitored are salient and represent distinct temporal discontinuities, such as lights and tones, as opposed to more subtle signals demanding of a higher order of perceptual cognitive analysis, in order to discriminate them from background noise. Monitoring for the former may indeed require little effort because the events themselves are sufficiently salient to call attention focussed elsewhere. However monitoring for the latter can be expected to impose considerable workload, if effectively done.

Second, analysis of the workload imposed by failure detection (or any task for that matter) must also account for the multi-dimensionality of human processing resources, the demand for which underlies the concept of workload (Wickens, 1979). This multi-dimentionality has been the focus of extensive research (e.g., Navon and Gopher, 1979; Wickens, 1980) and indeed represented a major theme underriding the recent NATO conference on workload (Moray, 1979). Our analysis of the workload in failure detection was predicated on the assumption that this dimensionality is defined in part according to stages of information processing (perceptual encoding, central processing, and response). Such a conception represents a somewhat simplified abstraction of a more complex dimensionality described by Wickens (1980), who in addition considers dimensions related to cerebral hemispheres of processing and modalities of input and response.

Our vehicle for analysis of the dimensions of workload imposed by the failure detection task was through the imposition of various concurrent loading tasks. Indeed throughout Experiment II, and the training phase of Experiment III, subjects participated in additional conditions in which they performed concurrently a subcritical tracking task (Jex, 1967). This task required them to manipulate a finger control with the left hand to stabilize an unstable element presented laterally in the center of the failure-detection display. Between conditions we were able to adjust the instability constant of the subcritical task, and thereby impose greater or lesser amounts of concurrent load on the failure detector operator.

In Experiment III we noted a major difference between MA and AU detection performance. While the former declined with the introduction of the concurrent loading task, the latter was totally unaffected (Wickens and Kessel, 1980). Detection in both modes was unaffected by the difficulty level of the critical task.

In interpreting these results, we proposed two alternative explanations for the differing effects in the two modes. Either AU detection imposed no workload (demanded no resources), or alternatively the load imposed by AU detection was placed on functionally different resources from those deployed in performance of the subcritical loading task. Adopting the latter hypothesis, we proposed that resources associated with perception and central processing were primarily responsible for AU detection, while those associated with the selection and execution of manual responses were primarily involved in the subcritical task, and were utilized for at least some component of MA detection (presumably processing of the proprioceptive information, employed in MA detection).

To test the hypothesis that both kinds of detection do in fact demand resources, but of a qualitatively different nature, a second set of MA and AU detection trials was conducted (using separate subjects) and a perceptual/cognitive loading task was imposed. Subjects heard a string of two digit numbers and were required to subtract 7 from each. They were to be prepared at any time, upon a probed cue, to provide the response. Probes were infrequent, so the response load of the task was low; yet performance of the mental processing of the task could be inferred from the accuracy of the response to the unpredictable probes.

Unlike the subcritical task, the mental arithmetic task did produce a reliable decrement in AU detection efficiency. AU detection does therefore demand resources, and diversion of these resources to the mental arithmetic task derogated performance. Of further importance was the fact that MA detection was not influenced by the mental arithmetic task nor was the accuracy of tracking in this mode. Our conclusions asserted that the impact of the loading tasks was upon the processing channels used to monitor the system, visual for AU detection, proprioceptive for MA detection. The former utilizes resources associated with perception and central processing, the same resources as those demanded by the mental arithmetic task; the latter utilizes response-related resources, coincident with the subcritical loading task.

A point of more general relevance here is that automation (as operationally defined here by the AU monitor) does not necessarily eliminate or even reduce the workload of the human operator, but merely shifts the locus of processing demands. While interference with other manual responses might be attenuated by automation, interference with cognitive tasks might well be increased.

EXPERIMENT V: AU DETECTION IN A DYNAMIC ENVIRONMENT

Experiment IV suggested that AU detection depended in its performance upon processing resources that were functionally separate from those heavily utilized in manual control of an unstable first order system. The present experiment was intended to extend this conclusion to a dual task environment in which the demands of the detection or the control task are changing dynamically over time. As in the preceding experiments, subjects performed the detection and the subcritical tracking tasks concurrently. However, in contrast to these studies, the task characteristics did not remain constant across the trial but varied in difficulty in quasi-random fashion. For one group of 10 subjects this variation was imposed upon the tracking task by changing the instability constant. For a second group the detection task was varied by changing the bandwidth of disturbance noise introduced into the control loop which the autopilot was directed to nullify. Greater levels of noise served to "mask" the immediate manifestations of failures, and rendered them more difficult to detect.

In each case the variable task was designated as primary. That is, subjects were explicitly requested to maintain performance on that task at a constant level, despite fluctuations in its apparent difficulty. Monitary payoffs reinforced these instructions. Following logic developed by Wickens and Tsang (1979), it was assumed that, if common resources were demanded by both tasks and subjects exercised voluntary control over the allocation of these resources, then the performance constancy required by the instructions could be maintained by borrowing resources from the constant (secondary) task as primary load increased, and returning these at the periods of reduced primary demand. Accordingly, such a policy should produce fluctuations in secondary task performance corresponding in phase to the imposed variations in primary task demand. This covariation in time can, in turn, be quantified by the linear coherence measure used in time-series analysis (Wickens and Tsang, 1979). The coherence measure which varies between 1.0 (perfect covariation) and 0 (no covariation) then should reflect the extent to which resources were exchanged between the primary and secondary tasks.

In contrasting the results of the two experiments, we observed that the mean coherence measure between primary demand and secondary performance was considerably greater for subjects for whom the tracking task was variable (and primary) than for subjects for whom the detection task was variable (mean coherence = 0.27 and 0.17 respectively). Neither of these values is remarkably high but it is apparent that subjects were less able to borrow resources from tracking to cope with increases in detection difficulty (reflected by the 0.17 measure) than the converse.

The extent to which each task varied with changes in its own demand is reflected by coherence measures computed between primary demand and primary performance. The mean coherence between detection performance and detection difficulty was high (p^2 = .82). That between tracking performance and difficulty was considerably lower (p^2 = .40).

These results are generally consistent with those of Experiment IV, in that they emphasize the fundamentally separate resources underlying both tasks (supported operationally by the low coherence measures of both tasks' difficulty with the other tasks' performance). However, the results go further to emphasize what might be described as the "fragility" of failure detection: On the one hand, when demanded by an increase in its own difficulty, failure detection performance cannot benefit from borrowed resources of tracking and suffers heavily. Yet this separation of resources is not symmetrical because, unlike tracking, failure detection performance does deteriorate as well when tracking demand is increased. This fragility may represent an intrinsic characteristic of monitoring/detection tasks in general.

CONCLUSIONS AND ACKNOWLEDGEMENTS

The current series of experiments present a fairly coherent picture of the failure detection task under investigation. Experiment I suggested the generalizability of these studies to environments with lower levels of expectancy. Experiments II and III fairly conclusively demonstrated the existence of MA superiority in failure detection, in the context of the paradigm employed. These results thereby suggest that consequent costs may be associated with design innovations which serve to remove the operator from the control loop. Naturally there will often be factors that override these considerations and will require that the operator be placed in the role of an autopilot supervisor/monitor. In this regard the implication of the transfer study is that a major benefit can accrue to system monitors, if they have received a prior period of manual interaction with the system that is to be under supervision.

The results of Experiments IV and V argue that both classes of failure detection, AU as well as MA, impose mental workload - of a qualitatively different dimension in each case - that is not trivial when performed effectively. The resources required for AU monitoring are however easily diverted to other tasks and suggest the "fragility" of this task.

It must be emphasized, in conclusion, that these results are of course paradigm-specific. They may not generalize to situations in which the failure manifestations are highly salient; or alternatively to system dynamics with long time-constants, or extreme complexity. In the latter cases, the option of direct MA participation is unavailable because of the complexity and precision of control required. Nevertheless the general lesson of maintaining manual involvement at some level, to update the internal model, must still be relevant, just as the workload and expectancy issues transcent well beyong the specific paradigm investigated here.

The research reported here was supported by grants from the Air Force Office of Scientific Research (Grant No. AFOSR 77-3380) and from the National Science Foundation (Grant No. NSF BNS 78-007860). The authors acknowledge the valuable assistance of Ms. Pamela Tsang and Dr. Russel Benel in the conduct of Experiment V.

REFERENCES

Curry, R., and Gai, E., 1976, "Detection of random process failures", in: "Monitoring Behavior and Supervisory Control", T. Sheridan and G. Johannsen, eds., Plenum Press, New York.

Earing, R., 1976, "Expectancy effects in adapting to abrupt control failures", Unpublished Masters Thesis, University of Illinois.

Ephrath, A.R., and Curry, R.E., 1977, "Detection by pilots of system failures during instrument landings", IEEE Transactions on System Man and Cybernetics, SMC-7: 841-848.

Jex, H.R., 1967, "Two applications of the critical instability task to secondary workload research", IEEE Transactions on Human Factors in Electronics, HFE-8: 279-282.

Keele, S.W., 1973, "Attention and Human Performance", Goodyear, Pacific Palisades, California.

Kessel, C., and Wickens, C., 1978, "Development and utilization of internal models in dynamic systems. A comparison of monitors and operators as failure detectors", University of Illinois Engineering-Psychology Laboratory, Technical Report EPL-78-2/AFOSR-78-5.

Lappin, J., and Disch, K., 1972, "The latency operating characteristic: I. Effects of stimulus probability", Journal of Experimental Psychology, 92: 419-427.

Levison, W.R. and Tanner, R.B., 1972, "A control theory model for human decision making", NASA-CR-1953.

McDonnell, J.D., 1966, "A preliminary study of human operator behavior following a step change in the controlled element", IEEE Transactions on Human Factors in Electronics, HFE-7: 125-128.

McRuer, D., and Jex, H., 1965, "A review of quasi-linear pilot models", IEEE Transactions on Human Factors in Electronics, HFE-6: 62-85.

Moray, N., 1979, "Mental workload: Its theory and measurement", Plenum Press, New York.

Navon, D., and Gopher, D., 1979, "On the economy of the human processing system: A model of multiple capacity", Psychological Review, 86: 214-255.

Posner, M., Nissen, M., and Klein, R., 1976, "Visual dominance", Psychological Review, 83: 157-171.

Ruffle-Smith, H.P., 1979, "A simulator study of the interaction of pilot workload with errors, vigilance and decision", NASA Technical Memorandum 78482.

Sadoff, M., 1962, "A study of a pilot's ability to control during simulated stability augmentation system failures", NASA TN D-1552.

Weir, D.H., and Phatak, A.V., 1966, "Model of the human operator response to step transitions in controlled element dynamics", Second Annual NASA University Conference on Manual Control, NASA SP-128: 65-83.

Wickens, C.D., 1979, "Measures of workload, stress and secondary tasks", in: "Human Workload: Its Theory and Measurement",: N. Moray, ed., Plenum Press, New York.

Wickens, C.D., 1980, "The structure of attentioanal resources", in: "Attention and Performance VIII", R. Nickerson, ed., Lawrence Erlbaum, Englewood Cliffs, N.J.

Wickens, C.D., and Kessel, C., 1979, "The effect of participatory mode and task workload on the detection of dynamic system failures", IEEE Transactions on Systems, Man, and Cybernetics, SCM-13: 24-34.

Wickens, C.D., and Kessel, C., 1980, "The processing resource demands of failure detection in dynamic systems", Journal of Experimental Psychology. Human Perception and Performance, in press.

Wickens, C.D., and Tsang, P., 1979, "Attention allocation in dynamic environments", University of Illinois Engineering-Psychology Laboratory, Technical Report EPL-79-3/-AFOSR-79-3.

Young, L.R., 1969, "On adaptive manual control", IEEE Transactions on Man-Machine Systems, MMS-10, No. 4: 292-351.

A MODEL OF HUMAN FAULT DETECTION

FOR COMPLEX DYNAMIC PROCESSES

Renwick E. Curry

Aviation Safety Research Office
NASA Ames Research Center
Moffett Field, California

INTRODUCTION

An understanding of the human as a monitor is particularly important in any man-machine system; whether by design or by default almost all systems require some sort of human monitoring and supervision to ensure proper operation. Research in the past several years has recognized the importance of the human monitor's role in manned-vehicle systems (e.g., Sheridan and Johannsen, 1976). There is a similar interest in the role and capabilities of the human monitor and operator for other processes as well (e.g., Edwards and Lees, 1974); this interest has been intensified by the "human factors" contribution to the Three Mile Island nuclear power plant accident. That the human should perform as a less-than-ideal monitor and operator in such situations came as no surprise to the many researchers in this field.

It is generally assumed that proper human intervention for faults, failures, and anomalies implies the accomplishment of three separate tasks:

1) Detection of the failure, fault or anomaly
2) Diagnosis
3) Remedial action

Improper intervention may occur in any one of these three stages. In this article we focus on the detection task, since it must be the first of the three to be performed even if it may be difficult to behaviorally separate detection from diagnosis. (Many people will not exhibit any sign that an anomaly has been detected until a diagnosis has been performed, at least to the extent of identifying a plausible alternate to the null hypothesis that everything is proceeding normally). Laboratory and simulation studies have investigated human failure detection. Gai and Curry (1976), and Curry and Govindaraj (1977) have examined the monitor's performance in detecting changes in the mean, variance, and bandwidth of random processes. Wickens and Kessel (1979), Young (1969), and Eprath and Curry (1977) compared the ability of passive monitors and active controllers to detect malfunctions of a process. Gai and Curry (1977) modelled pilot's monitoring performance for instrument failures in an aircraft simulator, while Kiguchi and Sheridan (1979) considered a method for evaluating information measures of abnormal states of a process. A control-theoretic view of monitoring models is described in Kleinman and Curry (1977), and Rouse (1977) proposed a more general decision making model.

These studies are an important first step in modelling the ability of a human to detect failures, but they are somewhat simplified in terms of the information and knowledge actually used by operators in complex systems. (The paper by Kiguchi and Sheridan (1979) is one exception). There is such a wide variety of cues and clues used by the human that they almost defy analysis, yet on closer examination they seem very plausible: sights, smells, sounds, touch etc. are all examples. There are prescribed emergency procedures to be used in the event of the smell of an electrical fire in the cockpit. A further example of the complexity of the available information is shown in Figure 1, which contains a portion of an aeronautical chart (an approach plate) and its legend. The pilot's knowledge of this information, or lack of it, will strongly influence the interpretation of events and displays within the cockpit. Note the many different types of altitudes ... whether it is referenced to sea level or the airport is obviously important; whether one should remain above, below, or at the indicated altitude is also contained in the cryptic information. A star symbol near the name of the control tower means that the tower does not operate continuously; a pilot who does not know this may assume that his radio is inoperative when his after-hours calls to the tower go unanswered.

Information and operator knowledge are just as important in process control. According to one observer of the Three Mile Island nuclear accident (Ferguson, 1980):

> "The real concern here is not that the operators "did not" but rather "why did they not" follow the written procedures. First, they had a poor direct indication of leakage, by design. While there were other indirect means and observations by which they could have inferred leakage from these valves, it is not necessarily reasonable to expect the required analyses and deductions under conditions that then

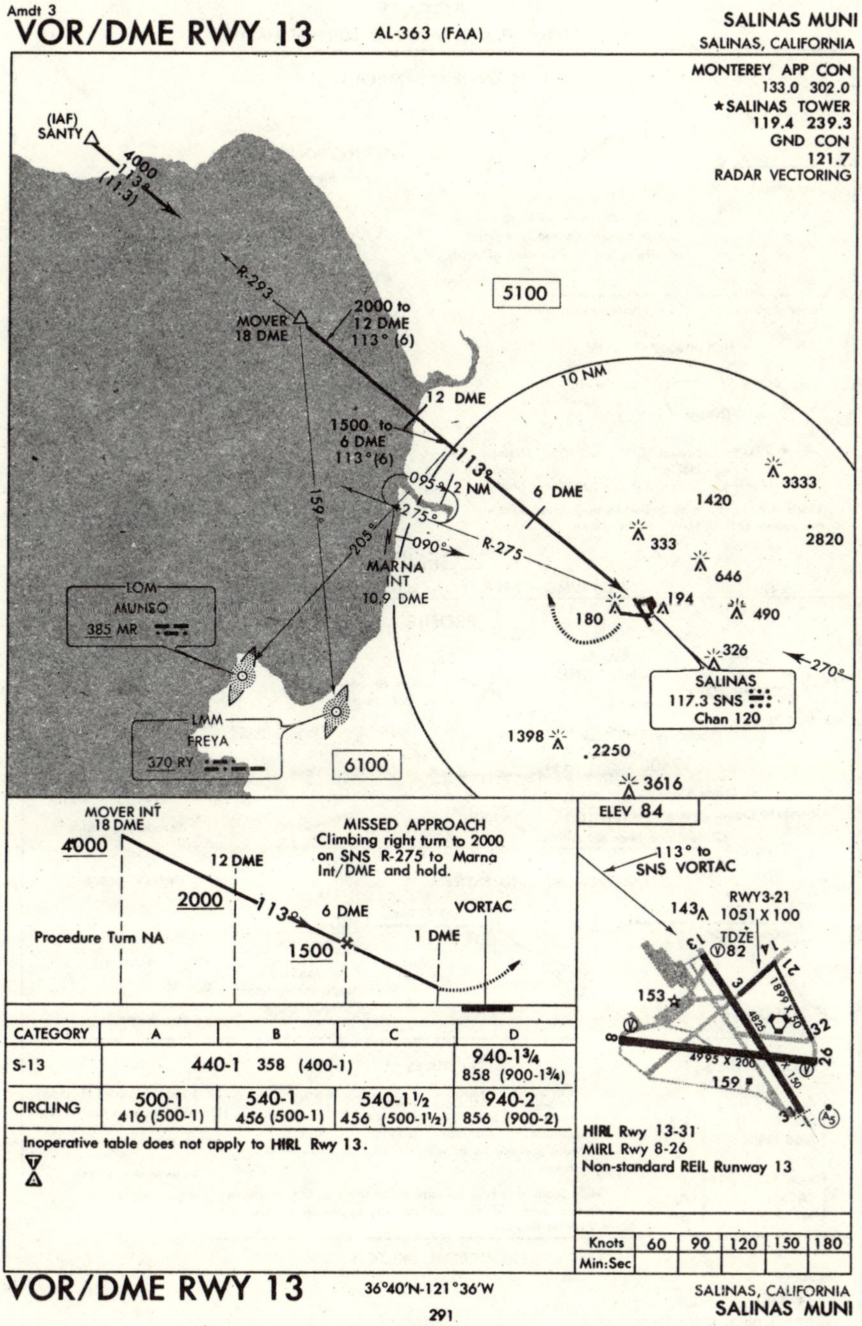

DO NOT USE FOR NAVIGATION

Figure 1. (a) Approach plate for aircraft landing

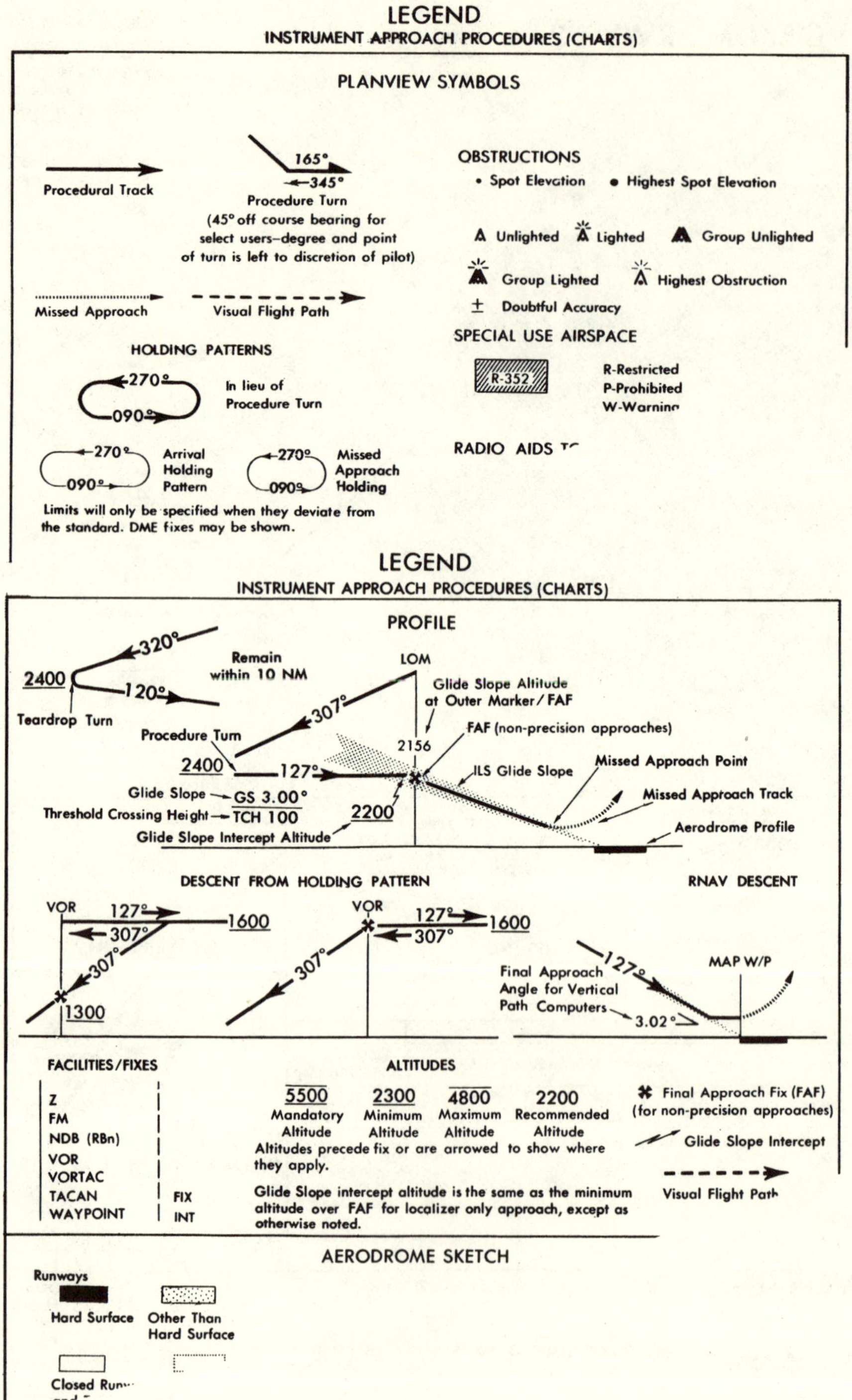

Figure 1. (b) Portions of Approach Plate Legends

existed in the plant. Second, they had been conditioned by their experience in this plant to expect leakage from these valves and high discharge line temperatures. And third, they had apparently developed the habit of ignoring this particular applicable procedure."

In summary, the detection of anomalies and failures in realistic, complex system requires that the human operator uses a wide variety of cues and an exceedingly complex internal representation or model of the process; the purpose of this article is to take one step in developing a model for human failure detection in such systems.

A MODEL OF HUMAN FAILURE DETECTION

Any model of human behaviour must be based on certain assumptions, either explicity or implicity stated. Hypotheses for manual control were proposed by Clement et. al. (1971), and were founded on several years of experimental research. These hypotheses were modified by Curry and Gai (1976) to reflect human monitoring behaviour and are summarized here:

- The human monitor (HM) uses a variety of internal models based on his experience.
- Satisfactory monitors show the characteristics of "good" inanimate failure detection systems.
- There is a (workload) cost to monitoring which can be alleviated by use of inanimate monitoring systems.

These working hypotheses are themselves a model of human monitoring - not necessarily complete, certainly not quantitative, but they do reflect the consensus of many researchers at this time.

The detection model proposed here assumes that the HM is evaluating the hypothesis that the processes is in normal operation. This evaluation of a single hypothesis is appropriate for the detection task and can be extended to the diagnostic task where alternate hypotheses are tested; see, for example, Rasmussen (1980), Rouse (1980), and the psychological aspects of hypothesis sampling (Brown, 1974).

A desirable feature of any detection model is the ability to account for a wide variety of information available to the human monitor of realistic systems. We assume that the attentional resources of the operator and the information content of the "displays" play a more important role than the specific sensory modality, but psychophysical limitations must be incorporated in the model. One concept that seems to allow for the combination of many different sources of information is the likelihood ratio. Let us assume that the HM has available to him a set of indicators at time t ($I_i(t)$). These indicators will be present and past readings of instruments, events or nonevents, odors, sounds, etc. The amount of information available to the HM will depend on his sampling strategy, the type of information, and the short term memory capacity, among other factors.

The Neyman-Pearson method of hypothesis testing suggests the following likelihood ration test as a reasonable measure

$$l(t) = \ln \frac{P(I_1(t),I_2(t),\ldots,I_m(t)|H1)}{P(I_1(t),I_2(t),\ldots,I_m(t)|H0)} \tag{1}$$

where H0 is the null or nominal hypothesis and H1 is the "best" hypothesis available from the current indicators. An upper bound to the numerator probability is 1, which assumes that each indicator is completely correct. We then have

$$l(t) = - \ln P(I_1,I_2,..,I_m|H0) \tag{2}$$

Using conditional probabilities, we may rewrite (2) as

$$l(t) = - \ln P(I_1|H0,I_2,I_3,\ldots) - \ln P(I_2|H0,I_3,I_4,\ldots) - \ldots \tag{3}$$

Assuming that the HM evaluates the indicators in the context of the null hypothesis only, equation (3) allows us to rewrite the log likelihood ratio as

$$l(t) = - \sum_{i=0}^{m} \ln P(I_i(t)|H0) \tag{4}$$

Evaluating each indicator in the context of the null hypothesis does neglect possible indicator interactions. However, there is ample evidence (Brehmer, 1980) that humans integrate information cues in a linear or additive manner.

The above equation is a central element in the model of human failure detection. It also provides a measure of how likely or unlikely each indicator is at any given time. It provides a measure of the detectability of specific failures and anomalies as reflected in the various indicators, a point to be discussed shortly.

Implicit in the model are the following considerations which must be addressed:

System Knowledge. The monitor's knowledge and experience with realistic processes is perhaps the most difficult to quantify. However, any analytic model of interaction with complex processes must come to grips with this problem since it is critical to the prediction of human performance. Models which lack structure representing the HM's knowledge of the system will not be useful in identifying sensitive areas where improper knowledge will lead to an improper assessment of the process.

Subjective Probability. The probabilities of the various indicator readings (conditioned on the nominal hypothesis) must be based on subjective probabilities. These in turn will be influenced by heuristics as well as the knowledge and system structure utilized by the HM. The last few years seen a substantial shift from normative/descriptive models of probability assessment to information processing heuristics (Slovic, Fischoff, and Lichtenstein, 1977). These heuristics and biases revolve around the concepts of representativeness, availability, and anchoring adjustment. While these deviations from normative behaviour must be considered in modelling human failure detection, it may be that they will have a secondary effect when considering the indicators of most realistic systems in most cases.

Attention Strategy. The attention strategy has a profound effect on calculating the likelihood ratio: information in any particular display, if utilized, can reduce subjective uncertainties and therefore increase the power of the likelihood ratio test of the null hypothesis; in the other extreme, if no information is sampled, then the uncertainties associated with the likelihood ratio must be based on the a priori knowledge.

Dynamic Systems. When the HM is observing the outputs of a dynamic system, it is assumed that the indicators will be processed according to an internal model of the dynamics of the process. "Good" failure detection performance suggests near optimal processing of the observations; Gai and Curry (1976, 1977) used this hypothesis to develop a Kalman Filter to represent the HM's processing of dynamic displays. The indicator for these investigations, which was influenced by the attention strategy, was the residual of the Kalman Filter (observed reading less expected reading). A similar model was used by Curry and Govindaraj (1977) for changes in process bandwidth and variance. These techniques are applicable for developing the indicators to be incorporated in the likelihood ratio if the dynamics appear to be an important part of the HM's processing. Space limitations preclude an adequate treatment here, but the references contain a more complete description. The major difference between the previous models and the likelihood ratio model is the incomplete internal model used by inexperienced HM.

Decision Strategy. An evaluation of the "internal" likelihood ratio as a decision variable for a single indicator was made by Gai and Curry (1976) with encouraging results. Thus we assume that the human monitor will declare (at least internally) that a failure or anomaly is present if any indicator in the likelihood ratio is above a level prescribed by the decision rule being used (e.g., a subjective expected utility). The choice of criterion level will influence the missed alarm rate and false alarm rate, as in conventional signal detection theory.

DETECTABILITY OF FAILURES AND ANOMALIES

The likelihood ratio concept can be used to evaluate the detectability of specific failures, and this may be the most important application of the likelihood model. Assume a specific sampling strategy so that the instantaneous likelihood ratio can be calculated. If the observer attends to indicators (say j and k at time t, then the observed likelihood ratio is

$$l^{o}(t) = - \ln P(I_j(t)|H0) - \ln P(I_k(t)|H0) \tag{5}$$

Averaging over the ensemble of trials using this previous sampling strategy, the average "observed" likelihood ratio is

$$E[l^{o}(t)] = - \sum_{i=0}^{m} P(A_i(t)) \ln P(I_i(t)|H0) \tag{6}$$

where $P(A_i(t))$ is the probability that the monitor will focus his attention on indicator "i" at time t. In the single-channel model of attention, the attention probabilities sum to unity; in a parallel processing model, they may sum to a number greater than unity. From a practical standpoint, it seems reasonable to assume that all indicators in the same modality have equal probability of attracting attention unless they are very low bandwidth signals relative to other signals; or unless previous experience suggests that they are prone to be overlooked due to a lack of conspicuity - a definite possibility in some failure flags on aircraft navigation instruments.

EXAMPLE CALCULATION

In this section we compute the likelihood ratio for a hypothetical example to show the types of analyses that may be made for specific failures and anomalies. For the process we assume the following:

> A vessel is to pressurized from ambient pressure (P0 atm gauge) to a final pressure Pf. The instantaneous pressure reading is available to the human monitor. Experience shows that when the pressure reaches approximately half its final value (P^*) a cooling system turns on to reduce the accompanying heat build up; temperature is not displayed. During the first portion of the process the normal pressure increases at a rate r_1 atm/sec, and at a rate r_2 atm/sec after the cooling system turns on (Figure 2). Experienced monitors are able to tell that the cooling system is turned on by the vibrations in the floor of the control room.

The example will examine the detectability of a failure of the cooling system to turn on; both experienced and inexperienced operators, with and without the use of an annunciator for the cooling system, will be considered.

<u>Pressure Rate Estimation</u>

In this example we will concentrate on the interactions of multiple sources of information rather than a model of dynamic information processing, although this can certainly be incorporated in the general model. It is assumed that the monitor derives pressure rate information from the pressure display.

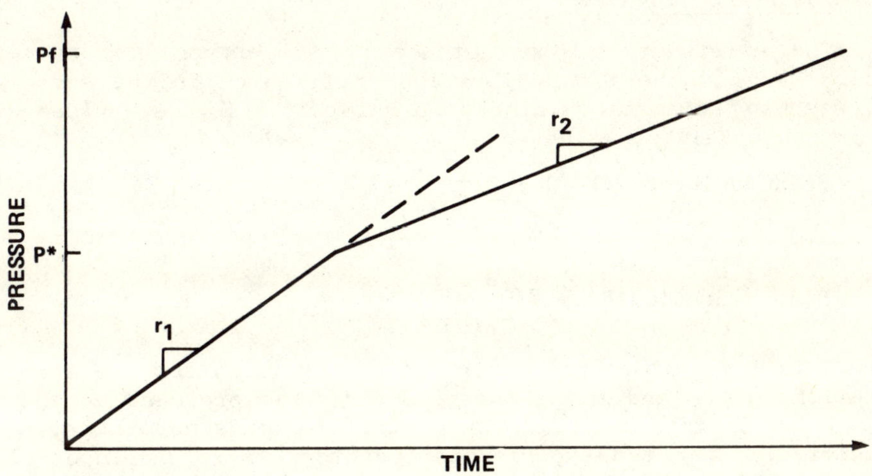

Figure 2. Pressure vs. time for example process

$$\hat{\dot{P}} = \dot{P} + v(t) \qquad (7)$$

where $\dot{P}$ is the actual pressure rate, and v(t) is the error in estimating pressure rate, assumed to be distributed as $N(0,\sigma_v)$. We assume that the estimated pressure rate is compared to the expected value (depending on the phase of the process) by the standard score

$$z(t) = \frac{\hat{P} - r_k}{\sigma_v} \tag{8}$$

This has a N(0,1) distribution during normal operation, and an $N(\bar{z}=(r_1-r_2/\sigma_v,1)$ distribution when the cooling system fails to start. The contribution to the average likelihood function may be calculated from the equivalent 1 degree of freedom chi-square variable

$$E[P(z|H0)] = P(z|r_1)P(r_1) + P(z|r_2)P(r_2) \tag{9}$$

$$= P(\chi^2(1)>1)P(r_1) + P(\chi^2(1)>1+\bar{z}^2)P(r_2)$$

Annunciator and Vibration

The existence or nonexistence of the annunciator indicator "cooler ON" and the floor vibration are treated in the same manner. Assuming that the monitors correlate the logical values with the pressure values, we have

$$P(\text{annunciator OFF}|P) = \Phi\left(\frac{P^* - P}{\sigma_{p(a)}}\right) \tag{10}$$

$$P(\text{no vibration}|P) = \Phi\left(\frac{P^* - P}{\sigma_{p(v)}}\right) \tag{11}$$

where Φ is the unit normal distribution function, and the standard deviations $\sigma_{p(v)}$ and $\sigma_{p(a)}$ correspond to the pressure uncertainties at which the vibration and annunciator cues actually appear, respectively; for the experienced operator, these should be close to the actual uncertainties.

Experience Level

The model allows the evaluation of different operator experience levels; we consider two levels of experience which effect the model parameters as shown in Table 1. We have assumed that the error in detecting pressure rate is primarily psychophysical in nature and the same for both monitors. The pressure uncertainty for the annunciator is lower for the experienced monitor. Both monitors place equal attention (P(A)=0.5) on the visual cues, and the vibration cue is attended to with P(A)=0 and .5 for the inexperienced and experienced monitors, respectively.

Table 1. Model Parameters

Display	Parameter	Experienced Monitor	Inexperienced Monitor
pressure	$(r_1-r_2)/\sigma_v$	2	2
annunciator	$\sigma_{p(a)}$	.05 Pf	.15 Pf
vibration	$\sigma_{p(v)}$	.05 Pf	not used

Results

In Figure 3 we have plotted the contributions to the likelihood ratio under the assumptions outlined above and a limitation on the subjective probabilities of (.001, .999). The knowledge of the experienced operator is reflected in the more rapid rise of the likelihood because he has less uncertainty about when the cooling system should turn on. His likelihood ratio achieves higher values because of the utilization of the vibration cue. The pressue display does not provide much information for either monitor in this case because the change in pressure rate is relatively small (2 standard deviations of the error in estimating

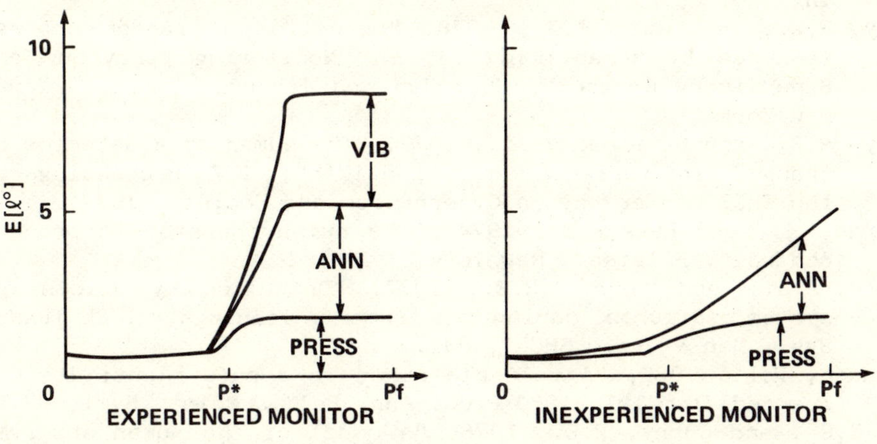

Figure 3. Average likelihood ratio for two monitors

rate). The plots were made under the assumption that all indicators are equally likely to be observed. If, for example, the annunciator is not readily visible from the operator's station, or if he is wearing cushioned shoes, then these indicators will have less contribution to the averages observed likelihood function. Also, correct instructions to the inexperienced operator about the vibration and the pressure range of the cooling system activation will increase his effectiveness.

SUMMARY

Human monitors of complex systems use a wide variety of information and internal models to assess the performance of the system. In this article we have proposed a model of failure detection which can be applied to these systems, but perhaps the greatest value of the model is in the evaluation of the detectability of failures and anomalies as a function of the utilized information and experience level of the monitor.

REFERENCES

Brehmer, B., 1980. "Models of diagnostic judgement", this volume.

Brown, A.S., 1974, "Examination of hypothesis-sampling theory, Psychol. Bull., 81, 773.

Clement, W.F., McRuer, D.T., and Klein, R., 1971, "Systematic manual control display design", AGARD Conference Proceedings No. 96.

Curry, R.E., and Gai, E.G., 1976, "Detection of random process failures by human monitors, in "Monitoring Behaviour and Supervisory Control", Sheridan and Johannsen, eds., Plenum New York.

Curry, R.E., and Govindaraj, T., 1977, "The human as a detector of changes in variance and bandwidth", 1977 Proceedings of the IEEE Conference on Cybernetics and Society, 1977.

Edwards, E., and Less, F.E., 1974, "The human operator in process control", Halstead, New York.

Eprath, A.R., and Curry, R.E., 1977, "Detection by pilots of system failures during instrument landings, IEEE TRANS. Sys., Man & Cyb., SMC-7, 841.

Ferguson, R.L., 1980, "A reliable plant is a safe plant; lessons learned from TMI, 1980 Proc. Ann. Rel. & Maint. Sym., 77.

Gai, E.G., and Curry, R.E., 1976, "A model of the human observer in failure detection tasks, IEEE Trans. Sys., Man. & Cyb., SMC-6, 85.

Gai, E.G., and Curry, R.E., 1977, "Failure detection by pilots during automatic landing; models and experiments", AIAA Jour. of Aircraft, 14, 135.

Kiguchi, T., and Sheridan, T.B., 1979, "Criteria for selecting measures of plant information with application to nuclear reactors, IEEE Trans. Sys., Man. & Cyb., SMC-9, 165.

Kleinman, D.L., and Curry, R.E., 1977, "Some new control theoretic models for human operator display monitoring", IEEE Trans. Sys., Man & Cyb., SMC-7, 778.

Rasmussen, J., 1980, "Models of mental strategies in process plant diagnosis", this volume.

Rouse, W.B., 1977, "A theory of human decisionmaking in stochastic estimation tasks, IEEE Trans. Sys., Man & Cyb., SMC-7, 274.

Rouse, W.B., 1980, "Experimental studies and mathematical models of human problem solving performance in fault diagnosis tasks", this volume.

Sheridan, T.B., and Johannsen, G., 1976, "Monitoring Behaviour and Supervisory Control", Plenum, New York.

Slovic, P., Fischoff, B., and Lichtenstein, S., 1977, Behavioral decision theory, Ann. Rev. of Psycol., 28,1.

Wickens, C., & Kessel, C., 1979, "The effects of participatory mode and task workload on the detection of dynamic system failures", IEEE Trans. Sys., Man & Cyb., SMC-9, 24.

Young, L.R., 1969, On adaptive manual control, IEEE Trans. Man-Mach. Sys., MMS-10, 292.

THE ROLE OF ATTENTION IN THE DETECTION OF ERRORS AND THE DIAGNOSIS OF FAILURES IN MAN-MACHINE SYSTEMS

Neville Moray

University of Stirling
Scotland

INTRODUCTION

Little or no direct experimental work exists on the role of attention in error detection and diagnosis. Therefore, this paper draws on well established approaches to the understanding of human information processing to suggest the direction in which may lie a model of the way the need to pay attention to many sources of information, and to the information received from those sources, gives rise to difficulties for the human operator in monitoring large automatic and semiautomatic systems.

Failures to detect or diagnose an abnormal situation in a richly connected multivariate man-machine-system may occur for many reasons. It may be that the relevant variable is not displayed, as in the case of the electromatic value at Three Mile Island, (IEEE Spectrum. Special Issue on Three Mile Island, 1979). It may be that the signal to noise ratio of a crucial signal is too low to allow the signal adequately to be sensed, as seems to have been the case with some of the communications which led up to the jumbo jet crash at Teneriffe (Human Factors Report on the Teneriffe Accident. Air Line Pilots Association, 1979. Washington, D.C.). It may be that the expectation of the human operators leads to a misinterpretation of information as again may have happened at Teneriffe. It may be that the human ignore information which is available and show "cognitive tunnel vision", concentrating on some variables to the exclusion of others, as in the Eastern Air Lines 401 crash in the Everglades, (Weiner, 1977). Or it may be that the human operator is swamped by too much information, as again at Three Mile Island, where one operator testified that he would rather have turned off all the warning lights, since they were providing no useful diagnostic information.

What features of behaviour and performance[x)] would lead one to ascribe failures to defective, or limited attention? What steps might be taken in man-machine-system design to effect the effects of limited attention?

It is unfortunate that much of the vast literature on human attention which have been accumulated from laboratory studies over the last twenty years is irrelevant to the study and understanding of large scale industrial systems. Most of the studies have been carried out using auditory signals, often near threshold, and usually with less than three messages present. Furthermore, usually sources of signals have been statistically independent. the experiments have used single trials and statistical independence of successive trials, and few studies with continuous and dynamic information have been made. Practice has usually been slight, and payoffs small. Rasmussen (1979) has contrasted such studies with the industrial setting:

> "Laboratory tasks tend to have a well-defined goal or target. Payoff matrices are artificial and of low value. The subject is controlled by the task. Task instructions arc specific. Task requirements are stable. Subjects are relatively untrained. By contrast in "real" tasks only a (sometimes vague) overall performance criterion is given and the detailed goal structure must be inferred by the operator ... The task may vary as the demands of the system vary in real time. Operating conditions and the system itself are liable to change. Costs and benefits may have enormous values. There is a hierarchy of performance goals. The operator is usually highly trained, and largely controls the task, being allowed to use what strategies he will. Risk is occurred in ways which can never be simulated in the laboratory".

It is useful to formalise slightly his description of the human operator confronted by a real system, by asking what are characteristic properties of such systems. I believe that the most important features, those from which a formal theory may be derived, are as follows:

[x)] By behaviour is meant observable states of the human: by performance observable states of the overall man-machine-system. See N. Moray, 1979 for a discussion.

1. There are multiple sources of information (state variables).
2. The state variables are characteristically dynamic, that is, functions of time, and usually continuous functions.
3. The variables are richly interconnected. Small groups of variables tend to be locally tightly coupled to each other, with the resulting "molecules" being less tightly coupled to other "molecules" than their internal couplings to each other.
4. There are many feedback loops
5. Values of state variables in different parts of a system are often highly correlated so that knowledge of one provides considerable information about the value of others.
6. Values of state variables are inherently ambiguous diagnostically. Because of the rich interconnectivity of systems the value of a single state variable is in general not sufficient to diagnose the cause of an abnormal state.
7. On the whole, given reasonable care in the design of displays, variables are presented with adequate S/N ratios.

Given these properties, how does man use attention to detect and diagnose errors?

SAMPLING BEHAVIOUR

Since most displays are visual and there is little reason to think that the use of other modalities will substantially improve things, the most obvious attentional limit is the rate at which sources of information can be sampled by visual scanning. Eye movements cannot be made at a rate significantly faster than 2 fixations per second and, as is well known, fixation is required for accurate pattern perception even if rates of change can be detected by the periphery of vision. Under normal conditions the dynamics of most industrial processes seem to be sufficiently low for the human scanning bandwidth to be adequate for monitoring most single variables, although there may be occasional moments when very rapid changes occur. However, even when the dynamics are low, the requirement to sample a large number of variables will cause difficulties, as we shall see, (Clement & Graham, 1968). We know that in many cases operators with long experience of a system adopt unconscious scanning patterns which can be very nearly optimal, (Senders, et al. 1966), although the sampling rate is frequently higher than would be predicted from the sampling theorem, (Clement, et al., 1968). The latter result is not surprising because a system which can rely on samples taken at 2W Hz in order to extract the information from a W Hz bandwidth source must have perfect memory for the last sample during the intersample interval, and humans do not. Both forgetting and interference from other information cause a loss of information. A number of studies have now found evidence that somewhere between 10 and 20 seconds quite significant loss of information occurs, even when only one signal has to be remembered. Richards and I, (Moray, et al., 1980) found that radar operators forget radar-like displays at a rate governed by the following equation:

$$\text{s.d. of memory of position} = a + b(t)^{1.5}$$

where $\underline{t}$ is the time since the last observation, and the constants depend upon the number of items to be remembered, and the amount and type of training of the observer; and Loftus, 1979, reports similar results for the recall of ATC messages. Hence if we assume with Senders that man samples in order to reduce his uncertainty, low bandwidth sources will tend to be sampled at intervals shorter than the Nyquist interval due to the loss of information from memory.

Further, some sources are more important than others. For example, Richards and I in the same study noted that the observer had a mean first passage time between his fighter and an enemy target aircraft which was far shorter than the mean first passage times to other echoes on the radar, although their optical properties were very similar. So although the forgetting rate must have been very similar, the operator was not prepared to tolerate as much uncertainty about the more important variables as about those less important, (his task being to direct his fighter to the neighborhood of the enemy aircraft). Indeed, the mean first passage times between fighter and target were so low that several fixations on each were made within the duration of one radar sweep, although no new information could have been added to the display, (Moray et al., 1980).

These ideas are summarised in Figure 1, which shows the time history of sampling for two variables, both of which must be monitored. The two time functions $\phi(x)$, $\phi(y)$ are bandlimited Gaussian functions with zero mean and bandwidths W_x and W_y Hz respectively. Their respective Nyquist intervals are shown as N^x_0, N^x_1 ... and N^y_0, N^y_1 ... where N_0 is the first observation made on the respective variable during the period of interest, and of course N^x_0, N^y_0 are not simultaneous. $\phi(x)$ is first sampled at O^x_t, and $\phi(y)$ at O^y_t. At those moments the observer's uncertainty U_x, V_y about their respective values is zero, but thereafter it

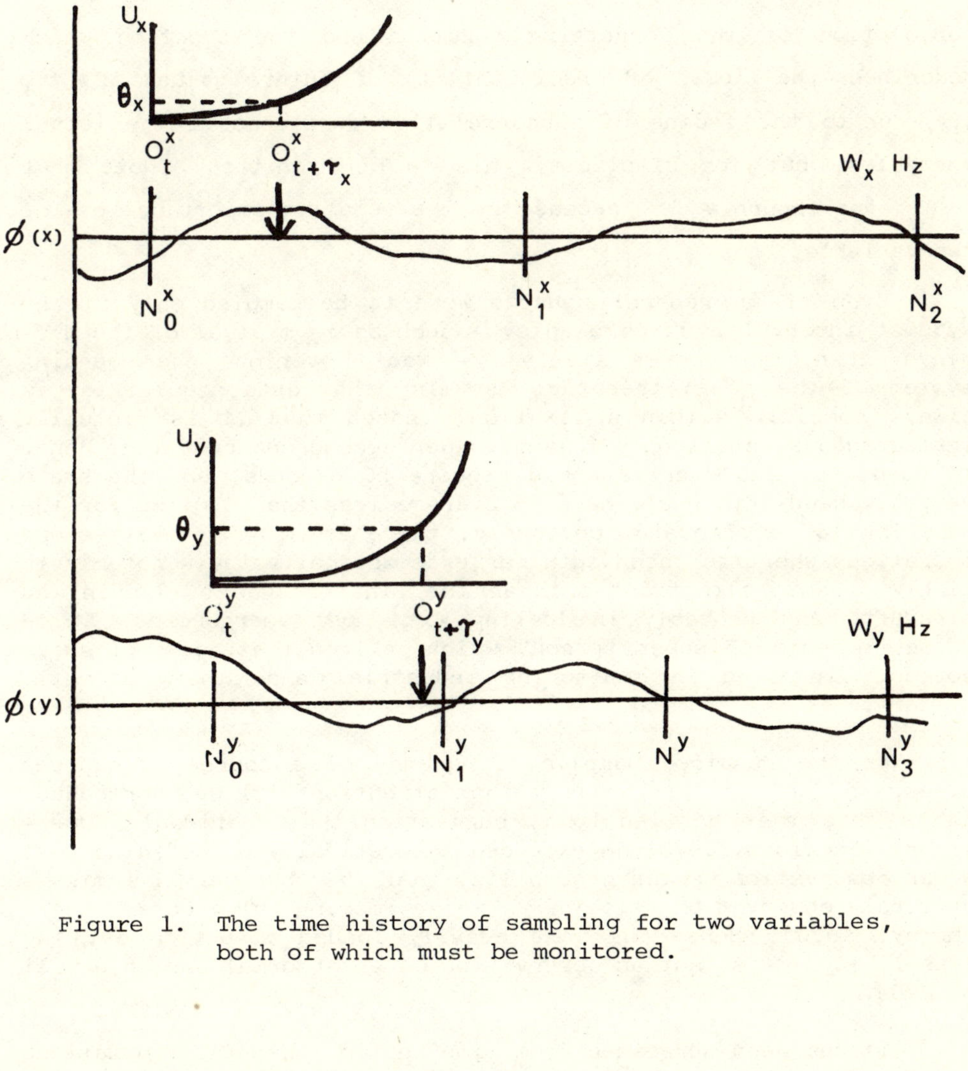

Figure 1. The time history of sampling for two variables, both of which must be monitored.

grows exponentially. The tolerable uncertainties are O_x and O_y, and $\phi(x)$ is a more important variable than $\phi(y)$. Hence $\phi(x)$ is next sampled (as the uncertainty function $U_x(t)$ grows due to forgetting) at $O^x_{t+\tau_x}$ and $\phi(y)$ at $O^y_{t+\tau_y}$, both intervals. (In fact the uncertainty function should really be represented as the convolution of the forgetting function and the function which described the loss of predictability of future values of $\phi(x)$, $\phi(y)$ due to the fading of subcorrelation functions of the latter variables, but for simplicity this is here omitted). Note that $\tau_x < \tau_y$ even though $W_x < W_y$ because of the relative importance of the two variables.

Even if in general sources were to be sampled only at the Nyquist interval a large display - such as is typical of the displays in nuclear power stations - would overload the scanning system of the human operator. Assuming that data acquisition is always complete within a fixation, indeed that it is virtually instantaneous, still only 2 samples per second can be taken. Hence to scan, say 100 sources would require 50 seconds, and the state variable bandwidth would have to average less than 0.01 Hz for the sampling to be adequate. Obviously, there are many activities and operations where the bandwidth far exceeds that value - notably in highly dynamic situations such as the final stages of landing an aircraft, and probably in dealing with most emergencies. It is these "moments of sheer terror" which enliven the "hours of sheer boredom" involved in monitoring industrial and other automatic systems.

As the required sampling frequency of a source or sources rises, so does the probability that attention will be overloaded. It is this that has led to my suggestion that Sheridan's Supervisor, (Sheridan, 1970) be extended theoretically by defining cost of an observation as the probability that the observer will miss a critical observation elsewhere while making his observation, (Moray, 1980). Hence the more heavily loaded attention is, the less likely it is that an observation of an abnormal variable will be made.

It has been suggested from time to time that the limitations of visual scanning may be overcome by using hearing as an extra input, but this hardly seems likely to help significantly. Although there have been situations in the laboratory where what appears to be parallel processing has been demonstrated, (Schneider and Shiffrin, 1977; Shiffrin and Schneider, 1977; Sanders, 1979) it is not likely to apply in industrial situations.

Deatherage (1972) suggests that four auditory signals is the maximum number which should be used, and there is much anecdotal evidence that pilots turn off auditory warnings during difficult phases of flight because they find them more distracting than helpful, (Weiner, 1977).

DATA ACQUISITION

The situation is made worse by the problem of data acquisition. It was mentioned above that sampling models have for the most part assumed that the data acquisition time was short compared with the interval between successive fixations. (But see Clement, W.F., et al. (1968) for an exception). In some studies, such as that by Senders et al. (1966), this can be seen to be so. But consider a case where perhaps because of a poor S/N ratio, or because of a very complex multidimensional display, the fixation duration required to extract "all" the information from the display is long compared with the sampling periodicity. A conflict must arise. The sampling algorithm will demand that the observer switch his attention to another part of the display, but the data acquisition algorithm will demand that he continues to observe the display currently fixated until he has extracted all the information it contains. This kind of conflict must occur, but has been ignored in models of sampling behaviour. It is possible - as is implied by the proposals of Jex and his co-workers for the re-design and amalgamation of displays for cockpits, (Clement, et al., 1968) that combining several instruments into one display could make matters worse for the monitor if the increase in fixation time is longer than the saving in scanning time. This could certainly be the case, for example, where a computer generated display is implemented which requires the operator to enter codes for required displays in a poorly designed keyboard in order for them to appear on a screen. It is by no means obvious that traditional keyboards or even some of the new designs will allow data to be accessed more rapidly than by glancing from one display to another or even walking about a control room. But reduction in the overall time per display must be an absolutely fundamental criterion for improving displays.

A plausible model for the data acquisition process which has independently been suggested at least by two people recently (Kvalseth, 1980; Moray, 1980) is sequential decision theory. This would assume that the observer sets criteria in terms of a priori probabilities and utilities for each observation, and continues to sample a source until enough evidence has been accumulated to satisfy the criterion for a decision as to whether the system is normal or not. The theory has also been used by Gai and Curry, 1976, in combination with optimal decision theory to model failure detection with considerable success. Kvalseth, T., 1980, found in an experimental test of the model that human operator performance correlated well with its predictions, although not perfectly. But it should be noted that any such model is likely to prove only approximate unless the observers are given very extensive practice, so that the payoffs and probabilities are incorporated into an unconscious internal model. There is good reason to think that when observers try to behave in accordance with instructions, rather than having had enough experience to behave automatically, suboptimal performance is almost always the role, (Moray, 1980). Gai and Curry, and Kvalseth, examined only single variable systems. But there is little doubt that a model on these lines is required, and in so far as the data acquisition algorithm restricts the rate at which sources can be scanned it will be a further source of limitation on the efficient distribution of attention.

The two mechanisms taken together, define a speed-accuracy tradeoff function for the observation of dynamic functions. If the emphasis is on accuracy then the data acquisition algorithm will take precedence in determining tactics, while if the absolute necessity to scan all variables in the minimum time is emphasised, the sampling algorithm will be given precedence, fixation duration will be driven down, and accuracy will suffer.

EFFECTS OF CORRELATION AMONG VARIABLES

The conclusion of the previous section is that a demand for accurate observation can lead to a failure to detect and diagnose errors, since in general such a demand will cause the operator to miss some moments at which he should have sampled certain variables, while a demand for speed will reduce the amount of evidence he has on which to base his decision. Let us now turn to an aspect of complex systems which in theory can offset this to some extent, but at the same time may give rise to other problems.

In general a large complex system is not uniformly interconnected. It can be regarded as being composed of a number of subsystems. At the lowest level individual variables are closely coupled only to a few others, and rather weakly to most. Hence one can, as it were, think of the correlation between variables, and the causal connections, as binding certain "atomic" variables into "molecules", within which there are high intercorrelations. This kind of structure suggests certain optimum tactics and strategies for examining the system as a whole. What follows is speculative: as far as I know there are no empirical investigations relevant to the claims to be made.

Let us begin by assuming that the observer has made a series of observations which have led him to believe that the system is functioning normally, and that he has just completed an observation. Which variable should he observe next assuming that all importance, bandwidth, etc. considerations leave the variables having equal priority? It would seem that he should not observe a highly correlated variable. Rather he should next sample a variable from a different "molecule". This is because high correlation already gives him considerable information based on the observation he has just made.

He can predict the value of highly correlated variables with a high a priori probability from that observation, and to observe the variables themselves will not gain him much in the way of further reducing his uncertainty. Hence an observation on a closely coupled variable is uneconomical in terms of the aggregation of evidence. The existence of such correlation will obviously compensate to some extent for the speed-accuracy problems mentioned earlier. Indeed it may well be worthwhile to make long, accurate observations of variables which occupy key positions in the correlational nexus, because so much can be predicted from them that it will be worth relying on the correlation rather than spending time on direct observation. It would seem then, that when the system appears normal, successive samples should be made from weakly correlated variables. But such a rule will mean that if an abnormal state arises in certain variables it may be a long time before they are observed. (In a similar way a strong internal model will help an overloaded operator by allowing him to predict future states of the system, provided that it stays normal. But in so far as he relies on predicting the future on the basis of recent past experience, and fails to make observations on the current values of state variables, abnormal readings will not be noticed. Thus we have the paradox that the better a system is known to the operator, the less likely he will be to discover that it is in an abnormal state).

If the latest observation made by the observer is an abnormal observation, however, the situation changes. Either the abnormality is arising in the variable which is examined, or that variable is being driven into an abnormal state by some other variable. In order to diagnose the cause of the abnormality it will be necessary to examine other variables. If no inputs to the variable are abnormal, then it must be the source of the abnormality. If some of its inputs are abnormal, either some other variable is the source of the abnormality or it is itself abnormal and there is a feedback loop from the variable to itself either directly or through other variables. In any case, it will be necessary to examine other variables, and because systems can usually be partially decomposed into subsystems, "molecules", it is more likely that the source of abnormality is highly correlated with an abnormal variable than that it is weakly correlated, since in physical systems correlation and causality are closely connected. Hence following an abnormal observation, highly correlated sources should be sampled.

This rule will have the effect of locking attention on to a small subset of variables until diagnosis is complete, and what we may call "cognitive tunnel vision" will result. Although no formal studies seem to have been made, there is ample evidence from accident reports to suggest that this does indeed happen. It can account for the otherwise inexplicable behaviour of the flightdeck crew on EAL 401, (Weiner, 1977), and the many other accounts where observers have completely ignored diagnostic data which was "staring them in the face" while concentrating on a small part of the system. It may indeed be a model for perceptual "set", and account for the inability of the operators at TMI to realise that the reason for the high water level in the pressurizer was not that the cooling system for the reactor was full but that it was empty.

DECISION CRITERIA

There is abundant evidence that attention can both affect the strength of signals that enter the nervous system, and also the observer's response criterion, (Moray, et al., 1976). The expectations of the observer, the probability of the events, the payoffs associated with the various outcomes and influence the interpretation put on a signal of a given strength when it is processed by the observer. As Broadbent, 1971, has put it, what the brain receives are states of evidence, not deterministic signals. The observer's pattern analysing mechanisms determine into what category an input will be placed and hence what he will perceive. Expected evidence is readily accepted as conclusive evidence for the state of the world, and is perceived as such that as a far lower S/N ratio than unexpected evidence. Such effects are well documented, and would account for the KLM captain's belief that he had received clearance for takeoff, and that the Pan Am aircraft had cleared the runway at Teneriffe. Most of us who have learned to programme computers will have experienced "invisible" programming errors which one can literally not perceive because of preconceptions about the place in the program that the error must be, ("I am certain the hardware is at fault"), but which can be instantly perceived by someone who has no preconceptions as to what to expect in the program. It may be that the unwillingness of human observers to accept evidence which would require them to change their hypotheses is related to response bias, although the topic has not been sufficiently investigated, (Taylor, 1975; Gai & Curry, 1978; Moray, 1980). Humans seem to want to confirm, rather than explore, the accuracy of their hypotheses.

PROPERTIES OF COMPLEX SYSTEMS WHICH MIGHT AID THE HUMAN OPERATOR

It is clear that certain design characteristics would aid the human operator to make more efficient use of his limited attention in the detection of errors and the diagnosis of their cause.

1. Minimise the number of displays. If it is absolutely imperative to display many variables, then displays should be developed which will allow several variables to be combined in a single display, with the caveats expressed earlier about the tradeoff between scanning time and data accessing time. It would seem worth exploring the possibility of directly displaying the probability of normality rather than the values of the state variables themselves.

2. Let the system monitor the monitor. With modern computers, and especially with displays where the human monitor must call up the variables which he wishes to examine, the answer to the classical question, "Quis custodet ipsos custodes?" is at last available. The system should keep a log of the intervals at which its variables are inspected, and should prompt the human if he has ignored some state variable for too long a time.

3. Minimise data acquisition time. Displays must be designed to reduce the time for which a given variable must be examined, since only by so doing can the observer be freed to examine other variables. This is a fundamental design requirement, providing that accuracy is maintained and mental workload is not unduly increased.

4. Use predictor displays. By giving the observer an estimate of the future states of the system he can use the time to his tactical best advantage, since he can detect periods when certain variables are likely to be in a normal state, and hence need not be observed. While not invariably better than standard displays predictor displays frequently reduce the load on the operator.

5. Make the system demand interrogation during diagnosis It is not at all clear how best this may be done, but there is ample evidence to suggest that a major problem during diagnosis is "cognitive tunnel vision". Some method must be found to prevent

the observer from becoming trapped in a subset of state variables. If the system can know which variables have been interrogated it should be able to detect changes in the observer's scanning pattern and prompt him at suitable moments. But failing that it may be necessary to divide functions among members of a team. A highly experienced operator working with a normal system uses a nearly optimal scanning strategy of whose properties he is characteristically unaware. When conscious scanning supervenes during error diagnosis the operator becomes far less efficient, (Moray, 1980). When a fault occurs in a large system it is usual for there to be a considerable delay before its effect spreads throughout the system. Hence it may be worth having one operator monitor the still normal parts of the system using his internal model efficiently, while another operator deals with the abnormal parts of the system.

TOWARDS A THEORY OF MONITORING AND DIAGNOSIS

Briefly, the following features seem to be needed in any theory which sets out to give an exhaustive account of the way in which human operators process information to detect errors and diagnose their causes.

1. Sampling algorithms - derived from the Sampling Theorem but modified by payoffs, importance, etc., probably in a way related to Sheridan's Supervisor Theory.
2. Data acquisition algorithms - probably using sequential decision theory.
3. Correlation sensitive algorithms - which allow the operator to make use of the structure of the system and to detect causality so as to generate tactics based on system structure.
4. Theory of Signal Detection - to account for the interpretation of data which is received.
5. Internal models of the system - acquired during practice and representing the operator's knowledge of the dynamics of the system, and which allow him to predict and extrapolate from present observations.
6. A theory of the generation of actions - not dealt with in this paper.

It goes without saying that while new technology offers outstanding opportunities to help the human operator, almost nothing is known about how best to use the new displays and data entry devices. Only when appropriate new ergonomics have been done will the theory and practice converge, and the design of systems, rather than components be possible.

REFERENCES

Broadbent, D.E., 1971, "Decision and Stress", Academic Press, New York.

Clement, W.F., Jex, H.R., Graham, D., 1968, "A manual Control-display theory applied to instrument landings of a jet transport." IEEE Trans MMS-9. 93-110.

Deatherage, B., 1972, "Auditory and other sensory forms of information presentation." in Van Cott, H.P. and Kinkade, R.G. (eds.). Human Engineering Guide to Equipment Design, U.S. Government Printing Office.

Gai, E.G. & Curry, R.E., 1976, "A model of the human observer in failure detection tasks." IEEE Trans. SMC-6, 85-94.

Gai, E. & Curry, R.W., 1978, "Perseveration effects in detection tasks with correlated decision intervals." IEEE Trans. SMC-8. 93-101.

"Human factors report on the Teneriffe accident." Air Line Pilots Association. 1979, Washington D.C.

IEEE Spectrum. Special issue on Three Mile Island, 1979.

Kvalseth, T., 1980, "A decision-theoretic model of the sampling behaviour of the human process monitor, experimental evaluation", Human Factors, 21, 671-686.

Loftus, G., 1979, "Short-term memory factors in Ground Controller/ Pilot Communications." Human Factors, 21, 269-181.

Moray, N., Filter., Ostry, D., Favrean, D & Nagy, V. 1976. "Attention to pure tones." Quarterly Journal of Experimental Psychology 38, 271-284.

Moray, N. (ed.), 1979, "Mental Workload: its theory and measurement." Plenum Press, N.Y. and London.

Moray, N., 1980, "Human information processing and supervisory control." Man-Machine Systems Laboratory Report, M.I.T.

Moray, N., Richards, M., and Lowe, J., 1980, "Final Report on the behaviour of fighter controllers." United Kingdom Ministry of Defence 2020/10ASA

Ostry, D., Moray, N., and Marks, G., 1976, "Attention, Practice and semantic targets." Journal of Experimental Psychology: Human Perception and Performance, 2, 326-336.

Rasmussen, J., 1979, in Moray, N. (ed.) "Mental workload: its theory and measurement." Plenum Press, N.Y. and London.

Sanders, A., 1979, "Some remarks on mental load. In Moray, N. (ed.) Mental workload: its theory and measurement." Plenum Press, N.Y. and London.

Senders, J.W., Elkind, J.I., Gringnetti, M.C., and Smallwood, R., 1966, "Investigation of the visual sampling behaviour of human observers." NASA CR-434.

Schneider, W. and Shiffrin, R.M., 1977, "Controlled and automatic processing I." Psychological Review, 84, 1-66.

Sheridan, T.B., 1970, "On how often the supervisor should sample." IEEE Trans. SSC-6, 140-145.

Shiffrin, R.J. and Schneider, W., 1977, "Controlled and automatic processing II." Psychological Review, 84, 127-190.

Taylor, F.J., 1975, "Finite fading memory filtering." IEEE Trans. SMC-5, 134-137.

Weiner, E.L., 1977, Controller Flight into Terrain Accidents. System induced errors." Human Factors, 19, 171-181.

EXPERIMENTAL STUDIES AND MATHEMATICAL MODELS OF HUMAN PROBLEM SOLVING PERFORMANCE IN FAULT DIAGNOSIS TASKS

William B. Rouse

University of Illinois at Urbana-Champaign
Urbana, Illinois 61801, U.S.A.

INTRODUCTION

One of the reasons often given for employing humans in systems is their supposed abilities to react appropriately and flexibly in failure situations (Johnson, Rouse, and Rouse, 1980). On the other hand, we seem to hear increasingly about incidents of "human error". The apparent inconsistency of these two observations can cause one to wonder what role the human should actually play. This question has led us to pursue a series of investigations of human problem solving performance in fault diagnosis tasks. Using three different fault diagnosis scenarios, we have studied several hundred subjects, mostly maintenance trainees, who have solved many thousands of problems. The results of these studies have led to the development of three mathematical models of problem solving behavior. The three tasks, results of the eight experiments, and the three models will be reviewed in this paper.

Besides trying to assess problem solving abilities, we have also invested considerable effort into studying alternative methods of training humans to solve fault diagnosis tasks. One issue that has particularly intrigued us concerns the extent to which humans can be trained to have general, context-free problem solving skills. From a theoretical point of view, it is of fundamental interest to know whether skills are context-free or context-specific. From a practical perspective, this issue is perhaps even more important in terms of training personnel to serve in multiple domains (e.g., to diagnose faults in a wide variety of systems). While our studies thus far have not provided a definitive answer to the context-free versus context-specific question, various bits and pieces of an answer have emerged and will be briefly discussed here.

The overall goal of this research is to determine an appropriate role for humans in failure situations and, to develop methods of training humans to fill that role. In this paper, we will attempt to show how our investigations are slowly but surely achieving this goal.

FAULT DIAGNOSIS TASKS

The three tasks that will be discussed here involve computer simulations of network representations of systems in which subjects are required to find faulty components. The three tasks represent a progression from a fairly abstract task that includes only one basic operation to another abstract task that includes two basic operations and, finally, to a fairly realistic task that includes several operations.

Task Number One

In considering alternative fault diagnosis tasks for initial studies, one particular task feature seemed to be especially important. This feature is best explained with an example. When trying to determine why component, assembly, or subsystem A is producing unacceptable outputs, one may note that acceptable performance of A requires that components B, C, and D be performing acceptably since component A depends upon them. Further, B may depend on E, F, G, and H while C may depend on F and G, and so on. Fault diagnosis in situations such as this example involves dealing with a hierarchy of dependencies among components in terms of their abilities to produce acceptable outputs. Abstracting the acceptable/unacceptable dichotomy with a 1/0 representation allowed the class of tasks described in this paragraph to be the basis of the task chosen for initial investigations.

Specifically, the task chosen was fault diagnosis of graphically displayed networks. An example is shown in Figure 1. These networks operate as follows. Each component has a random number of inputs. Similarly, a random number of outputs emanate from each component. Components are devices that produce either a 1 or 0. All outputs emanating from a component carry the value produced by that component.

A component will produce a 1 if: 1) All inputs to the component carry values of 1, 2) The component has not failed. If either of these two conditions are not satisfied, the component will produce a 0. Thus, components are like AND gates. If a component fails, it will produce values of 0 on all the outputs emanating from it. Any components that are reached by these outputs will in turn produce values of 0. This process continues and the effects of a failure are thereby propagated throughout the network.

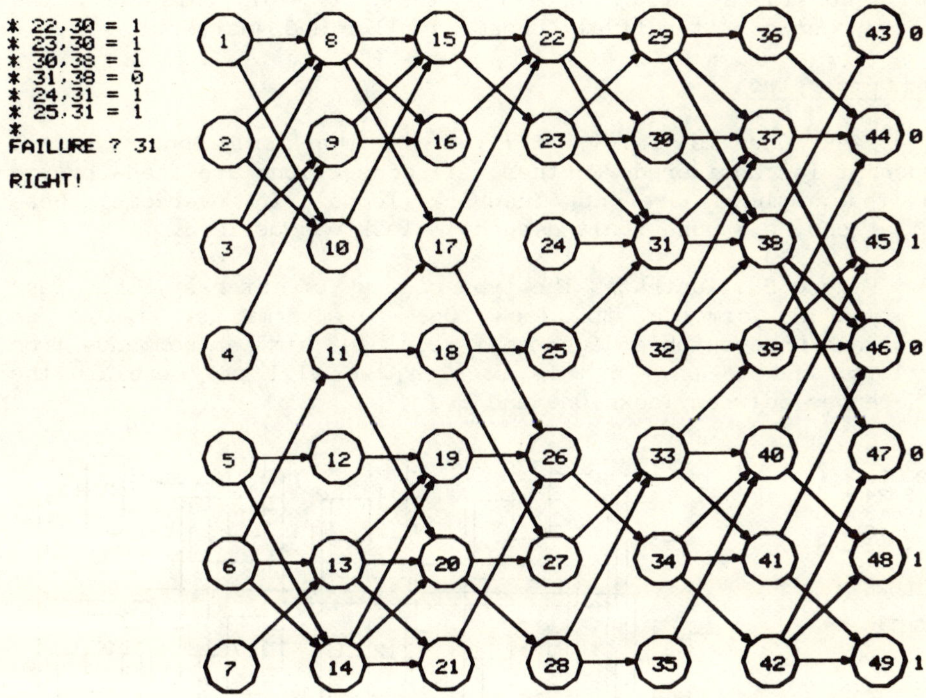

Figure 1. An Example of Task One

A problem begins with the display of a network with the outputs indicated, as shown on the righthand side of Figure 1. Based on this evidence, the subject's task is to "test" connections between components until the failed component is found. The upper lefthand side of Figure 1 illustrates the manner in which connections are tested. An * is displayed to indicate that subjects can choose a connection to test. They enter commands of the form "component 1, component 2" and are then shown the value carried by the connection. If they responded to the * with a simple "return", they are asked to designate the failed component. Then, they are given feedback about the correctness of their choice. And then, the next problem is displayed.

In the experiments conducted using Task One, computer aiding was one of the experimental variables. The aiding algorithm is discussed in detail elsewhere (Rouse, 1978a). Succinctly, the computer aid was a somewhat sophisticated bookkeeper that used the structure of the network (i.e., its topology) and known outputs to eliminate components that could not possibly be the fault (i.e., by crossing them off). Also, it iteratively used the results of tests (chosen by the human) to further eliminate components from future consideration by crossing them off. In this way, the "active" network iteratively became smaller and smaller.

Task Number Two

Task One is fairly limited in that only one type of component is considered. Further, all connections are feed-forward and thus, there are no feedback loops. To overcome these limitations, a second fault diagnosis task was devised.

Figure 2 illustrates the type of task of interest. This task is somewhat similar to Task One in terms of using an acceptable/unacceptable dichotomy, requiring similar commands from subjects, and so on. In this section, we will only explain the differences between Tasks One and Two.

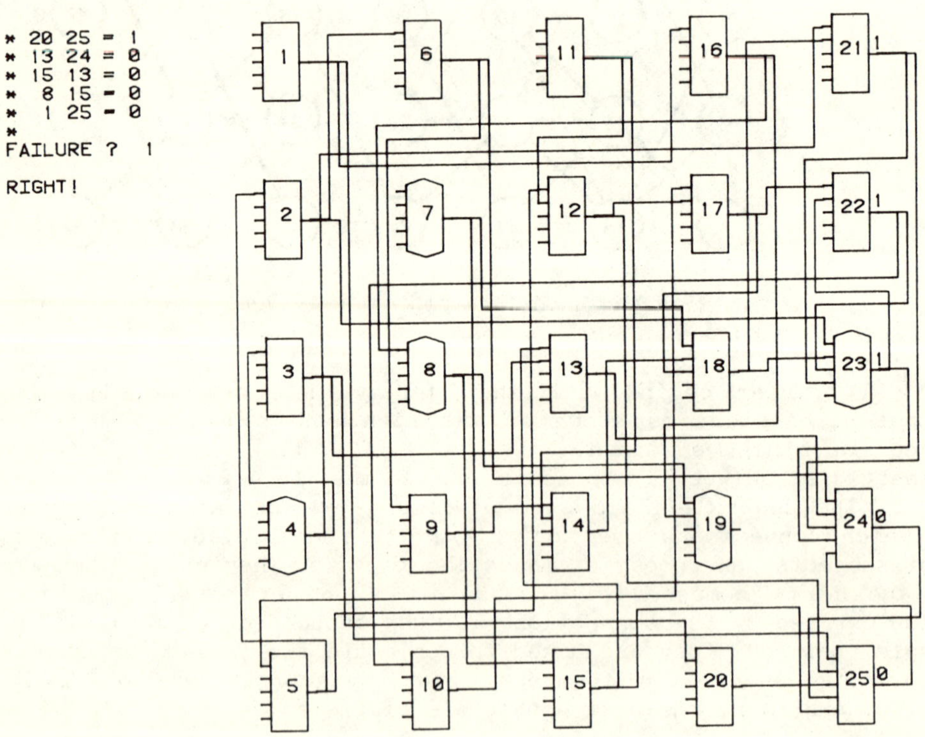

Figure 2. An Example of Task Two

A square component will produce a 1 if: 1) All inputs to the component carry values of 1, 2) The component has not failed. Thus, square components are like AND gates. A hexagonal component will produce a 1 if: 1) Any input to the component carries a value of 1, 2) The component has not failed. Thus, hexagonal components are like OR gates. For both AND and OR components, if either of the two conditions is not satisfied, the component will produce a 0.

The overall problem is generated by randomly connecting components. Connections to components with higher numbers (i.e., feed-forward) are equally likely with a total probability of P_{FF}. Similarly, connections to components with lower numbers (i.e., feedback) are equally likely with a total probability of $P_{FB} = 1-P_{FF}$. The ratio P_{FF}/P_{FB}, which is an index of the level of feedback, was one of the independent variables in the experiments to be discussed later. OR components are randomly placed. The effect of the ratio of the number of OR to AND components was also an independent variable in the experiments.

Task Number Three

Tasks One and Two are context-free fault diagnosis tasks in that they have no association with a particular system or piece of equipment. Further, subjects never see the same problem twice. Thus, they cannot develop skills particular to one problem. Therefore, we must conclude that any skills that subjects develop have to be general, context-free skills.

However, real-life tasks are not context-free. And thus, one would like to know if context-free skills are of any use in context-specific tasks. In considering this issue, one might first ask: Why not train the human for the task he is to perform? This approach is probably acceptable if the human will in fact only perform the task for which he is trained. However, with technology changing so rapidly, an individual is quite likely to encounter many different fault diagnosis situations during his career. If one adopts the context-specific approach to training, then the human has to be substantially retrained every time he changes situations.

An alternative approach is to train humans to have general skills which they can transfer to a variety of situations. Of course, they still will have to learn the particulars of each new situation, but they will not do this by rote. Instead, they will use the context-specific information to augment their general fault diagnosis abilities.

The question of interest, then, is whether or not one can train subjects to have general skills that are in fact transferrable to context-specific tasks. With the goal of answering this question in mind, a third fault diagnosis task was designed (Hunt, 1979).

Since this task is context-specific, we can employ hardcopy schematics rather than generating random networks online such as used with Tasks One and Two. A typical schematic is shown in Figure 3. The subject interacts with this system using the display shown in Figure 4. The software for generating this display is fairly general and particular systems of interest are completely specified by data files, rather than by changes in the software itself. Thus far, we have concentrated on various automobile and aircraft systems and, in particular, powerplant systems.

Task Three operates as follows. At the start of each problem, subjects are given fairly general symptoms (e.g., will not light off). They can then gather information by checking gauges, asking for definitions of the functions of specific components, making observations (e.g., continuity checks), or by removing components from the system for bench tests. They also can replace components in an effort to make the system operational again.

Associated with each component are costs for observations, bench tests, and replacements as well as the a priori probability of failure. Subjects obtain this data by requesting information about specific components. The time to perform observations and tests are converted to dollars and combined with replacement costs to yield a single performance measure of cost. Subjects are instructed to find failures so as to minimize total cost.

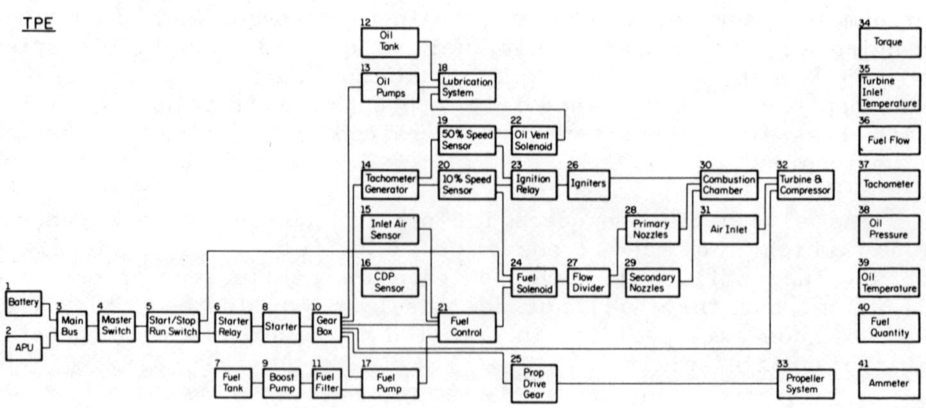

Figure 3. An Example of Task Three

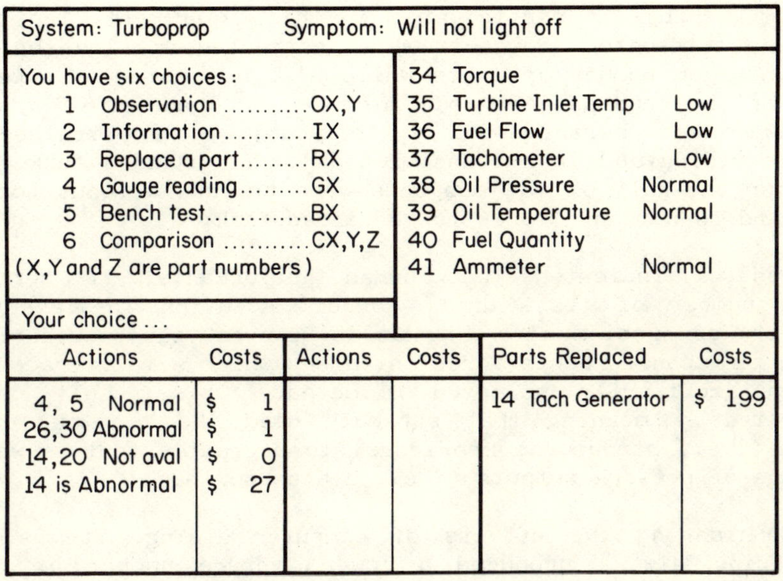

Figure 4. Display for Task Three

Because the software developed for this task is very general, we feel that it will be used quite extensively for future investigations. In recognition of this flexibility, it seemed appropriate to devise an acronym. We concluded that an excellent acronym was FAULT which stands for Framework for Aiding the Understanding of Logical Troubleshooting.

EXPERIMENTS

Using the above tasks, eight experiments have been completed. In this section, we will quite briefly review the statistically significant results of these experiments.

Experiment One

The first experiment utilized Task One and considered the effects of problem size, computer aiding, and training. Problem size was varied to include networks with 9, 25, and 49 components. The effect of computer aiding was considered both in terms of its direct effect on task performance and in terms of its effect as a training device (Rouse, 1978a).

Eight subjects participated in this experiment. The experiment was self-paced. Subjects were instructed to find the fault in the minimum number of tests while also not using an excessive amount of time and avoiding all mistakes. A transfer of training design was used where one-half of the subjects were trained with computer aiding and then transitioned to the unaided task, while the other one-half of the subjects were trained without computer aiding and then transitioned to the aided task.

Results indicated that human performance, in terms of average number of tests until correct solution, deviated from optimality as problem size increased. However, subjects performed much better than a "brute force" strategy which simply traces back from an arbitrarily selected 0 output. This result can be interpreted as meaning that subjects used the topology of the network (i.e., structural knowledge) to a great extent as well as knowledge of network outputs (i.e., state knowledge).

Considering the effects of computer aiding, it was found that aiding always produced a lower average number of tests. However, this effect was not statistically significant. Computer aiding did produce a statistically significant effect in terms of a positive transfer of training from aided to unaided displays for percent correct. In other words, percent correct was greater with aided displays and subjects who transferred aided-to-unaided were able to maintain the level of performance achieved with aiding.

Experiment Two

This experiment utilized Task One and was designed to study the effects of forced-pacing (Rouse, 1978a). Since many of the interesting results of the first experiment were most pronounced for large problems (i.e., those with 49 components), the second experiment considered only these large problems. Replacing problem size as an independent variable was time allowed per problem, which was varied to include values of 30, 60, and 90 seconds. The choice of these values was motivated by the results of the first experiment which indicated that it would be difficult to consistently solve problems in 30 seconds while it would be relatively easy to solve problems in 90 seconds.

This variable was integrated into the experimental scenario by adding a clock to the display. Subjects were allowed one revolution of the clock in which to solve the problem. The circumference of the clock was randomly chosen from the three values noted above. If subjects had not solved the problem by the end of the allowed time period, the problem disappeared and they were asked to designate the failed component.

As in the first experiment, computer aiding and training were also independent variables. Twelve subjects participated in this experiment. Their instructions were to solve the problems within the time constraints while avoiding all mistakes.

Results of this experiment indicated that the time allowed per problem and computer aiding had significant effects on human performance. A particularly interesting result was that forced--paced subjects utilized strategies requiring many more tests than necessary. It appears that one of the effects of forced-pacing was that subjects chose to employ less structural information in their solution strategies, as compared to self-paced subjects. Further, there was no positive (or negative) transfer of training for forced-paced subjects, indicating that subjects may have to be allowed to reflect on what computer aiding is doing for them if they are to gain transferrable skills. In other words, time pressure can prevent subjects from studying the task sufficiently to gain skills via computer aiding.

Experiment Three

Experiments One and Two utilized students or former students in engineering as subjects. To determine if the results obtained were specific to that population, a third experiment investigated the fault diagnosis abilities of forty trainees in the fourth semester of a two-year FAA certificate program in power plant maintenance (Rouse, 1979a).

The design of this experiment was similar to that of the first experiment in that Task One was utilized and problem size, computer aiding, and training were the independent variables. However, only transfer in the aided-to-unaided direction was considered. Further, subjects' instructions differed somewhat in that they were told to find the failure in the least amount of time possible, while avoiding all mistakes and not making an excessive number of tests.

As in the first experiment, performance in terms of average number of tests until correct solution deviated from optimality as problem size increased. Further, computer aiding significantly decreased this deviation. Considering transfer of training, it was found that aided subjects utilized fewer tests to solve problems and that they were able to transfer this skill to problems without computer aiding. A very specific explanation of this phenomenon will be offered in a later discussion.

Experiment Four

Experiment four considered subjects' performance in Task Two (Rouse, 1979b). Since the main purpose of this experiment was to investigate the suitability of a model of human decision making in fault diagnosis tasks that include feedback and redundancy, only four highly trained-subjects were used. The two independent variables included the level of feedback and the ratio of number of OR to AND components in a network of twenty-five components.

The results of this experiment indicated that increased redundancy (i.e., more OR components) significantly decreased the average number of tests and average time until correct solution of fault diagnosis problems. While there were visible trends in performance as a function of the level of feedback, this effect was not significant. The reason for this lack of significance was quite clear. Two subjects developed a strategy that carefully considered feedback while the other two subjects developed a strategy that discounted the effects of feedback. Thus, the average across all subjects was insensitive to feedback levels. One of the models to be described later yields a fairly succinct explanation of this result.

Experiment Five

The purpose of this experiment was to investigate the performance of maintenance trainees in Task Two, while also trying to replicate the results of experiment three. Forty-eight trainees in the first semester of the previously noted FAA certificate program served as subjects (Rouse, 1979c).

The design involved a concatenation of experiments three and four. Thus, the experiment included two sessions. The first session was primarily for training subjects to perform the simpler Task One. Further, the results of this first session, when compared with the results of experiment three, allowed a direct comparison between first and fourth semester trainees.

The second session involved a between subjects factorial design in which level of feedback and proportion of OR components were the independent variables. Further, training on Task One (i.e., unaided or aided) was also an independent variable. Thus, the results of this experiment allowed us to assess transfer of training between two somewhat different tasks.

As in the previous experiments, Task One performance in terms of average number of tests until correct solution deviated from optimality as problem size increased and, the deviation was substantially reduced with computer aiding. However, unlike the results from experiment three, there was no positive (or negative) transfer of training from the aided displays. This result led to the conjecture that the first semester students perhaps differed from the fourth semester students in terms of intellectual maturity (i.e., the ability to ask why computer aiding was helping them rather than simply accepting the aid as a means of making the task easy).

On the other hand, Task Two provided some very interesting transfer of training results. In terms of average time until correct solution, subjects who received aiding during Task One training were initially significantly slower in performing Task Two. However, they eventually far surpassed those subjects who received unaided Task One training. This initial negative transfer and then positive transfer is an interesting phenomenon which we hope to pursue further.

Experiment Six

This experiment considered subjects' abilities to transfer skills developed in the context-free Tasks One and Two to the context-specific Task Three (i.e., FAULT). Thirty nine trainees in the fourth semester of the two-year FAA certificate program served as subjects (Hunt, 1979).

The design of this experiment was very similar to previous experiments except the transfer trials involved FAULT rather than the context-free tasks. The FAULT scenarios used included an automobile engine and two aircraft powerplants. Both Tasks One and Two were used for the training trials. Overall, subjects participated in six sessions of ninety minutes in length over a period of six weeks.

Despite difficulties with considerable variability among subjects and problems, the results supported the hypothesis that context-free training can affect context-specific performance. For two of the three powerplants used with FAULT, it was found that training with the computer-aided version of Task One reduced cost to solution, mainly because expensive bench tests were avoided and more cost-free information was gathered.

Experiment Seven

The purpose of this experiment was to replicate experiment six using first semester rather than fourth semester maintenance trainees. Sixty trainees participated. The design of the experiment was very similar to experiment six except that only Task One training was used. Further, one of the aircraft powerplant scenarios was changed to allow inclusion of a more sophisticated system.

Since, as noted earlier, the analyses of the results of experiment six were plagued by a very substantial degree of inter-subject and inter-problem variability, it was decided to employ more fine-grained measures for analyses of the combined results of experiments six and seven (Hunt and Rouse, 1980). One of these fine-grained measures involved partitioning subjects' suboptimality (i.e., expenditures greater than optimal) into those due to errors and those due to inefficiency. Another measure was the expected information gain (in bits) per action. A third measure reflected the subjects' allocation of expenditures among observations, bench tests, and unnecessary replacements.

Use of these fine-grained performance measures led to quite clear conclusions. Fourth semester trainees who had received aided training with Task One were consistently able to achieve significantly better performance on the power plant problems, especially for problems involving less familiar power plants. The results for first semester trainees were mixed with a substantial positive transfer of aided training indicated by two performance measures and a slight negative transfer of training indicated by the third measure. For all trainees it was found that their suboptimality in terms of inefficiency could be attributed to their focussing on high cost, low information gain actions (i.e., bench tests and replacements) to a much greater extent than the optimal solution.

Experiment Eight

The purpose of this experiment was to evaluate the transfer of training with Task One, Task Two, and FAULT to real equipment (Johnson, 1980; Johnson and Rouse, 1980). Thirty-six fourth semester trainees participated as subjects. Each subject was allocated to one of three training groups. One group was trained using a sequence of Task One and Task Two problems. Another group was trained using FAULT. The third group, the control group, received "traditional" instruction including reading assignments, video taped lectures, and quizzes. The transfer task involved five problems on two real aircraft engines. In addition, subjects later (approximately one month) transferred to a completely unfamiliar context involving an autopilot system which was simulated using FAULT. This final observation was included to evaluate the hypothesis that context-free training would be superior for unfamiliar contexts.

Performance measures for the real equipment problems included overall cost, a performance index based on a fine-grained analysis of each action, and an overall rating by an observer. Results indicated that traditional instruction was only superior if explicit demonstrations were provided for the exact failures to be encountered (i.e., three of the five real equipment problems). Otherwise, there were no significant differences among the three training methods. Thus, context-free training was as useful as context-specific training as long as the instruction was general in nature (i.e., did not provide "cookbook" solutions for particular problems).

Considering correlates of real equipment performance, it was not surprising to find that the score on a quiz concerning the exact problems to be encountered was the best predictor. Beyond that, performance using Task One and Task Two was a good predictor of real equipment performance while performance using FAULT was by no means as useful. This result provides evidence for supporting the use of the context-free tasks as fault diagnosis aptitude tests.

The results for overall performance (i.e., cost) on the autopilot problems indicated that context-free training was not superior to context-specific training. However, preliminary analyses using the fine-grained performance measures utilized for analyzing the results of experiments six and seven indicate that context-free training may be superior when viewed on a fine--grained level. The analyses will soon be completed (Johnson, 1980).

MODELS OF HUMAN PROBLEM SOLVING PERFORMANCE

The numerous empirical results of the experimental studies discussed above are quite interesting and offer valuable insights into human fault diagnosis abilities. However, it would be more useful if we could succinctly generalize the results in terms of a theory or model of human problem solving performance in fault diagnosis tasks. Such a model might eventually be of use for predicting human performance in fault diagnosis tasks and, perhaps for evaluating alternative aiding systems. More immediately, a model would be of use in focusing research results and defining future directions.

Fuzzy Set Models

One can look at the task of fault diagnosis as involving two phases. First, given the set of symptoms, one has to partition the problem into two sets: a feasible set (those components which could be causing the symptoms) and an infeasible set (those components which could not possibly be causing the symptoms). Second, once this partitioning has been performed, one has to choose a member of the feasible set for testing. When one obtains the test result, then the problem is repartitioned, with the feasible set hopefully becoming smaller. This process of partitioning and testing continues until the fault has been localized and the problem is therefore complete.

If one views such a description of fault diagnosis from a purely technical point of view, then it is quite straightforward.

Components either can or cannot be feasible solutions and the test choice can be made using some variation of the half-split technique. However, from a behavioral point of view, the process is not so clear cut.

Humans have considerable difficulty in making simple yes/no decisions about the feasibility of each component. If asked whether or not two components, which are distant from each other, can possibly affect each other, a human might prefer to respond "probably not" or "perhaps" or "maybe".

This inability to make strict partitions when solving complex problems can be represented using the theory of fuzzy sets (Rouse, 1980). Quite briefly, this theory allows one to define components as having membership grades between 0.0 and 1.0 in the various sets of interest. Then, one can employ logical operations such as intersection, union, and complement to perform the partitioning process. Membership functions can be used to assign membership grades as a function of some independent variable that relates components (e.g., "psychological distance"). Then, free parameters within the membership functions can be used to match the performance of the model and the human. The resulting parameters can then be used to develop behavioral interpretations of the results of various experimental manipulations.

Such a model has been developed and compared to the results of experiments one, two, and four (Rouse, 1978b, 1979b). Two particularly important conclusions were reached. First, the benefit of computer aiding lies in its ability to make full use of 1 outputs, which the human tends to greatly under-utilize. Second, the different strategies of subjects in experiment four can be interpreted almost solely in terms of the ways in which they considered the importance of feedback loops.

It is useful to note here that these quite succinct conclusions, and others not discussed here (Rouse, 1978b, 1979b), were made possible by having the model parameters to interpret. The empirical results did not in themselves allow such tight conclusions.

Rule-Based Models

While the fuzzy set model has proven useful, one wonders if an even simpler explanation of human problem solving performance would not be satisfactory. With this goal in mind, a second type of model has been developed (Pellegrino, 1979; Rouse, Rouse, and Pellegrino, 1980). It is based on a fairly simple idea. Namely, it starts with the assumption that fault diagnosis involves the use of a set of rules-of-thumb (or heuristics) from which the human selects, using some type of priority structure.

Based on the results of experiments three, five, and six, we have found that an ordered set of twelve rules adequately describes Task One performance, in the sense of making tests similar to those of subjects 89% of the time. Using a somewhat looser set of four rules, the match increases to 94%. For Task Two, a set of five rules resulted in a 88% match. We have also found that the rank ordering of the rules is affected by training (i.e., unaided vs. aided).

The insights provided by this model led to the development of a new notion of computer aided training. Namely, subjects were given immediate feedback about the quality of the rules which the model inferred they were using. They received this feedback after each test they made. Evaluation of this idea within experiment six resulted in the conclusion that rule-based aiding was counterproductive because subjects tended to misinterpret the quality ratings their tests received. However, it appeared that ratings that indicated unnecessary or otherwise poor tests might be helpful.

Models of Task Complexity

It is interesting to consider why some fault diagnosis tasks take a long time to solve while others require much less time. This led us to investigate alternative measures of complexity of fault diagnosis tasks (Rouse and Rouse, 1979).

A study of the literature of complexity led to the development of four candidate measures which were evaluated using the data from experiments three and five. It was found that two particular measures, one based on information theory and the other based on the number of relevant relationships within the problem, were reasonably good predictors ($r = 0.84$) of human performance in terms of time to solve Task One and Task Two problems. The success of these measures appeared to be explained by the idea that they incorporate the human's understanding of the problem and specific solution strategy as well as the properties of the problem itself.

All of the modeling results noted above were based on problems involving Tasks One and Two. We have also tried to apply these models, especially the rule-based model, to describe human performance using FAULT. Our success here has been limited by what Rasmussen (1981) would call a shift from topographic to symptomatic search strategies. In other words, once subjects shift from a context-free to context-specific situation, they attempt to use rules that map directly from the symptoms to the solution. In many cases, this mapping process can be adequately described by a rule-based model. However, not infrequently it appears that subjects utilize what might be termed highly context-dominated rules, based on their past experiences, that have relatively little relevance to the particular set of symptoms with which they are dealing. While such rules can often be described after their use has been frequently observed, it is very difficult to predict their existence beforehand. More work needs to be devoted to this problem.

CONCLUSIONS

Within this paper, we have reviewed three fault diagnosis tasks, eight experiments, and three models of human problem solving performance in fault diagnosis tasks. The empirical results, as well as the results of interpreting the performance comparisons with the mathematical models, lead to several interesting conclusions. Human performance increasingly deviates from optimality as problem size increases, particularly because of the human's inability to utilize information about what has not failed. Forced-pacing makes this deviation more pronounced. In situations where feedback loops are predominant, some subjects appear to discount the importance of the loops to the extent that performance is substantially degraded. The time required to solve a fault diagnosis problem, in terms of the complexity of the problem, is related to the human's understanding of the problem as well as the intrinsic properties of the problem.

General, context-free skills learned with computer aiding in one task can be successfully transferred to another context-free or context-specific task, particularly if trainees are sufficiently mature intellectually and motivated to study how the computer helps them. Positive transfer of training can be explained as a reordering of priorities within a set of basic problem solving rules. This reordering appears to enable trainees to utilize the structure of the problem to a greater degree and thereby make more efficient tests in the sense of achieving greater reductions of uncertainty per unit cost. This effect is most pronounced for context-free or less familiar context-specific tasks. Performance on context-free tasks is, nevertheless, highly correlated with performance on familiar real equipment tasks.

Our current efforts in this research area are being directed at three particular issues. We are continuing the transfer of training studies with particular emphasis on assessing how different degrees of intelligence embedded in FAULT might affect transfer of training, especially to real equipment. In an effort to expand our scope of applications beyond automobile and aircraft maintenance, we are now studying how the ideas discussed in this paper might be applicable to shipping operations and process control situations. Finally, on a more theoretical level, we are continuing to pursue issues related to problem representation and modeling how humans understand fault diagnosis tasks.

ACKNOWLEDGEMENTS

This research was mainly supported by the U.S. Army Research Institute for the Behavioral and Social Sciences under Grant No. DAHC 19-78-G-0011 and Contract MDA 903-79-C-0421. Some of the earlier work reported here was supported by the National Aeronautics and Space Administration under Ames Grant NSG-2119. The author gratefully acknowledges the contributions of his colleagues R.M. Hunt, W.B. Johnson, S.J. Pellegrino, and S.H. Rouse.

REFERENCES

Hunt, R.M., "A Study of Transfer of Training from Context-Free to Context-Specific Fault Diagnosis Tasks", MSIE Thesis, University of Illinois at Urbana-Champaign, 1979.

Hunt, R.M., and Rouse, W.B., "Problem Solving Skills of Maintenance Trainees in Diagnosing Faults in Simulated Powerplants", University of Illinois at Urbana-Champaign, July 1980.

Johnson, W.B., "Computer Simulations in Fault Diagnosis Training: An Empirical Study of Learning Transfer from Simulation to Live System Performance", PhD Thesis, University of Illinois at Urbana-Champaign, 1980.

Johnson, W.B., Rouse, S.H., and Rouse, W.B., An Annotated Selected Bibliography on Human Performance in Fault Diagnosis Tasks, Alexandria, VA: U.S. Army Research Institute for the Behavioral and Social Sciences, Report No. TR 435, 1980.

Johnson, W.B., and Rouse, W.B., "Computer Simulations for Fault Diagnosis Training: From Simulation to Live System Performance", Proceedings of the 24th Annual Meeting of the Human Factors Society, Los Angeles, October 1980.

Pellegrino, S.J., "Modeling Test Sequences Chosen by Humans in Fault Diagnosis Tasks", MSIE Thesis, University of Illinois at Urbana-Champaign, 1979.

Rasmussen, J., "Models of Mental Strategies in Process Plant Diagnosis", this volume.

Rouse, W.B., "Human Problem Solving Performance in a Fault Diagnosis Task", IEEE Transactions on Systems, Man, and Cybernetics, SMC-8, No. 4, pp. 258-271, April 1978. (a)

Rouse, W.B., "A Model of Human Decision Making in a Fault Diagnosis Task", IEEE Transactions on Systems, Man, and Cybernetics, SMC-8, No. 5, pp. 357-361, May 1978. (b)

Rouse, W.B., "Problem Solving Performance of Maintenance Trainees in a Fault Diagnosis Task", Human Factors, Vol. 21, No. 2, pp. 195-203, April 1979. (a)

Rouse, W.B., "A Model of Human Decision Making in Fault Diagnosis Tasks that Include Feedback and Redundancy", IEEE Transactions on Systems, Man, and Cybernetics, Vol. SMC-9, No. 4, pp. 237-241, April 1979. (b)

Rouse, W.B., "Problem Solving Performance of First Semester Maintenance Trainees in Two Fault Diagnosis Tasks", Human Factors, Vol. 21, No. 5, pp. 611-618, October 1979. (c)

Rouse, W.B., Systems Engineering Models of Human-Machine Interaction, New York: North-Holland, 1980.

Rouse, W.B., and Rouse, S.H., "Measures of Complexity of Fault Diagnosis Tasks", IEEE Transactions on Systems, Man, and Cybernetics, Vol. SMC-9, No. 11, pp. 720-727, November 1979.

Rouse, W.B., Rouse, S.H., and Pellegrino, S.J., "A Rule-Based Model of Human Problem Solving Performance in Fault Diagnosis Tasks", IEEE Transactions on Systems, Man, and Cybernetics, Vol. SMC-10, No. 7, July 1980.

SYSTEM COMPLEXITY, DIAGNOSTIC BEHAVIOUR, AND REPAIR TIME: A PREDICTIVE THEORY

Joseph G. Wohl

The MITRE Corporation, Box 208
Bedford, MA 01730

INTRODUCTION

Existing maintainability prediction-evaluation procedures are in need of certain major refinements, namely; 1) ability to include design characteristics which are fixed during the earlier stages of the design process; 2) more precise and sensitive measurement of the troubleshooting process; 3) a basis for relating data taken under laboratory conditions to data taken under field conditions; and 4) the development of indices based on an explanatory model which includes functional relationships based on measurable parameters. Recent work by Rouse et al (1978a, 1978b) has begun to address items (2) and (4) through controlled experiments in the laboratory. The present paper provides an explanatory model and relates it to equipment design parameters and to repair time data for operational equipment taken under both laboratory and field conditions.

BACKGROUND

The basis for this paper arose out of two separate but related activities. The first summarized the results of experimental studies on the effects of packaging design on equipment repair time (Wohl 1961). These studies indicated that repair times taken under laboratory conditions were exponentially distributed, in contrast to data for many other equipments taken under field conditions, which were found to be more or less lognormally distributed. One study in which repair time data for the same equipment (an FPS-20 radar) were taken in both laboratory and field environments (Kennedy 1960) indicated an exponential distribution for the laboratory data but a non-exponential distribution for the field data. This work suggested that a model of the interaction process between man and machine in a maintenance situation might be developed. It also indicated that such a model would have to account for differences between maintenance environments.

The second activity involved a major weapon system program in which an unusual opportunity arose to control the collection of field maintenance data on a number of major electronic subsystems including sensors, computers, and navigation equipment and representing a wide spectrum of circuit functions, technology, packaging, modularization, and self-test capability. Highly motivated personnel were trained to record all aspects of the maintenance process under actual field conditions. The data were easily corrected for such factors as waiting for spare parts, administrative time, etc., and the resulting data as reported here represent active repair time; i.e., time during which either diagnosis, test or replacement was actually in process.

Active repair time data were taken on the same equipments for different maintenance teams, different installations, and different times of the year. No statistically significant differences could be related to these conditions. The data were therefore combined and summarized in the form of graphs of their cumulative frequency distributions (CFD). When plotted on lognormal paper, the data indicated clearly that the distributions were not lognormal. The data were then replotted on cumulative Weibull distribution paper, on which a CFD of the form

$$F(t) = 1-e^{-(t/\theta)^{\beta}}$$

plots as a straight line. The result for representative equipment is shown in figure 1 (similar plots were obtained for all seven of the equipments for which data were obtained). The data clearly do not represent exponential distributions, which would plot as straight lines of slope $\beta = 1.0$.

DISCUSSION

The characteristic shape of the seven plots suggests the existence of two quite separate causal processes, one responsible for those repair times below 2 hours and the other for those above. Regarding the first process, 65 to 80 percent of all repairs have obviously been completed within 2 hours, represented by the early high-slope portions of the plots. Such repairs are facilitated by the usual methods of maintainability design such as built-in test equipment, self-test and self-diagnostic capability, good maintenance manual design, and good equipment packaging and modularization practice.

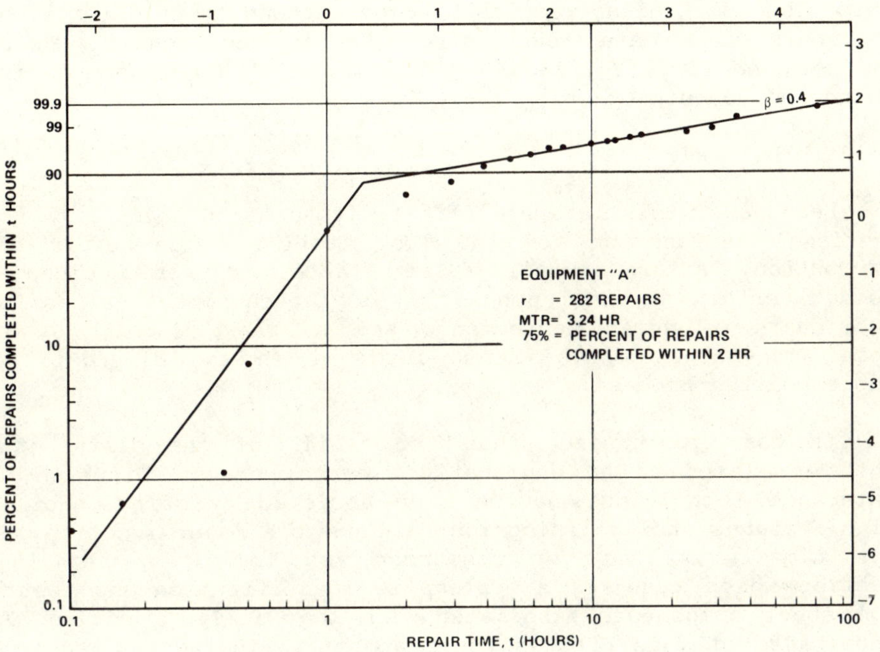

Figure 1. Cumulative Frequency Distribution (CFD) of Repair Time for Equipment "A," Plotted on Weibull Probability Paper

Regarding the second process, it would appear that in the remaining 20 to 35 percent of the repairs these maintainability design features failed to produce the intended results. Those field repairs which took longer than 2 hours to complete took much longer, which appeared to be due solely to difficulty in fault diagnosis.

The high slope of this initial portion ($\beta > 1.0$) suggests an accelerating process in which the probability of locating the failed component during an interval Δt increases with t, most likely due to increasing knowledge and rapid reduction of ambiguity. On the other hand, the low slope portion ($\beta < 1.0$) suggests a decelerating process in which the probability of locating the failed component during an interval Δt decreases with t, possibly due to increasing rarity of the failure mode, increasing difficulty in symptom interpretation, or increasing complexity of the circuit segments being examined.

In any case it is clear that it is this long "tail" of the distribution that contributes in an overriding manner to the mean repair time measure: For each of the seven equipments exemplified by figure 1, the arithmetic mean repair time was between 3 and 6 hours, even though 65 to 80 percent of the repairs were completed within 2 hours. Similar results have been found to hold for almost all repair time data taken under field conditions, with one exception; namely, flight-line maintenance of highly modularized airborne equipment for which $\beta = 1.0$.

INITIAL HYPOTHESIS

Because of this leverage effect, it appeared logical to seek a rational explanation for the long "tail" of the repair time distribution. After several false starts, it was decided to consider certain physical properties of the equipment as possible determinants of human performance and to seek a model of the special man-machine interaction involved in the diagnostic or troubleshooting process.

It was hypothesized that the "tail" of the distribution might be related to an enumerative search process in which the maintenance technician, having been assisted by maintainability design features in localizing the failure to a major function, now finds himself on his own resources, so to speak, and in a situation which requires a systematic identification and test of all suspect components associated with that failed function. The determinants of this process would appear to include at least the following:

a. The number of leads directly associated with the (unknown) faulty component; hence, the number of electrical junction points (defined by Kirkhoff's law) which might to a first-order approximation exhibit out-of-tolerance readings (direct connections only).

b. The accessibility of electrical junction points as test points.

c. The total number of components directly connected to all of the tests points having out-of-tolerance readings (i.e., the set of suspect components).

d. The capability for logical inference on the part of the technician.

This list includes only first-order effects; i.e., it does not take into account either indirect series effects or reflected effects of failed components upon remote junction points in the same circuit. Nor does it account for reliability differences among suspect components. On the other hand, it is independent of the technology or the packaging approach being used, being equally applicable to old tube-type and transistorized equipment as well as modern MSI and LSI equipment and to poorly as well as highly modularized equipment.

To summarize, the search process must begin with tests and/or out-of-tolerance indications. These result in the identification of a set of suspect components which then becomes the set of items to be searched. The search process is enumerative in the sense that each of the suspect components in turn must be isolated and tested until the failure source(s) is (are) located. Since the order in which the suspect components are isolated and tested may depend as much upon accessibility as upon logical inference or reliability weighting, treatment of the search process as random and uniform appears justified as a first approximation. The expected number of components searched per failure will, under this assumption of a uniform random distribution, be half the number of suspect components for that failure. However, we are interested in the characteristics of the distribution over many failures; and the size of the set of suspect components will differ for each failure. It is this random variable which appears to be the key to maintenance time prediction.

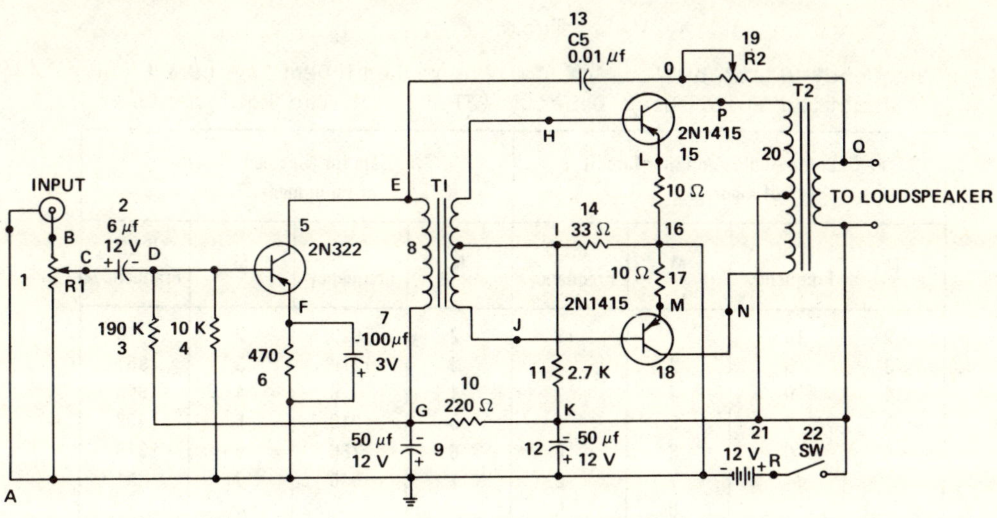

Figure 2. 12 Volt Audio Amplifier

To a first approximation, the frequency distribution of suspect components can be computed from a simplified network analysis. The approach is illustrated in figure 2 using a simple audio circuit for ease of understanding. The two types of "nodes" previously noted are identified in figure 2: <u>components</u> with leads emanating from them, and <u>electrical junction points or nodes</u> with leads attached to them. Each component is attached to as many electrical junction points as it has leads, and vice versa. Thus, when a component with N leads fails, then (to a first approximation) N junction points may indicate out-of-tolerance readings (voltage, resistance, frequency, bit stream, waveform, and so forth). In the absence of logical inference (again, to a first approximation), all of the components connected to all of these out-of-tolerance junction points are now suspect, and hence candidates for the enumerative search process.

Thus, under the simplified hypothesis, the primary determinants of the number of steps in this process for any given equipment are twofold:

a. The frequency distribution of component leads (i.e., number of components with N leads attached).

b. The frequency distribution of junction point leads (i.e., the number of junction points with M components attached).

Table 1

Frequency Distributions of Component Lead Density (CLD) and Junction Point Lead Density (JPLD) for Two Equipments

a. Data for Audio Amplifier Circuit of Figure 2				b. Data for Nineteen Circuits of Equipment "A"			
N	CLD Frequency	M	JPLD Frequency	N	CLD Frequency	M	JPLD Frequency
2	15	2	11	2	5926	2	1370
3	5	3	3	3	1278	3	951
4	0	4	2	4	0	4	660
5	2	5	1	5	210	5	458
6	0	6	0	6	786	6	318
		7	0	7	546	7+	31
		8	0				
		9	0				
		10	0				
		11	1				
Total	= 22	Total	= 18	Total	= 8746	Total	= 3788
$\overline{N}$	= 2.5	$\overline{M}+1$	= 3.05	$\overline{N}$	= 2.5	$\overline{M}+1$	= 3.28
$I = \overline{NM} = 5.12$				$I = \overline{NM} = 5.70$			

Again, as an illustration, the frequency distributions of N and M associated with the circuit of figure 2 are tabulated in table 1. For convenience, we have called these component lead density (CLD) and junction point lead density (JPLD). Their associated density functions will be called $P_N(i)$ and $P_M(i)$ respectively.

The number of suspect components (i.e., candidates for enumerative search) is then determined by a convolution of these two functions. If the time for isolation and test of a suspect component is also a random variable defined by a density function P(t), then the cumulative distribution F(t) of total time for the enumerative search process can be computed by convolving the three functions $P_N(i)$, $P_M(i)$, and P(t).

A closed-form solution for F(t) was obtained assuming as an approximation that $P_N(i)$ and $P_M(i)$ are geometric distributions, and P(t) is a Poisson distribution. One of the interesting results of the solution was a circuit complexity index I, defined as follows:

$I = \bar{N}\bar{M}$, where $\bar{N}$ is the expected value of $P_N(i)$
$\bar{M}$ is the expected value of $P_M(i)$ (corrected for i-1).

This result can be shown to be true for any distributions of $P_N(i)$ and $P_M(i)$; it simply represents the expected number of components connected to a given component. Alternatively, it represents the suspect set of components less one. In addition, λ is defined as the expected value of the diagnostic rate (i.e., $\lambda = 1/\tau$ where τ is the average test time per component).

TEST OF HYPOTHESIS

Some sample results were plotted for various values of I and λ. While the Weibull plots were all straight lines, their slopes ($\beta = 1$, representing exponential distributions) did not match those of the field data for Equipment "A" or for any of the other seven equipments. However, as will be discussed later, they did match the laboratory data.

These results were reviewed in the light of several possible interpretations of the field data, for which $\beta < 1$:

a. One possibility is that the maintenance technician may experience an increasing complexity of circuit segments as he steps through a sequence of tests. But the likelihood of this occurring is zero on the average, as long as the steps in the sequence are random and the circuit itself does not change during the trouble-shooting process.

b. A relatively rare failure mode, such as a cold solder joint or a hairline crack in a printed circuit, could indeed produce a situation in which the maintenance technician would spend an inordinate amount of time in testing. The mechanism involved here is that the failure in effect produces a new "component" (e.g., the cold solder joint) which is not immediately recognized as such.

c. However, the simplest interpretation is that, on the average, successive diagnostic tests on components simply take increasingly longer times, either for the reason given in "b" above or because the technician tends to do the easiest tests first or works on the most accessible components first, or finds it increasingly difficult and more time-consuming to interpret the observed symptoms as the successive components he tests are found to be "good".

IMPROVED HYPOTHESIS

The latter interpretation can be represented in the following way. Assume that the distribution of times required to test the nth component (i.e., during the nth step following a system failure) is given by:

$$P(t) = \lambda_n e^{-\lambda_n t} \quad \text{(modified Poisson assumption)},$$

where $1/\lambda_n$ is the average time required to test the nth component. The new hypothesis assumes that λ_n simply decreases or, equivalently, that the average component test time increases geometrically with n; that is,

$$\lambda_n = \lambda_o a^n$$

This relationship reflects a simple linear ordering of suspect components by increasing difficulty of diagnostic interpretation: as successive suspect components are eliminated from consideration as the failure source, diagnostic interpretation takes increasingly longer.

A closed-form solution for F(t) was not possible, but an approximation was developed whose error is proportional to $(1-a)^3$ and is negligible for all cases of interest. Plots were obtained for various combinations of values for I, λ, and a. Note that all three of these parameters are capable of experimental observation.

TEST OF IMPROVED HYPOTHESIS

While it was not possible to obtain independently measured values for either λ or a from the field data described herein, it was possible to compute the complexity index I for one of the equipments for which repair time data was available (Equipment "A"). The circuit diagram for this equipment was subjected to a detailed analysis to obtain the frequency distribution for component lead density (CLD) and junction point lead density (JPLD). A total of 8746 components and 3788 junction points from 19 separate circuits in the equipment was included in the analysis. The results are shown in table 1. Under the assumption that the frequency distributions are Poisson, the complexity index is then given by $I = \overline{N}\overline{M} = 5.7$ for Equipment "A".

A family of curves for F(t) was computed for various values of the parameters I, a, and λ. The repair time data for Equipment "A" was then plotted on the set of curves representing $I = 5.7$. For repair times greater than 1 hour, the best visual fit of the data, as shown in figure 3, was obtained for $a = 0.85$ and $\lambda = 3.33$ hr. Due to the laborious nature of the effort involved, Equipment "A" was the only equipment subjected to a detailed accounting of junction points.

To obtain an indication of the generalizability of this result, the repair time data for a number of other equipments were also plotted against the same family of curves, using appropriate values of I, a, and λ to achieve a best visual fit for repair times greater than 1 hour. CFD's for six additional weapon system equipments were obtained and plotted with results similar to that shown in figure 3.

In addition, available data from previous maintainability research experiments were reviewed and plotted. These are summarized in figure 4. Curves 1 and 2 in figure 4 represent repair time distributions for two different designs of a radar simulator taken in a controlled laboratory environment, with curve 1 representing the results of improved packaging for maintainability (Wohl 1961). Curves 3 and 4 represent the results of laboratory and field maintenance respectively for the FPS-20 search radar (Kennedy 1960).

It is clear that for the laboratory environment, an excellent fit would be obtained with the parameter value $a = 1.0$ and appropriate values of λ, thus corroborating the simpler Initial Hypothesis as a special case of the more general Improved Hypothesis. Note that as equipment is simplified for ease of maintenance (i.e., as in the repackaging of equipment 1 into equipment 2) average component test times ($\tau = 1/\lambda$) become shorter. Simpler equipments (i.e., 1 and 2 vs all others) exhibit the same effect. Also, when moving from the field environment into the laboratory, the parameter "a" changes from a value of 0.85 to a value of 1.0. This may be associated with increased ability for logical inference or increased opportunity for task concentration (i.e., less distraction) in the laboratory environment.

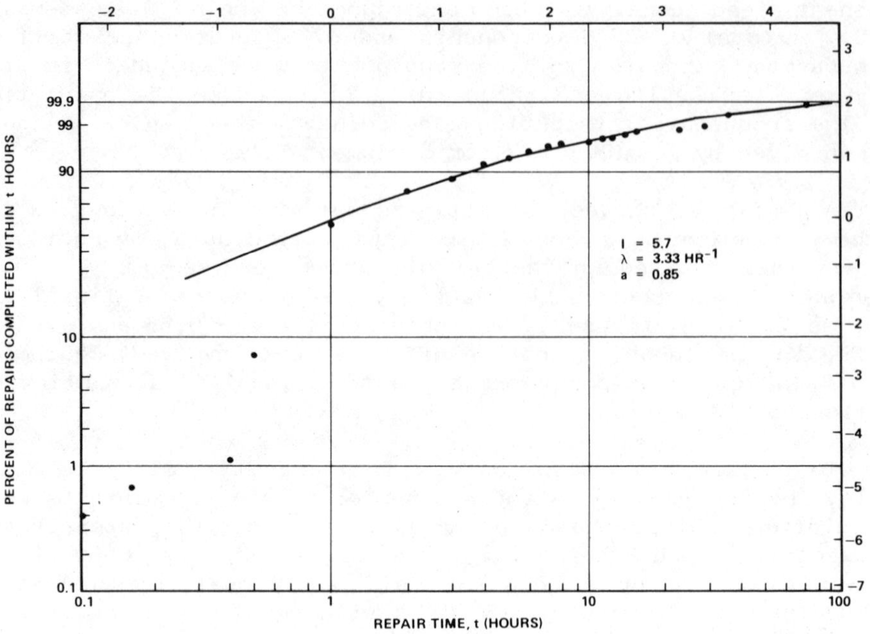

Figure 3. CFD Data for Equipment "A" Plotted Against F(t) as Computed from Improved Hypothesis (Appendix C)

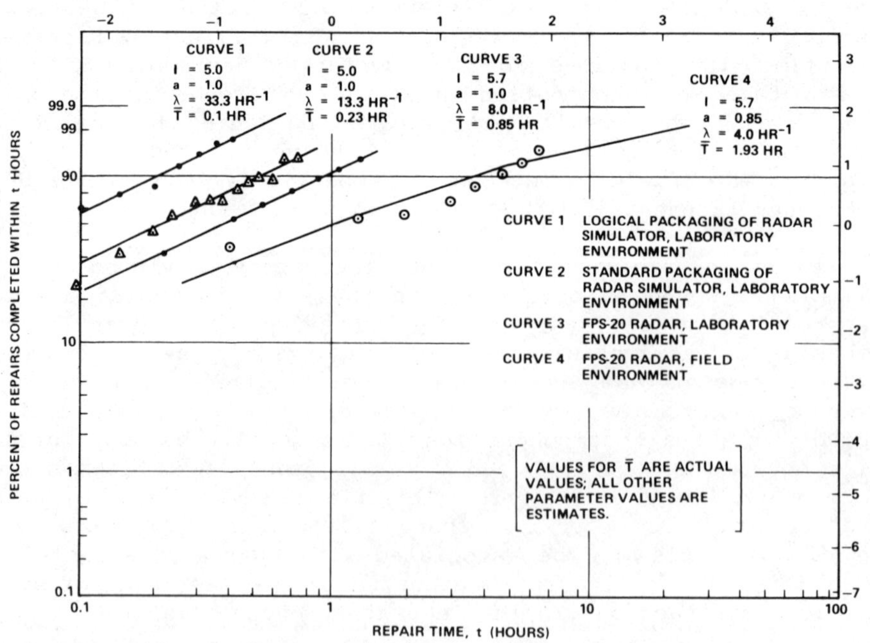

Figure 4. CFD Data for Radar Simulator and for FPS-20 Radar Plotted Against F(t) as Computed from Improved Hypothesis (Appendix C)

The expected value of total diagnostic time (defined by $\bar{T} = \int_0^\infty t\, d(F(t))$ was then calculated and was found to have the form $\bar{T} = \tau f(I,a)$. Specifically, there are three cases of interest:

(1) For $a = 1$,

$$\bar{T} = \tau\left[\frac{1}{I(1 - \frac{1}{I})^2}\right]\left\{\frac{1}{2}\ I^2\left[1 - (\frac{1}{I})^2\right] - \ln I\right\}$$

(2) For $a \le (\frac{I - 1}{I})$,

$$\bar{T} = \infty$$

(3) For $1 > a > (\frac{I - 1}{I})$

$$\bar{T} = \tau\left[\frac{1}{I(1 - \frac{1}{I})^2}\right]\left\{\frac{-a^2(2 - a)\ln\left[\frac{\frac{1}{I} - 1 + a}{a}\right] - a\ln I - Ia(1 - a)(1 - \frac{1}{I})}{(1 - a)^2}\right\}$$

It is clear from cases (1) and (3) above that the general relationship

$$\bar{T} = \tau G(I,a)$$

may be useful in predicting $\bar{T}$ for various types of equipment under both laboratory and field conditions.

Using (1) or (3) as appropriate for each equipment, values of $\bar{T}$ as predicted by the Improved Hypothesis were calculated. Table 2 summarizes all of the parameter values for ten sets of equipment. The last two columns in table 2 contain the computed and measured values of mean repair time, $\bar{T}$. Assuming the two sets of values to be independent, the coefficient of correlation as calculated from the data in table 2 using the product-moment method was $r = 0.97$.

However, the best-visual-fit method of estimating the parameters I, λ and a would tend to increase the degree of

Table 2

Values of I,a,λ,τ and $\bar{T}$ for Ten Equipments

Equipment	Environment	Sample Size	Parameter Estimated by Best Visual Fit of Data to F(t) Curves				Mean Active Repair Time, $\bar{T}$ (hrs)	
			I	a	λ (hrs^{-1})	τ (hrs)	Predicted	Measured
A	F	282	5.7	0.85	3.33	0.30	2.88	2.99
B	F	91	7.5	0.90	5.0	0.20	1.91	3.23
C	F	112	5.7	0.85	2.0	0.50	4.80	5.15
D	F	257	7.5	0.87	3.0	0.333	5.79	5.94
E	F	135	5.0	0.85	3.0	0.333	2.20	2.48
F	F	45	7.5	0.90	5.0	0.20	1.91	2.27
G	F	61	7.5	0.90	2.5	0.40	3.82	4.72
1	L	–	5.0	1.00	33.3	0.03	0.097	0.10
2	L	–	5.0	1.00	13.3	0.075	0.244	0.23
3	L	73	5.7	1.00	8.0	0.125	0.406	0.85
3	F	42	5.7	0.85	4.0	0.25	2.40	1.93

$\bar{T}$ = mean active repair time, hours
I = complexity index
λ = average diagnostic rate per component, hr^{-1}
τ = average test time per component, hours = 1/λ
F = data taken under field conditions
L = data taken under laboratory conditions

dependency among the variables. Hence, the correlation must be less than 0.97. Independent measures of the parameters I and λ for each equipment would be required in order to validate the theory and determine the true correlation.

IMPLICATIONS FOR DESIGN AND TRAINING

Reducing the complexity index can have a marked impact on the shape of the repair time distribution. In addition, the value of λ effectively shifts the entire family of F(t) curves along the horizontal axis. Thus, if we can assume that the parameter "a" remains relatively invariant among various types of equipments for a given maintenance environment, at least as a first approximation, the values of I and λ can be established as design goals to insure that mean repair time falls below some criterion value. These design goals can then be achieved through appropriate circuit partitioning, modularization, and test point access.

In addition, for a given equipment configuration it may be found that the level and type of training may produce differential effects on both parameters λ and a.

While the estimated values of all the parameters appear reasonable, a controlled experiment would be required to validate the Improved Hypothesis and to establish specific relationships among I, λ, a, level of expertise of the repairmen, type of maintenance environment, and mean active repair time. The laboratory setup of Rouse et al (1979b) would lend itself to such an experiment if distractions such as those encountered under field conditions could be introduced.

CONCLUSION

The Improved Hypothesis appears to provide a useful explanation of the underlying man-machine interaction process in diagnosing electronic equipment malfunctions. It provides a direct basis for predicting not only mean active repair time but the entire "tail" of the cumulative frequency distribution for active repair time, using a minimum of assumptions. It is consistent with the notion that diagnostic time can be expected to vary radically with interconnectivity among components; i.e., with circuit and equipment complexity. Since the parameters I and λ are circuit-design related, it can be used at the earliest stages of design as an additional input into the circuit partitioning process. It appears applicable to prediction of design impacts at line, field, and depot maintenance levels. The theory may also be useful from a training standpoint. Training of various types (e.g., generalized problem solving versus generalized troubleshooting versus equipment-specific troubleshooting) can be assessed with respect to their differential effect on the parameters a and λ if the maintenance environment is held constant. Depending on the results of such an assessment, training techniques and associated costs could be influenced. Further experimental verification under controlled conditions appears justified, considering the potential impact on military standards for maintainability design and prediction as well as on training.

ACKNOWLEDGEMENT

I am indebted to Mr. David Gootkind of The MITRE Corporation for the mathematical developments, for computational assistance, and for stimulating discussions of the relationship between theory and data.

REFERENCES

Kennedy, H. and Retterer, B.L., 1960, "Maintainability Measurement and Prediction", in Electronic Maintainability, Vol. 3, F.L. Ankenbrandt, ed., Engineering Publishers, N.J.

Military Standardization Handbook, No. 472, 1966, Maintainability Prediction, MIL-HDBK-472.

Rouse, W.B., 1978a, "Human Problem Solving Performance in a Fault Diagnosis Task", IEEE Transitions on Systems, Man and Cybernetics, Vol. SMC-8, Nc. 4, pp. 258-271.

Rouse, W.B., 1978b, "A Model of Human Decisionmaking in a Fault Diagnosis Task", IEEE Transactions on Systems, Man and Cybernetics, Vol. SMC-8, No. 5, pp. 357-361.

Rouse, W.B., 1979a, "A Model of Human Decision Making in Fault Diagnosis Tasks That Include Feedback and Redundancy", IEEE Transactions on Systems, Man and Cybernetics, Vol. SMC-9, No. 4, pp. 237-241.

Rouse, W.B., and Rouse, S.H., 1979b, "Measures of Complexity of Fault Diagnosis Tasks", IEEE Transactions on Systems, Man and Cybernetics, Vol. SMC-9, No. 11, pp. 720-727.

Wohl, J.G., 1961, "Research Data on Maintainability", IRE Transactions on Human Factors in Electronics, Vol. HFE-2, pp. 112-113.

MODELS OF DIAGNOSTIC JUDGMENTS

Berndt Brehmer

Department of Psychology
University of Uppsala
Box 227
S-751 04 Uppsala, Sweden

The diagnostic process may be divided into three stages: data collection, information integration, and feedback (Kahneman & Tversky, 1979). The present paper is concerned only with the second of these stages: the information integration stage. This is not because the other stages are unimportant, but because most of the research on human judgment has been aimed at this stage. One reason for this may be that it provides a particular challenge, for whereas the other two stages in the diagnostic process are overt and public, the information integration stage tends to be covert and private: the diagnostician often is not able to describe how he arrives at at his judgments (e.g., Hoffman, 1960).

The challenge to research, then, is this: How do we describe a mental process of which the person himself is not aware? Fortunately, there is a simple solution to this problem. This solution takes as its point of departure that although a person may not be able to describe how he makes his judgments, he is nevertheless able to make them when he is given the relevant information. If we know what judgments a person has made, and the information upon which the judgments are based, it is possible to construct a model that relates the judgments to the input information. If our model is successful, it will produce the same judgment as the person does when given the same information. The model is then a simulation of the mental process, and we will be

---

Preparation of this paper was supported by grants from the Swedish National Defense Institute and the Swedish Council for Social Science Research.

able to learn about this covert and private process by studying the overt and public simulation.

The problem, then, is to chose an appropriate model for this simulation. Psychologists studying human judgment processes have found that linear statistical models, such as multiple regression, are useful in this context. This is not the place to discuss the reasons for this particular kind of models; I have done so at length elsewhere (Brehmer, 1979). It is sufficient to point out two important features of these models. First, they can be used for describing systems that contain uncertainty. Thus, they can be used, not only to describe the mental process involved in diagnosis but also the diagnostic tasks. That is, the same model can be used to describe both the person and the task, and this enables us to compare them. This is important, because it makes it possible to determine the extent to which the person uses the information available for his judgments in an adequate way. This reveals how well the person performs, and will suggest how he would have to change to improve.

A second advantage of linear models is that they describe the process in terms that are readily understood by the person being analysed. This is, of course, a necessary condition if we want the person to use the information to change his mental system to improve his judgments.

Linear models have now been used to analyse diagnostic judgments in a wide variety of circumstances. The subjects studied include stock brokers, clinical psychologists and physicians (see, e.g., Slovic & Lichtenstein, 1971 for a review). In an as yet published series of studies, we have used linear models to study a variety of judgments by psychiatrists, psychologists and nurses.

These studies are carried out in a series of standardized steps. The aim of the first step is to set up the judgment task. In this step, the subjects for the study are interviewed to ascertain what information they require for making the judgments under investigation. For example, if the purpose of the study is to investigate how the psychiatrists make judgments about the risk that a patient will commit suicide, each psychiatrist will first be asked what things he or she will need to know about the patient to decide whether or not the patient will commit suicide. These interviews will yield a list of symptoms, most of which are usually common to all of the subjects, although some, of course, may be idiosyncratic. In the second step, a subset of the symptoms on the list are selected. If the purpose of the study is to analyze the group of subjects, this list may comprise those that are common to all of the subjects. It is, however, also possible to work with each individual subject's list of symptoms, doing all analyses on a single subject basis. Whatever the approach the next step is to construct a set of "patients" by combining the symptoms according to some set of principles. This step is critical for the success of the study in two respects. Firstly, the cases resulting from the combination of symptoms must be credible. Therefore, the set of cases should be screened by the prospective subjects. Secondly, the method used for constructing the set of cases determines what analyses may be performed. If the symptoms are combined orthogonally, it is possible to use analysis of variance procedures, but if the symptoms are correlated, multiple regression procedures must be used.

This has some consequences when it comes to testing for deviations from additivity in the combination of the information from the various symptoms (see Anderson, 1968), and for the indices of weight that are used, see below. The "patients" are then given to the subjects who make a judgment about each case. In most studies, subjects judge each case twice, so that it becomes possible to ascertain the reliability of their judgments. The judgments are then analysed for each subject separately by analysis of variance, multiple regression or some other variety of the linear model. This analysis then yields information about five important aspects of the judgment process:

1. The symptoms actually used by each subject. This is shown by the presence or absence of significant main effects for the symptoms in the analyses.

2. The relative weights given to the symptoms. This is shown by the weight indices calculated, such as the beta weights in the multiple regressions, or the ω^2 calculated from the analysis of variance results.

3. The form of the functions relating the judgments to each symptom, i.e. whether a symptom is used in a linear way, so that the higher the value of symptom, the higher the judgment, or whether it is used in a nonlinear way, e.g., so that there is an optimum value, which leads to a high judgment, while departures from this optimum value in either direction leads to lower judgments. This information can, of course, be obtained only when the symptoms are quantitative.

4. How the subject integrates information from different symptoms into a judgment. He may integrate the information additively, e.g., by adding or averaging it, or configuratively, i.e. in such a way that the weight given to one symptom varies with the value of another symptom. Alternatively, he may use a multiplicative, rather than additive, rule. This aspect of the process is assessed by first examining the analysis results for significant deviations

from additivity, and if there are such deviations, determining the exact form of the non-additive rule used by the subject.

These four aspects of the judgment process, i.e., what symptoms are used, their relative importance, the functional relations between each symptom and the judgments, and how the rule used for integration information from the symptoms into unitary judgments, are aspects of the process that are readily understood by a subject, so these aspects of the process are thus easily communicated to the subject. This, however, is not true of the fifth aspect of the process revealed by this kind of analysis:

5. The reliability, or consistency, of the process. This is the extent to which the subject uses the same rule from case to case. It is shown by the error variance in his system as estimated, for example, by the test-retest reliability of his judgments, i.e., the correlations between the judgments made at two different occasions, or by the residual, or error variance. Whereas the analysis of actual judgment regularly shows that the process is inconsistent, people seem to have no subjective awareness of this.

The results of studies using this methodology are easy to summarize because the results are essentially the same, regardless of what kind of subjects have been studied or what kinds of judgments have been investigated. There are four main results.

The first of these is that the form of the judgment process tends to be very simple. It is simple in two aspects. Firstly, very little information seems to be used for the judgments. A person may ask for ten different symptoms for making his judgments about suicide risk, but the results of the analysis of his actual judgments then shows that he uses very few, usually only two or three. Secondly, the process is simple in that it tends to be additive rather than configural. Few studies have yielded any deviations from additivity, and when such deviations have been found, they are usually small and of little systematic importance in the process.

A second result from studies of judgment is that the process is generally found to be inconsistent. The subjects in these studies usually do not seem to use exactly the same rule from case to case, and when the same case is presented a second time, the judgment may differ considerably from what it was the first time.

A third finding in judgment studies is that there are wide individual differences in judgments. The correlation between the judgments made by two subjects for the same cases is often quite low, even though the subjects are experts with years of experience in their field. In part, this disagreement in judgments is due to lack of consistency; when the processes that produce the judgments are not perfectly reliable, the judgments cannot, of course, be perfectly correlated. However, not all of the disagreement is due to lack of reliability. There are also systematic differences between subjects in many cases. They may differ both in which symptoms they use, and in the relative weights they give to the symptoms they use.

A fourth important finding is that the subjects are not very good at describing how they make their judgments. When a model of the process is constructed from subjective descriptions of the process, the judgments produced by this model usually do not correlate very highly with those actually made by the subject (see, e.g., Hoffman, 1960).

Two results are of particular importance here. The first is that the process lacks consistency. Inconsistency of this kind seems to be a general feature of cognitive faced with tasks that contains uncertainty. It is not a simple matter of lack of reliability, for the degree of consistency is systematically related to the nature of the judgment task. Two aspects of the judgment task are especially important: its predictability and its complexity.

As for predictability, as defined, for example, by the multiple correlation between the cues and the variable to be judged, the results show that the degree of reliability in the subject's cognitive system varies monotonically with the predictability of the task. Thus, that the higher the predictability of the task, the more consistent the subjects tend to be. This is true of various laboratory tasks, as well as of judgments collected from experts performing tasks with which they have years of experience (Brehmer, 1976). The explanation for this result is not known, but its implication is clear: the characteristics of human judgment processes depart in systematic ways from what is required for optimality according to statistical decision theory.

As for complexity, the results suggest that as the complexity of the judgment task increases, the consistency decreases. Thus, when the subjects are required to use information from few symptoms, they are more consistent than when they have to use many symptoms, and when the task requires the subjects to use nonlinear rules, they are less consistent than when they have to use linear rules (Brehmer, 1971). To some extent, the subjects improve with training, but these effects seem to be rather limited (Brehmer, 1979b). These results have been interpreted to mean that, in some respects, judgment is like a motor skill. Just because a person knows what rule to use for his judgments, it is not certain that the judgments will actually follow this rule, and the subject will, of course, not detect that his judgments do not follow the rules he intend to use, (Brehmer, Hagafors, & Johansson, 1980).

The second general result of importance in this context is that there are wide individual differences, also among experts. At first, this may be surprising, because we would expect that experts, having essentially the same kind of experience, ought to have learned about the same things. Thus, they ought to make their judgments in the same way. Analyses of the circumstances under which experts have to learn show, however, that the possibilities of learning anything from the kind of the experience provided in these settings is rather limited (Brehmer, 1980). In short, the argument is that to learn from experience in these settings, people have to have hypotheses relevant to their task. These hypotheses must be statistical hypotheses, because the tasks often contain a large measure of uncertainty. But people generally do not employ these kinds of hypotheses, and they are therefore not able to profit from experience as they should. When subjects are given statistical hypotheses, their ability to use them is severely limited, presumably because they cannot process the amount of information needed to test these statistical hypotheses in an adequate way (Brehmer,1979c). To learn the same thing from experience, people would have to have adequate hypotheses, and use them correctly. Since they do not, it is no longer surprising that they do not learn the same thing.

These and other results (see Brehmer, 1980 for a review) show that when the task has some complexity, and when it requires the subjects to handle relations which contain some uncertainty, experience does not guarantee good judgment. The alternative would be to teach people to make good judgments. However, great problems are involved in trying to teach judgment. We have already remarked that people seem to have only limited ability to describe how they make judgments. Consequently, it is very hard for an expert to teach a novice how to make judgments in the same way he does. The problems are further exacerbated because the teacher will not understand what the pupil is doing. Consequently, the pupil is likely to be left to his own devices, and he will have to acquire whatever expertise he can get from his own experience. This means that he will have to learn on the basis of largely unintelligible remarks from his older and more experienced colleagues, and from the feedback he may receive from the system he is trying to learn. Such feedback, when provided at all, may be very rare, thus providing little information. Furthermore, the feedback will contain error, thus making it hard to use for learning. These considerations show that to teach judgment, we cannot rely on the traditional approaches to teaching. A new approach is needed.

A NEW APPROACH TO TEACHING JUDGMENT

The most important problem in teaching judgment is to provide adequate feedback to the learner. Since diagnostic tasks are often probabilistic in nature, the feedback provided contains error, and in addition, the feedback actually provided is often very infrequent and it may occur after a considerable delay, it may not be of much use. Furthermore, it is not particularly informative even under the best of circumstances. This is because it gives only indirect information about what is to be learned. The feedback usually informs the learner only whether he was right or wrong, or,at best, about the direction of his error. It does not tell him why he made an error. Therefore, he must use the feedback information to infer why his judgment was correct or why it was not correct. This may lead to problems. A typical diagnostic task requires the person to learn relations between symptoms and judgments, and single instances of outcome feedback telling the subject that he was wrong does not inform the learner how he should change the relations between the symptoms and his judgments. If the task is probabilistic, an error may not even mean that he should change the relations between the symptoms and his judgments. Having a teacher may not help much because the teacher faces the same problem as the learner: he has to infer what was wrong with what the pupil did, since the pupil may not be able to tell him exactly what he did.

A second problem in teaching judgment is to create a good description of the task to be learned. For many judgment tasks, there may exist no objective account of the tasks; all the relevant knowledge about the task is in the heads of experts, and the experts may not be able to describe what they know in such a way that it can be used to teach a person to make judgments. Consequently, to obtain the knowledge needed, it may be necessary to analyse judgments of experts to determine what they know. An approach to the solution of this problem has already been outlined in this paper. We know turn to problem of providing feedback.

As a step towards the solution of this problem, we (Hammond & Brehmer, 1973) have developed a computer based system. This system presents information to the learner in the form of a series of cases. For each case, the learner makes a judgment. After a sufficient number of cases, the system performs an analysis of the judgments of the learner, and then displays the results graphically on a screen. These displays allow the learner to compare the characteristics of his cognitive system with those of the task. Thus, the system will display the relative weights given to the symptoms by the learner next to the weights he should use, so that the learner is informed of any discrepancies. Furthermore, the system displays the functional relations between each cue and the judgments together with the correct functional relations, so that the learner may compare his way of using the cue with the correct way. The system also provides information about the consistency of the judgments. Furthermore, it is also possible to display individual cases for which the learner has made especially grave errors for discussion and analysis with a teacher.

The system, then, provides exactly the kind of information needed for learning a judgment task. It does not require the learner to infer how he should change, but shows exactly what changes are needed in the parameters of his cognitive system. As might be expected, this system leads to rapid learning also when the task has considerable complexity (e.g., Hammond, 1971).

The actual experiments in using the system for training in practical situations is, as yet, rather limited, although some attempts have been made. On theoretical grounds, the system has considerable promise, and as it is tried out in new circumstances, we will know more about its practical usefulness.

APPLICATIONS TO OTHER SYSTEMS

The general approach to diagnostic judgment described in this paper is of course, developed mainly for handling the problems related to psychological and medical diagnosis. It has not been developed for the problem of trouble shooting or assessment of mechanical and electronic systems. There is at least one important difference between the diagnostic problem facing a psychiatrist and that facing an engineer. This is that the system with which the engineer is concerned has been created by other engineers. The system facing the psychiatrist on the other hand, has been created by forces that are unknown to him. This makes a difference. For the electronic or mechanical system, plans and specifications are available, so that it is possible, at least in principle, to find whatever is wrong with the system when it is not functioning properly. For the task facing the psychiatrist, there is no such guarantee that he will find the real problem with the patient. Thus, the psychiatrist works under genuine uncertainty, but the engineer does not.

However, it seems that the difference between the task facing the engineer and the psychiatrist may be diminishing as the complexity of the technical systems increase. As these systems become more complex (say on the order of a nuclear plant or a computer) it no longer seems to be possible to predict the systems perfectly, and it becomes harder and harder to decide when the system is functioning properly. The task facing those who take care of these systems thus seems to approach that facing psychiatrists, or others who take care of systems not created by man. This suggests that the cognitive processes of engineers in these tasks would become similar to those of physicians and psychologists, and that the general approach developed for the study of diagnostic judgment in these areas would become applicable also to the tasks facing the engineer.

REFERENCES

Anderson N.H. A simple model for information integration. In R. P. Abelson, et al. (Eds.), Theories of cognitive consistency: A source book.Chicago: Rand McNally, 1968.

Brehmer, B. Subjects's ability to use functional rules. Psychonomic Science, 1971, 24, 259-260.

Brehmer, B. Note on clinical judgment and the formal characteristics of clinical tasks. Psychological Bulletin,1978, 83, 778-782.

Brehmer, B. Preliminaries to a psychology of inference. Scandinavian Journal of Psychology, 1979, 20, 193-210 (a).

Brehmer B. Effect of practice on utilization of nonlinear rules in inference tasks. Scandinavian Journal of Psychology, 1979, 20, 141-149(b).

Brehmer, B. Note on hypotheses testing in probabilistic inference tasks. Scandinavian Journal of Psychology,1979, 20, 155--158(c).

Brehmer, B. "In one word: Not from experience". Acta Psychologica, 1980. In press.

Brehmer, B., Hagafors, R., and Johansson, R. Cognitive Skills in judgment: Subjects' ability to use information about weights, function forms, and organizational principles. Organizational Behaviour and Human Performance,1980. In press.

Hammond, K.R. Computer graphics as an aid to learning. Science, 1971, 172, 903-908.

Hammond, K.R., and Brehmer, B. Quasi-rationality and distrust: Implications for international conflict. In L. Rappoport and D. Summers (Eds.). Human Judgment and Social Interaction. New York: Holt, Rinehart, and Winston, 1973.

Hoffman, P. J., Paramorphic representation of clinical judgment. Psychological Bulletin, 1960, 57, 116-131.

Slovic, P., and Lichtenstein, S. Comparison of Bayesian and regression approaches to the study of information processing in judgment. Organizational and Human Performance, 1971, 6, 649-744.

MODELS OF MENTAL STRATEGIES IN PROCESS PLANT DIAGNOSIS

Jens Rasmussen

Risø National Laboratory
DK-4000 Roskilde, Denmark

INTRODUCTION

It has long been recognised that operators' misidentification of abnormal system states plays a significant role in major system break-downs (see e.g. Cornell, 1968, Rasmussen 1968) and the problem has been subject to much discussion since the Three Mile Island incident. Consequently, there has been a rapid increase in efforts to design computer-controlled man-system interfaces which are effective in supporting operators' identification of the actual operating state of the system.

Operators' actions upon a plant must always be based on an identification of the operational state of the system. To identify the state means to give it a name, to label it in terms which, depending upon the situation, will refer to the functional state of the system; to the cause of this functional state; or directly to the related control action. Such state identification is always present as an important element of operator activities, e.g. to check whether the system is ready for intended operations or to confirm that actions have brought the system to the proper target state. The boundary between such routine identification and identification of abnormal states, i.e., diagnosis, is ill-defined and depends very much on the operators' prior experience.

Identification of a system state which has not previously been experienced must be performed as an identification of the actual, internal anatomy and function of the system from the observed behaviour. This is a very complex task which, however, in process plant control is simplified by the fact that the plant will be known to have functioned properly, and therefore the identification can result from a search to locate a change with reference to knowledge of the normal state or function. The term diagnosis for this kind of identification is in some respect misleading. The general meaning of "diagnosis" is a determination of the cause of some observed symptoms, which is not necessarily the case in our context. The ultimate purpose of diagnosis in process plant control is to link the observed symptoms to the actions which will serve the current goal properly. Depending upon the situational context, the object of the diagnostic search may vary: to protect the plant, search may concentrate on patterns of critical variables related to stereotyped safety actions; to compensate for the effect of the change, search for alternative functional paths bypassing the effect of the change will be appropriate; to restore normal state, search in terms of the initiating physical change - the cause - is necessary. In consequence, the diagnostic task implies a complex mental process which is very situation and person dependent.

To support the design of an effective man-machine interface, a description of the operators' mental processes is necessary.

DESCRIPTIONS OF OPERATORS' DIAGNOSTIC STRATEGIES

The mental activity of operators in a diagnostic task is very difficult to analyse and describe in detail due to the strong dependence on details of the actual situation as well as the operators' skill and subjective preferences. Fortunately, to support systems design we do not need detailed process models of the mental activities which <u>are used</u> by the operators. System design must be based upon higher level models of the structures of effective mental processes which the operators <u>can use</u> and their characteristics with respect to human limitations and preferences so that operators can adapt individually and develop effective strategies. Rather than descriptions of the course and content of actual mental processes, we need descriptions of the structure of possible and effective mental processes.

In the system design context, a description of mental activities in information processing concepts is preferable since it is compatible with the concepts used for design of the data processing and control equipment. For this purpose, the human data processes can be described in terms of data, models and strategies. <u>Data</u> are the mental representations of information describing system state which can be represented on several levels of abstraction which in turn specify the appropriate information coding for the data presentation. <u>Models</u> are the mental representation of the system's anatomical or functional structure which can be related directly to the appropriate display formats. <u>Strategies</u> here are taken as the higher level structures of the mental processes, and they relate goals to sets of models, data and tactical process rules. The different possible strategies must be characterized in terms referring to mental limitations and emotional preferences in order to be able to judge the effect of subjective and emotional performance criteria upon the actual performance. From a systems design view-point, a separation of the descriptions of the functional, data processing aspects and the subjective value aspects of the performance is advantageous, but identification of the proper factors for describing their interaction is necessary. See fig. 1.

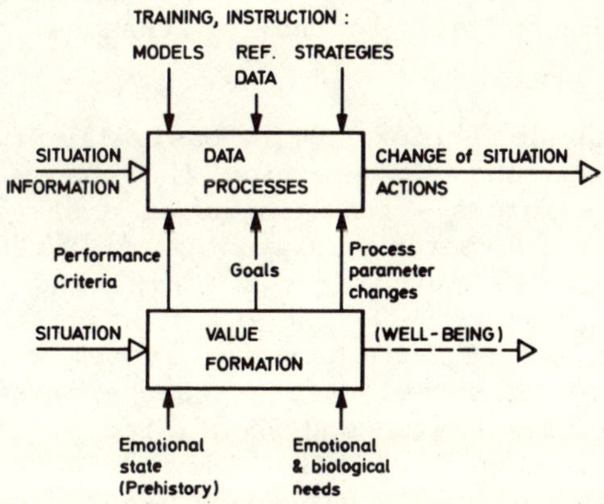

Man as a system component. Design of systems depends on descriptions of man and machines which are compatible in structure and concepts. For automated systems, information processing concepts are natural choices for integrated functional design. Functional properties of man depend, however, on emotional features of work situation.

System as man's work environment. Consideration during design of subjective values and preferences demands a description of work situation in psychological terms, relating features of the situation to subjective values and emotional states.

Two separate descriptions are then needed for compatibility with engineering and psychology. Parameters and variables suitable for description of their interaction must be found. Descriptions of human mental functions typically depend on situation analysis and information process models. Descriptions of subjective values and preferences typically depend on factor and scaling analysis and emotional state models.

Fig. 1. Descriptions of related human functions and values.

Within this framework the actual performance of operators can be described and analysed using tactical rules to describe the control of the detailed processes within a formal strategy as well as the causes of the frequent shifts between the formal strategies which take place in response to variations in the fit to the immediate performance criteria. During real life performance such shifts occur frequently when difficulties are met in the current strategy or information is observed which indicates immediate results from another strategy. Since the shifts are controlled by detailed person and situation dependent aspects, they result in a very individual course and content of the mental processes in each case and, consequently, in great difficulties when generalizing is attempted from detailed mental process models. Our approach has been to derive descriptions of the structure of formal strategies from the analysis of a number of verbal protocols from varying diagnostic task situations. This study has led to a kind of typology of diagnostic strategies and an identification of some of the performance criteria guiding the operators' choice of strategy.

TYPOLOGY OF DIAGNOSTIC SEARCH STRATEGIES

In general, the diagnostic task implied in supervisory systems control is a search to identify a change from normal or planned system operation in terms which refer the controller to the appropriate control actions. The controller can in this context be a man and/or an automatic control system. Several different strategies are possible for this search and, within the individual strategies, the object of search and the domain in which the search is performed will depend on the specific situation and immediate intention of the controller (Rasmussen, 1978).

The diagnostic search can be performed in basically two different ways. A set of observations representing the abnormal state of the system - a set of symptoms - can be used as a search template in accessing a library of symptoms related to different abnormal system conditions to find a matching set. This kind of search will be called symptomatic search. On the other hand, the search can be performed in the actual, maloperating system with reference to a template representing normal or planned operation. The change will then be found as a mismatch and identified by its location in the template. Consequently, this kind of search strategies has previously been named topographic search (Rasmussen and Jensen 1974). The difference between the two kinds of search procedures is related to a basic difference in the use of the observed information. Every observation implies identification of an information source and reading the content of the message. By symptomatic search, reference to the identity of system state is obtained from the message read; by topographic search, reference is taken from the topographic location of the source, while the messages are subject to good/bad judgements which are used for tactical control of the search.

Topographic Search

The topographic search is performed by a good/bad mapping of the system through which the extent of the potentially "bad" field is gradually narrowed down until the location of the change is determined with sufficient resolution to allow selection of an appropriate action. The domain in which the search is performed will vary. The search can be performed directly in the physical system but, in most cases, the search is a mental operation at a level of abstraction which depends upon the immediate goal and intention of the controller and upon the form of the reference map or model available. Also the resolution needed for the final location depends upon the actual circumstances.

The topographic strategy is illustrated by the information flow graph of fig. 2. The main elements of the strategy which will be considered in more detail are the model of system used to structure the search; the kind of data used to represent the actual, failed plant state and the normal, reference state; and finally, the tactical process rules used to control the search sequence.

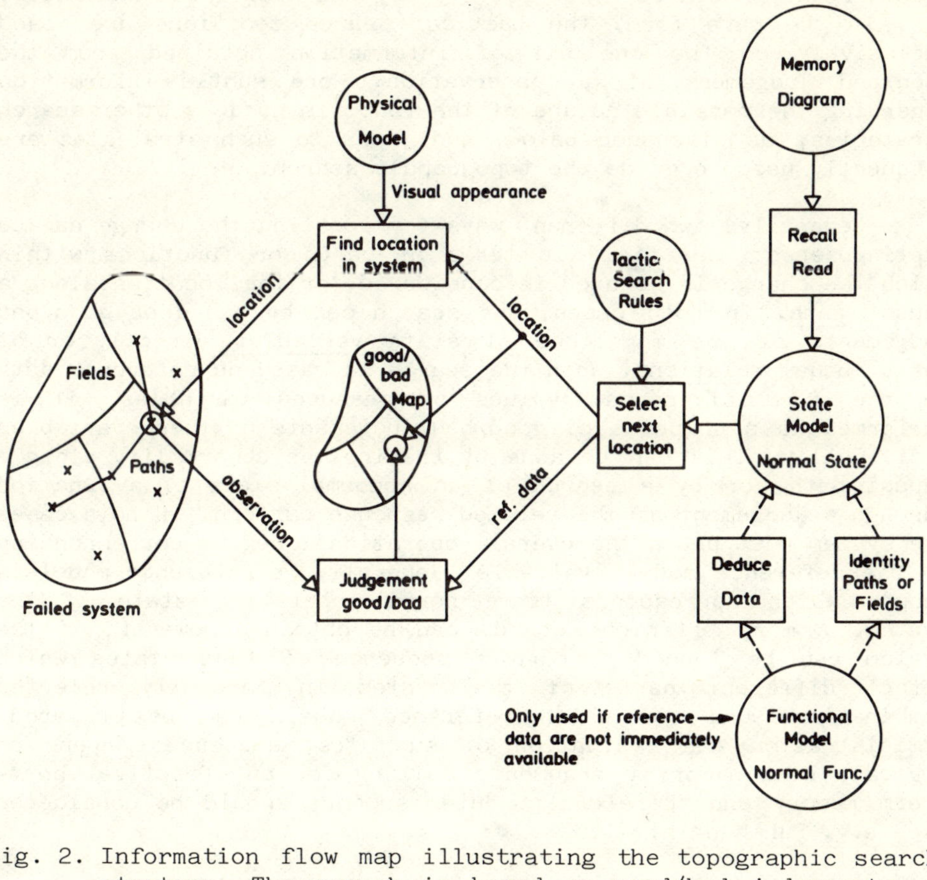

Fig. 2. Information flow map illustrating the topographic search strategy. The search is based on good/bad judgements of variables along a path or of patterns related to a field.

The topographic search is performed as a good/bad mapping of the system which results in a stepwise limitation of the field of attention within which further search is to be considered. The search depends on a map of the system which gives information on the location of sources of potential observations for which reference information is available for judgements. The map is a model which may identify the potential sources of observations relative to the topology of the physical system itself, of its internal anatomical or functional structure, or of its external purposes. The search sequence is based on a set of, often heuristic, rules serving to limit the necessary field of attention. If different external functions can be related to separate internal parts or subsystems, a good/bad scan of external functions effectively identifies the internal field for further search. If a faulty input/output relation is found, the related causal route should be searched, e.g. by the half-split heuristic, etc. In the pure form, the tactical search decisions are based exclusively on the one bit of information obtained from the good/bad judgement of an observation. More subtle information regarding the possible nature of the fault is used by other search strategies, as discussed below, and jumps to such strategies are frequently used to guide the topographic search.

Generally, two different ways for locating the change can be distinguished: the field in terms of parts or functions within which the change is located is considered, or the location along a causal path. In both cases, the search can be based on good/bad judgements of the magnitude of state variables directly or of their mutual relations. When the search is based on reference data in the form of normal values of measured variables, it is performed as a sequence of good/bad judgements of these variables individually. This can be done by tracing the abnormality along a causal path, or by a search of an abnormal state of a function through a judgement of the related response pattern. In both cases the system must be in the overall operational regime corresponding to the reference model available. Generally, a reference model is chosen which corresponds to a normal operating state of the system. A more efficient search can be obtained however, if the system can be forced through a sequence of test states which affect different parts of the system in carefully selected combinations and for which reference models can be prepared. Administration and evaluation of such test sequences depend on logical, combinatorial arguments calling for an effective short-term memory and therefore computer support should be considered (see e.g. Furth et al. 1967).

Being based on a long string of simple judgements, this search strategy based on variables individually is effective for human diagnosis if supported by a suitable map with reference data. However, the tactical decisions depend upon assumptions about the direction of causality which may break down due to intrinsic and structural feedback effects and thus lead to a cul-de-sac. In this respect, strategies which are based on judgement of the relationships in sets of observations are superior. If boundaries can be identified around system parts or functions for which the input-output relations are subject to constraints from known laws and relation, such as mass or energy conservation, component characteristics, etc., then the state of the field within the boundary can be found from a test of these relations. The reference for the search will then be more independent of feedback effects and of variations in operational level than is the case when magnitudes of variables are judged directly. This is because judgements are performed on data relations representing system properties rather than magnitudes of variables representing system states.

The tactic rules control the use of a map of the system. This map may depict the structure of the system at several levels of abstraction (Rasmussen, 1980) and the proper level for the search depends upon the goal or intention of the diagnostician, as discussed above. However, the most effective level may also change during the search. Our verbal protocols, e.g., recorded by electronic engineers during trouble-shooting of computer systems, repeatedly demonstrate reference to the structure of external functions at the beginning of a search while the reference subsequently moves through the levels of information flow structures; of elementary information processing units; of electrical signal flow paths; and finally ends down at the level of electronic components, their characteristics and physical locations. As attention is "zooming in" during search, there is a simultaneous need to move to more concrete, physical levels of search. In case of industrial process plant, a corresponding shift of level of abstraction can be effective, if the role of information flow structures in the above example is replaced by the use of mass and energy flow maps as a representation of the overall system function. (For details, see Lind 1980).

It is important for effective topographic search that observations or measured data are available in a form corresponding to the level of the reference model in use. This means that physical variables measured in the system are only directly compatible with the needs at the lowest level of abstraction; for judgements at higher level of abstraction corresponding to information or energy flow, the measured data must be transformed accordingly. If a human diagnostician must do this himself, a significant mental load is added to the task. Efficient search depends upon availability of all the information accessible in the system in codes compatible with the map at each of the levels used. For example, for high level search in terms of energy and mass flow structures, all available information such as physical variables (temperatures, pressures, flows), states of valves and switches, and status of supplies, should be converted to flow and storage level information exclusively. This means that the same measured data will be needed in several different combinations and transformations; a need which can be matched perfectly well by computers.

The information available in observations is used rather uneconomically by topographic strategies, since they depend only upon good/bad judgements. Furthermore, they do not take into account previously experienced faults and disturbances. Therefore, switching to other strategies may be necessary to reach an acceptable resolution of the search or to acquire good tactical guidance during the search. However, the topographic search is advantageous because of its dependence upon a model of normal plant operation - which can be derived during design or obtained by data collection during normal operation. Therefore, consistency and correctness of the strategy can be verified and, since it does not depend on models of malfunction, it will be less disturbed by multiple or "unknown" disturbances than strategies based on disturbance symptoms.

Symptomatic Search

Symptomatic search strategies are based on the information content of observations to obtain identification of system state, instead of the location of the information source in a topographic map. The search decisions are derived from the internal relations in data sets and not from the topological structure of system properties. In principle, a search is made through a set of abnormal data sets, "symptoms", to find the set which matches the actual observed pattern of system behaviour. The reference patterns can be collected empirically from incidents of system maloperation or derived by analysis or simulation of system's response to postulated disturbances. Furthermore, reference patterns can be generated on-line if the controller has a functional model available which can be modified to match a current hypothesis about the disturbance.

When the diagnosis is performed in the data domain by a search through a library of symptom patterns, it has no logical relation to system function. The result is directly the label of the matching symptom pattern which may be in terms of cause, effect, location or appropriate control action directly. Depending upon the structure of the controller and its memory, the search can be a parallel, data driven pattern recognition, or a sequential decision table search as illustrated by fig. 3.

Pattern recognition plays an important role in human diagnosis; it can efficiently identify familiar system states and disturbances directly, but it is also used frequently during e.g. topographic search to guide the tactical decisions. Recognitions are then typically based on more fuzzy or general reference symptoms in terms of generic fault patterns referring to types of functions or physical parts, such as noise characteristics, instability, or forms of non-linearity.

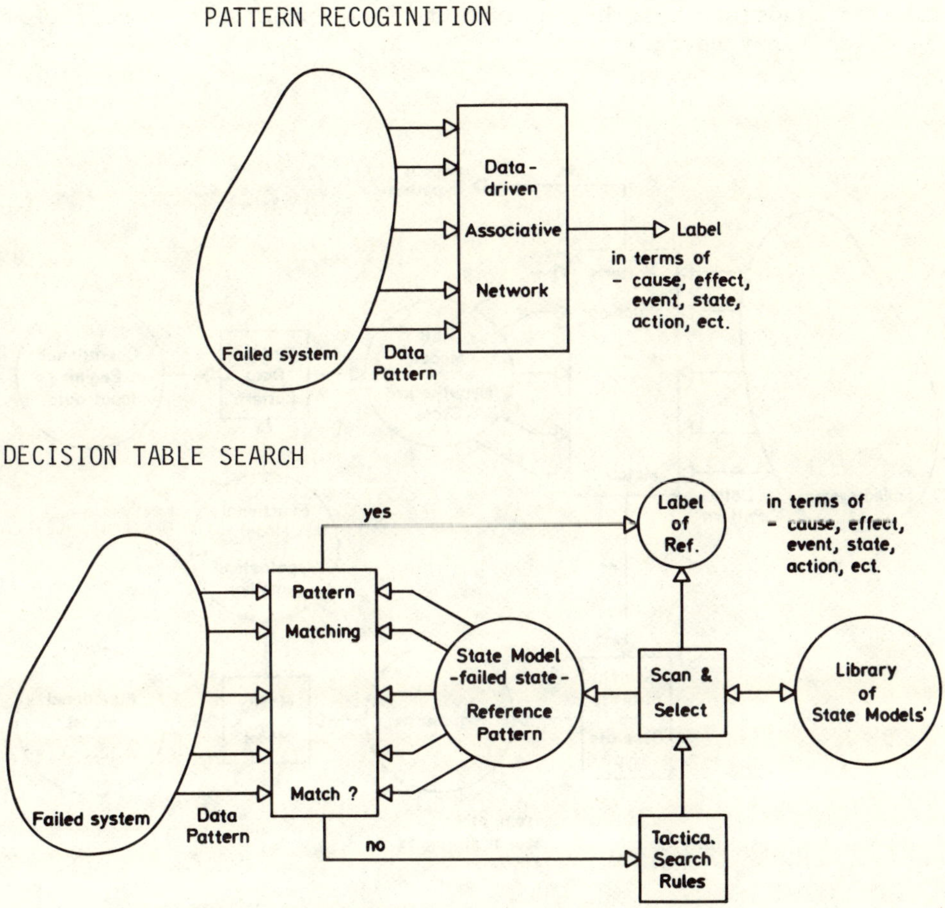

Fig. 3. Information flow maps for symptomatic diagnosis based on pattern recognition or search through a library of symptoms.

Decision table search depends upon a set of tactical rules to guide the search which can be based on probability of occurrence, a hierarchical structuring of the attributes (like Linné's generic system for botanical identification), or functional relations, stored as fault trees, etc. Human diagnosticians probably would use decision tables for verification of more ambiguous recognitions. Decision tables have been used for plant monitoring by Berenblut et al. (1977) and Lihou (this volume).

If a search is based on reference patterns generated "on-line" by modification of a functional model in correspondence with a postulated disturbance, the strategy can be called search by hypothesis and test. The efficiency of this search depends upon the tactics of generating hypotheses. Typically, in human diagnosis, hypotheses result from uncertain topographic search or fuzzy recognitions.

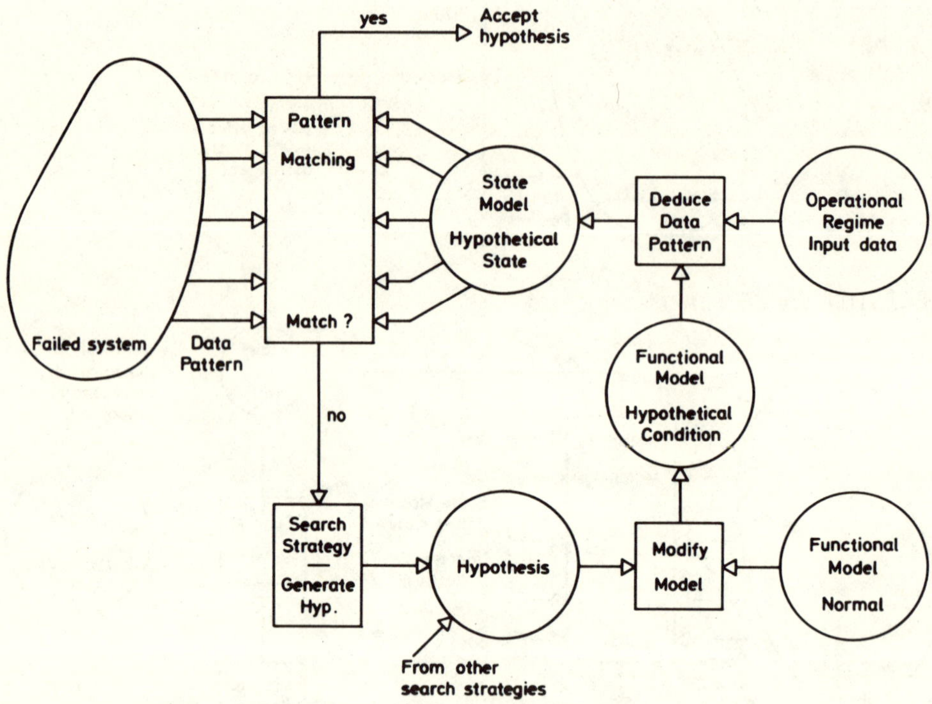

Fig. 4. Information flow map for symptomatic search by hypothesis and test. The figure illustrates conceptual test. In practice, the test may be performed by correcting the system and test for normal performance instead of modifying the model as shown.

Symptomatic search is advantageous from the point of view of information economy, and a precise identification can frequently be obtained in a one shot decision. One serious limitation will be that a reference pattern of the actual abnormal state of operation must be available, and multiple faults and disturbances must be considered. This means that reference sets must be prepared by analysis or recorded from prior occurrences. Or, the reference sets can be generated on-line by means of a functional model of the system which can be modified on occasion to simulate the abnormal system state in accordance with the current hypothesis.

Simulation of process plant malfunctions in this context puts some special demands upon the necessary functional model. The model must be able to simulate overall system states outside the operating ranges which normally are considered in models used for system design and control studies. Such models are typically structured as a network of relations amongst physical variables, and these relations must be known for the relevant states of maloperation. Furthermore, to update the model according to a postulated fault, the correspondence between the physical, material change which constitutes the fault and the effect in terms of changes of structure and relations within the model must be known. Since the demand for accuracy in this kind of simulation is rather low, an acceptable solution of these two conditions will generally be to discretize the quantitative variables into a few ranges with reference to normal or critical values, and to arrange the corresponding relations into sets labelled in states of physical parts or components. The result is a model which resembles the type normally used for human cause-and-effect reasoning; a model structured in objects which have properties and interact by events. At present, this is typically the model used for automatic alarm or disturbance analysis introduced to support operators in process plant diagnosis (Dahll et al., 1976, Taylor et al., 1977, Bastl et al., this volume).

The mental load in using a model of the system's abnormal state to generate a set of symptoms to test a hypothesis is so high that very often a human diagnostician will prefer to test the hypothesis by correcting the system in accordance with it, and to test whether its response pattern hereby turns into a normal or desired pattern. This strategy is generally effective for trouble-shooting in a workshop environment and to test correcting actions for minor disturbances in a plant control room. For major disturbances, however, the strategy will not be effective, and support of operators by computers during test of hypothesis by model modification and functional arguments must be considered.

The intrinsic differences between the various diagnostic strategies and the effect of shifts in strategies used by subjects appear typically in diagnostic experiments.

Shepherd et al. (1977) studied the effects of different ways of training operators for plant diagnosis and found that operators trained by the rules obtained from experienced operators were superior in the diagnosis of not previously encountered faults compared with operators trained in plant theory. These in turn were superior to operators trained by practising diagnosis. The results are reproduced by Duncan (this volume), fig. 6. These differences can be readily explained, since the different training methods support the use of different strategies, i.e., topographic search, hypothesis and test and recognition, respectively.

Topographic search in abstract flow structures is very similar to the rule-based search in "context free" networks described by Rouse et al. (1980). Compare the structures of the networks of Rouse (this volume), fig. 1, and Lind (this volume), fig. 8. It is interesting to note the observation of Rouse that the rule-based model describes the context-free strategies reasonably well but breaks down in context depending experiments. This is probably because the context initiates shifts to symptomatic strategies which are depending upon the individual subject's prior experience.

MAN-MACHINE INTERACTION IN DIAGNOSIS

In future industrial control rooms, the diagnostic task will be performed by a complex interaction between the plant's control computers and its operating staff. The solution of the problem of allocating data processing functions to operators and computers must be resolved by a careful consideration of the mental load put on the operators when the strategies for the computer's part of the task are chosen. In case of plant maloperation, the work situation of the operator implies a multi-dimensional resource/demand fit. The mental load of the operator will very much depend upon the possibilities left for him to resolve resource/demand conflicts by selecting a proper strategy for the task. The dimensions characterizing the demands from different strategies are shown in table 1. The strategies are often complementary in several respects. For example, good/bad mapping by topographic search can be based on a long but rapid stream of simple

STRATEGY / PERFORMANCE FACTOR	TOPOGRAPHIC SEARCH	RECOGNITION	DECISION TABLE	HYPOTHESIS AND TEST
TIME SPENT	-	LOW	-	-
NUMBER OF OBSERVATIONS	HIGH	LOW	-	LOW
DEPENDENCY ON PATTERN PERCEPTION	-	HIGH	-	-
LOAD UPON SHORT TERM MEMORY	LOW	LOW	HIGH	HIGH
COMPLEXITY OF COGNITIVE PROCESSES	LOW	LOW	-	HIGH
COMPLEXITY OF FUNCTIONAL MODEL	LOW	-	-	HIGH
GENERAL APPLICABILITY OF TACTICAL RULES	HIGH	-	-	LOW
DEPENDENCY ON MALFUNCTION EXPERIENCE	LOW	HIGH	-	LOW
DEPENDENCY ON MALFUNCTION PRE-ANALYSIS	-	-	HIGH	-

Table 1 The table illustrates the dependence of different factors of a performance criterion upon the choice of diagnostic strategy.

judgements. It is uneconomic in its use of information, but relies on simple data processing and low short term memory load. On the other hand, search by hypothesis and test is information economic, but data processing is complex and memory load high.

The performance criterion is clearly a complex factor which is not merely a simple question of the mental load or the amount of input information present in the control room. The criterion underlying the choice of strategy in an actual situation will be subjective, since the resources, and the value assigned to a margin in the different dimensions of the fit, will vary among individuals. In addition, emotional factors such as curiosity, excitement in risk taking etc. may influence the choice. The performance criteria behind the choice of a strategy are neither explicitly known to, nor consciously used by operators in actual situations. For highly skilled persons our work on electronic trouble-shooting indicates an immediate relation between different task situations and specific routine strategies. They were, however, able to identify the relations and explain them rationally during interviews. Furthermore, skilled persons seem to have a fixation with respect to preferred strategies in specific situations - this is in a way a tautology since skill can be defined by lack of the hesitation involved in conscious selection of proper strategies.

The conclusion of this discussion is that the performance criteria which are guiding human choices have to be inferred from studies of actual performance. For performance in complex man-machine cooperation in computer-based industrial control rooms, experimental studies are at present necessary. However, some guidance can be obtained from studies in other contexts. In our study of electronic trouble-shooting (Rasmussen and Jensen 1974) and diagnosis in computer systems, we have found some guiding principles which correspond to the findings in other studies of strategies (Bartlett 1958, Bruner et al. 1967) and from analyses of error reports from process plants (Rasmussen 1979).

The principal rule behind the strategies seems to be to choose the way of least resistance. Instead of making overall plans for the search, the tendency is to make rapid or impulsive decisions all along the search based only upon the information observed at the moment. This means that as long as simple, familiar routines are felt to give progress, there is little tendency to withdraw for recapitulation of previous results or considerations of more complex functional or causal arguments. At the same time, there is "a point of no return" in the attention applied at the instant of a decision. Information may be clearly available and observed after a decision without any effect even if it clearly contradicts the decision taken. These tendencies make strategies which demand inferences from several observations very unreliable if the data are available in sequence, since they often will give diagnostic references individually. The same effect can be seen in reports from industrial accidents. In complex abnormal situations, a set of abnormal indications are likely to be interpreted as a coincidence of familiar occurrences rather than a pattern related to a complex unknown plant state (Rasmussen, 1979).

The basic performance criterion behind the choice of the "way of least resistance" is to minimize the load upon short term memory. This criterion seems to be acting so strongly that it often overrules other resource/demand conflicts. A very important aspect in the design of computer support of operators in diagnostic tasks will then be to relieve his short term memory and this can be done in several ways: simultaneous presentation of information which should be used together; relieving the operator of secondary tasks such as memorizing functional or anatomic information; conversion of data to the domain of the search; generation of reference data, etc. Removal of such secondary tasks decreases the number of opportunities for associations which can lead the operators off the path of his current strategy.

Effective computer support of operators in complex diagnostic tasks depends on the willingness of operators to accept the support offered by the computer. Experience with computer aided decision making in other fields (Halpin et al. 1973) shows that acceptance depends upon the reliability and trustworthiness of the support, and upon the extent to which the operator understands and accepts the strategy used by the computer, and is able to follow its course of arguments. This means that the diagnostic task cannot be divided into separate operations which are allocated to either the computer or the operator. Ideally, both operator and computer must be able to perform the diagnosis to some extent. Automatic computer diagnosis is necessary in case of automatic protective functions and, in general, to guide the operator by localizing disturbances in properly formatted displays, e.g. in the form of topographic maps. The computer diagnosis should not, however, be carried further than consistency can be verified. On the other hand the operator must also be able to perform the diagnosis. He must be able to understand and monitor the computer's results, and he must be able to supplement the computer diagnosis by his knowledge of prior cases, extraordinary operational conditions, abnormal equipment states, etc. This in turn means that information must be available to the operator at a level of abstraction and in a format compatible with his preferred strategy. This strategy depends on the actual plant situation, the operator's task and his previous experience, and ideally an "intelligent" computer should be able to infer the strategy from the operator's request for information on the basis of its own diagnostic results.

For illustration, some typical roles of operators and computers in different diagnostic task situations which we have chosen for experimental evaluation of computerized control rooms are shown in table 2. Examples of display formats related to these roles are published elsewhere (Goodstein and Rasmussen, 1980; Goodstein, this volume) together with the detailed criteria for man-machine interface design which are used for the current experiments.

CONCLUSION

The present trend in the introduction of the process computer in the design of man-machine interfaces has been towards presentation of measured variables on visual display units as bar graphs and/or mimic displays together with attempts to unload the operator by alarm analysis and reduction. The development is influenced by the fact that the variables measured and the alarms introduced are still depending upon the choice of equipment suppliers rather than upon an overall analysis of the total system's functions and the resulting tasks of its operators.

In the optimal computer-based design, however, the sharp distinctions between the functions of alarm and safety systems, of control systems and of operators will disappear, and a key role of the computer will be as a partner of the operator in higher level supervisory control. This means that many computer functions in the man-machine co-operation will be very situation and task specific. To be practically feasible, this implies that the strategies and programs at the higher levels of the hierarchy of production optimization and disturbance control should be standardized and therefore to some extent must be plant and equipment independent. This is only possible if they are based on concepts at a device-independent level of abstraction, and this has supported our interest in generalized diagnostic strategies based on mass and energy flow structures.

It is questionable whether this new optimum can be reached by a step by step modification of existing design tradition without crossing unacceptable performance valleys. A jump to a new optimum has to be based on a proper hypothesis which preferably should be tested experimentally. This is the aim of a co-operative Scandinavian project on Control Room Design and Operator Reliability of which the present work is a part. The project has been partly sponsored by the Nordic Council of Ministers.

REFERENCES

Barlett, F., 1958, "Thinking. An Experimental and Social Study" (London: UNWIN).

THE GOAL OF DIAGNOSIS	OPERATOR'S TASK AND STRATEGY	COMPUTER'S SUPPORT OF OPERATOR'S TASK AND STRATEGY	COMPUTER'S AUTOMATIC TASK AND STRATEGY
PROTECT PLANT	Verification of automatic safety actions and consequent plant state. Recognition and decision tables.	Presentation of patterns of critical variables in direct relation to standard/reference values by visually perceptible patterns. Tabulation of safety action sequences and related state patterns.	Monitoring of critical variables; automatic actions according to stored decision tables.
PROTECT OPERATION	Identify endangered critical variables; check patterns against normal/specified.	Initiate operator's attention and action. Present critical variables in patterns related to normal state for rapid visual search.	Monitoring of measured variables; derivation of "normal", reference data; detect discrepancies.
	Identify disturbed function, e.g. mass/energy flow paths. Topographic search in flow structures.	Guide operators by messages about informative displays to attend to. Present maps of flow paths with state of flows related to normal, for direct visual search.	Automatic topographic search in mass/energy balance structures for identification of faulty/disturbed balances. Convert all measured information into flow and level information related to balances.
	Select proper control facility to counteract disturbance.	Indicate possible control points on flow maps. Present secondary and normally not used paths and control points on demand. Support verification of decisions.	Store design data base. Simulate system's response.
LOCATE INITIAL DISTURBANCE OR FAULT	Locate source of disturbance by topographic search in causal (mass/energy/information flow) network.	Guide operators by messages about which informative displays to attend to.	Automatic topographic search in mass/energy flow structures and balances for identification of faulty/disturbed balances.
		Present maps of flow structures with state of flow and reference to normal for direct visual search.	Convert measured data from all sensors into flow and level information compatible with energy/mass flow representation. Arrange for hierarchy of display coverage.
	If necessary, increase resolution by symptomatic search on component/equipment level or	Support operator's visual recognition of symptoms by suitable graphic presentation of patterns of data related to parts and components. Assist operator's test of hypothesis by presentation of stored symptoms in suitable patterns.	Collect and store reference data representing normal state. Simulate abnormal operation according to operator's hypothesis - cause/consequence modelling. Store decision tables for test of data against known symptoms.
	Use preplanned test sequences to identify faulty components.	Support evaluation of test results by presenting results and combinatorial conclusions of tests.	Decision table evaluation of combinatorial conclusions of test results.

Table 2. Illustrative example of operator and computer roles during different diagnostic task situations in a process plant control room.

Bastl, W. and Felkel, L., "Disturbance Analysis Systems" (this volume).

Berenblut, B.J. and Whitehouse, H.B., 1977, "A Method for Monitoring Process Plant Based on a Decision Table Analysis", The Chemical Engineer, March 1977, pp. 175-181.

Bruner, J., Goodnow, J.J. and Austin, G.A., 1956, "A Study of Thinking", Wiley, New York, pp. 330.

Cornell, C.E., 1968, "Minimizing Human Errors", Space Aeronavtics 1968, vol. 49, March, pp. 72-81.

Dahll, G. and Grumbach, R., 1976, "On-Line Analysis of Abnormal Plant Situations", presented at OECD Halden Project Meeting, Sanderstölen, Norway, March 1976.

Duncan, K.D., "Training for Fault Diagnosis in Industrial Process Plant" (this volume).

Furth, E., Grant, G. and Smithline, H., 1967, "Data Conditioning and Display for Apollo Prelaunch Checkout", Test Matrix Technique, Dunlapp and Associates; NASA, N-68-12531.

Goodstein, L.P. and Rasmussen, J., 1980, "Man-Machine System Design Criteria in Computerized Control Rooms", "ASSOPO 80", IFIP/IFAC Conference, Trondheim, Norway, June 1980.

Goodstein, L.P., "Discriminative Display Support for Process Operators" (this volume).

Halpin, S.M., Johnson, E.M. and Thornberry, J.A., 1973, "Cognitive Reliability in Manned Systems, IEEE Trans. on Reliability, vol. R-22, no. 3, August 1973, pp. 165-169.

Lihou, D.A., "Aiding Process Plant Operators in Fault Finding and Corrective Action" (this volume).

Lind, M., 1980, "The Use of Flow Models for Automated Plant Diagnosis (this volume).

Rasmussen, J., 1969, "Man-Machine Communication in the Light of Accident Records", Int. Symp. on M-M Systems, Cambridge, IEE Conf. Records No. 69 (58-MMS, vol. 3).

Rasmussen, J. and Jensen, A., 1974, "Mental Procedures in Real Life Tasks: A Case Study of Electronic Trouble Shooting", Ergonomics, 17, no. 3, pp. 193-307.

Rasmussen, J., 1978, "Notes on Diagnostic Strategies in Process Plant Environment", Risø-M-1983.

Rasmussen, J., 1979, "On the Structure of Knowledge - a Morphology of Mental Models in a Man-Machine System Context", Risø-M-2192.

Rasmussen, J., 1979, "What Can Be Learned from Human Error Reports", in: Duncan, K., Gruneberg, M. and Wallis, D. (Eds.), Changes in Working Life, John Wiley & Sons (Proceedings of NATO International Conference on Changes in Nature and Quality of Working Life, Thessaloniki, Greece (to be published).

Rouse, W.B., Rouse, S.H. and Pellegrino, S.J., 1980, "A Rule-Based Model Human Problem Solving Performance in Fault Diagnosis Tasks", IEEE Trans. on Sys, Man and Cybernetics (to appear).

Rouse, W.B., "Experimental Studies and Mathematical Models of Human Problem Solving Performance in Fault Diagnosis Tasks" (this volume).

Shepherd, A., Marshall, E.C., Turner, Ann and Duncan, K.D., 1977, "Diagnosis of Plant Failures from a Control Panel: A Comparison of Three Training Methods", Ergonomics, 20, pp. 347-361.

Taylor, J.R. and Hollo, E., 1977, "Experience with Algorithms for Automatic Failure Analysis", in: J.B. Fussel and G.R. Burdick (Eds.): Nuclear Systems Reliability Engineering and Risk Assessment, SIAM, Philadelphia.

MATHEMATICAL EQUATIONS OR PROCESSING ROUTINES?

Lisanne Bainbridge

Department of Psychology
University College London
London WCIE 6BT, England

INTRODUCTION

Within the context of this conference, we want to know the factors which affect human ability to detect and diagnose process failures, as a basis for console and job design. Understandably, human factors engineers want fully specified models for the human operator's behaviour. It is also understandable that these engineers should make use of modelling procedures which are available from control engineering. These techniques are attractive for two reasons. They are sophisticated and well understood. They have also been very successful at giving first-order descriptions of human compensatory tracking performance in fast control tasks such as flying. In this context they are sufficiently useful for the criticism, that they are inadequate as a psychological theory of this behaviour, to be irrelevant for many purposes. Engineers have therefore been encouraged to extend the same concepts to other types of control task. In this paper we will be considering particularly the control of complex slowly changing industrial processes, such as steel making, petrochemicals and power generation. We will find that while control engineering methods may be adequate to give a first-order description of human control behaviour in very simple tasks, they do not contain mechanisms with sufficient potential to account for all aspects of behaviour in more complex situations. I will try to discuss the types of mathematical model which appear to best fit some of the data from these tasks. I will also describe some of the other types of behaviour for which a mathematical representation seems insufficient or inappropriate.

In order to make these comparisons I will take some of the basic concepts in some typical engineering approaches and compare them with data from human control studies. As many engineering models distinguish between input processes and control decisions, these are discussed separately. Unfortunately, as studies of human control behaviour show that these distinctions between obtaining information, making decisions (and setting control targets) are often not clear, this way of organising the paper gives a rather distorted representation. It also entirely omits some central aspects. The paper will therefore start with a discussion of the overall nature of process control tasks, and of some of the richer aspects of behaviour for which mathematical equations do not provide the types of concept needed.

I would certainly accept that most current psychological models cannot be used effectively in solving engineering problems, and that, as they are not fully specified, they have little intrinsic value for engineers. I would like, however, to argue that the most effective response to this is not to make more sophisticated control theory models, but to put more work in developing more powerful models of different types. It must also be admitted that it is only possible at the moment to give a very superficial indication of the wealth of problems which need to be dealt with, and of the mechanisms which might be useful in more fruitful models of human process control.

THE GOALS OF PROCESS CONTROL

A simple expression of the process operator's goal would be that he should maintain the process output within tolerance around the target level. This description may be appropriate to the control of single variables, as in Veldhuyzen and Stassen's (1977) model of ship steering. In many processes, however, the operator's task is more like that of a ship's pilot deciding which direction the ship should go in, and dynamic control is a relatively minor aspect of the task. Umbers (1979b), in a study of gas grid control, found an average of 0.7 control actions per hour. Beishon (1969), interviewing a bakery oven controller, while he was doing his job, found that 11% of his skilled decision making was concerned with control and observation. These investigators both found that goal- or target-setting was a major part of the operator's activity; and that it influences task behaviour in such an integral way that it cannot be dealt with in models which are distinct from models of control decisions. In general, studies suggest that the operator is concerned with goals at several levels. He converts goals which are not fully specified into an operationalisable form. He predicts targets which are affected by varying product demand or process behaviour. He plans sequences of future behaviour, by the process and himself. Such sequences may be evaluated against multiple goals, and generation and evaluation of these sequences may be difficult to distinguish. He may also be able to generate entirely new behaviour in unfamiliar situations. We therefore need to understand and be able to model all these types of behaviour.

Future Targets

Frequently the general goals specified by management are inadequate as guides for controlling the plant, and the operator has to convert them to sub-goals. For example, Beishon (1969) found that the time/temperature combinations at which cakes could be baked were not fully specified in the job description, and had been learned by the operator. Such knowledge could be modelled, at a first-order level, by table-look-up, and the learning by parameter trimming.

Many process tasks are more complex than this, as the control target varies on-line as a function of product demand or stages in production. In some industries management takes the decisions at this level. In others, predicting targets may be part of the operator's task. Umbers (1979b) studied gas grid controllers, who ensure that sufficient gas is available to meet demand over the day. It is inefficient to control the gas supply by feedback from demand levels, as 2-4 hours notice is needed to increase production using cheap methods, and the pressure of stored gas provides a "free" force for distribution, so a major part of the task is to predict how much gas will be used that day and have it available beforehand. Umbers concluded that observational techniques provided inadequate information on the controller's methods, so he recorded verbal protocols, asking the controllers to "think aloud" while doing the task. Verbal protocols are often analyzed by producing flow diagrams which would generate the same sequence of activity as in the verbal report. Therefore, to accept many of the following examples in which verbal reports are used as data one has to accept that an operator can give useful and valid reports (see Bainbridge, 1979), and that some of the operator's behaviour can be described validly in terms of program-like routines. Umbers found that 28% of the verbal reports were concerned with predictions of gas demand. A computer makes a demand prediction based on the 4 previous same days of the week. The operator can revise and override this estimate, by finding records of a previous day which better match the present day than the ones used by the computer, allowing for the weather, national holidays, etc. Such judgements and comparisons would be most easily modelled by conditional statements. The knowledge of factors influencing demand can presumably be modelled in the same way as other types of knowledge of dynamic relations. This is discussed later.

The future target may be a function not only of product demand but also of process behaviour. The writer (Bainbridge, 1974) has studied allocation of power to 5 electric-arc steel-making furnaces, with a limit on the amount of power which could be used in a ½ hr. period. Steel-making furnaces go through a sequence of stages which require different amounts of power. Therefore, for example, if too much power is being used now, but shortly one of the furnaces will change to a stage using less power, then this may compensate for the present over-usage, and there is no need for action by the operator. Goal setting and control strategy are here combined. Protocols from these controllers show that they make a list of future furnace changes, and keep in mind particularly those which will happen in the current ½ hr. Generating the time of the next stage change in each furnace is done by arithmetic, using displayed and reference data. Ordering these events in time requires a simple sort procedure. Simple arithmetic could be used to model the process of checking whether a future event will compensate for a present power usage which is unacceptable by average criteria. However this would not be a good model for all controllers. The protocols suggest that some operators assess the effect of an event in relative rather than numerical terms. Also an arithmetic account does not model the operator's understanding of why this anticipation is a good strategy.

Another important aspect of this behaviour is that the operator makes the list of future events during a general review of the process state made when the control state is acceptable. The list is then immediately available for reference when power usage is unacceptable. The process of building up and referring to this background information (or "mental picture") about the present process state plays an important part in control behaviour. These examples suggest that control activity is generally oriented to the future. The operator can only make such predictions if he has adequate knowledge of the process dynamics. See also the discussions of sampling, working storage and control action choice.

Planning Future Sequences of Behaviour

Allowing for the future can be still more complex. In Beishon's (1969) bakery study, much of the operator's thinking was concerned with planning future activity. For example, he had to choose the best sequence in which to bake the waiting cakes. To minimise time and power usage, the cakes should be baked in a sequence of increasing or decreasing temperatures. Again there is no clear distinction between goal-setting and control. It would seem that a simple sort routine for arranging cakes in order of required conditions would give a simple description of what the operator does, though not necessarily of how he does it. The actual situation is more complex however, as he is controlling 3 ovens, so assignment of cakes to each will depend on present conditions and capacity, requiring a multidimensional evaluation.

As well as planning the best sequence of behaviour for the process, the operator also has to plan his own sequence of actions. In simple tasks with one display and one control, the operator's control actions are all on the same variable. The sequence of actions to obtain a step change: "turn control up to start controlled variable moving towards target, turn it down to stop the movement at the target, turn it to a level which will maintain output on target" has been observed in many tasks, e.g., Crossman and Cooke (1962). Passage from one type of activity to another can be modelled by dividing the phase-plane into different areas, as by Veldhuyzen and Stassen (1977). The operator however also has to time his display and product sampling. Interleaving the potential different types of activity requires a complex multidimensional evaluation. This will be affected by personal as well as production goals; for example, the operator may do one activity after another because they are close together on the process rather than because this is the optimum timing of the second action from the point of view of process operation. Beishon comments that planning gives rise to activities, the reasons for which are not obvious from the current context and are therefore not identifiable unless one has been discussing with the operator what he is doing and thinking about.

All these types of planning involve the operation of multiple goals. Some preliminary points about this will be made in the next sections.

Problem Solving

The previous examples show that the operator is evaluating possible behaviours against several goals which operate simultaneously. There are parallel goals, e.g., baking cakes, using minimum time and fuel, and an acceptable level of operator effort. Some of these may be implemented by sub-goals, e.g., cakes are baked with minimum time and fuel if they are put in the oven which is nearest to the required temperature. Thus within one goal there may be multiple levels of goal organisation which cannot necessarily be described as a hierarchy (cp. Bainbridge, 1975). Therefore, to understand their effects on behaviour, we need to understand the nature of different types of goal structure, and this is something about which we know very little.

This complex structure of goals is involved in the generation of new types of behaviour in unfamiliar situations. This occurs both in problem solving, in learning (see "Acquisition of the Internal Model"), and in on-line choice between known behaviours in a specific known situation (see the next section). Problem solving involves generating new behaviours which will meet the task goals. To be brief it is easiest to give a non-process-control example. If the goal is to get a nail in, a simple response to this goal is to find a hammer and hit the nail in. However, we do not only have simple and direct goal-response links. If we cannot find a hammer then we look for another object which can be used with the same effect. To be able to do this, each goal must have associated with it a specification of the characteristics required of a "response" which will implement it, and instructions about how to check for them. Each response must have associated with it information about the type of results it can provide. Winograd (1972) has described an interesting attempt at a program which works in this way. The above points therefore suggest that for the process operator to generate new behaviours, his process knowledge must be specified in a goal-oriented way, which can also be searched in a goal-directed way to find new sequences of process behaviour which will lead to the control goals. A model of such behaviour would be necessary for understanding how an operator reacts to system failure.

On-Line Behaviour

It is perhaps easier to model how multiple goals affect the choice between already available strategies from moment to moment, to suit particular on-line conditions. The writer, (Bainbridge, 1974), in the power allocation study, found that the operator had one routine for finding the target power usage for the rest of the ½ hr., and other routines for choosing the best action to make. These routines choose which of the 5 furnaces to reduce the power supply to, total level of power usage being the primary control goal. The first routine finds the furnace which started melting last so least fuel will be wasted by allowing it to cool. The second routine finds which melting furnace is least loaded with metal, for the same reasons, and so on. Each routine therefore considers actions which are increasingly refined from the point of view of meeting the secondary control goals of maximising steel throughput and quality. Suppose the operator judges that the level of power usage indicates an action will be needed soon. He uses the first action choice routine, which indicates a good, though not necessarily the best, furnace to be changed. He then samples the control state again. If action is urgent he can make the action chosen on a simple basis, if it is less urgent he uses the next routine, which refines the action choice, and so on.

There are two ways, in this example, in which the urgency of the control state affects the choice of next best behaviour. When action is urgent, the primary goal of power control overrides the secondary goals and action is chosen on the simplest basis. In addition, he does not always return to check the control state at the end of each of the control choice routines. One might model this by returning, at the end of each routine, to some "executive" which decides what would be the most effective way for the operator to spend the next few moments. In the unacceptable-power--usage context there is a choice between refining the action choice, sampling the power usage, or making a control action, and the choice between these is a function of the urgency of the control state at the last sample. Utility "calculations" are presumably used in these "executive" decisions, but play only a partial role within a more complex structure.

Umbers' (1979b) study of gas-grid control gives another example. He found 40% of the verbal reports were taken up with decisions about whether action was necessary, compared with 3.5% concerned with selecting a suitable action. Again the controller's task was to find a balance between the various costs of changing the levels of gas generation, considering various goals which are secondary to the main one of meeting gas demand. The verbal report analysis suggested that the different costs were combined together in some categorising process which involves a subjective scale which may differ between operators.

Evaluation

Frequently the operator's choice between alternatives is modelled by an Expected Value calculation as in decision theory. The example from Umbers raises the question whether the operator's decisions are in this specific, numerical form. The EV calculation does require the assumptions that numerical utilities are known to, and used by the operator. Operators however do not necessarily have full cost information. Paternotte (1978) found, from interviews with operators and management, that operators controlling 8 distillation columns had very little information about costs. This was deliberate policy. However, if one assumes that the operator must use some costs in making decisions, then if he is not given information about costs, he must assign his own, which may lead to idiosyncratic actions. Hopefully, one could give correct cost information to an operator without adding to his difficulties.

Verbal report evidence, as in the example from Umbers *op cit*, suggests that the operator usually works by making relative comparisons, e.g., *a* is better than *b*, where *a* and *b* are categorical values. This implies that decisions might be modelled by a mechanism which can operate on an ordinal scale, rather than the ratio scale required by the calculation of EV functions. This "linguistic" handling of utilities might be better represented by "fuzzy" methods, cp. preliminary discussion by Efstathiou and Rajkovic (1979).

The discussion so far has implied that the generation of alternatives and their evaluation are in two separate stages. When generating complex sequences one quickly comes to the problem that generating all the present alternative behaviours using an algorithm is an unrealistically large task. Heuristics therefore have to be used to suggest the most useful alternatives to consider first. Heuristics of course are a form of evaluation, so it becomes difficult to separate generation from evaluation.

Conclusion

In general this section suggests that some of the main problems in understanding and modelling the process operator's behaviour lie in representing his orientation to the future, and his generation of the future sequence of subgoals by which the main goals in a complex task will be achieved. The main studies described have found it appropriate to describe parts of the operators' behaviour using information processing flow diagrams. These are described only at a first-order level of accuracy in the papers quoted, but on the whole it would not be difficult to represent them using a conventional numerical computer programming language plus list-processing facilities. It should therefore not be too large a problem to give a fully specified description of the operators' activities.

One should of course remember that this discussion has by no means captured the complexity of the goals in any real situation. The operator acts on the process to control it, he may also act on it to maintain his skill, prevent boredom, have some fun, or to impress or confuse the onlookers. Perhaps we do not need to have a model which accounts for all of these to have a model which is useful for practical purposes. However, it should be evident that goals are important components of complex process control activity and require a major research effort.

INPUT PROCESSES

Most models of human input processes in control tasks have been concerned on one hand with human input limitations, with the accuracy of readings taken by an operator and the lag before he makes his response, and on the other hand with the interval between his samples and the problems of sampling several variables. The purpose of taking a sample is usually assumed, i.e., to compare it with a target value. The discussion in the previous section has suggested that the notion of a control target is not necessarily simple, so presumably making a comparison with it is not simple either. Also, for example, comparisons are often not made by arithmetic but by pattern recognition, particularly but by no means only in check reading, and this can have important implications for optimum instrument design. However, these aspects have not received much accurate analysis, although there are many sensible design recommendations, so the discussion here will concentrate on concepts which do appear in several models, and on comparing these concepts with actual operator behaviour, to suggest aspects which need to be included in a fuller understanding of how the process operator takes in information.

Human Response Lag

The lag (reaction time) between taking in information and reaching a decision has an important influence on the quality of human control of quickly reacting "processes" such as aircraft, but becomes of progressively less direct importance when the process response lags are much longer than the human lags. However there are some points which need to be kept in mind.

Kok and van Wijk (1978, p. 6) state that "the human information processing time (can be modelled) by a single pure time delay which is equal for all observed outputs independent of its modality (visual, auditory, etc.)". This assumption may be an acceptable approximation in models of fast tracking tasks, in which one well-designed display is used and neuromuscular control is the major limiting factor in performance. It can be positively misleading in slow process control for several related reasons. Decision making time is usually much longer than the 150-200 msec assumed in fast tracking models, as more complex distinctions are involved. As we have already seen, coming to a decision may involve lengthy information considering procedures for which a simple stimulus-response representation could be inappropriate. The variability in reaction time as a function of display and task is not a small fraction of the average RT, but reaction times in some situations can be several times longer than in others (e.g., Teicher and Krebs, 1974). If one does not consider all the factors which can affect reaction time, such a quality of display, compatibility, and console layout, one can easily forget the major influence which interface design can have on process control performance.

Accuracy of Display Readings

Many modellers assume that the human operator is an inaccurate reader of displays, for example Kok and Wijk (1978, p. 6) say that "the human uncertainties due to scanning or sampling (can be modelled) by a normal white observation noise on each of the observed variables". Indeed the human sensory threshold can be considered as adding "noise" to the reading obtained (as in any other sensing device) and identifying a stimulus does involve a decision. The threshold is not constant but is a function of location of stimulus, e.g., on retina or skin, and of display design, as mentioned by Baron and Kleinman (1969). It is also a ratio (to a first-order level of description) of the sensory adaptation level, and is affected by the utilities of misses, false alarms, etc. However, it seems that introducing noise into models of the human operator has had the main effect that a Kalman filter is then needed to remove the noise (whose characteristics have been carefully chosen to be those a Kalman filter can deal with) as the model control equations acting on the sensory measures cannot handle stochastic inputs. One can argue both that the human operator may use "inaccurate" categorised display readings for good reason, and also that sometimes when noise has to be added into control engineering models of the human operator to make them match his control behaviour, this may be due to limitations in the model rather than to limitations in the operator. One might also mention that Paternotte's (1978) operators complained that their control task was difficult because the instruments and controls were not accurate. (See the section on Reliability of Input Measures).

When the human process operator makes categorised display readings these will be inaccurate by engineering standards, but the display may be read in this way because the reading is being used, not for feeding into an action-size calculation, but as a determinant of decision making for which a categorised value is more appropriate. In the writer's steel-making study (Bainbridge, 1971) the control state (power usage level) was evaluated into three types: alright, above/below, increase/decrease action needed. These three categories have different implications for the next best behaviour as discussed earlier. To describe this simply as a noisy display reading would be to misrepresent what the operator is doing with the information.

In a different way, "noise" may be introduced into modelling to account for a mismatch between model and human performance, i.e., the "remnant". In some cases the need for this might be taken as a measure of the inadequacy of the model, with a larger remnant indicating that less of the human operator's performance has been accounted for by the model. It may of course be necessary to include noise in a human operator model, particularly when his knowledge of the process is only adequate to predict its behaviour to within statistical or categorised limits. I consider that noise should be used in models with great care, as it has a more general danger. Many models of the human operator represent him as a rather simple control device with noise. While the originators of such models may hold that such a model has only the status of a first-order description of the human behaviour, it is very easy to fall into thinking that such a model is a valid representation of actual human control mechanisms. One can then easily infer that the human operator is a rather poor quality simple control device best replaced by something which has been properly designed. As this type of model does not contain any components by which one could begin to represent the things which the human operator is particularly good at, such as pattern recognition, planning and problem solving, it could be easy to forget, when using such models, that human operators do have this sort of potential.

Intersample Interval and Process Knowledge

Most operator models are concerned with the task of sampling at a rate such that the behaviour of the variable being sampled can be reconstructed, though it is now accepted that there are several other reasons why the operator might sample, such as failure detection, which would lead to different optimum strategies (see e.g., Kleinman and Curry, 1977). We need to discuss what does determine the human operator's intersample intervals. We can also ask whether this type of notion is sufficient to account for all human sampling behaviour, particularly in multi-variable process control tasks and how much the operator can remember about a complex process state. (See the section on Multi-Variable Sampling).

The simplest sampling decision is whether to sample or not. Sheridan (1976) describes a model of supervisory control in which a sample measure gives an estimate of the system state, this indicates an appropriate action which has an expected value. The model selects the measure (including not sampling) which maximises this expected value. The Expected Value calculation requires knowledge of the distribution of process states likely to occur, and of the precision of the various measures available. While it is unlikely that assessments of the optimum time to sample could be made without these types of knowledge, it would be optimistic to assume that they are perfect. (Knowledge about measurement precision will be discussed in the next section).

We can note that an input track to be followed in tracking tasks and a process output to be controlled, are equivalent in the sense that both are behaviour of the external world over time, about which information is needed as a basis for action choice. Presumably one can therefore use related models for knowledge which give the ability to predict a future track and knowledge which gives the ability to predict rather than sample future process behaviour. However, on the whole, different types of models have been used to represent knowledge in these two tasks. When following an input track (without preview) it is usually assumed that the human operator's sampling follows sampling theory, or its later developments, i.e., that he knows the statistical properties of the track. (Preview allows the human operator to use more complex types of input processing such as pattern recognition, e.g., Sheridan, 1966). In sampling process outputs, the signals are more redundant when the operator also has knowledge about the process dynamics and his own behaviour. Baron and Kleinman (1969) allow for sampling as a function of system dynamics and controller gains. Carbonell (1966) explicitly allows for the effects of control actions, while he and Kleinman and Curry (1977) mention the effect of correlations between instruments on sampling. In discussing monitoring behaviour Sheridan (1976) suggests that knowledge of process states which are likely to occur is obtained by extrapolation from trends in the present observation. Several types of control model assume that the human operator has knowledge of the process dynamics in the form of a deterministic internal model, which is used to predict the effect of an action on process behaviour, as the basis for choosing the best control action. Presumably there is no reason why this model could not also be used to predict process behaviour as a result of the present control settings, as a basis for determining the best next sampling time.

One might then suggest that either type of model, statistical or deterministic, or any possible intermediate combination, could be the basis for determining sampling intervals for both tracks and process outputs. For example, human sampling of input tracks has usually been tested with random inputs, so it is hardly surprising that a random model is sufficient to fit the behaviour. Other studies, however, show that the operator can also learn deterministic properties of the track, e.g., he can follow pure sine waves (see e.g., Pew 1974) which will influence both his need to sample and his strategy for following the track. Presumably the operator can learn about, and develop his own model for the behaviour of the statistical properties of external world by some ability to extract the correlations and conditional probabilities in its behaviour. He might learn about its determinacies by noting that conditional probabilities are close to 1, or by some different type of mechanism such as pattern recognition. On the other hand, the operator's internal model of the process is usually not perfect but may be partly statistical (see later on). For example, the process behaviour may fluctuate as a function of factors not under the operator's control, such as changes in the environment or in the quality of input materials. If the operator has not been able to learn these dependencies he would react as if these are random fluctuations in process output, about which some statistical properties are known. We therefore need much more research on how human operators learn and use their knowledge of non-random inputs before we can predict how their sampling of such inputs will differ from that predicted by statistical models.

Knowledge of process dynamics can also be important when the human operator is monitoring the operation of automatic controllers for failure. This is only simple to monitor in steady-state control, when the target is stationary. If the automatics/computer are controlling trends or step changes, then the operator needs to know the trajectory of target performance against which to compare actual behaviour. We can ask how the human operator knows this trajectory, and whether he needs to control the process himself in order to know it. Brigham and Laios (1975) found that operators did not learn how to control a process by watching automatics do the task.

Failure detection requires that the operator's sampling should be a function of failure probabilities, rather than the probabilities of normal events. The operator should therefore sample displays with low signal probabilities. It is well known (e.g., Macworth, 1970) that human beings cannot attend effectively in directions in which nothing is happening, for periods longer than ½ hr. This implies that monitoring automated control is like other watch-keeping situations, and requires rapidly rotating shifts for optimal behaviour.

The discussion so far has implied acceptance of the notion that decisions about when to sample are distinct from other types of task decision. The use of the same process knowledge in action choice and behaviour prediction would suggest that these two decisions may be closely interrelated. Actual studies of process behaviour suggest that sampling decisions may also be more complex, and may be affected by different types of mechanisms. As related earlier, the writer's study (Bainbridge, 1974) indicated that deciding to sample was part of a more general "executive" decision about the operator's best next behaviour. The operator "alternated" between a routine concerned with sampling the control state and ones for increasing refinement of action choice. The sampling model that this is most like is Carbonell's (1966) queueing model, but this is more complex since all the operator's activities, not just his sampling, are "queueing" for consideration. The effect of such a mechanism is that the time between samples is determined by the length of the other routines. However the sampling interval is also a function of control state urgency, as this affects how many other routines are allowed to intervene before the operator returns to checking the control state. This will be the case whenever the operator has to choose between many possible activities and sampling is just one component.

Reliability of Input Measures

We can ask whether the operator's knowledge of the reliability of the measures available to him is independent of the knowledge he uses in predicting future process behaviour. If the operator acquires his knowledge of the process by learning about correlations and conditional probabilities in its behaviour, then there is no way in which he could distinguish noise in the instrument measures from noise in the process behaviour, unless independent evidence is available about one or the other. Noticing unreliability in the instruments is not different from noticing unusual behaviour in the process as a whole. Both can only be done if independent evidence, or a model in some general sense of knowledge of what should happen, is available for comparison with what is actually happening. An effective model of the process is necessary before the operator can diagnose that one part of the process is not behaving as it should do given the readings of other related factors. Sheridan (1976) contrasts his model of human supervisory control, in which sensory measures and control parameters are trimmed and the process model left constant, with a "conventional Kalman-Bucy estimator control system" in which the process model is trimmed and the sensory measures are left constant. It may be more realistic to assume that knowledge of these two aspects develops together.

With both the instruments and the process, given sufficient experience, the operator should be able to learn about this unreliability. This knowledge too would presumably be in the form of statistical properties or determinacies (e.g., levels of uncertainty about process behaviour, and knowledge of the types of things which can go wrong) which could be incorporated with other aspects of the operator's internal model of the process. Normally such knowledge is effective and useful. There are at least two ways however in which it can be misled. Once an instrument has been identified as unreliable the operator may diagnose peculiar changes on it as due to instrument failure rather than as something wrong with the process, and so fail to diagnose a process failure. This of course does not indicate that the operator is no good at assessing likelihoods, but underlines the importance of instrument maintenance. On the other hand, judgements that the process is behaving correctly may change relatively quickly given contrary evidence, which can be a problem if this contrary evidence is unreliable! Bartlett (1943) found that fatigued pilots blame system malfunction when things go wrong. Bartlett inferred that the tired pilot was not implementing the size and timing of actions that he should have done, though he thought that he was. Consequently the aircraft behaved in an unexpected way, and the pilot attributed this to failure in the aircraft rather than to himself. This again would support the notion that it is difficult for the operator to distinguish different parts of the process behaviour without independent evidence.

Multi-Variable Sampling and Working Storage

The usual approach to modelling human sampling of multi--variable processes is to describe this behaviour using developments of sampling theory. Senders (1964), studying sampling of randomly varying displays, found that attention was allocated according to the probability of an event on a display, although Hamilton (1969) has found that this occurs only when the signal rates are relatively high. One can suggest that statistical or deterministic knowledge of dependencies between process variables would increase the redundancy of signals and so reduce the sampling rate necessary. This is still to argue within the same framework of notions about sampling, however, while the writer's (Bainbridge, 1974) study suggests that something rather different may be going on. In the power control task the operator sampled many variables which were not relevant to the primary goal of controlling power usage but were relevant to the choice of an action which would best meet secondary control goals. These secondary variables were, in this task at least, not sampled at a rate such that the pattern of changes in their levels could be reconstructed. In fact such a notion would be inappropriate here as changes in these variables were step-changes occurring at fairly predictable times. Also there was a much larger number of variables than could be remembered perfectly between samples (though the form in which some variables are remembered is an important part of efficient decision making, see below).

It seems that these secondary variables are sampled in two contexts. Their levels are checked during the action choice routines, so sampling of these inputs is not independent of control choice. They are also sampled during the general process reviewing which the operator does when he is not under action pressure (as discussed earlier under Future Targets). This is one example of the general checking which operators have been seen to do during quiet periods in many studies. The writer's analysis suggests that the operators are not simply or only checking control loops which they do not normally have time to get round to. The items about the process which are remembered after making this review, are not, on the whole, the variable values as originally sampled (see Bainbridge 1974, 1975). The sampled values may be stored in a pattern different from the one in which they are displayed which is more convenient for future reference, e.g., furnace stages are displayed by furnace, but are remembered as lists of furnaces in the same stage. The more important items remembered are the results of previous thinking, for example the time of the next power demand change, and the best action to make then. Here the operator is not storing raw data about the process state, and the present "process state" as seen by the operator is not only its behaviour at this particular moment but also includes its and his potential behaviour. This suggests that the operator is building up a background "mental picture" of the process state which will enable him, when he does need to make an action, to make a wise decision within the overall context of immediately accessible and relevantly organised information about total plant behaviour, rather than simply in response to one variable which has been found to be out of tolerance. The on-line development of this background memory, and its operation in decision making, have strong implications for manual take-over from automated systems. Modelling the development and operation of this memory requires devices which are not available in most programming languages but which could be mimicked by them.

Summary

The previous section concentrated on important aspects of complex manual process control behaviour which are missing in simple models. In this section we have again argued that simple representations of the human operator's information processing limits give a misleading impression of his potential contribution to control tasks. The need for an operator to sample may depend on his statistical or deterministic knowledge of the process, and knowledge about process behaviour may be difficult to distinguish from behaviour of the instruments or his own muscle dynamics. In more complex tasks the operator may not sample at predetermined times or intervals, but at the end of a sequence of other activities, when sampling is the next behaviour with the highest expected value. He may also review the task variables in a way which is structured by the task decisions rather than the process dynamics, to build up a working store of the whole state of the process for use in later decision making. Such analyses again support the need for information processing representations of at least some aspects of manual control behaviour, and suggest that much further research is needed to understand the nature of the operator's temporary "mental picture" of a particular process state and long-term "mental model" of the overall process dynamics, and their relation to his sampling behaviour. As these are both aspects of the operator's knowledge, his sampling may vary with experience. This is another topic about which we know very little, though Umbers (1976, p. 224) found that trainees sampled and then decided what to do, while experienced operators sampled with a purpose.

CONTROL ACTION CHOICE

Many engineering models of control action choice are concerned with the size and timing of actions. However, many of the examples given here have already implied that action choice is often more complex than simply aligning control change to error. For example, in Umbers' (1979b) task, the multiple goals to be met are being considered within a decision about whether an action is necessary. This section will be primarily concerned with the operator's knowledge of process dynamics. Many engineering models of control of single variables assume that the controller has perfect knowledge of the process behaviour, and is an ideal feedback controller. It might be more realistic to say that he exercises great ingenuity in controlling given the information at his disposal. In models which assume that the operator has perfect knowledge, this knowledge is used to predict the effects of the actions available, as a basis for choosing the best. We can ask whether these assumptions, which will be discussed separately in the next sections, are wrong in detail, or wrong in kind as a way of accounting for all types of manual process control behaviour. Within the discussion of the operator's knowledge we will consider what form it may take, and what may affect its acquisition.

The Form of the Operator's Process Knowledge

Kok and van Wijk (1978, p. 6) start their operator modelling from several assumptions, including: "The human operator has a perfect understanding of the system to be controlled, the statistics of the system disturbances, the interrelation between the observed output and the system state (the display parameters), the task to be performed, and his own inherent limitations" This assumption greatly simplifies the task of modelling, as one can include the known process dynamics in the model without asking any questions about the form in which they should be represented (and avoiding the interesting point that operators are able to control processes for which the dynamics cannot be fully specified). Of course, not all engineering models of the process controller do assume that he has perfect knowledge; some interesting examples are the fuzzy-set controllers e.g., Mamdani and Assilian (1975). However, as several models do make this assumption we need to discuss the extent to which this is a valid notion, and the distortions of understanding which it might lead to. In many cases it is essential to be able to represent the operator's non-perfect knowledge as a basis for valid predictions of his behaviour. Detailed evidence on control choice suggests that it is possible to control without full knowledge of process dynamics, that correct open-loop actions can be made by experienced operators with only a very primitive type of process knowledge, and that process knowledge may sometimes differ in form from a simple description of input/output relations.

Control behaviour in several studies suggests that a very primitive level of knowledge is sufficient for control. Beishon (1967) and Paternotte (1978) both found that operators controlled by moving the process output in the required direction in small steps. Both investigators suggest that this occurs because the operator has poor knowledge of the process dynamics. This method of control, which is possible but not efficient, requires knowledge only of the direction and (approximate) gain of control movement, and of the time to wait before checking for an action's effect. Paternotte's operators were controlling 8 distillation columns. In interviews the operators said that it was impossible, with the existing controllers, to make actions as accurately as desired, that accurate control was useless because of inaccurate instrument readings, that the precise effects of control actions on quality values were unknown, and that lack of information concerning quality forced careful strategies. Evidently these operators are trying to control within a high level of uncertainty about the process behaviour measures given by the instrumentation. One may infer that inefficient control occurs because the nature of the console design, process or task makes it difficult to acquire the higher levels of knowledge about process behaviour which are necessary for more sophisticated control strategies.

When the operator does predict process behaviour this may also, in some tasks, be a simple statement about direction of change rather than a numerical specification of what will happen. In Cooke's (1965) study only 10-15% of statements about the present state were in relative rather than numerical form, but predictions were not numerical. In protocols collected from operators starting up turbines for electric power generation (Rasmussen and Goodstein, personal communication) the predictions were mainly simple, e.g., "it'll soon be up". However, Umbers (1976) found that predictions were numerical. It is rash to generalise to the reasons for this difference from so few examples, but one might point out that the operators making trend predictions were working from analogue displays and predicting the process behaviour, while Umbers' operators were working with digital information and predicting control targets.

Studies of experienced operators who do exert efficient control (e.g., Crossman and Cooke, 1962; Cooke, 1965) suggest that they may choose their actions without considering alternatives, by a process which is not available to verbal description. Acquisition of knowledge about actions appropriate to a given control context does not require predictions from a perfect internal model, but can be acquired from experience of correlations between action and effect. It might be misleading to assume that this knowledge is in the form of a very simple input-output look-up table however. Crossman and Cooke op cit found, by measuring correlations between control actions and various dimensions of control state, that it was only valid to describe an inexperienced controller as working by feedback since the correlations decreased with practice. From other data they concluded that the experienced operator used mainly open-loop control. This correlational learning, which enables the experienced operator to control without trial-and-error, gives a primitive form of process knowledge, without a separate specification of the nature of and reasons for process dynamics which can be used and discussed independently of doing the task.

These types of example suggest that the operator acquires these simple forms of knowledge by parameter trimming, though this may not be represented in the operator in specific numerical terms. However, analyses such as Cooke's (1965) suggest that such parameter trimming can be only a component of the learning, rather than the whole or even a major determinant of its development. Cooke's verbal protocols, and other data collected from university students controlling a simple one-dimensional process, the temperature of a water bath, suggest that control strategies are also based on hypotheses about how the process works. Some relatively simple propositions about its behaviour are combined together to make predictions. An alternative way of expressing this would be to suggest that the operator must start with some hypotheses about the "equations" which it is appropriate to trim, for example realising that it is necessary to take lag into account, e.g., (Cooke op cit, p. 365):

> "I think I'll take it off a bit when it gets up to about 75 because I don't want it to overshoot the mark (85) and I imagine it will still have some effect on the water inside the tube some time after the heating has stopped."

Some of the other propositions which Cooke's students mentioned were that heating was faster than cooling, sampling continuously was not necessary, and various control strategies were available. The students mentioned these points as they realised them from their experience of trying to control the process. (It was possible for a student to mention one of these points but for it not to lead to a revision of his control strategy, and vice versa). Such propositions are not automatically assumed by a beginner operator, but are acquired by training and experience. Development of a sufficient set of these propositions is necessary for adequate control. Some people may not be able to acquire such propositions from unaided experience, or may not be able to implement the more complex control which they imply. For example Crossman and Cooke (1962) found that the control performance of low intelligence people trying the water-bath task showed that they understood the notion of proportional control but were not taking lag into account.

Such analyses suggest that human learning of process behaviour does not start with a complete equation of the appropriate order, in which the parameters are then trimmed by learning, but that learning also involves acquiring the appropriate components of the "equations" before they can be trimmed. At a more complex level of modelling one would also have to account for the way in which increasingly sophisticated knowledge of process dynamics leads to the generation of new (to the operator) and more effective control strategies. (See also sections on Problem Solving and Predicting Action Effects). One assumes that such "propositional" knowledge of the process is at a "higher level" than the simple correlational learning described above, precisely because it may have the potential to be used for working out what to do in unfamiliar situations, which would not be possible given knowledge only of an "if this, do that" form.

Most of the above points about the nature of the operator's internal model have been inferred from his task performance, and sometimes from his verbal reports. The problems of studying the form of this model directly are very large. Presumably it is only studiable by the classic technique of making models for the operator's internal processes and testing their performance against his. An interesting example of this is given by Jagacinski and Miller (1978), who consider parameter estimation, to fit a model to behaviour in the usual tracking task, is too multi-dimensional to be successful, so they use a simpler step-change task. They fit an equation to the operator's performance, but admit that they have no information about how the operator has actually represented the system dynamics internally. This is probably an ultimate limit to any modelling of mental processes. However one should still be able to do useful work by testing, via models, a richer range of ideas about the nature of the operator's internal processes.

Acquisition of the Internal Model

The last section suggested that the operator's process knowledge can be at different levels of sophistication. If his potential for control is a function of knowledge then the quality of control is affected by anything which affects this knowledge. Studies suggest that the main influences, which all interact in their effect, are interface design, experience or training (see Crossman and Cooke, 1962), and the more general working conditions. Some aspects of these will be mentioned briefly.

The interface, and the operator's interactions with it, can affect the extent to which he can notice correlations between variables and learn the properties of their behaviour over time. One example comes from a study by Wickens and Kessel (1979). They assume that the effectiveness of a previously learned internal model can be measured by the speed and accuracy of detecting that system dynamics have changed. They find that this detecting is done better by people who have controlled manually than by those who have learned about the process by monitoring automatic controllers. The inference is that direct interaction with the process in manual control allows better learning of process dynamics. (This has strong implications about manual take-over from automated systems). If one accepts that primitive knowledge of process dynamics is mainly in the form of knowledge about directions and size of changes, then this emphasises the importance of compatibility between directions of change in displays, controls and process. Task load, the amount of work to be done, can affect moment to moment use of strategies of greater or less complexity (see earlier). For example Paternotte and Verhagen (1979), studied control of a simulated distillation column run at several speeds. The operators commented that they changed from feedforward to feedback control at higher task speeds. Task load may also affect opportunities for longer-term learning about process behaviour. For example Paternotte (1978), in his study of operators controlling 8 distillation columns argued that the operators used a "small changes with feedback" strategy because they had to divide their attention, so it was easiest to use a simple strategy with standard actions. One could also argue that 8 columns is above the number of processes which an operator could keep track of separately. He would therefore not be able to learn the individual characteristics of the 8 columns, which were not the same, so he would not have the knowledge from which he could generate control choices specific to particular columns.

Another more general aspect of working conditions would be the division of decision-making responsibility between operators, foremen and management, or between operators in a team, which would affect a particular operator's opportunities to experience parts of the process dynamics.

Comparing Alternative Actions

Control models often assume that the operator's process knowledge is used to predict the effects of alternative actions, as a basis for choosing between them. This is an attractive idea, but the actual data on manual control suggest as usual that the human operator works in ways which are sometimes simpler and sometimes more complex than this. We can discuss the two aspects separately: does the operator compare alternative actions? and does he predict the effect of actions in choosing his behaviour? Observational data on manual control can only show what the operator finally decided to do, not what other possibilities he considered while making the decision. Some information on this is available from verbal reports, although it should be remembered that there are occasions when several possibilities slip through one's mind much too quickly to be reported on. With this proviso on interpreting verbal protocols, when one looks for evidence of comparing actions one finds that this happens in two contexts: when comparing the effectiveness of past actions, and when predicting possible sequences of action in complex tasks with many degrees of freedom.

Past exemplars are used as patterns for effective behaviour now (see the discussion of Umbers', 1979b, gas-grid controllers), or a previous lack of success is used to suggest the way to revise behaviour to try this time, e.g., (Cooke, 1965):

> "It seems that this time I got up to the required range rather quickly but at the expense of the fine control keeping it there. It first of all went up too far, then it went below as in the first trial but nearly so bad. The second trial I seemed to take more time getting up there but once I got there I stayed there better than this time at any rate".

A complete model for this type of behaviour would need to contain a memory which could retain "episodes" rather than simply numbers, plus comparison processes which could also suggest new strategies to try. Again this example suggests that some operators revise their control strategy by more complex cognitive processes than would be represented by "parameter trimming".

There are two examples of situations in which experienced operators do make predictions about alternative behaviours. Umbers' (1976) operators predicted the ways in which gas demand might develop, and therefore the need for actions later in the shift e.g., (p. 326):

> "We'll be looking at it hourly and then we'll decide later whether it's necessary to increase during this shift or whether to make provision for it between 6 and 8 o'clock this morning".

The clearest example of comparing several predicted actions does not come from a control task. Smith and Crabtree (1975) collected verbal protocols from well-practised people during a laboratory task in which items had to be scheduled through a sequence of machines with different capacities. The task was to optimise the routes used. This is a task with a large number of degrees of freedom of action, and the people predicted and compared up to 3-4 alternatives.

Predicting Action Effects in New Situations

These examples lead one to ask whether the notion that control choice is made by predicting and comparing the effect of alternatives is inappropriate. Certainly the notion that control is oriented to the future has been mentioned frequently already. In particular, predictions about the future have been discussed in relation to future control targets, possible sequences of behaviour, and process sampling. Umbers (1979a) lists studies which find evidence for open-loop or feed-forward control. The data on predicting and comparing the effects of actions suggest, however, that this is not done by experienced operators in standard situations.

Predictions may appear as a reason for behaviour, e.g., (from Cooke, 1965) "I'm turning it down to 90, which will make it go down because it's below boiling point". Presumably knowledge about the effects of different actions is also used in the original development of a good control strategy. For example, in the furnace power control task, the knowledge that cutting power to a furnace which is at a particular stage of making a quality steel will disrupt the quality of its output leads to the strategy of cutting power to other furnaces. This information is no longer mentioned when the operator uses the strategy, e.g.:

> "I shall have to cut (furnace) E a bit, it was the last to come on, what is it making by the way? ... E make stainless, oh that's a bit dicey ... I shall not have to interfere with E then",

but the information may be available when the operator is asked to explain his behaviour when he is not actually doing the job, e.g.:

> "If a furnace is making stainless, it's in the reducing period, obviously it's silly, when the metal temperature and the furnace itself is at peak temperature, it's silly to cut that furnace off".

(These extracts come from the same operator in the same session). This suggests that predicting and comparing actions may be done primarily during the development of new behaviour in unfamiliar situations, which would occur particularly during learning, or when something has gone wrong with the process. Here is an extract from some operators having difficulty with starting up a turbine (Rasmussen and Goodstein, personal communication):

Operator A	I don't think anything will happen if we run it all the way up
Operator B	yeah, we're repeating ourselves, right as soon as we come up past a given point there then we can't
Operator A	that's the question
Operator B	yeah, but that was the one which alarmed wait and see when it comes over 15, what will it do
Operator A	we won't get any "reverse program" will we?
Operator B	no, no
Operator A	so we can continue
ALARM	there it goes again

These predictions seem to be made using conditional statements which include fairly simple propositions about process behaviour. The types of conditional statement which occur can be analysed, as a basis for beginning to understand this type of behaviour. These are basically of two types, about conditions which affect variable values (including the effect of actions) and about conditions on the use of actions. This use of process knowledge can be compared with Rasmussen and Jensen's (1974) study of electronic maintenance technicians, in which they found that the technicians used a simple general strategy appropriate to many instruments rather than using functional knowledge specific to a particular one. Laboratory studies which test diagnosis, given minimal information about random links, may be appropriate for investigating this type of diagnosis. The above anecdotal examples from process operators do however suggest that they may use functional knowledge. This would be more appropriate in process operators as they are concerned with one process with meaningful relations between the parts, of which they have a great deal of experience, and they need to find out how to control the process despite there being something wrong with it, as well as finding out what is wrong.

This evidence therefore suggests that operators do their trouble shooting by thinking through conditional statements which include simple dynamic knowledge about the process, mainly in the form of directions of change. This is related to the points made about Cooke's (1965) finding that sufficient strategies are based on sufficient simple, mainly cause-and-effect, propositions about process behaviour. It is also related to the points made on problem solving in the section on control goals. This would imply that control strategies are the result of goal (i.e., required behaviour) oriented search through conditional propositions about potential process behaviour. This would suggest that the adequacy of problem solving/trouble shooting would depend on the adequacy of these propositions, and of the procedures used in searching through them. This is something that we need to understand much more fully if we want to aid operators in their trouble-shooting. Rigney and Towne (1969) present a simple model for maintenance technicians' activity sequences which is of this type.

Summary

The operator's knowledge of process dynamics may not be in the form of control equations, but may be the result of simple correlational learning, which could lead to, or be related to, conditional propositions about general aspects of process behaviour. The effectiveness of the operator's "internal model" will depend on his opportunities for interaction with the plant, which emphasises the importance of interface design and training, and has implications for manual take-over. In choosing his control actions the operator may recall previous control attempts. He may predict the effects of an action as a justification for that action, or in trying to work out what to do in unfamiliar situations. His ability to do this will depend on the form and content of his process knowledge, and modelling this type of behaviour may require sophisticated models of cognitive activity.

GENERAL CONCLUSIONS

This paper has attempted to review the usefulness of control theoretic models, developed for fast tracking tasks, in describing manual control of slowly changing industrial processes. In many cases it seems that slowly changing tasks allow different types of cognitive processes to be used, or the task complexity requires different responses. The paper has not described the full complexity of process control behaviour, as would be evident for example, from reading Edwards and Lees (1974). However, there are still several major themes which have required much cross-referencing in a discussion divided into sections on goals, inputs and output decisions, particularly as these three aspects of behaviour are not necessarily clearly distinguished. The operator's mental or internal model depends on his interactions with the task and is basic to his potential performance. Task decisions are also a function of his "mental picture" or knowledge of the present process state. Complex cognitive activity may be involved in deciding which of the available behaviours is most appropriate in a given multi-dimensional situation, or in generating new behaviours in unfamiliar situations.

The question remains however, whether there are models which can be developed to a level of rigour which would be attractive to engineers. It was suggested earlier that much of the cognitive activity could be modelled by existing computer programming languages. Such programs could be used to predict qualitative aspects of behaviour, the results of decisions. They would not automatically produce quantitative predictions about time and accuracy. This would require parallel calculations, for which we have not yet really got adequate data. For some of the more sophisticated notions which have been mentioned briefly, neither the concepts nor the performance data are yet available. Perhaps some suitable concepts are emerging from Artificial Intelligence, as for example reviewed by Johannsen and Rouse (1979), though often their concepts represent logically possible ways of doing complex tasks, rather than ones which have been tested for parallels with human performance. Essentially we need a great deal more sophisticated analysis of performance in complex tasks, from which human operator mechanisms can be induced.

REFERENCES

Bainbridge, L., 1971, "The Influence of Display Type on Decision Making Strategy", in: "Displays. Conference Publication No. 80", Institution of Electrical Engineers, London.

Bainbridge, L., 1974, "Analysis of Verbal Protocols from a Process Control Task", in: Edwards and Lees, op cit.

Bainbridge, L., 1975, "The Representation of Working Storage and Its Use in the Organisation of Behaviour", in: "Measurement of Human Resources", W.T. Singleton and P. Spurgeon, eds., Taylor and Francis, London.

Bainbridge, L., 1979, "Verbal Reports as Evidence of the Process Operator's Knowledge", Int. J. Man-Machine Studies, 11:411.

Baron, S., and Kleinman, D.L., 1969, "The Human as an Optimal Controller and Information Processor", IEEE Trans. Man-Machine Syst., MMS-10:9.

Bartlett, F.C., 1943, "Fatigue Following Highly Skilled Work", Proc. Roy. Soc. B., 131:247.

Beishon, R.J., 1967, "Problems of Task Description in Process Control", Ergonomics, 10:177.

Beishon, R.J., 1969, "An Analysis and Simulation of an Operator's Behaviour in Controlling Continuous Baking Ovens", in: "The Simulation of Human Behaviour", F. Bresson and M. de Montmollin, eds., Dunod, Paris, reprinted in: Edwards and Lees op cit.

Brigham, F.R., and Laios, L., 1975, "Operator Performance in the Control of a Laboratory Process Plant", Ergonomics, 18:53.

Carbonell, J.R., 1966, "A Queuing Model of Many-Instrument Visual Sampling", IEEE Trans. Hum. Fact. Electron., HFE-7:157.

Carbonell, J.R., Ward, J.L., and Senders, J.W., 1968, "A Queuing Model of Visual Sampling: Experimental Validation", IEEE Trans. Man-Machine Syst., MMS-9:82.

Cooke, J.E., 1965, "Human Decisions in the Control of a Slow Response System", Unpublished D.Phil. Thesis, University of Oxford.

Crossman, E.R.F.W., and Cooke, J.E., 1962, "Manual Control of Slow-Response Systems", in: International Congress on Human Factors in Electronics, Long Beach, California, reprinted in: Edwards and Lees op cit.

Edwards, E., and Lees, F.P., 1974, "The Human Operator in Process Control", Taylor and Francis, London.

Efstathiou, J. and Rajkovic, V., 1979, "Multi-Attribute Decision Making Using a Fuzzy Heuristic Approach", IEEE Trans. Syst., Man, Cybern., SMC-9: 326.

Hamilton, P., 1969, "Selective Monitoring in Multisource Monitoring Tasks", J. Exper. Psychol., 82:34.

Jagacinski, R.J., and Miller, R.A., 1978, "Describing the Human Operator's Internal Model of a Dynamic System", Human Factors, 20:425.

Johannsen, G., and Rouse, W.B., 1979, "Mathematical Concepts for Modelling Human Behaviour in Complex Man-Machine Systems", Human Factors, 21:733.

Kleinman, D.L., and Curry, R.E., 1977, "Some New Control Theoretic Models for Human Operator Display Monitoring", IEEE Trans. Syst., Man, Cybern, SMC-7: 778.

Kok, J,, and van Wijk, R., 1978, "Evaluation of Models Describing Human Operator Control of Slowly Responding Complex Systems", Delft University Press.

Mackworth, J.F., 1970, "Vigilance and Attention", Penguin, Harmondsworth.

Mamdani, E.H., and Assilian, S., 1975, "An Experiment in Linguistic Synthesis with a Fuzzy Logic Controller", Int. J. Man-Machine Studies, 7:1.

Paternotte, P.H., 1978, "The Control Performance of Operators Controlling a Continuous Distillation Process", Ergonomics, 21:671.

Paternotte, P.H., and Verhagen, L.H.J.M., 1979, "Human Operator Research with a Simulated Distillation Process", Ergonomics, 22:19.

Pew, R.W., 1974, "Human Perceptual-Motor Performance", in: "Human Information Processing: Tutorials in Performance and Cognition", B.H. Kantowitz, ed., Erlbaum, New York.

Rasmussen, J., and Jensen Aa., 1974, "Mental Procedures in Real-Life Tasks: A Case Study of Electronic Trouble Shooting", Ergonomics, 17:293.

Rigney, J.W., and Towne, D.M., 1969, "Computer Techniques for Analysing the Micro-Structure of Serial-Action Work in Industry", Human Factors, 11:113.

Senders, J.W., 1964, "The Human Operator as a Monitor and Controller of Multi-Degree of Freedom Systems", IRE Trans. Hum. Fact. Electron., HFE-5: 2.

Sheridan, T.B., 1966, "Three Models of Preview Control", IEEE Trans. Hum. Fact. Electron., HFE-7: 91.

Sheridan, T.B., 1976, "Toward a General Model of Supervisory Control", in: "Monitoring Behaviour and Supervisory Control", T.B. Sheridan and G. Johannsen, eds., Plenum Press, New York.

Smith, H.T., and Crabtree, R., 1975, "Interactive Planning", Int. J. Man-Machine Studies, 7:213.

Teichner, W.H., and Krebs, M.J., 1974, "Laws of Visual Choice Reaction Time", Psych. Rev., 81:75.

Umbers, I.G., 1976, "A Study of Cognitive Skills in Complex Systems", Unpublished Ph.D. Thesis, University of Aston in Birmingham.

Umbers, I.G., 1979a, "Models of the Process Operator", Int. J. Man-Machine Studies, 11:263.

Umbers, I.G., 1979b, "A Study of the Control Skills of Gas Grid Control Engineers", Ergonomics, 22:557.

Veldhuyzen, W., and Stassen, H.G., 1977, "The Internal Model Concept: An Application to Modelling Human Control of Large Ships", Human Factors, 19:367.

Wickens, C.D., and Kessel, C., 1979, "The Effects of Participatory Mode and Task Workload on the Detection of Dynamic System Failures", IEEE Trans. Syst., Man, Cybern., SMC-9: 24.

Winograd, T., 1972, "Understanding Natural Language", Edinburgh University Press.

TASK ANALYSIS AND ACTIVITY ANALYSIS IN SITUATIONS OF FIELD DIAGNOSIS

Jacques Leplat

Laboratoire de Psychologie du Travail
de l'E.P.H.E. ERA. CNRS. n° 236
41 rue Gay Lussac, 75005 Paris

INTRODUCTION

A review of publications on diagnosis shows that only few of them relate to systematic field studies. Experiments in laboratory or simulation conditions are more frequent, though their relation to work situations is not always really examined. This relation gives rise to difficult problems (Leplat, 1976, 1978); it is necessary, however, in order to justify experimental studies, installation of systems, realisation of work supports, and training of operators. The necessary relation between field situations and laboratory conditions shows various forms (Rouse, 1979), considering for instance the representative quality of experimental tasks or the general character as opposed to the specific nature of diagnosis skill. One of the major difficulties of such a study certainly results from the analysis of field situations because of undeniable practical difficulties and also, because of lack of adequate theoretical outlines to guide such an analysis. Our present contribution will present a summary of such an outline, using examples to illustrate its necessity.

We shall first establish the distinct character of work situations, then develop two basic concepts: task and activity. They are not particularly original but, in our opinion, we are far from having demonstrated the full interest of distinguishing and co-ordinating them. This will be aimed at in the present study, stressing the necessary role of field analysis in every diagnosis study, particularly when it is undertaken in view of possible applications. No quantitative justification will be sought, our concern being concentrated on outlooks for studies and guidelines of analysis.

CHARACTERISTIC FEATURES OF WORK SITUATIONS

The essential differentiation between features of work situations and experimental conditions in the laboratory shall first be reviewed. The first and fundamental distinctive feature is given by the origin of situations. Work situations are neither born in an investigator's mind nor are they generated for the purpose of analysis. They are generated in order to realize a special production under determined conditions. The goals assigned to the operator and his work conditions are generally defined without consulting the investigator. His possibility of intervention is therefore limited, because of technical or economical or management contraints. The reference system of the operator is not always clearly set and clearly defined. Work instructions are often deficient, moreover, they do not necessarily fit with what the operator is actually doing. The latter is not passive in regard to his task. Particularly, he may use information not foreseen by the designer of the installation, making inferences from system regularities or from information on the functioning of systems controlling the one he is working on. The longer he works on the installation, the easier is it for him to get this information.

Features of the situation may vary largely, depending on the operator's knowledge level and on the type and level of training he has acquired. In work situation studies especially, one might meet old, very experienced operators who have automatized some stages of their activity, giving rise to problems unknown to the experimenter. The latter, working generally on new subjects, still in the training stage, meets other difficulties. Consequently, we assume that work analysis - here the analysis of diagnosis - shall define both the task and the kind of activity it requires.

ELEMENTS FOR ANALYSIS OUTLINE AND METHODOLOGY

Analysis of diagnosis situations, as well as of work situations in general, requires some essential distinctions which we shall now discuss. Under situation we define the "task-subject" system and its environment. Task shall be defined as the goal to achieve under some circumstances (Leontiev 1976).

The goal is indicated theoretically in the instructions. In the case of diagnosis, it does not mean modification of the initial states but their attribution to a class, using mostly mental operations. All elements and characteristic features taken into account in order to achieve the goal are considered as conditions. The set of all conditions is sometimes defined as the "device" (Hoc 1980). In process control, for example, the device will consist of components and the functioning instructions of the installation initiated for detecting the origin of the incident.

To execute the task, that means to achieve the determined goal, the device shall be used according to a procedure which may be put sometimes in terms of an algorithmic model. This procedure determines all the paths leading from the initial stage to incident identification.

The operator reacts to the task with an activity. This activity might be described on the basis of observable records (e.g., observed or manipulated objects). But in the case of diagnosis, activity is essentially mental. It may be assumed that activity is regulated by an internalized procedure, becoming thus what we shall call a <u>process</u>. In this way, process determines the rules of activity generation, indicating how and when device properties shall be treated in order to achieve the goal. Properties are recoded at the perception or representation level.

Activity analysis thus reveals which conditions are actually taken into account by the operator to execute the task. It becomes possible to state the deviations (Vergnaud, 1968) between the task as prescribed in the instructions, and the effective task, realised actually and fitting with the subject's activity (Figure 1). This is pointed out in several field studies on work (Faverge, 1955, 1966, Herbst, 1974, Leplat and Cuny, 1979). The deviations may concern conditions of execution, goals and/or procedures, (Figure 2). Their identification is an important stage in the investigation. They demonstrate in fact a maladaptation of "man-task" coupling, requiring besides its statement, an analysis of its origin and consequences. It may appear, for example, that the operator does not intervene at the moment foreseen by the instructions and that he takes into account signs appearing

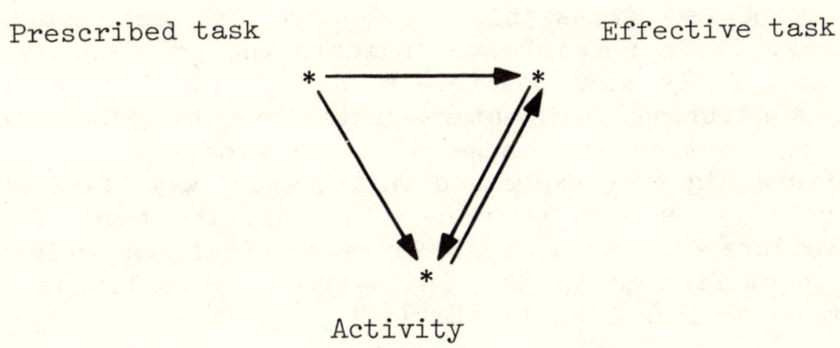

Figure 1 : The tree concepts involved in alaysis of a work situation.
Arrows show the dependence relations

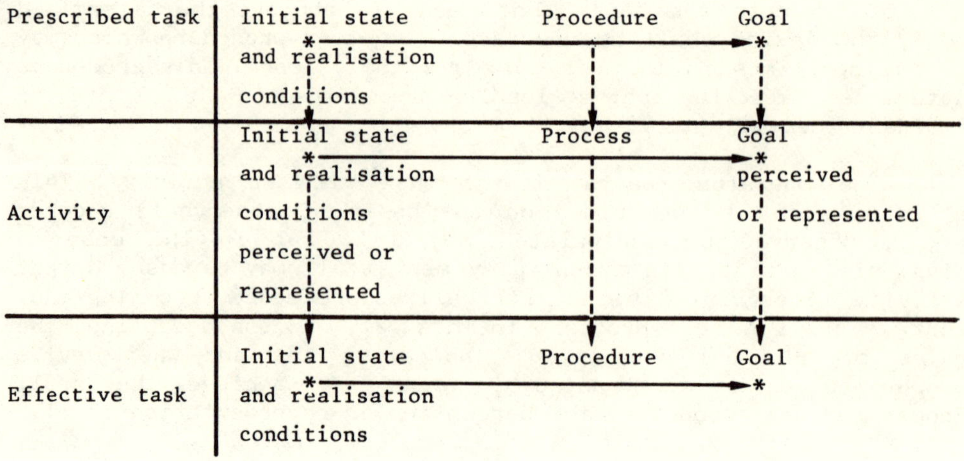

Figure 2 : 3 Analysis levels and relations between concepts

directly in the output product instead of indications on his console, being thus able to apply an abbreviated but not always efficent procedure.

Deviations are also presented and discussed in Landa (1974) with respect to the distinction he introduces between "algorithmic prescription" and "algorithmic process".

1. Task expression and optimal task

The task is expressible in a more or less explicit and detailed way. It is possible to indicate the goal solely and, on the contrary, it is also possible to describe with high precision execution conditions and intermediate targets. Thus the task assigned to operator in charge of identification of an installation failure might be expressed in a summary way "find origin of the failure". It is also possible to detail the task "first find out if failure is mechanical or electrical in origin, then continue identification", the indication for continuation being possibly more or less clearly stated.

An explanation of task T may be presented in symbolic way as a structure of sub-tasks (t_i):

$$T = \{t_1, t_2, \ldots\ldots t_k\}$$

Each sub-task is related to a sub-goal; its realisation being a condition for execution of the next task. It shall be noted, in fact, that realisation of a given task under specific circumstances might require only some of tasks t_i. A more precise expression of what often forms the algorithm of execution could be suggested by the distinction between procedure and condition (s) of execution in the tasks (that is the approach employed by Lyapounov and Shestopal, cited by Landa, 1974, p. 55).

Division into sub-tasks requires clarification of a procedure for organising and determining the sub-tasks (Figure 3). Depending on the result of one elementary procedure (+ or –), one or another of the following sub-tasks will be executed, the result being really the condition of execution for this sub-task.

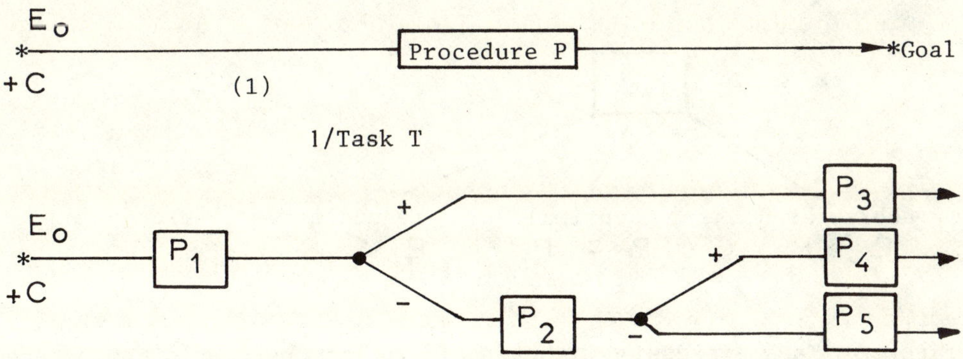

(2)

2/ Division into sub-tasks t_1, t_2, t_3, t_4, t_5

Figure 3 : Theoretic model of a diagnosis task defined globally (1) or more detailed (2). A sub-task is represented by e—[P_i]—s e represents entry and realisation conditions, s output or result obtained by the procedure realised. The result becomes entry and condition for realisation of next task.

Each sub-task may be divided in more elementary sub-tasks. Thus $t_i = \{t_{i1}, t_{i2}, \ldots . t_{ik}\}$ and so on.

A task is immediately executable for a subject and completely determined by him if he has previously acquired the ability to realize the corresponding procedures; that means, if he has the process available (or internalized procedures). If the process has not been elaborated and the tasks or sub-tasks cannot be executed, the subject must undertake their acquisition, and thus the task becomes a problem for him.

An <u>optimal task for a subject</u> could be defined as a task description that is just precise enough to make task execution possible for him. A more rough or summary description would oblige him to acquire abilities he does not possess yet. A more precise and thus redundant description (Figure 4) would be useless, when concerning elementary, already mastered tasks, or disturbing, when

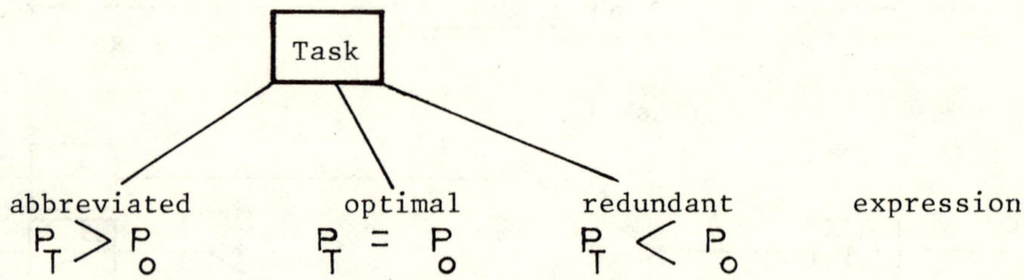

Figure 4: Task expression with respect to the operator's competence.

P_T procedure prescribed for task realisation

P_O the largest procedure internalized by the operator

$>$ more summary than $<$ more detailed than

the structure of distinguished sub-tasks does not agree with those required by the activity; though both $t_1, t_2, \ldots t_K$ and $t'_i, t'_2 \ldots t'_k$ may make it possible to realize the same global task T. In fact, in the latter case, acquisition of a new method would be required, which is not necessary if the final goal or the goals relative to the optimal task are exclusively pertinent.

The optimal task varies for the same subject during training. It becomes more and more summary in the sense that sub-tasks become fewer and larger, and expressions of the global task become progressively more allusive. A psychological problem occurring at this point is to know if the subject will be able to perform elementary tasks again, perhaps to explain them verbally or for embedding as elements of other complex tasks.

On the other hand, in a sense, analysing the activity by a subject means determining task T and related elementary tasks t_i; that is, the units of treatment available for the subject for execution of task T and their organisation.

2. Description of optimal task and regulation of activity

Description of the optimal task for a subject at a given stage leaves open the problem of control of his activity. To cite an extreme case, the optimal description for an experienced operator will be simply an utterance of the task goal (find origin of the failure). But the method of control might be quite varied in this task. In remote-control, for example, the process of diagnosis may be based on the identification of displayed signals or on knowledge of the functioning rules of the system. Definition of the mode of activity regulation, which is correlated with the subject's competence, cannot be deduced solely from observable data collected on one and the same task. It should be based on different methods; variation of "normal" execution conditions, verbal clarification, error analysis etc. ... This shall allow one to determine the level of functioning of the operator at a given moment. Rasmussen (1979) recently provided useful discussions of this point, particularly concerning diagnosis.

A way to deal with this problem is to determine the way operators define and organise sub-tasks, particularly sub-goals. They may be organised as a construction of specific sequences (on our figure 3, for example, (t_1, t_3), (t_1, t_2, t_4), (t_1, t_2, t_5) or according to formal features of the installation (example: dichotomous strategies). Their definition may be also be based on knowledge of the functioning rules of the system, derived using different principles (e.g., progressive/regressive).

3. "Task-activity" dialectics

Task analysis and activity analysis are closely related. Activity analysis assumes some previous knowledge of the task, allowing one to assign at least a provisional meaning to observed behaviours. But task analysis and description assume knowledge of the features of the subject's activity, otherwise the task defined does not agree at all with the one actually executed, the danger being the greater, the more finely the task is described in terms of sub-tasks. Consequently, work analysis frequently proceeds in a spiral movement; better task approximation leads to better activity approximation, and inversely. Progressively the analyses can enrich each other until reaching the desired quality of precision.

The necessity of this connection was stressed by Newell and Simon (1972) who wrote in particular:

"Demands on the task environment and psychology of subjects (...) should never reduce the reader into thinking that as a psychologist he should be interested only in the psychology of the subject. The two aspects are in fact like figure and ground - although which is which depends on the momentary viewpoint" (p. 55).

ROLE OF TASK ANALYSIS IN ACTIVITY ANALYSIS

Based on the theoretical outline thus far defined, we shall now show how data obtained by task analysis may be used to acquire better knowledge of activity, and also, what are the limits of this utilization. Study of goals and study of procedures shall be successively examined in this regard.

1. Prescribed and effective goals

Gaps between the prescribed and effective task may involve goals or sub-goals. The first important origin of gaps is a function of the permitted tolerance. Some operators accept larger deviations of the system than those officially stated; e.g., they intervene only when indications deviate appreciably from the zone considered as "normal". Analysing the origin of these allowances may conclude that the cause is simple ignorance. It may also result from a knowledge of the installation's functioning, showing for instance that, all other conditions being considered, the gap should decrease rapidly without intervention (De Keyser, 1972). Gaps can reveal a strategy.

Gaps are frequently multi-dimensional and the importance attached to each dimension is not always clarified, besides it may vary depending on operators. In his study of a rolling mill installation, Delahaut (1966) observed that criteria applied by subjects did not agree with those of factory management. The latter gave priority to urgent orders - an order concerns different-sized ingots - without considering the need for frequent modification of controls thus implied and its disadvantages. On the contrary, some operators tried to reduce the amount of control modifications so as to facilitate the work, without considering order urgency. Gaps between the two tasks were partly justified by the characteristics of the operators; the latter, ex-rolling-mill operators themselves, took more in account the work of this type of operator, as technical conditions of the installation's functioning.

2. Standard procedure and activity

The task analysis may generate procedures. The formal procedures may be compared to those actually realized by the subject and thus represent a model of the process. Studies by Rigney (1969) clearly showed the utility of the method for diagnosis investigation. The author elaborated the optimal procedure for searching for failure on basis of the failures/symptoms matrix. He then compared this procedure to the one really applied by operators and calculated the difference between them using various indicators. Gaps reveal the properties of the activity, e.g., "the technicians operate with incomplete and inaccurate hypothesis sets most of the time while trouble-shooting, just as they operate with lowgrade symptom-malfunction relationship information" (p. 428). But these properties are negatively defined, and, even if it is not uninteresting to know that technicians do not behave according to Bayes' model, it would be still more interesting to know the model according to which they do behave.

It is possible to try to reduce gaps using a model; the way the author does, involves substituting subjective probabilities for objective probabilities. It could also be possible to introduce a variable accounting for the cost of control process (difficulties of tests, time and ease of access). Examination of model predictions - based on the hypothesised task features - would allow one to learn something about data resulting from the subject's activity and could thus help to determine the characteristic features of the organisation of the latter.

However, this method is limited, particularly by the fact that it assumes the task realized by operator as being the task prescribed, especially, that the reference system is the same. Further, the model leads one to collect certain types of observation which are not necessarily the most revealing as to the process applied by the subject. Thus a procedure model of the Bayes' type would neglect all aspects concerning the perceptual configuration of system components which, as it is known, plays a role.

Task analysis and models are very useful but appear limited if not enlightened by activity analysis, as we shall see in the following chapter.

ROLE OF ACTIVITY ANALYSIS IN TASK DEFINITION

We have seen how activity analysis generates the definition of the task really executed by the operator. Task knowledge is necessary for correct analysis of operators' performance and errors. Without knowledge of the actual system on which the subject works, it is not possible to foresee the effect of installation modification and of work conditions in general. This problem is fundamental in some work situations which often reveal that the task is poorly defined or sometimes even little known by staff management. Cuny (1977) thus observed during investigation on a process in the chemical industry that official notes and necessary instructions defined rules neither applied nor appliable since their authors strongly understated the complexity of the situation presented to the operator (p. 12). We shall try to show using two points how activity analysis may lead to reconsidering the initial task conception.

1. Identification of the reference system

When an experimenter presents a task to a subject, the experimental device is well defined and the experimental methodology is sufficiently well elaborated as to avoid artifacts and to assert that the task proposed by the experimenter is the task of the subject as well. That is not the case in work situations when the task assigned to operator is surrounded by a complex environment, frequently related to other tasks, more or less connected to them. In order to understand the operator's activity, it is necessary to recognize the actual system on which he works. We might thus also find an answer to the question asked by Rasmussen (1979); how far back should the search for error causes go?

It was demonstrated that, at such a workplace, operators may work on a different system than the one assigned in the prescribed task. The differences may have various origins. They may result, first, from an enlarged reference system. De Keyser (1972) found several examples of the type, particularly in process control in the iron and steel industry. She showed that operators in search of causes of malfunctions of an installation, "tracked far back in the process" going beyond the officially assigned limits of their work. In the same way, Assenheim (1969) showed in work concerning quality control in a glass factory, that the "work horizon" of inspectors varied strongly. Some of them concentrated on sorting, often considering their work as part of a large system including preceding working places. The second type saw their role in terms of diagnosis, with results influencing preceding stages of production, in order to lead to corrections on the output product.

The extent of the system considered seems to depend largely on the intervention possibilities the system allows. This idea is expressed in a recent paper by Rasmussen (1979): "The failure of a component is generally accepted at the level where replacement is convenient" (p. 3). Investigating the diagnosis tasks of the chief operator in a data control system room, Michard (1978) observed: "If some parts of system require, for technical reasons e.g., distance, a verification by other operators, the subject tends to eliminate these parts from logical diagnosis units on which he is working" (p. 69), which may produce setbacks in his diagnosis work. In studies on safety systems with operators trained for recognition of accident causes with the help of error trees, Dogniaux (1978) could show that the size of constructed error trees diminished with distance from the time of training. The author attributed it to the fact that operators did not mention any more the factors on which they knew they had no possibility of intervention. In this case, experience leads rather to reduce the reference system.

In comparison, in the task of breakdown diagnosis based on a synoptic display, De Keyser and Piette (1970) showed an increase of the field of variables in proportion to the operators' competence.

The extent of the reference system depends thus on several factors working in different directions and their combined effects are not always easy to predict.

2. Delimitation of the considered temporal field

The subject's activity is characterized by size and by extent of the system taken into account but also by its temporal span. Activity analysis often showed the importance of what was sometimes called the "temporal horizon" in activity regulation. The "temporal horizon" may imply two dimensions: the future and the past one. As for the past, for example, it is known that an important factor of diagnosis tasks is determined by the occurrence frequency of different breakdown types. The frequency is reconstructed by the subject integrating more or less of his previous experience.

Operators may also take into account the systematic variations shown by the indications, and then extrapolate them. As noticed by De Jong and Koster (1974), the subject constructs a "phenomenal limit curve". According to the authors, when the curve approaches the tolerance limit, the operator makes a control action. We are here in the presence of a variable determining control occurrence, the so-called sampling time. Thus, the temporal dimension of the system is essential for task definition. The task may change completely with the size of this dimension.

As for the future dimension, experimental studies, in particular by English psychologists working with Bartlett, revealed the importance of anticipation phenomena, showing that subjects' reaction is not solely aimed at correction of the present situation but also at generating conditions which would eliminate error causes in future. In studies on air trafic control (Leplat, Bisseret, 1965), it could be demonstrated that besides recognizing risks of immediate conflicts, operators also identify those which might occur in the more or less distant future, allowing themselves time for intervention. The intervention concerns thus not only solution of the particular conflict itself, but attempts to deal with a situation with minimized probability of new conflicts occurring.

In a study of observations of synoptic displays in the iron and steel industry, De Keyser and Piette (1970) found two types of activity related to the extent of the temporal horizon. For the first type, diagnosis and interventions are short term, of symptomatic and palliative nature, while in the second case, diagnosis and interventions are long term and aim for stabilization of later functioning.

Thus analysis of diagnosis activity reveals the importance of the temporal dimension, often neglected in formal task definition. This dimension can interfere according to various modalities and knowledge of the modalities often influences directly the efficiency of training and ergonomics action.

CONCLUSION

The differentiation and relationships between task and activity analysis, as proposed in the present text, demonstrate the richness as well as the difficulty of field investigations on diagnosis situations.

Some important conclusions may be stressed:

- Caution against an exclusively technically-minded attitude: This means over-emphasis of task analysis by the technical expert considering his explicit or implicit activity model identical to the model of the operator.

- Caution against anticipated simulation: Taking into account situation features without determining the features pertinent to the operator's activity at the considered moment. Simplifications and limitations of the reference system itself, introduced by simulation tasks, may result in eliminating fundamental task features, and it is necessary to evaluate the consequences of this reduction. Also, the subjects' competence represents an important factor, in particular, one should always bear in mind that the activities used during acquisition may considerably deviate from those displayed by experienced operators. Simulation emphasizes the importance of coordinating task and activity analysis.

- The concept of the optimal task developed above may have direct applications in training, planning and conception of diagnosis aids. This item is close to the conclusions drawn by Duncan (1975).

- On a theoretical level, the analysis above approaches some aspects of those proposed by Soviet psychologists, influenced by Leontiev (1976) and Galperine (1966) in connection with action generation and with the notion of the basis of orientation. This perspective may be further developed. The investigations clearly set problems relative to activities elaboration and planning, which cannot be solved without the help of psychology and the theoretical outlines it suggests. This help is essential for pertinent approaches to experimental problems (Leplat, 1976), within which simulation is a specific case.

REFERENCES

Assenheim, G., "Etude d'un Systeme au Travers d'un Centre Privilégié,: un Poste de Controle en Cristallerie", Le Travail Humain, 32, p. 1-12, 1969.

Cuny, X., "Analyse de l'Activité de Régulation dans une Situation de Conduite de Processus". Doc. inter. du labo de Phych. du Travail, 1977.

Delahaut, J., "Le phénomene de Régulation au Niveau de l'Entreprise", in L'Ergonomie des Processus Industriels, J.M. Faverge et al. (cf. infra) p. 61-82, 1966.

de Jong, J.J., Koster, E.P., "The Human Operator in the Computer-controller refinery", in E. Edward and F.P. Lees, The Human Operator in Process Control. Taylor and Francis, London, p. 196-205, 1974.

de Keyser, V., Piette, A., "Analyse de l'Activité des opérateurs au Tableau Synoptique d'une Chaine d'agglomération", Le Travail Humain, 33, 2, p. 341-352, 1970.

de Keyser, V., "Fiabilité et Expérience", in Fiabilité et Sécurité Luxembourg C.E.E. ch. II, p. 79-140, 1972.

Duncan, K.D., "An Analytical Technique for Industrial Training" in Measurement of Human Resources, Taylor and Francis, London, Ed.W.T. Singleton and P. Spurgeon, p. 131-150, 1975.

Dogniaux, A., "Approche Quantitative et Qualitative d'un Probleme de Sécurité Industriel", J. of Occup. Accidents, 1,4, 311-330, 1978.

Faverge, J.M., "L'Analyse des Processus Industriels", Ed. de l'Institut de Sociologie, 1966.

Galperine, P., "Essai sur la Formation par Etapes des Actions et des Concepts" in Recherches Psychologiqies en URSS Editions du Progres Moscou, p. 114-132, 1966.

Herbst, P.G., "Socio-Technical Design", Tavistock Publications, 1974.

Hoc, J.M., "Articulation Entre Description de la Tache et Caractérisation de la Conduite", Bulletin de Psychologie, 1980, 34, 4-11, p. 207-212.

Landa, L.N., "Algorithmization in Learning and Instruction", Educational Technology Publications, Englewood Cliffs, New Jersey, 1974.

Leontiev, A., "Le Développement du Psychisme", Ed. Sociales, Paris, 1976.

Leplat, J., Bisseret, A., "Analyse des Processus de Traitement de l'Information chez le Controleur de la Navigation Aérienne", Bull. du CERP, 14, 1-2, p. 51-68, 1965.

Leplat, J. "Testing Hypotheses in Situations not Designed by the Experimenter", Studia Psychologica, 18,2, 117-124, 1976.

Leplat, J., Cuny, X., "Eigenschaften und Funktionen von Repräsentationen bei der Prozesskontrolle", in F. Klix, K.P. Timpe, Arbeits- und ingenieur-psychologie und Intensivierung, p. 61-67, 1979. V E B Deutscher Verlag der Wissenschaften Berlin, 1979.

Michard, A., "Représentations Opératives et Modeles de Processus dans les Taches de Diagnostic", These, Université de Paris VIII, 1978.

Newell, A., Simon, H.A., "Human Problem Solving" Prentice Hall, Inc. Englewood Cliffs, New Jersey, 1972.

Ombredane, A., Faverge, J.M., "L'Analyse du Travail", Paris P U F, 1955.

Rasmussen, J., "What Can be Learned from Human Error Reports", in Duncan, K., Gruneberg, M., and Wallis, D., (eds.): Changes in Working Life. John Wiley & Sons. (Proceedings of the NATO International Conference on Changes in the Nature and Quality of Working Life, Thessaloniki, Greece, 1979).

Rigney, J.W., "Simulation of Corrective Maintenance Behavier", in La Simulation du Comportement Humain, p. 419-428, Dunod, Paris, 1969.

Rouse, W.B., "Problem Solving Performance of Maintenance Trainees in a Fault-Diagnosis Task", Human Factors, 21,2, p. 195-204, 1979.

Vergnaud, G., "La Réponse Instrumentale Comme Solution de Problemes: Contribution", These de Doctorat de 3e Cycle, Paris, 1968.

MODELS AND EXPERIMENTAL RESULTS CONCERNING THE DETECTION OF OPERATOR FAILURES IN DISPLAY MONITORING

Barry H. Kantowitz and Raymond H. Hanson

Human Information Processing Laboratory
Department of Psychological Sciences
Purdue University, West Lafayette, Indiana 47907, USA

INTRODUCTION

This paper deals with the human operator's ability to monitor visual displays. We shall discuss operator failures to detect specified signals or changes in display information. This is indeed a limited topic since the mere detection of a signal is always followed by some operator action. If an operator could perform as a perfect detector but failed to initiate corrective action there would be no practical benefit from this ideal detection. Yet, it is also clear that detection per se is a necessary prerequisite for operator action. An operator who fails to detect a deviation from some controlled process will never be able to control effectively. It is our hope that a partitioning of life into detection versus later stages of information processing will allow the ultimate emergence of a detailed model of all of operator perception, cognition and action. The research discussed here will be only the "front end" of the more complete model.

Accordingly our first effort must be to carefully distinguish that behaviour we call "detection" from other related aspects of operator behaviour. While any definition is to some extent dictated by fiat, we believe it is important for us to articulate explicitly how we structure the detection task, if only so that others can disagree. From this we shall proceed with a brief overview of five classes of models that have been widely used to describe and/or explain human behaviour in detection tasks. Although we firmly believe that models are best stated in terms of formal systems like mathematics and computer programs, our emphasis will not be upon equations. Instead, we shall try to focus on the assumptions about the human operator that are necessary to render the formalisms appropriate: e.g., linearity, superposition, memory capabilities and so forth. Finally, we shall offer a simple taxonomy in order to organize the empirical results from experiments that have obtained data from warm, breathing humans. Indeed, we shall reserve the term "experiment" to describe only those reports that contain such data; reports that describe the behaviour of a computer used to try out various parametric combinations of particular models will be called "simulations" rather than experiments. We think it is only fair to warn the reader of our prejudices from the onset. As psychologists we are concerned primarily with the behaviour of humans required to detect a change of state in a visual display. The behaviour of normative and optimal models of the detection task per se, while interesting and important, does not concern us nearly as much as the behaviour of humans. Perhaps an analogy may clarify this distinction. In computer science research can be organized into a continuum ranging from artificial intelligenge (AI) to computer simulation of human behaviour. While very efficient algorithms can be formulated for AI solutions to cognitive problems like playing chess, many of these algorithms are utterly inappropriate as descriptions of human behaviour since they require processing resources the human lacks. Our interest would be in those heuristic solutions that could be attained within human capabilities. So, if we appear to slight some very clever models of detection, we ask the reader to remember that our focus is upon human behaviour rather than model behaviour.

DEFINING THE DETECTION TASK

It is easy to list examples of detection tasks. We suspect that if we asked participants to sort cards containing task descriptions there would be general agreement ($r = .83$) about which tasks on the list were detection tasks. But such an inductive approach may not be as helpful as attempting a formal definition. For example, it is also easy to get agreement about which particular sets of mappings from stimulus to response are more compatible than others, but this exercise does not lead directly to a theoretical understanding of S-R compatibility.

Our first definition of detection is relatively informal. Detection is the ability to notice or perceive a change in a set of stimuli. While at first glance such a definition may appear satisfactory, it fails to specify a behaviour that will allow some external observer to verify the "ability to notice or perceive". While it is perhaps possible that detection states might be monitored by purely physiological measures derived from brainwaves or heart rates, thus eliminating the need for overt behaviour, these psychophysiological variables derive their validity from correlation with an external observable response and so fail to avoid the issue of overt behaviour. In short, one can never know if a stimulus change has been detected without compelling the observer to perform some overt action. It is this action requirement that complicates the definition of detection. To anticipate our argument, we now state that the more complex is this required action, the more likely is the human operator to depart from a state of "pure detection".

One of us has argued elsewhere (Kantowitz, 1974) that the process of response selection is an important and crucial aspect of human information processing. Fascinating as response selection is, for our present focus upon detection it represents an unwanted complication of the detection process. An ideal detection task would require no response from the human. But this theoretical ideal is rather impractical. So we must settle for paradigms that minimize response selection. One scheme for calculating the amount of response selection imposed by a task uses the Shannon information metric (Kantowitz, 1974). A task that has but a single response has no information in the technical sense and so is the best we can do in defining a practical detection task. To the extent that more than one response is available to the human operator, we are departing from pure detection tasks and complicating the psychological processes forced upon the operator.

Perhaps some concrete examples may help make this point more salient. Imagine an operator faced with a visual display consisting of ten green squares. If any square turns red the operator is required to detect this stimulus change and to respond by pushing a large button labelled STOP. Since there is but a single response, this situation is close to an ideal detection task. There is no response information, although there is stimulus information of up to 10 bits depending upon the probability distribution over the stimulus set. (The next paragraph explains why more than one stimulus was used in this example). Since there is more information in the stimulus set than in the response set, this situation is an example of a many: 1 stimulus-response mapping. There is virtually no diagnosis or response selection imposed upon the operator. Once a change in the stimulus set is detected the requisite action has already been pre-selected by the task structure.

Now let us complicate the situation by requiring separate response buttons for each stimulus square. (This is, of course, formally identical to the case with one square and one button). This is a 1:1 stimulus-response mapping. While response information is greater than zero, it will usually not exceed stimulus information. Thus this type of detection task requires some diagnosis for the operator must not only detect a stimulus change but also must decide what response to make. Note that the "simplest" detection task--one red/green square and one response button--falls within this 1:1 mapping rather than the many:1

mapping discussed above. Yet, we believe that the many:1 task is a better exemplar of detection in general. It is only the degenerate case of the 1:1 mapping that is as simple. But even tasks with 1:1 mappings may go beyond our classification as detection tasks. For example, a 1:1 mapping with a continuous (analog) signal and continuous response set is more complex than detection. Such a tracking task is better considered an example of control rather than of detection since a large component of the task has to do with the psychological processes of response selection and execution. Thus, we define a detection task partially by exclusion: detection tasks require detection and little (ideally, nothing) else. Finally, a situation can exist where many responses are appropriate once a stimulus change has been detected. This 1:many mapping is far too complicated to be called a detection task and will not be discussed further.

In summary, it should be clear that, the title of this paper not withstanding, we cannot study a pure detection task. Any task of practical significance will be contaminated, to some degree, by response processes. Furthermore, our use of stimulus-response mappings to describe different degrees of detection, should make obvious the close relationship between practical detection tasks and the choice reaction tasks studied by the Dutch physiologist Donders more than a century ago.

A QUARTET OF MONITORING MODELS PLUS SOME PSYCHOLOGY

Before discussing any models of display monitoring we shall first mention the relationship between detection and monitoring. The two terms are closely related but not identical. Any monitoring task requires detection. But the response processes that are a contamination in a pure detection task are cheerfully accepted as a necessary evil in the monitoring task. So we define the monitoring task to include both detection and action. While the monitoring task may not require the operator to initiate a control action to correct an error, or to diagnose or interpret a set of error messages, it does require an overt response indicating that an error or stimulus change was detected.

This section briefly surveys four classes of model drawn from engineering that have proved useful in describing the monitoring task: queueing models, control theory models, signal detection theory and information theory. Each of these models has a sound mathematical base and the virtue of clear, precise prediction. Since entire tomes have been written to describe each of these models, it would be presumptuous to attempt any detailed comparisons among the four models. (We shall assume the reader to be generally familiar with these models). Instead, we will briefly note some of the assumptions of each and then compare the entire quartet to a fifth class of models drawn at least as much from psychology as engineering. This fifth class generally lacks the elegance and often the formal properties of the quartet but has nonetheless proved useful to psychologists studying human behaviour.

Queueing Models

The queueing class of models is considered first because it appears to lend itself most readily to the monitoring task where the operator must attend to more than one display or more than one stimulus element within a display. The basic constructs of queueing theory (Cox & Smith, 1961; Kleinrock, 1975) such as the number of customers (tasks to be serviced), rates of arrival, costs of delay, time required to service a task, and so forth can be directly applied monitoring tasks as performed by machines, humans (Carbonell, 1966) or both (Rouse, 1977).

The typical queueing model reported in the pages of the IEEE Transactions on Systems, Man and Cybernetics does not have a simple analytic solution; often no analytic solution is possible. So experimenters rely upon computer simulation to generate predictions of their models. Although one is often compelled to make strong mathematical assumptions -- e.g., Gaussian signals, random walks, exponential distributions -- to provide analytic solutions, there is less need for this when using computer simulations. Nevertheless, most simulations make the same kinds of a priori mathematical assumptions. Note that such mathematical assumptions differ from process assumptions such as describing how dynamic priority in a queue is achieved, that is, under what conditions (if any) can a later customer be serviced prior to a customer already in the queue.

Process assumptions in published queueing models are logical and orderly being based upon the expected behaviour of a rational, if not ideal, servicer. For example, a rational queueing model would service a customer if available rather than look for additional new customers. It would be highly inefficient for a device that required appreciable time to check on customer availability to seek out new customers (that might not exist) in preference to serving a customer at the head of the queue. Yet, our research (to be described later) has found that humans exhibit precisely this type of irrationality. If there is a "dead space" in the task requiring service such that it could be serviced now or the server could take a chance that service might be deferred without cost, humans tend to look for other customers that are nearer to the upper limit of the dead space, even though on the average this strategy results in poorer performance. In fact, humans may "foolishly" refuse to service any available customer in a queue until the entire queue has been scanned even though such a strategy increases memory load, often beyond the processing resources of the human operator.

A common engineering approach to this problem is to resolve the discrepancy between model and human behaviour by noting that the human is stupid. Of course, few would put the distinction so baldly. The modal way to handle this is to called the model an example of optimal performance. Hence, the human need not feel offended at being less than optimal since, as we all know, nobody is perfect. But this approach puts the cart before the horse. Unless one is concerned primarily with the task per se, rather than the human's performance of that task, it makes more sense to start from the human's behaviour. There are many examples in the psychological literature where exhaustive search works better for the human, even though a rational machine might prefer to terminate searching when an object of interest is discovered (e.g., Sternberg, 1967).

The point of this objection to most queueing models is that psychologists search for consistency in human behaviour rather than rationality. Impeccable principles of logical analysis often fail to capture the essentials of human behaviour. Human behaviour is not always transitive, commutative and linear. Since a queueing model is a set of rational assumptions, it may be quite difficult to discover which particular subset is at fault when human behaviour does not coincide with model behaviour. This difficulty will not be solved by replacing a model with a more complex and sophisticated model, even if the new model fits better. The opposite approach of removing assumptions so that small subsets can be falsified is a better way to study human behaviour. Such a strategy is not popular since it implies leaving situations with high face validity -- e.g., at least having laboratory displays look something like a tiny airplane cockpit display -- in order to exercise greater control in laboratory situations that offer greater potential for a theoretical understanding of human behaviour but fail to have immediate face validity -- e.g., what does a Donders' choice reaction have to do with Three Mile Island? (This point cannot be expanded here but see Kantowitz (1980) for a detailed discussion in the context of interfacing theoretical psychology and applied human factors).

Another serious difficulty with queueing models lies in the adroit way they finesse (or ignore) the detection process. How does the model know that a customer is available for service? Most models either assume that this detection process is automatic and error-free or assume some probability distribution over available customers. Now, we have nothing against probability distributions, although we have never invited one home for dinner. They are essential components of almost all models. But invoking a probability distribution does not mean that a process is now understood. In fact, the opposite is more likely. Distributions are used when we are ignorant. (This is one reason why the precise distribution chosen is seldom of great import and is therefore selected to maximize convenience). They allow us to continue our modelling without being hung up in the details of one tiny portion of our model. We have no objection to this common practice so long as we realize that distributions posited by fiat are necessary reifications that help the gears and cogs of our models turn. But each such arbitrary distribution represents a problem that will have to be solved eventually. When we are concerned with human behaviour, the detection process is far too important to ignore. Queueing models will not provide outstanding descriptions of human behaviour until they are elaborated to explicitly account for the detection process.

Information Theory

Although many cognitive psychologists believe that information theory has not lived up to its billing and cannot offer a sound foundation for the explanation of human behaviour, we are in the minority of experimental psychologists who stoutly maintain that information theory is an extremely useful tool (Kantowitz, 1975). The best-known application of information theory to human monitoring is the model of Senders (1964). Like the queueing model, it assumes the operator to be rational. In this case rationality implies knowledge of the Shannon 2WT sampling theorem. Thus, the bandwidth of the signals should determine how an observer fixates successive displays. Senders experimental procedures were particularly well chosen to study the detection process since the same response was used for all display dials; that is, the stimulus-response mapping was many:1. Hence, it was not surprising that his empirical results were well fit by his model. But even so there were small systematic differences between data and model. Moray (1979) has convincingly explained these as due to the human's inability to accurately match extremes of observed statistical distributions. Such conservatism is a well-known feature of human decision making.

Since Senders reported data only about fixation times and transitions his model cannot be completely evaluated. Studies reporting operator failures and detection times are also needed. But Senders has demonstrated that information theory is a good start for a description of human monitoring. Of course, the information metric itself is consistent with several internal information-flow models within the human and the success of the model in predicting fixations does not guarantee the correctness of the internal information-flow postulated by Senders. A far more detailed examination of human data will be required before firm conclusions can be reached.

Signal Detection Theory

The theory of signal detection (Green and Swets, 1974) has been so widely adopted by experimental psychologists studying perception that its origins in electrical engineering are in danger of being forgotten. It too has an optimal aspect, the theory of the ideal observer, that the human observer cannot attain. Although it provides an excellent description of the detection of single signals, it is only recently that problems of attention and multiple signal sources have been attacked within this framework.

The most serious limitation of signal detection models is that they have been elaborated primarily for situations with defined observation intervals. A two-dimensional matrix can be used to organize these models with the columns being observation interval defined or undefined. The rows would be signal-response mapping: one signal and one response, n signals and one response, n signals and n+1 responses. With the recent addition of a model for detection and recognition (Green and Birdsall, 1978) the defined observation interval column of our matrix would have an entry in every cell. But the undefined observation interval would have a model only for the case of a single signal (Lucas, 1967; Watson & Nichols, 1976). While one might hope that this single case could be extended to the next two blank cells, so far we have been unable to discover published models for undefined observation intervals with more than one signal. Since the observation interval for the typical monitoring task is undefined, or at least self-defined by the observer and not the experimenter, this is a serious limitation in applying the theory of signal detection to monitoring tasks.

Control Theory

The importance of control theory as a model of human behaviour can be seen in a recent issue of Human Factors (Volume 19, 4) devoted entirely to control theory. In particular, the optimal control model (Kleinmann, Baron, and Levinson, 1970) has generated much excitement, although many of us (especially aging psychologists) find its mathematics to be obscure. The basic assumption of any optimal controller is that the system adjusts its insufficient resources by minimizing some particular criterion. When the controller is a human, we have not been able to discover precisely which internal criterion is subject to this minimization process, especially since other parameters in the model (e.g., temporal constants and noise parameters) can interact in ways that are not intuitively obvious. The solution of having the experimenter tell the operator what objective criterion is to be minimized merely ignores this difficulty since there are no guarantees that the human can accomplish this. Any check necessarily invokes the entire model making it hard to assess the validity of any particular submechanism within the optimal control mode.

As psychologists with modest mathematical training, we find it virtually impossible to evaluate one part or stage of this model in isolation. Since the model is so complex, when it fails to fit data, it may be difficult to know which component of the model is responsible. While this difficulty has been discussed in the literature, the complementary problem has not: When the model is good at predicting data it is hard to know which part is correct. The ability of a model to generate correct predictions, while essential, does not of itself prove that the internal information flow assumed by the model is correct (MacKay, 1956). Given enough parameters, we can fit anything -- even an elephant! Although there have been faint efforts to explore the possible isomorphisms between stages of the optimal control model and stages of human information-processing theory (Pew and Baron, 1978), these efforts have been so rudimentary that no evaluation is yet possible. The model is a linear system and the human is not. For example, Langolf, Chaffin and Roulke (1976) have shown that no linear control model can explain the details of human voluntary movement. Yet given the mathematical complexity of the linear model, a non-linear optimal control model is too horrible to contemplate. And the linear model certainly is not without success. So, we must reluctantly conclude that it is too early to pass judgement on the optimal control model as a theoretical construct that explains human behaviour. The complexity of the model has outstripped our ability to evaluate it.

Comparing Quantitative and Qualitative Models

Two important conclusions can be drawn about the quartet of quantitative models discussed above. First, all are adequate as models of the detection or monitoring task, but none are outstanding as models of the human. Second, our discussion of them was able to proceed quite reasonably with little reference to data. (Again we remind the reader that the term "data" refers to empirical results garnered from live humans and not to computer simulations). The models are so well developed that interesting articles can be published solely to discuss how the models behave under different circumstances (e.g., Rouse, 1977). This is a remarkable state of affairs when such models are proffered to explain human behaviour. Engineers, who by training are accustomed to precise and complex models -- the world is either over damped, under damped, or critically damped -- may find the need to give priority to human data hard to understand. Indeed, this may only confirm their belief that the typical psychologist is incapable of precise models because he flunked calculus as a freshman. But there is no psychological reality in the naked mathematical equation. It is only after detailed consideration of human data, and not before, that many psychologists are willing to write formal models. And so the fact that meaningful discussion about models can be conducted without reference to data suggests to the psychologist that the models are too complex. Indeed, it has been often decried by some of the very best theoretical psychologists that our ability to test models has lagged far behind our ability to create them (Broadbent, 1971, p. 5). And so while we admire these models for the ingenious creations they are, and while we support the necessity for formal statements of models, we feel it is first necessary to examine data before generating elaborate models that often cannot be reduced to simpler and more falsifiable forms.

A more psychological approach is to search our common assumptions of classes of models that have been successful in describing human behaviour. A model is but an analogy and is only a first approximation towards a theory. (See Kantowitz and Roediger, 1978, Chapter 1 for a discussion of the functions of different types of theoretical statements in explaining human behaviour).

These assumptions can be found in psychological models of attention. Some representative models from which appropriate lists of assumptions may be gathered include the limited-capacity model (Broadbent, 1971), the variable-capacity model (Kahneman, 1973), response conflict models (Kantowitz, 1974), hybrid models (Kantowitz and Knight, 1976), and multiple-resource models (Navon and Gopher, 1979). But these assumptions in turn rest upon operational definitions and converging operations that allow assumptions to be translated into testable hypotheses. This is a sticky business, especially in the area of attention where there are many pitfalls for the unwary researcher attempting to pin down a relatively simple concept like "capacity" (Kantowitz, Note 1) let alone a more complex concept like "mental workload" (Moray, 1979). And so our concern for high-level theory compels us to focus upon data, the topic of the remainder of this paper.

EMPIRICAL RESULTS FROM HUMANS MONITORING DISPLAYS

While it is easy to organize a set of empirical results according to the type of model to which the data have been applied, we shall not use such a taxonomy. In order to emphasize the importance of data per se and of the procedures as viewed by the human operator who may be unfamiliar with any model of monitoring behaviour we shall instead focus upon two independent variables that have been manipulated in studies of monitoring: signal rate and event predictability. By signal rate we mean the frequency of occurrence of stimuli or changes in a stimulus display per unit time. We shall arbitrarily partition experiments into those that have slow signal arrival rates on the order of one event per minute or slower, and rapid rates where stimuli occur more often than one per minute. Thus, signal rate is an index of the temporal density of signals. Event predictability is divided, like Caesar's Gaul, into three parts: random processes, partially predictable processes and completely predictable events. Note that this tripartite division refers only to the type of stimulus and not to its temporal uncertainty. Hence an experiment with only one known signal that occurred at random times would be classified as a completely predictable event.

Rapid Signal Rates

Due to space limitations we shall select only one or two experiments in each of the following categories. These should be sufficient to allow the reader to understand how the task taxonomy works.

Predictable events. A complex experiment performed by Wickens and Kessel (1979) is a good example of monitoring for a predictable event. The basic task required the operator to either track manually or to monitor (autopilot condition) a track composed of two nonharmonically related summed sinusoids. The system failure which operators had to detect was a change in the system dynamics from a mixed velocity and acceleration system to one that was almost entirely second-order dynamics. In order to acquaint subjects with the higher acceleration in the control dynamics in the failure condition, they received two practice trials with only these system dynamics. When the system failed it always shifted into this second-order mode. Failure detection was better for all subjects in the monitoring task relative to the tracking task. However, Wickens and Kessel quite properly recorded detection latencies as well as hits and false alarms. The improved detection in monitoring was at the cost of increased detection latencies. Wickens and Kessel concluded that when both detection accuracy and latency were considered, overall performance was poorer in the monitoring condition. This conclusion, while not necessarily incorrect, may be questioned. The major basis for their conclusion was a bi-variate function plotting accuracy against latency. Relying solely upon the direction of change in two dimensional space as experimental conditions change can be a dangerous practice since it involves implicit assumptions about the reliability and particularly the scaling properties (in a psychometric sense) of the plotted variables (Kantowitz and Knight, 1976b). This potential problem can be especially acute when one variable has both upper and lower bounds -- the area under an ROC function must fall between .50 and 1.0 -- while the other variable (latency) has no upper limit except that imposed by the experimenter terminating the observation interval. An explicit model of speed-accuracy trade-off (Pachella, 1974) is required before one can state that any given increase in detection accuracy was offset by an increase in latency. Furthermore, a more appropriate statistical data analysis would have been a multiple analysis of variance of detection accuracy and latency together, rather than separate univariate analyses.

The preceeding discussion was, of course, based only upon a single-task data. Wickens and Kessel also included dual-task conditions and ancillary data analyses that bolstered their conclusion about the superiority of the manual control condition and these data and analyses have not been reviewed here. It is interesting to note that their finding of little effect of increasing secondary task difficulty combined with a strong effect of adding the secondary task is entirely consistent with a hybrid information-processing model proposed by Kantowitz and Knight (1976a).

Partially predictable events. An example of a partially predictable event was studied in Kvålseth (1979). His display was a trifle unusual since it could not be seen by subjects until they made what psychologists call an observing response (Holland, 1963) overtly requesting the next display value. Such a request had an explicit cost as did allowing the display to exceed some critical value. The categorization of this type of experiment depends upon the forcing function. With a random forcing function any observation X_i would give no information about X_{i+1}. With a large step-function in the process such that X_{i+1} could always exceed the critical value regardless of X_i this process would be classified as completely predictable since the limit could not be approached incrementally. Thus, it makes most sense to study this process with a partially predictable forcing function that gradually moves the process towards the limit of critical value. Kvålseth found that in his monitoring task subjects' sampling (or observing) behaviour did not change with the critical value or the ratio of observation costs to costs of exceeding the limit; this is, of course, yet another instance of the sub-optimal behaviour we have come to expect from humans. However, subjects who were controlling rather than monitoring the process tended to sample less often as the cost ratio increased. (The critical value was not an independent variable in the control experiment). Kvålseth attributed the sub-optimal behaviour in part to the memory load imposed by the observing-response procedure. As in the Wickens and Kessel experiment, passive monitoring lead to less optimal behaviour than did controlling, although the model did much better than the subjects in both experiments.

A similar experiment in a multi-task environment has been recently completed by us at the Purdue Human Information Processing Laboratory. Subjects monitored either five or ten displays simultaneously and responded once a critical value was exceeded on any display based upon a 1:1 mapping. The display information was always visible. The most important independent variables were the rates at which individual displays approached the same critical value, and augmented display information that identified the rates at which the digital displays were changing. Superior performance on high-rate displays was gained at the cost of worsened performance on the easier low-rate displays. Addition of a reaction time loading task in a second experiment did not alter this trade-off. Since display values were continuously available these results are more readily explained by attentional limitations than by memory limitations of the type suggested by Kvålseth.

Random processes. Our example of a random process is a monitoring experiment reported by Theois, Brelsford and Ryan (1971). They presented a binary sequence of zeros and ones generated by two independent processes. The first string had probability P_1 of each of the first i events being one, and the second juxtaposed string had probability P_2. The subjects' task was to estimate the point in the sequence at which the underlying probabilities changed from P_1 to P_2. Since the transition trial varied randomly on successive presentations, this is a random process. Of course, an optimal model exists based upon maximum likelihood estimates and, of course, subjects did not behave optimally. However, humans became more optimal as the difference between P_1 and P_2 increased.

CONCLUSIONS

Optimal models are not optimal for describing human behaviour. The human is not an optimal monitor. The human approaches optimal models only in the simplest experimental situations and even then does not approach too closely. Our understandable tendencies to complicate experimental situations to see how well the optimal model will do only further befuddle the human monitor. If we wish to understand human behaviour rather than model behaviour we need to develop simpler experimental paradigms even if they are beneath the dignity and capability of an optimal monitoring model.

REFERENCES

Broadbent, D.E., "Decision and Stress". New York: Academic Press, 1971.

Carbonell, J.R., "A Queueing Model of Many-Instrument Visual Sampling". IEEE Transactions in Human Factors in Electronics, HFE-7, 157-164, 1966.

Cox, D.R. and Smith, W.L., "Queues". London: Methuen, 1961.

Green, D.M. and Birdsall, T.G., "Detection and Recognition". Psychological Review, 85, 192-206, 1978.

Green, D.M. and Swets, J.A., "Signal Detection Theory and Psychophysics. Huntington, N.Y.: Krieger, 1974.

Holland, J.G., "Human Vigilance". In D.N. Buckner and J.J. McGrath (Eds.), Vigilance: A symposium. New York: McGraw Hill, 1963.

Kahneman, D., "Attention and Effort", Englewood Cliffs, N.J.: Prentice Hall, 1973.

Kantowitz, B.H., "Double Stimulation". In B.H. Kantowitz (Ed.), Human information processing: Tutorials in performance and cognition. Hillsdale, N.J.: Lawrence Erlbaum Associates, 1974.

Kantowitz, B.H., On the Beaten Track", Contemporary Psychology, 20, 731-733, 1975.

Kantowitz, B.H., "Interfacing Engineering Psychology and Human Information Processing", In E.A. Fleishman (Ed.), Human performance and productivity. Hillsdale, N.J.: Lawrence Erlbaum Associates, 1980.

Kantowitz, B.H. and Knight, J.L., "Testing Tapping Timesharing, II. Auditory Secondary Task". Acta Psychologica, 40, 343-362 (a), 1976.

Kantowitz, B.H. and Knight, J.L., "On Experimenter-Limited Processes", Psychological Review, 83, 502-507 (b), 1976.

Kantowitz, B.H. and Roediger, H.L., "Experimental Psychology, Chicago: Rand McNally, 1978.

Kleinman, D.L., Baron, S., and Levinson, W.H., "An Optimal Control Model of Human Response, Part I: Theory and Validation", Automatica, 6, 357-369, 1970.

Kleinrock, L., "Queueing Systems, New York: Wiley, 1975.

Kvålseth, T.O., "A Decision-Theoretic Model of the Sampling Behaviour of the Human Process Monitor: Experimental Evaluation, Human Factors, 21, 671-686, 1979.

Langolf, G.D., Chaffin, D.B., and Foulke, J., "An Investigation of Fitts' Law Using a Wide Range of Movement Amplitudes, Journal of Motor Behaviour, 8, 113-128, 1976.

Lucas, P.A., "Human Performance in Low-Signal-Probability Tasks". Journal of the Acoustical Society of America, 42, 158-164, 1967.

MacKay, D., "Towards an Information-Flow Model of Human Behaviour, British Journal of Psychology, 47, 30-43, 1956.

Moray, N., Ed., "Mental Workload", New York: Plenum, 1979.

Navon, D. and Gopher, D., "On the Economy of the Human-Processing system". Psychological Review, 86, 214-255, 1979.

Pachella, R.G., "The Interpretation of Reaction Time in Information-Processing Research", In B.H. Kantowitz (Ed.), Human information processing. Hillsdale, N.J.: Lawrence Erlbaum Assoc., 1974.

Pew, R.W. and Baron, S., "The Components of an Information Processing Theory of Skilled Performance Based on an Optimal Control Perspective". In G.E. Stelmach (Ed.),

Information processing in motor control and learning. New York: Academic Press, 1978.

Rouse, W.B., "Human-Computer Interaction in Multitask Situations", IEEE Transactions on Systems, Man and Cybernetics, SMC-7, 384-392, 1977.

Senders, J.W., "The Human Operator as a Monitor and Controller of Multidegree of Freedom Systems". IEEE Transactions in Human Factors in Electronics, HFE-5, 2-5, 1964.

Sternberg, S., "Two Operators in Character Recognition: Some Evidence from Reaction-Time Measurements", Perception & Psychophysics, 2, 45-53, 1967.

Theois, J., Brelsford, J.W. and Ryan, P., "Detection of Change in Nonstationary Binary Sequences", Perception and Psychophysics, 9, 489-492, 1971.

Watson, C.S. and Michols, T.L., "Detectability of Auditory Signals Presented without Defined Observation Intervals", Journal of the Acoustical Society of America, 59, 655-668, 1976.

Wickens, C.D. and Kessel, C., "The Effects of Participatory Mode and Task Workload on the Detection of Dynamic System Failures, IEEE Transactions on Systems, Man and Cybernetics, SCM 13, 24 34, 1979.

TOWARDS A THEORY OF QUALITATIVE REASONING ABOUT MECHANISMS AND ITS ROLE IN TROUBLESHOOTING

John Seely Brown & Johan de Kleer
XEROX PARC
Cognitive and Instructional Sciences
3333 Coyote Hill Road
Palo Alto, California 94304

INTRODUCTION

One of the intriguing properties of many expert troubleshooters is their ability to diagnose systems that they have never seen before. These experts construct their own qualitative causal models of how a system functions given a description of its structure and what it is supposed to do. What is the basis of this skill? Towards answering this question, we have been investigating properties of the "mental models" that such experts might be constructing for representing the underlying mechanisms of a system, i.e. their understanding. Some of these properties are apparent. Firstly, they involve a qualitative understanding of how the system functions in contradistinction to the analytic or quantitative models often taught to engineers. Secondly, these models appear to have many of the properties of a simulation which is, metaphorically speaking, "run" in the mind's eye or what might be called an envisionment. From running such an envisionment the expert can discover and/or encode much of a system's underlying causality.

Our approach for studying the nature of these models is not to directly probe those of an expert but rather to hypothesize what such a mental model might be and then implement it on a computer in order to see if it can be used to answer the same kinds of questions that an expert can answer using his model. Our goal in such an enterprise is not to discover any one model for any one system but rather to find a formal representation which is capable of expressing the underlying causal models of a wide class of machines be they electronic, hydraulic, mechanical or steam systems. But, any representation scheme rich enough to capture such a wide class of devices is also capable of supporting numerous representations of the same device many of which are so ad hoc and brittle that they can be used for virtually nothing that was not anticipated when they were created - hardly characteristic of the robust understanding of an expert. Our concern, therefore, has been to discover a set of principles for constraining a causal representation of a system to ones that will be i) maximally "robust" in their ability to facilitate answering questions about how the systems functions and malfunctions and ii) "learnable" in that the representation of how the system functions does not, in any way, presuppose the very mechanism that it is trying to describe - an issue that we will return to throughout this paper.

MOTIVATIONS FOR A QUALITATIVE THEORY OF MECHANISM

The motivation for developing a theory of mechanism for circuits stems from our observations of using SOPHIE (Brown and Burton 1975) as an aid to teaching troubleshooting. SOPHIE, a system for teaching electronics, relied on a numerical circuit simulator to answer students' questions and to critique their hypotheses about what might be wrong. For example, SOPHIE, could answer such questions as - "What would happen if R22 were shorted?" (with respect to a given circuit being studied by a student) with an answer such as - "The output voltage would be .623 volts. In a working circuit it is 9.93 volts." Although SOPHIE could be used quite effectively for teaching trouble-shooting, we discovered two interrelated shortcomings: (1) SOPHIE could not give adequate explanations for its deductions, and (2) the students rarely learned a more global understanding of a circuit's behaviour (e.g. as a feedback system) from using it.

Since SOPHIE employed a numerical circuit simulator (in conjunction with some problem solving heuristics) to answer questions about the circuit, it was often unable to give explanations for its deductions. In the above example, SOPHIE, having deduced that the output voltage drops to .623 volts from 9.93 volts when R22 is shorted, cannot give a causal explanation as to why it dropped. The student would more profit from something like "If R22 shorts, Q4 can no longer amplify and thus all the output current must be supplied by Q3 which cannot provide enough current to the load." Circuit simulators, especially ones using relaxation techniques, cannot provide such an explanation. This limitation becomes even more serious in troubleshooting, since one often needs to infer what could possibly explain the observed symptoms - a task which is computationally prohibited with a numerical simulator. Thus, the original SOPHIE was itself incapable of efficient troubleshooting.

From observing students using this system, it became all too apparent that they were acquiring only one aspect of the skill of troubleshooting: using knowledge of circuit components to localize faults. However, isolating faults that required expertise in understanding of how the components interacted with each other proved extremely difficult even after extensive training with the system. This observation raised two questions: (1) what were the mental models of how and why a given device worked that experts seem to have which we expected our students to tacitly acquire, and (2) what kinds of explicit explanations could SOPHIE give that would facilitate the meaningful acquisition of these mental models.

For an intelligent computer assisted instructional system (ICAI) to be successful it must have an explicit theory of how students actually learn troubleshooting in order to guide its dialogue with the student. This, in turn, suggests that the instructional system must have explicit models of a student's partial understanding of how a device works. A quantitative simulation of a device's behaviour provides an unlikely candidate to base such partial models on since, for a numeric simulation to run at all, it must be complete with respect to the agreed upon grain size.

However, not every qualitative simulation model which meets the above requirements is useful. It may, at first, seem that it would be very simple to build a qualitative model for a particular circuit, but appearances are misleading. From the mental model viewpoint, we are not interested in teaching the student one qualitative model of a particular circuit; rather, we want the student to learn a collection of models that is adequate for a class of circuits. Therefore the particular model constructed must obey some principles relating it to models of other circuits. Ideally, we would like the student to learn how to construct his own qualitative models of how a circuit functions from examining just its schematic and its operating specifications. Likewise, from the ICAI viewpoint, we would like there to be some easy way of taking a circuit and producing a qualitative model for it; it would be very frustrating if it required months of effort to construct a viable qualitative model of a given circuit. That is, we would like the qualitative models for a circuit to be constructed as directly as the quantitative models which use numerical circuit analysis.

Building a qualitative model of a circuit that is useful in troubleshooting raises even more difficulties. The qualitative model must be able to deal accurately with all possible faults which might produce completely unexpected behaviour. Also, to be able to rely on the qualitative model for troubleshooting, we must be sure that it accurately models behaviour even in those regions where it has never been tested. This requires that the qualitative model obey certain principles. Since the candidate model is going to be used to instruct the student, there must be some reason to believe that the student can acquire it - i.e. that it is learnable. Thus, although it seems easy to construct a qualitative model for a circuit, the above examples illustrate the extreme difficulties involved. This paper does not attempt to present a theory powerful enough to support the kind of circuits SOPHIE is capable of dealing with but, instead, tries to outline some basic principles that need to be understood before a general theory can be developed for handling more complex devices.

MODELS AND PRINCIPLES

This paper can be viewed from two perspectives. The first pertains to the training community in terms of their interests in discovering ways to teach a person how to operate, maintain or diagnose a general class of systems. Here, the task is to teach these skills in a way that makes them highly transferable thus enabling a person trained for one system to transfer easily his understanding to a similar system without requiring major retraining. A typical device used for such training is a generic simulator which attempts to portray the underlying mechanisms and cause and effect relationships that are in common to the entire class of systems. In designing generic simulators, the need for having a high fidelity simulation of a particular system is no longer of major concern. Instead, the central issue is achieving a "cognitive fidelity" which involves portraying the abstractions that unify the individual members of the class into a prototype that illuminates the mechanisms in common to the whole class. Much of what follows is an attempt to show how to construct a simulation of a particular system that has many of the desired properties of a generic simulation. In fact, ideally one might hope that a generic simulation has a nearly direct match to the mental models of an expert.

This paper may also be of interest to the cognitive science community in that it attempts to clarify what constitutes a robust and learnable representation of how a given system functions. There are two aspects to this issue. The first involves constructing a robust qualitative simulation of the machine. The second concerns using this simulation in order to infer the information needed for constructing a causal description (network) of how it functions.

A prerequisite for this latter task is having a language for expressing the various kinds of causal relationships found in machines. For this, we draw heavily upon the pioneering work of Rieger by creating a variant of his epistemology for representing the cause and effect relationships comprising a system's mechanisms. This epistemology will be defined by specifying a set of causal primitives along with a set of constraints for connecting the primitives together into possible causal descriptions.

Determining what constitutes a robust qualitative simulation turns out to be rather involved. However, determining a set of principles that captures our intuitions about what makes one simulation more robust that another will also provide insights into the process of constructing an understanding of how a system functions. In fact, one of the interesting aspects that emerged from this work was the close connection between the properties that make a simulation robust and those that make it learnable.

Two basic principles for characterizing the critical properties of what makes a simulation robust and learnable will be discussed in this paper. The first principle is called the causality principle which, very briefly, requires that each event in the simulation has an explicit cause. Qualitative simulations that have this property can be easily run, for example, in the mind's eye. Intuitively, this principle also facilitates learnability in that the cause and effect relationships that unfold in the running of the simulation can be easily inverted and thus used to discover what aspects of the simulation model need to be revised when the simulation, itself, fails to produce the desired behaviour or predictions. Thus, the causality principle enables one to cope with the "assignment of credit/blame" problem which requires identifying which steps in a chain of inferences are responsible for a particular undesirable outcome. This problem in one form or another has plagued nearly all learning systems.

The second principle is called the "no function in structure" principle. It requires that the rules for specifying the behaviour of the constituent parts of the overall system can in no way refer, even implicitly, to how the overall system functions. Satisfying this principle ensures that the behaviours of each of the system's parts can be totally represented and understood in a context free manner - that is, independent of the system it is embedded within. The simulation then shows how the behaviours of the individual parts become combined in order to produce the behaviour of the overall system. Failing to adhere to this principle is similar to having the explanation of how the systems functions already presuppose some aspect of how it functions.

Violating the "no function in structure" principle also limits one's ability to use the simulation for making predictions about how the system might function if a given component were modified or faulted. In particular, since modifying a structural aspect of the system could radically alter its underlying mechanism, then those rules which specified a part's behaviours by implicitly relying on this mechanism might no longer be valid thus making predictive aspects of the simulation highly questionable.

METHODOLOGY

Our approach has been to study a class of devices which on the one hand are intuitively understood by nearly everyone but on the other hand are sufficiently complex to raise many problematic issues - some of which were unanticipated by us. The device we have chosen is a simple doorbell or electromechanical buzzer. It is sufficiently complex to stress any current theory of causal mechanisms including our own and, as we shall see, finding an internally consistent qualitative simulation of the buzzer that also satisfies both the causality principle and the various variants of the "no structure in function" principle turns out to be surprisingly difficult. Similarly, since the buzzer is a common household device, the reader will likely be able to judge the subtleties involved in finding a principled simulation of it as well as appreciate how that simulation can be used to answer a wide variety of questions. Granted that the need for defining a set of principles for constraining the representation of a system might have been more convincing if we had chosen a more complex paradigmatic example, the chance of having readers follow and even anticipate the various pitfalls that unfold as we explore its potential models would have been radically decreased. Again, let us stress that the buzzer, is, itself, of no fundamental importance. Nor is any one device, no matter how complex it is! Rather, we are striving for a useful, consistent and systematic representation of how a device works with the hope that finding such a model will enable us to uncover a set of principles, if not a process, for determining what constitutes an ideal, qualitative understanding of its mechanisms.

TECHNICAL ISSUES[1)]

What form should such a representation take and what should we expect to be able to do with it? As with much of this paper, let us proceed by example. Pictured below is a diagram of the buzzer.

Its functioning seems to be easily described by a small set of rules basically asserting: "The clapper-switch of the buzzer closes, which causes the coil to conduct a current, thereby

1) Some of the following sections are derived from our paper "Mental Models of Physical Mechanisms and their Acquisition" which appears in "Cognitive Skills and their Acquisition", Erlbaum, 1980.

generating an electromagnetic field, which in turn pulls the clapper arm away from the switch contact, thereby opening the switch, which shuts off the magnetic field, allowing the clapper arm to return to its closed position, which then starts the whole process over again." Indeed, we could easily create a set of formal rules that would produce the above ad hoc description, but does that description produce a useful understanding?

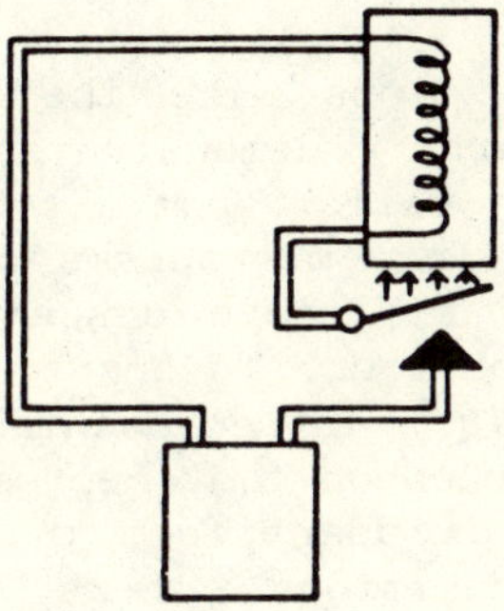

Figure 1: Buzzer

This question can be answered using a definition of robustness of a mental model: a model is robust with respect to its device structure if the questions that can be asked about the device structure can be answered correctly (or correctably). The device structure implicitly defines the terms of a descriptive language, and the questions using those terms concern the ramifications of perturbing the device structure or the component attributes. The ad hoc description of how the buzzer is supposed to buzz does not provide an understanding that can address these possible questions.

For example, listed below are some typical questions that one might expect a person who has a thorough, qualitative understanding of the buzzer's mechanism to be able to answer without recourse to analytic models of its components nor to differential equations, etc.

Table of Questions about the Buzzer

a) What happens if we reverse the leads of a battery (i.e. change the polarity of the battery in the circuit)?
b) What happens if we switch the leads of the coil?
c) What happens if we remove the battery from the circuit?
d) What happens if we short the switch contact?
e) What happens if we make the switch arm lighter (or even weightless)? For example, what happens qualitatively to the

frequency of the vibrator or the size of the gap when such a change is made?

f) What happens if we place a light-weight permanent magnet on the clapper arm? Does the frequency of the buzzer change? Does it matter in which way the battery is hooked up?

g) What happens if we put two clapper switches in series (or parallel)? If one buzzes with a lower frequency than the other when subjected to the same magnetic field, then what happens when they are placed in series?

Attempting to answer questions like the above - some of which are admittedly quite hard - demonstrates the surprising richness of inferences that can follow from a totally qualitative understanding of the underlying mechanisms of the buzzer. Similarly, the inadequacies of the above description of how the buzzer functions become quite obvious. To answer these and other, unanticipated questions just from a representation of how the buzzer works places quite a burden on that representation and the processes that interpret it. But apart from the failure of that description to answer such questions, there is a principled objection: a great deal of that description already presupposes how the buzzer works.

For example, the statement "the clapper-switch closes causing the coil to conduct a current" presupposes a source of current or battery in the circuit. Furthermore, it presupposes that the switch, coil and battery are all arranged in a very particular way and that when the switch closes, it necessarily passes current. Switches in general do not pass current, they only pass current if they are connected to batteries and wires in particular configurations. Although it may be true that for this buzzer that the switch passes current, we can only say so because we already understand how it works.

The desired level of understanding is this: given a description of how the switch operates in all possible circuits, use that along with similar descriptions of the other parts of the buzzer to explain how and why the switch passes current as it closes. In general, the problem is how to connect the functioning of the device (i.e. its ringing) with the structure of the device (i.e. the buzzer) without presupposing how the device functions.

A CAUSAL MODEL FOR THE BUZZER

The fundamental question we are addressing in this paper is given that the behaviours of magnets, batteries, coils, etc. are understood generally and sufficiently, how do the understandings of the behaviours of the individual pieces combine to explain the composite behaviour of buzzing. We will adopt the following simple syntax to formalize the behaviours of the individual parts. A component can be in one of a number of different states, each of which is characterized by some rules. Each such state represents a distinct region of behaviour for the component in question (e.g. a switch can be open or closed). The rule for a state consists of a definitional part (e.g. the battery is disconnected) and a conditional which makes a test (e.g. if the coil is not pulling) to determine whether some consequence applies (e.g. the switch will become closed). The general form for a component model is:

```
<COMPONENT>: <STATE1>: (<DEFINITIONAL-PART>)*, (IF <TEST>, <CONSEQUENCE>)*
             <STATE2>: (<DEFINITIONAL-PART>)*, (IF <TEST>, <CONSEQUENCE>)*
```

"(...)*" indicates that "..." may occur an arbitrary number of times. The <definitional-part> can be used in two ways: the first concerns its use as a criterion for determining whether the component is in a given state and the second concerns its use as an imperative: given that the component is declared to be in a particular state (criterion) then statements made in the <definitional-part> are asserted to be true (imperative) about the component's behaviour. These assertions can then be examined by an inferential process to determine their ramifications. In simple cases these ramifications can be determined by examining the tests of the current state of every component model.

The models for the buzzer part are:

```
SWITCH : OPEN:
           battery is disconnected,
           if coil is not pulling, switch will become CLOSED.
         CLOSED:
           battery is connected,
           if coil is pulling, switch will become OPEN.

COIL : ON:
         coil is pulling,
         if battery is disconnected, coil will become OFF.
       OFF:
         coil is not pulling,
         if battery is connected, coil will become ON.
```

In order to combine these descriptions of the behaviours of the individual parts to determine the behaviour of the overall buzzer, we set out to construct a simulation. Starting with some arbitrary[2] initial state for each of the components (i.e. a cross-product or composite state), their definitional parts are asserted and then each test is examined to determine the conse-

[2] footnote next page.

quences of those assertions being true. From these consequences the next state is determined. The discovery of each next state can be considered as a time increment. The time elapsed between increments is variable since increments refer to interesting events rather than the passage of some fixed unit of time. The determination of the next composite state can be very complex. The consequences of the prior composite state may indicate that multiple components will change state in the next time interval. In the actual physical system it may be critical that one component change state before another but, in a qualitative simulation, the time scales for each of the models are not inherently comparable, since they do not utilize a uniform time metric. The exact ordering of the transitions cannot be directly determined without considering the overall purpose of the device and other nonlocal properties that require complex inferencing schemes. A second complication stems from the qualitative nature of the inputs and other internal parameters. In this case it might be ambiguous whether some threshold in the test was passed, resulting in various components optionally changing state. Such a situation can force the consideration of parallel envisionments[3].

The transition table is:

	SWITCH	COIL	BATTERY	COIL
0	CLOSED	ON	connected	pulling
1	OPEN	ON	disconnected	pulling
2	OPEN	OFF	disconnected	not pulling
3	CLOSED	OFF	connected	not pulling
4	CLOSED	ON	connected	pulling
5	...			

2) For more complex devices this choice cannot be arbitrary since some of the composite device states may be either contradictory or inaccessible - requiring several initial states. A composite state is contradictory if the definitional parts of two of its component states make contradictory assertions.

3) However, many of these difficulties can be avoided by detecting and eliminating self-contradictory device states and invoking other deductive machinery. It is the responsibility of the simulation process P1 to identify the ambiguities, but it is also the responsibility of P1 to prune as many of the resulting envisionments as possible based on local considerations.

In the initial state (t = 0), the coil's being on causes the switch to change state from closed to open. If the switch is open, the battery is disconnected by definition. In the next state (t = 1), the battery's disconnection causes the coil to change state from on to off. If the coil is off, it is not pulling - by definition. The construction continues. The simulation can also be represented as a state transition diagram:

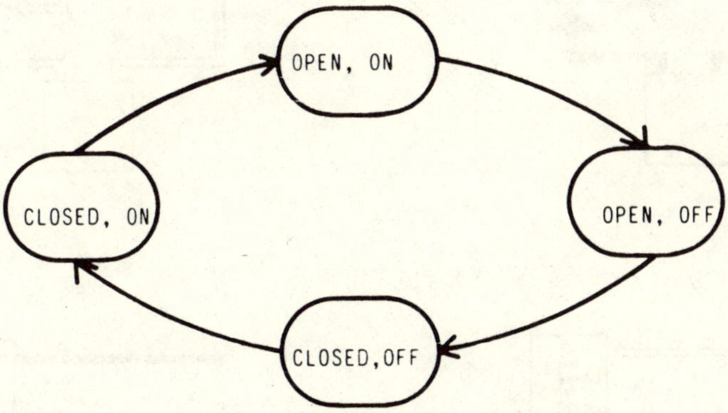

Figure 2: State-Transition Diagram for the Buzzer

The above table can also be used to construct a series of snapshots of the buzzer functioning over time, see figure 3.

NO FUNCTION IN STRUCTURE & THE LOCALITY PRINCIPLES

The buzzer model, although it explains the vibration, has so much of the knowledge of how the buzzer works embedded in the models of the individual components that it provides little insight into how the buzzer, as an interacting collection of components, functions. Implicit within each of the models is that it is connected to other constituent components comprising a buzzer. The models name other components and their internal states directly, thereby presuming that they are part of a buzzer. That the components are physically connected by wires and magnetic fields is often ignored. For example, the model for the switch assumes its connection to a battery and that when it opens it will prevent current from flowing and disconnect the battery from the coil. The model for the coil assumes that it is not the current flowing through it, but the fact that some battery is connected to something which enables the magnetic field to exist. Thus, much of the essential functionality of the overall buzzer has been embedded within its constituent parts. There is little chance that these same models would be useful for understanding a buzzer which was constructed even slightly differently. For example, a buzzer having two clappers hooked in series or parallel is sufficiently different for these models to be of little use[4].

[4] footnote next page.

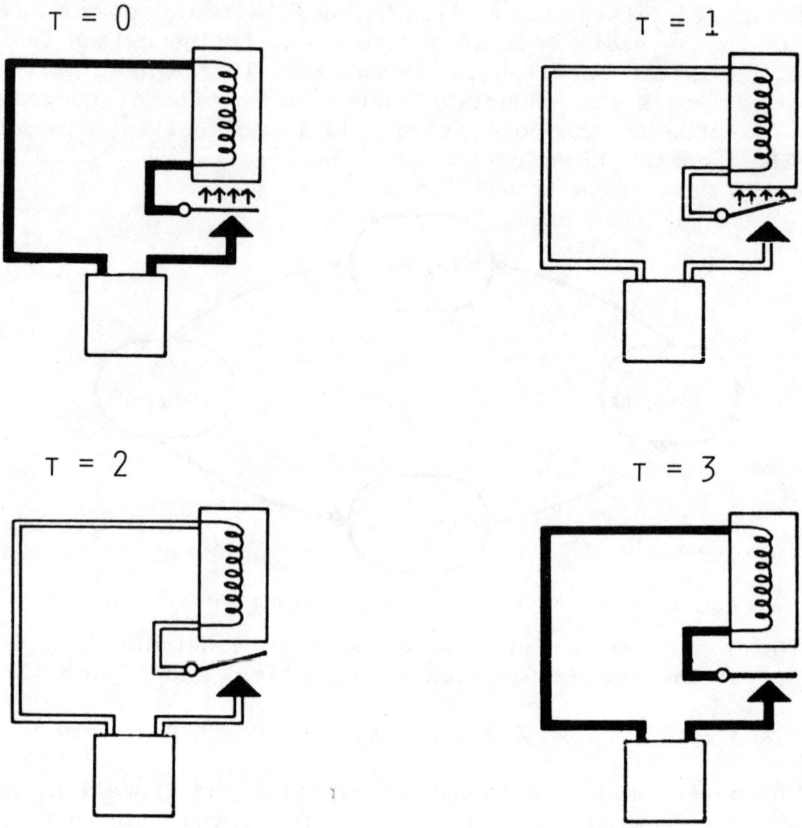

Figure 3: The Functioning of the Buzzer

Syntactic restrictions can be placed on the models of the individual components to ensure that they reference local quantities. Such locality restrictions help to avoid gross violations of the "no function in structure" principle The first locality principle demands that there is no reference to other components in rule consequences. The previous models' rules met that principle. A similar locality principle can be used to restrict the definitional aspect of component rules. But if a locality principle were also enforced on the tests of component rules, there would be no way for models of different components to

4) Notice that in this case contradictory states exists: if one clapper is open and the other is closed the rules make contradictory assertions about whether the battery is connected. Or if that were resolved by having better switch models, the next state of the coil is ambiguous.

ever interact. Thus, there is no way to avoid some nonlocality. To extend the locality principle to all three parts of a rule forces the introduction of connections as entities independent of the component models.

CONNECTIONS AND DEVICE TOPOLOGY

In order to avoid nonlocality in the component models we need to draw a distinction between the models for components and the method by which these models communicate. We introduce connection as a simple stateless model which is primarily a placeholder for information. Although we make the internal state of a component inaccessible to other components and connections, models for components and connections communicate by sharing information. For example, both the model for a specific wire (e.g. wire 2) and its adjacent part (e.g. coil) will know the current flowing from the wire into the coil. The simplest model for a wire is one which consists of the knowledge of the current through it, and it shares this information with the components on either end of the wire. The only information that is shared by connections are attributes which are related to the actual physics by which components interact (e.g. voltage, current, pressure, force, flow, etc.). The "no function in structure" principle also applies to connections. For example, it requires every wire in the buzzer to be modelled in the same way and also every attribute of the same type (e.g. force) to be treated in the same way (e.g. obey Newton's law). The component-connection distinction allows us to model the effect of the coil on the clapper switch: the coil being on (state of a component) causes a strong magnetic flux (attribute of a connection) which causes the clapper-switch to open (state of a component).

Formally, it can often be arbitrary as to which parts of the buzzer are components and which are connections. All of the buzzer models presented thus far implicitly assume wires and magnetic fields to be connections. This is not necessary. We could have modelled a wire as a component having certain properties but then we would have had to introduce an end-of-wire connection to attach the wire component to the switch component. The determination of which parts should be modelled as connections and which as components can be quite subtle since a connection assumes that the model describing its behaviour is extremely simple. For example, it would have been, in principle, impossible to model the switch as a connection since a switch has internal state. If we had decided the switch was a connection, we would never be able to construct an adequate model for the buzzer.

The device topology of a physical device consists of a description of the components of the device and their connections:

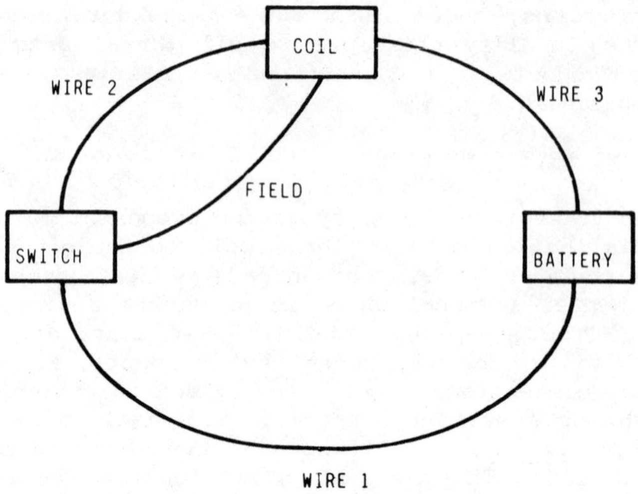

Figure 4: Device Topology of the Buzzer

Although a specific model for a specific component is permitted to know about the current in a specific wire, the class-wide or prototype is not. Therefore the prototype model can only express its potential connection to some wire. It refers to information that it might possibly share with connections as variables. When the prototype model is used in a specific device these variables need to be replaced with device-specific quantities, since that is the only way two component models can communicate.

A prototype model for the switch is:

```
Information terminals: i1,i2,f1
SWITCH : OPEN:
           i1 <- 0, i2 <- 0
           if f1 = 0, SWITCH will become CLOSED.
         CLOSED:
           i1 <- 1, i2 <- 1
           if f1 = 1, SWITCH will become OPEN.
```

A variable such as "i1" is intended to indicate the value of some attribute (e.g. a current of one ampere) which can then be communicated among models. Unlike component states which do not change until acted upon by some other model, variable values are direct results of components being in particular states and thus are only considered valid as long as the components which caused them do not change state. If a variable value is changed by some model it may not be changed again until the component that originally caused the variable value changes state[5]. The semantics of "a <-b" is $a[t+1] = b[t]$ for each time t in which the rule is valid. Or informally, a's value gets b's value. In the above model, b is a constant and the effect of "a <-b" is to set a's value to that constant. For example, as long as the switch is open, the rule "i1 <-0" applies, thus i1 is set to zero immediately after the switch becomes open and cannot change until after the switch ceases to be open. The actual amount of time elapsed moving from t to t+1 is arbitrary - it can be infinitesimally small or extremely large. Our convention is that time t+1 refers to the next interesting event after time t; thus the time elapsed moving from t to t+1 has no relation to the time elapsed moving from t+1 to t+2.

We now have a well-defined precise way of moving from a device topology to a set of device specific models which does not violate the "no function in structure" principle. For each node in the device topology where a connection attaches to a component, a new unique quantity must be invented; then, for each component and connection, a copy of its prototype is made with the information terminals replaced with the appropriate circuit specific quantity. This process ensures that the component models are local since the only quantities a model can reference are those that are associated with the connections which are adjacent to it in the device topology. Thus many of the violations of the "no function in structure" principle are avoided.

One possible set of specific models for the buzzer are:

```
SWITCH : OPEN:
             I1 <- 0, I2 <- 0
             if F1 = 0, switch will become CLOSED.
         CLOSED:
             I1 <- 1, I2 <- 1
             if F1 = 1, switch will become OPEN.

COIL : ON:
           F1 <- 1
           if I2 = 0, coil will become OFF
           if I3 = 0, coil will become OFF.
       OFF:
           F1 <- 0
           if I2 = 1, coil will become ON
           if I3 = 1, coil will become ON.
BATTERY: I1 <-> I3.
```

---

5) If this is violated; for example, if one component changes the current to one ampere to two, the model for the composite device is inconsistent.

A REPRESENTATION FOR CAUSAL ATTRIBUTION

Rieger & Grinberg (1977) have developed a system for representing cause-effect knowledge about physical mechanisms. A significant contribution of their work is an epistemology for the functionality of a physical mechanism. For example, their representation distinguishes between a state-change and a tendency to produce a state-change, and allows a tendency to cause a state-change but not vice versa. However, the state of an object may enable a tendency to exist which causes a state change. This same effect is manifested in our models: the only way one part can affect another is through connections.

The prior analysis of the buzzer can be used to construct a crude representation similar to Rieger's. Events are represented by nodes of which there are only two types: (1) state changes which represent a change in a component state, and (2) tendencies which represent attributes of the connections being forced to some value. There are four types of links which represent the causal relationships between these two types of events.

The first link is enablement:

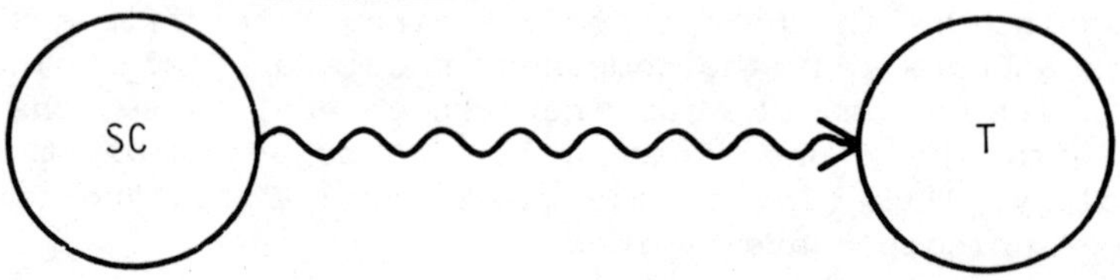

Figure 5: Enablement

The tendency T is a direct result of some component changing to a particular state SC. Since, the component being in that state SC is what enables tendency T to exist. For example, if the switch changes to the open state, the tendency of the current is toward zero.

The second link is cause:

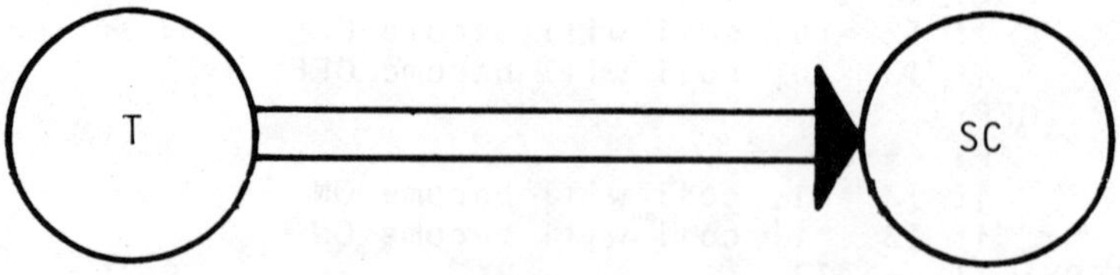

Figure 6: Cause

The tendency T is forcing the particular component to change state. For example, if the magnetic field starts the tendency of pulling, the switch will change state from closed to open (SC).

The third type of link is propagate:

Figure 7: Propagate

The laws for connections derive T2 from T1, thus propagating values through the device topology. For example, if the current through the switch is zero, the current through the battery also becomes zero.

The final type of link is antagonism:

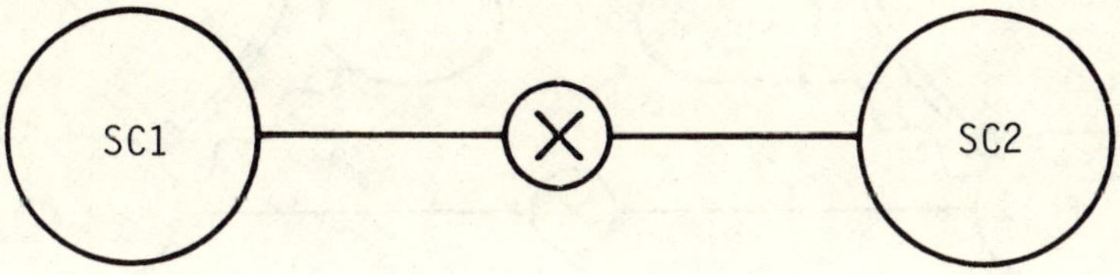

Figure 8: Antagonism

Both state changes SC1 and SC2 cannot hold simultaneously and are therefore termed antagonistic. For example, the switch state open is antagonistic to the switch state closed.

A representation for the functioning of the buzzer represented in this epistemology is shown in Figure 9.

This representation is constructed from the transition table produced by the last bell model. Every edge which represents an attribute value being successfully tested for a state transition is represented by a cause link. Every edge where a definitional rule is used is represented by an enablement link. Every edge where a connection law is used to derive a new attribute value from an old one is represented by a propagate link. Note that this representation for function cannot be constructed from the transition table alone. To identify the origin of the changes in the table and their respective type requires referring back to the particular component rules that produced the change.

In order to determine the behaviour of the buzzer over time one can just step through the links in figure 9 reading the sequences of values at successive time increments.

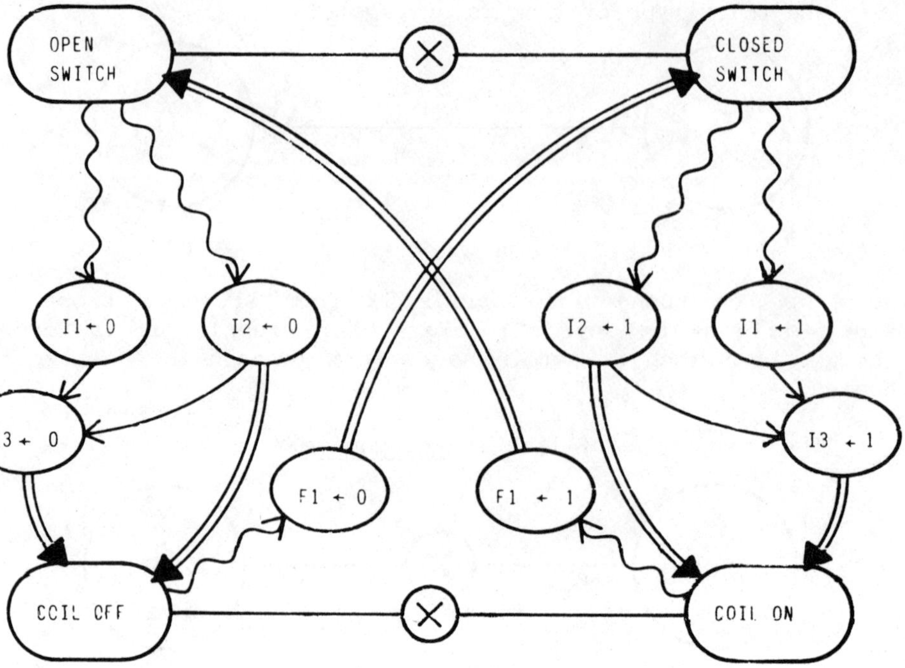

Figure 9: Causal Attribution of Buzzer

CONCLUSIONS

Although it seems easy to construct a qualitative model which successfully mimics the behaviour of some correctly functioning physical system, it is very difficult to construct a model which is also accurate when the system contains some fault. Often the model which mimics the working behaviour, rather than predict no behaviour at all, will also predict some functioning when the device is faulted, but the predictions will not be correct. Such a model is of little utility for the maintainer who must troubleshoot the system, or for the operator who must successfully operate the system when catastrophes occur. However, if the models used obey the principles stated in this paper, they will be faithfull in modelling faulty as well as correct system behaviours. These are the kinds of models which we would like maintainers and operators of systems to possess and ICAI systems to impart.

REFERENCES

Brown, A.L., 1976, "Qualitative Knowledge, Causal Reasoning and the localization of Failures," Artificial Intelligence Laboratory, TR-362, Cambridge: M.I.T.

Brown, J.S. and R. Burton, 1975, "Multiple Representations of Knowledge for Tutorial Reasoning" in D. Bobrow and A. Collins (Eds.) Representation and Understanding, New York: Academic Press.

Brown, J.S., A. Collins and G. Harris, 1977, "Artificial Intelligence and Learning Strategies", B.B.N. Report 3634, I.C.A.I. Report 6, Cambridge: B.B.N. (Also in Learning Strategies, New York: Academic Press, 1978).

de Kleer, J., 1979, "Causal and Teleological Reasoning in Circuit Recognition". Artificial Intelligence Laboratory, TR-529, Cambridge: M.I.T.

Stevens, A. and A. Collins, 1978, "Multiple Conceptual Models of a Complex System", B.B.N. Report 3923, Cambridge: B.B.N.

Rieger, C., and M. Grinberg, 1977, "The Declarative Representation and Procedural Simulation of Causality in Physical Mechanisms", Proceedings of the Fifth International Joint Conference on Artificial Intelligence, pp. 250-255.

SYSTEM DESIGN AND OPERATOR SUPPORT

CHAIRMAN: H. G. STASSEN

SECRETARY: H. TALMON

SYSTEM DESIGN AND OPERATOR SUPPORT

Henk G. Stassen

Department of Mechanical Engineering
Delft University of Technology
The Netherlands

CHAIRMAN'S REVIEW AND DISCUSSIONS OVERVIEW

INTRODUCTION

After reviewing the 10 papers to be presented at this session we can roughly cluster them into three groups. To start with, we have three papers which all together cover almost the entire area:

Johannsen's paper on "Fault Management and Supervisory Control of Decentralized Systems": a paper which explains the modern and future developments in control theory and in computer hardware and software development.

Lees's paper on "Computer Support for Diagnostic Tasks in the Process Industries": a paper that in a very elegant way reviews a number of possibilities to assist the human operator in a diagnostic task.

Pau's paper on "The Application of Pattern Recognition to Failure Analysis and Diagnoses": a profound paper which covers the field almost entirely.

As a second group of papers, we can mention those papers where the accent of the paper is laid upon the method of designing support systems for the human operator, and where the applications are mainly used as an illustration to emphasize the strength of the method described. To this group of papers we count the papers on flow models of Lind and of Goodstein, and that on disturbance analysis systems by Bastl and Felkel.

Finally, we see a set of papers with a strong emphasis on the application, and where particular aspects of human functioning in relation to the design of the system are elucidated. To this set we include the remaining papers of Syrbe, Dellner, Lihou, and Gaddes and Brady.

SURVEY PAPERS

In supervising processes, two different supervision situations should be recognised: (1) Steady state control where the human operator has to run the plant under normal conditions in some kind of optimal way, and (2) non-steady state control where the supervisor has to start up or shut down the plant or where he has to retune the plant setting due to severe disturbances, failures of equipment, etc. According to Sheridan (1976) the role of the supervisor here is to operate in four modes, planning, teaching, monitoring and intervening in the programmed operations of the computer. In each of these modes, the computer can take over certain tasks from the human; the basic problem is that of the task allocation between human and computer or process. The question seems to be: "Which information at what moment in which way should be presented about the computer or process to the human, and vice versa?"

In this conference, the topic of interest is centered around human detection and diagnosis of system failures, a task for humans and assistive support systems in non-steady state control situations. The understanding of these topics requires a very broad approach where, besides the successfully used control theory in steady state control (Kok and van Wük, 1978) detection and decision theory, information theory and pattern recognition certainly will be very helpful tools in understanding human behavior. In this context, Johannsen reviews in a nice way what we may expect in the near future from progress in control theory and in computer science. The arguments for the existing trend to decentralisation and regarding progress in multilevel control of large scale systems are well taken; present theory and instrumentation may be able to handle the problems. However, to what extent this theory and currently available instrumentation when combined with the effort put into increasing human and system reliability will be able to distribute the "intelligence" between human and computer seems to me the major problem to be studied. Certainly the computer can be a very helpful tool in fault management situations; the discussed concepts of failure coverage, tolerance and redundancy management are essential for human failure diagnosis, but the basic problem can be summarized by the question: To what level can we automate the plant with regard to the decision sharing between computer and human, and at what level should the human supervisor be able to intervene in future, decentralized multi-level controlled processes? Not discussed by

Johannsen, but probably of equal importance is the question: At what level of automation should we aim in order to balance between performance of human and system and mental workload of the human supervisor under all, steady as well as non-steady state, control situations?

The future role of computers in fault diagnosis is also discussed by Lees where he states "The problem of alarm analysis in real time may be seen as one aspect of a generic problem - that of fault propagation in process plants". Certainly, the concept of fault propagation is a most important item and it is a pleasure to see in what an excellent way Lees reviews the different methods developed to create fault propagation structures. His two methods (1) the process alarm network method and (2) the process fault tree method are excellent examples of the importance of his ideas. In discussing both methods, it is interesting to see what the problems and limitations of the methods are. In particular, the point is made that "Much of the fault propagation process can be modelled using models which describe the normal operation of the units. In most cases, however, it is necessary to include in the model of the unit a description of particular mechanical faults. It may be necessary to include faults even in a very simple model". The important question then arises: What should we know in terms of models about normal operation and non-normal operation of each individual unit, and what do we actually know beforehand?

Finally, Pau adds to this session a very interesting dimension by indicating the relation between the failure analysis and diagnosis and the pattern recognition methodology. He says: "The main idea is that the way in which diagnostics data are analysed, and detection decision later taken, is very similar to pattern recognition procedures with learning, feature extraction, and classification". In his excellent review paper covering a broad area, he shows the limitations as well as the potential possibilities of these techniques for failure diagnosis. The pattern recognition methods looks to be very promising, although Pau in his section on application states: "Any specific implementation is a unique blend of the specific aspects and methods surveyed in the previous section". Hence his statement implies that while general methods to solve failure diagnosis problems can be developed one is unable to generate general solutions for implementation due to the specific character of each implementation?

PAPERS ON METHODS

Several authors have focussed their attention on newly developed methods for failure diagnosis. In his paper, and in particular in the appendices, Lind illustrates the use of flow models for automated plant diagnosis. This method, which is more or less related to bondgraphs, shows how unknown plant information can be derived on the basis of conservation laws. In a second paper on this topic, based on Rasmussen's three-level model of human behavior, Goodstein illustrates how with this method displays can be developed for detection and identification of process disturbances. One of his major goals in this paper is to show the significance of two complementary concepts, i.e. field of attention and level of abstraction.

Fault trees and cause-consequence diagrams are well-known methods in the diagnosis of failures (see for example Lees's paper). The cause-consequence diagram is an extension of the fault tree or, to be more specific, it is a combination of the fault tree with failure mode and effects analysis. The Disturbance Analysis System, as discussed by Bastl and Felkel, is based on these cause-consequence diagrams and can be considered as a simulation of the process to be supervised based on (very) simplified process models. Presently this promising method is under consideration at several institutes, and has contributed already significantly to the safety of nuclear reactors. In the discussion we have contributions by three participants, i.e. Dr. Mancini, Mr. Wreathall and Dr. Hanes who will comment on this subject.

APPLICATION PAPERS

As pointed out, four papers are more or less application oriented, which, of course, does not say that attention is not paid to the methods used. Syrbe states in his paper clearly that, due to the increase in complexity, errors in technological systems can not be avoided by perfection alone. He argues that one has to develop fault tolerant systems. After a nice discussion of what fault tolerant means, he shows an example of a large scale application in the steel industry. Also Dellner discusses the importance of fault tolerant systems; in particular he is interested in modelling the human's role in fault tolerant systems. From this he draws conclusions for display design, system testing and maintenance. He illustrates his paper with examples of an electric utility power generation plant, an electronic switching system and an autopilot. Lihou presents a method to help operators in fault finding, based on operability studies. By means of cause equations and symptom equations he suggests a rather difficult looking method to assist the human operator in his tasks; however, the method could be easily described with the help of Boolean calculus. The paper is illustrated with an example from the chemical industry. Finally Gaddes and Bradly discuss the System Approach Manual Maintenance project for avionics, a paper which well illustrates to what results one might come just by working systematically.

SUBJECT	REVIEW			METHODS						APPLICATIONS			
AUTHOR / KEYWORD	JO-HANN.	LEES	PAU	LIND	GOOD-STEIN	BAS.+ FELK.	MAN-CINI	WREAT-HALL	HA-NES	SYR-BE	DELL-NER	LI-HOU	GADD.+ BRADY
DECENTRALISATION	*												
TASK ALLOCATION		*									*		
LEVEL OF INTERVEN.											*		
THREE-LEVEL HUM. BEHAVIOR MODEL					*								
INTERNAL MODEL	*	*				*	*	*	*				
FAULT COVERAGE	*												
FAIL. TRANSPAR.	*	*											
REDUND. MANAG.	*												
FAULT PROPAGAT.		*											
FAULT TOLERANCE	*									*	*		
RELIABILITY/SAFETY	*					*							
PATTERN RECOGN.			*										
FAULT TREE CAUSE CONSEQ. DIAG. SYMPTOM EQ.		*				*	*	*	*			*	
FLOW MODELS				*	*								
DIST. AN SYST.						*	*	*	*				

Table 1. List of keywords referring to failure diagnosis with the related papers.

CONCLUDING REMARKS

In table 1 the major aspects discussed in the papers are listed by means of keywords. It is of interest to see that much effort is laid upon the methods to be developed to help operators. In terms of Rasmussen's three-level human behavior model, the methods are intended to take over human tasks related to either skill-based and rule-based behavior or to make it possible for the human to act only as a problem solver using knowledge-based behavior. Whether these trends are welcome or not cannot be completely foreseen at this moment.

DISCUSSION SESSION: SYSTEM DESIGN AND OPERATOR SUPPORT

The papers have been discussed in the same order as they have been reviewed; that is, first the more general papers, then the papers on methods, and finally the application oriented papers. In addition to the paper on Disturbance Analysis Systems (Methods) three short contributions were presented for which we have included summaries.

In the presentation and the ensuing discussion of Johannsen's paper, one of the most important topics seemed to be the transparency vs. reduction of displayed information controversy: Key information should not be hidden.

The discussion following Lees's presentation clarified the point that the results of on-line fault tree analyses should be a display of alternative hypotheses combined with aids for the interpretation, rather than the display of a unique solution.

The current pattern recognition methods, as they were discussed by Pau, yield a resolution of some 15 concurrent patterns but during the discussion it became clear that sequential pattern recognition potentially may lead to a solution for much more complex situations. Another point of interest referred to the difference between human and automatic pattern recognition. Some of the most prominent applications can be found in the field of quality control where the need for human inspection rates may be reduced to the order of a few percent while improving, overall performance.

Before summarizing the interesting discussion on the pro's and contra's of the different methods presented, we will start with summaries of short extra contributions.

Presentation by Dr. Mancini

Dr. Mancini's contribution on Event Sequences and Consequence Spectrum, a Methodology for the Probabilistic Transient

Analysis of System Physics and Logics, was as follows:

The usual approaches for probabilistic accident evaluation do not satisfactorily take into account the dynamic aspects of the random interaction between the "physics" of the transients and the "logic" (states) of the systems. Indeed current probabilistic techniques separate the system reliability analysis from the real dynamic development of an accident. Actually, the accident moves in a specific direction according to the values assumed by physical parameters such as temperature, pressure, etc. (which control the intervention of protection and/or mitigating systems) and to the occurrence of logical events, such as lack of intervention of the demanded systems, delays, partial failures, etc., which can occur at random times.

An adequate methodology must indeed consider the possibility that a given initiating event will trigger, in its temporal progression, new logical events which, in turn, can act on the physical evolution of the accident in a continuous dynamic interactive process. Moreover, such methodology must take into account the statistical spread of physical parameters relevant for the transient behaviour.

The next few paragraphs briefly review a new general methodology (ESCS/LAM: Event Sequences and Consequence Spectrum by the LAM Technique) for the connection of the logical to the physical aspects in order to investigate the physically possible temporal accident sequences and their consequence spectrum, evaluating at the same time the corresponding probability of occurrence.

The basic concepts of the proposed technique can be summarized in the following items:

- The nominal behaviour of a system (either steady state or operational transient) corresponds to the nominal behaviour of its components.
- Incidental occurrences arise from failure or degradation events which change the component nominal behaviour.
- Each component is described by analytical relationships involving the physical variables which account for the component behaviour in all its possible states.
- Uncertainty distributions of physical parameters are taken into account by approximating continuous distributions with histograms and dealing with them in the same way as with any other component state.

The set of analytical relationships, characteristic for the component, is controlled by parametric logical operators which select the appropriate relation for the component behaviour under each given particular state.

In this way the system behaviour, both under nominal and under incidental conditions, is synthetically described by a set of parametric equations. As a consequence, all incidental occurrences can be identified without limiting the analysis to preselected abnormal transients or accidents by the automated generation of all possible sets of logical events.

The "consequence" of a given sequence (in terms of occurrence of a certain temperature, pressure, stress, and so on) is immediately obtained by the numerical solution of the corresponding equation set. The associated probability follows from the determination of the probability of those sequences which cause a system physical variable to range within a given interval.

Presentation by Mr. Wreathall

The following summary describes the contribution of Wreathall on Disturbance Analysis Systems (DAS). He posed and answered four questions.

1) Are they necessary? DAS concepts have developed as a result of three or four convergent streams. Firstly mathematical methods for the representation of dynamic physical processes have developed in 1950's, 1960's and early 1970's in such a way that system modelling engineers feel that they can model the world - literally! Secondly and strongly related to mathematical progress has been the development of high speed low cost - low weight computing enginees. Thirdly, integrationist methods of analysis, such as reliability analysis, have grown which enable different technical domains to be spanned so systems engineers and scientists feel easy roaming about in everyone-else's backyard. But, is it necessary? In Taylor's work on Abnormal Occurrence Reports in 1975 he reported the relative impact of operator errors in the overall frequency of errors is 12% (Taylor, 1975). If we look at the subsidiary causes within that overall group, lack of recognition of fault was 7% (of the 12%) and miscognition was 2% of the 12% - i.e. 0.24%. Nearly 50% of the operator errors were simply omitting steps in procedures, etc. Whilst these fault groupings may be not very explicit or necessarily accurate to 1 or 2%, they feel about right to me. Three Mile Island started because someone left out a step in a procedure and left the isolating valves which coupled the auxilliary feed system to the main feed system closed. The state of control room ergonomics was very poor. Any one with five minutes common sense applied to the problem could see that. That in itself does not mean to me a direct need

for DAS. Sensible engineering with finite resources says the problems giving greatest impact in the operation first. I submit that is not the cognitive area.

2) Does DAS work? I suppose it must, but how or where does it? Felkel seems to be the only author willing to quote evidence that DAS does achieve something, as opposed to assuming it does. This relates to the EPRI/C-E simulator connection. I do not know the details, but the evidence in the paper seems to imply DAS systems may well speed up the operator's learning process for his empirical(?) knowledge of plant behavicur, i.e. "creating a mental model". I do not infer anything else from the work. Maybe that is all that one does need. Now taking Jens' model (fig. 1) (Rasmussen, 1980), can one infer what DAS is doing here? The DAS models being talked about are representations of known predictable associations between plant processes. These are the parameters pertaining to rule-based behaviour - the middle ground in the "other" Rasmussen figure (2) - i.e. DAS is not contributing much to the elimination of "rare big" accidents. I do believe it could improve availability in plants and marginally reduce the risk - remember the relative impact of cognitive operator mistakes.

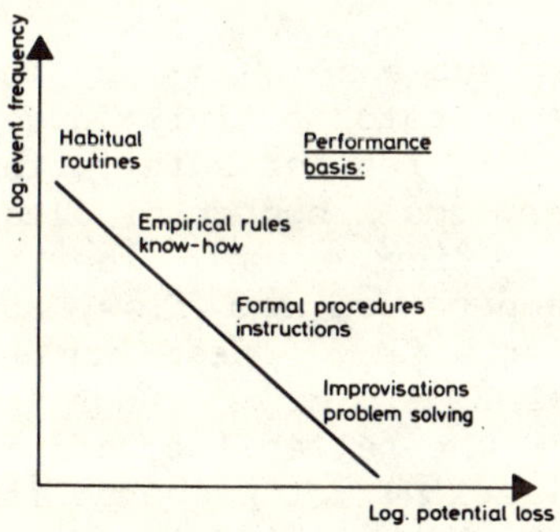

Fig. 1. The basis of human performance depends on the frequency of call for action, and the acceptable frequency of an event will depend upon the potential loss implied.

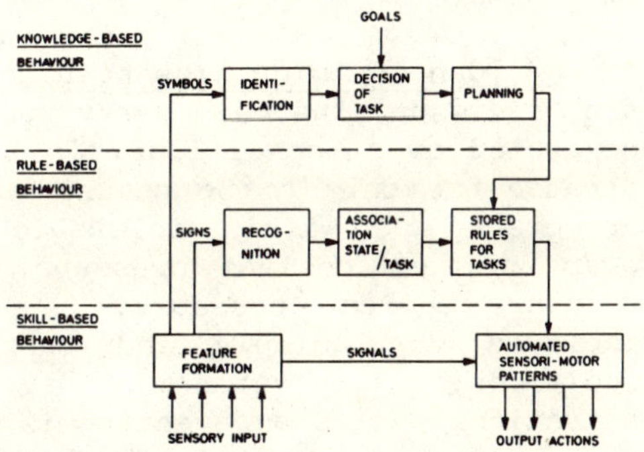

Fig. 2. Schematic illustration of different domains of the control of human performance.

Reproduced from Rasmussen (1980)

3) Can it make things worse? Bearing in mind that DAS systems presently talked about use models representing physical processes within plants, we have a real difficulty. We have incomplete knowledge of basic physical processes within nuclear power plants. That is critical when considering the effectiveness of post trip cooling. You may ask how can we operate and claim to be safe if this is so. This is done by bounding the problem, making pessimistic assumptions, but the behaviour thus modelled is of little use to form diagnosis models. For instance, the small

LOCA in TMI design studies was assumed to be bounded in terms of cooling sufficiency by pressurised fault or major LOCA studies. That was true - the heat transfer areas and heat-sink ratings are adequate for minor LOCAs, it's just that the plant behaviour in terms of parameter excursions were not what the designers expected. System engineers and scientists must allow that the real world may be very different from what they expect. Designers are sometimes wrong, knowledge is lost. The small LOCA problem was raised by TVA at least 1-2 years before TMI. ACRS, NRC and others had all exchanged letters about it. They were just lost inside B&W (see Kemeny).

Going back to Bob Taylor's work on classes of errors and their sources, design errors are 35% of the total - the dominant cause, and 25% of those are due to effects <u>unknown</u> at time of design. Also 13% of all errors which would most probably be missed in any DAS model are installation, maintenance or fabrication errors. So - the operators cannot rely on this aid; for maybe 10-20% of all faults the model is deficient. Any operator that I have ever met would simply pull the plug on anything so often in error. Remember the DAS is supposed to be reducing this 0.24% of all faults!

<u>4) Can anything else be done</u>? Taking nuclear safety as the point of issue, the main worry seems to be operator activity under unexpected event conditions. To be assured of reactor safety, the operator has only to be assured that a very small number of plant parameters are within tolerable bounds - for example reactor power, primary coolant pressure, bulk temperature and flow, and secondary coolant pressure, temperature and flow. These can be displayed <u>very</u> simply and <u>very</u> reliably. For this range of conditions, the operator must have faith in the reliability of his indications - it's an event he is (by definition) not prepared for and he must have something to be sure about. Maybe the information in this area must be sensed, transmitted and displayed by diverse means. The Hinkley Point B control room has been modified at the operators' request to provide reliable and diverse indication of successful safe post-accident operation. One is tempted to indulge in instant design, but I believe that safety displays on control room desks should be simple - analogous with the basic "T" of aircraft pilot instruments. The operator has knowledge entirely outside the level of models discussed. For instance, the operator may suddenly get failure of several different subsystems. He will know that say all the cubicles are in a single room so maybe local plant damage is the cause (fire, flood, etc.) or it may be more subtle - at one power plant in the US, it was known that cleaners in equipment areas caught switches with their brooms, etc. I will be impressed when we can get DAS working to sort that out!

Presented by Dr. Hanes

Finally we will give some notes by Hanes of a study performed under the direction of John Gallagher of Westinghouse Electric Comp. His presentation provided an overview of the results of an EPRI-funded study to determine the scope and feasibility if a plant-wide Disturbance Analysis and Surveillance System (DASS). The feasibility was determined by developing a process for designing the DASS and, thereby, for estimating the costs of designing and building the DASS. The scope was determined by developing DASS functions which would assist the operator in the improvement of plant safety and availability. In addition the process included some estimator of the potential benefits of these DASS functions in the form of relative improvements in safety and increased plant availability.

During the presentation, it was pointed out how the signals driving and information displayed in this conceptional DASS were different than in the GRS-STAR and Halden DAS approaches. The relation between Rasmussen's skill, rule and knowledge based behaviour model and operator information needs were discussed.

Discussion (continued)

Some controversy during the general discussion of the different DAS(S) methods presented centered around the following items.

- The availability/safety trade-off, a factor that seems to be rather unquantified.
- The influence of DAS upon operator errors is yet unclear; problems may occur during unforeseen events.
- There still appears to exist serious problems in modelling process knowledge.

Comparing the previous European DAS concept described in Felkel's paper with the new American DASS proposals, as for instance illustrated by Hanes's priority list of DASS-functions, it appears that functions like "subsystems configuration monitoring" and "system state identification by data integration" have gained highest priority while "cause determination" and "advice on proper action" have moved down below the 7th priority. This ordering of the DASS functions was argued for plant availability reasons; for plant safety the priority order was slightly different.

Furthermore it was mentioned during the discussion that:

- Remembering the work by MIT that decision making by committee can be very poor under stress, consider that because the computer is an aid and not a master, the man plus the computer is actually a committee.

- Work at Connecticut University was mentioned as potential evidence that DAS can corrupt the decision process. In those studies, the computer prompting with 100% correct analyses reduced the speed and accuracy of the operator's diagnosis.

Lind's flow model gave rise to some principal disagreement concerning the aspects of the dynamics used in his method, but it turned out that the dynamics were used for the state estimation rather than the prediction.

Finally, a comment of Whitehouse referring to Lind, Goodstein, Felkel, Mancini, Wreathall, and Hanes presentation said: "The "Anticipator System" developed by Hawker-Siddeley is intended to (1) warn the operator that the plant is moving towards an unwanted state, (2) present the operator with the results of the designer's analyses to give the maximum information available for him to base his plan of action, and (3) review plant operating data at a resolution depending on the plant state for subsequent analysis, particularly of un-analysed events. The pattern recognition is by the use of Decision Tables, which are normally prepared by the designer. Techniques are available to reduce the size of these tables to manageable levels: Disaggregation of the plant into a number of units, and treating the connections separately is a powerful way. Currently the tables are prepared manually: They can earlier be designed to extract as much information as required from existing instrumentation, or to determine what measurements are required in order to identify particular events which are required to be signalled".

The presented algorithms (Bastl, Felkel, Lihou, Lees and Whitehouse) were debated in terms of their correctness in relation to possible non-additivity of symptoms in case of multiple failure situations. Correctness has even been suggested to be more important than completeness. This topic deserves more attention.

REFERENCES

Kok, J.J., R.A. van Wük, 1978, "Evaluation of Models Describing Human Operator Control of slowly Responding Complex Systems", Delft, DUP, pp. 1-235.

Rasmussen, J., 1980, "Some Trends in Man-Machine Interface Design for Industrial Process Plants", Risø-M-2228, March 1980.

Sheridan, T.B., 1976, "Toward a General Model of Supervisory Control", in: T.B. Sheridan, G. Johannsen, "Monitoring Behaviour and Supervisory Control", NY, Plenum Press, pp. 271-281.

Taylor, J.R., 1975, "A Study of Abnormal Occurrence Reports", Risø-M-1837, September 1975.

FAULT MANAGEMENT AND SUPERVISORY CONTROL OF

DECENTRALIZED SYSTEMS

Gunnar Johannsen

Forschungsinstitut für Anthropetechnik (FAT)
Königstr. 2, D-5307 Wachtberg-Werthhoven
F.R. Germany

INTRODUCTION

Performance and reliability of complex technical systems (large-scale systems) may often be improved if decentralized hierarchical control concepts are used. Examples for such systems are transportation systems, air traffic control systems, complex vehicle guidance and control systems, power plants and networks, chemical plants, and manufacturing networks. All of these systems have taken or will take advantage of advanced automation using intelligent computers.

Some remarkable developments related to decentralized systems have taken place in automation technology, (Syrbe and Will, 1977; Avizienis, 1978), and control theory, (Athans, 1978), during the last years. The two main areas of activity are the design and the extension of theories and techniques on the one hand and the development of new hardware and software for their implementation on the other hand. These developments are far from being terminated.

As examples for hardware and software, microprocessors and microcomputers, distributed computer systems, fault-tolerant digital systems, higher programming languages for process computers (e.g., PEARL), and interactive, computer-generated process-control displays are to be mentioned, (Syrbe and Will, 1977; Avizienis, 1978). Microcomputers have started to change the concepts of automation from centralized to decentralized control where this seems to be appropriate. Cathode-ray tubes (CRT) and color-TV displays allow the human operators to get the information they need as a manager for supervisory control of the automated processes. On the basis of these technological improvements, it seems to be possible to design future systems with a distribution of intelligence between computers and human operators, between different levels of control, and between different subsystems within one level. Decision-making and control tasks are then to be allocated to the decentralized intelligent subsystems.

Interestingly enough, the development of control theory towards decentralization has occurred so far somewhat independently from that of the mentioned computer technology. A collection of papers representing the state-of-the-art in decentralized control has been published in a special issue of IEEE, (Athans,1978). A very brief review will be given in the next section of this paper.

Both developments, i.e., those of control theory and computer technology, have reached the state of a relatively high maturity so that more effort should be devoted, now, to merging these two together. The result would be the implementation of advanced control laws with decentralized microcomputers. At the same time, a better consideration of human-computer interaction for supervisory control and fault management of these systems is called for.

Some ideas how to approach this ambitious integration and some preliminary results are presented in this paper. In subsequent sections, the notion of supervisory control and shared decision-making responsibility between human operators and computers as well as the system reliability aspects and some possibilities of designing interactive fault-management systems are discussed.

DECENTRALIZED SYSTEMS AND APPROPRIATE CONTROL THEORY

Decentralized systems are complex systems which can be divided into subsystems with a certain degree of autonomy. One has to distinguish between local and functional decentralization. Examples for local decentralization, i.e. often spatial distribution, are given in chemical plants, interconnected power systems, and C^3 (command, control, and communications) systems. The functional decentralization allows a separation into subsystems with different dominant time scales, e.g., the separation between fast attitude motion and slow overall vehicle motion in vehicle guidance.

Decentralized control theories are needed for controlling these systems. As Athans points out in his Guest Editorial, (Athans, 1978), most theories for decentralized control so far proposed are extensions of the current tools associated with centralized control. If future large-scale systems are to be operated more efficiently, more interdisciplinary effort is necessary to design completely new theories. Therefore, Athans asks for the interaction between control theorists, operations researchers, communications engineers, and computer scientists, (the author of this paper would add human factors specialists). The new theories should bring about new concepts of solutions, new definitions, e.g., for optimality with specific consideration of reliable operations, as well as a better understanding of the value of information for decision making. Although Athans' assessment of the situation shows some interesting directions for future research, it is too negative on the whole with respect to the relatively high maturity of the theories published in the same special issue, (Athans, 1978).

Before applying decentralized control concepts, one has to realize that they are not always the best. Sometimes, centralized control is better as this is depending on the systems dynamics. Therefore, the solution of an application problem has to start as usual with a systems analysis before working on the control synthesis. Decentralized control may be best if a highly complex system is present, i.e. in the extreme, no way exists of finding a proper model for the whole system. The systems analysis has also to find out in which way the system can be decomposed for decentralized control, e.g., with respect to time scale or spacial distribution.

Model building and especially model simplification are of essential importance for the control of large-scale systems. Because of the complexity of these systems, a simplification of their structure is advantageous or even necessary before the control synthesis can be commenced. The existing techniques allow a decomposition and order reduction of the technical system, (Athans, 1978; Litz, 1979).

A very suitable tool for decomposing complex systems is the perturbation method, (Sandell, Jr. et al., 1978; Kokotovic, O'Malley, Jr., and Sannuti, 1976; Khalil and Kokotovic, 1978). With weak coupling between the subsystems, the following systems description is given:

$$\begin{bmatrix} \dot{\underline{x}}_1(t) \\ \dot{\underline{x}}_2(t) \end{bmatrix} = \begin{bmatrix} \underline{A}_{11} & \epsilon\underline{A}_{12} \\ \epsilon\underline{A}_{21} & \underline{A}_{22} \end{bmatrix} \begin{bmatrix} \underline{x}_1(t) \\ \underline{x}_2(t) \end{bmatrix} + \begin{bmatrix} \underline{B}_1 & \underline{0} \\ \underline{0} & \underline{B}_2 \end{bmatrix} \begin{bmatrix} \underline{u}_1(t) \\ \underline{u}_2(t) \end{bmatrix} \tag{1}$$

where ε is a small positive parameter. For $\varepsilon = 0$, both subsystems are completely decoupled. Then, a completely decentralized control structure can be built up (see Figure 1). If a restricted amount of information links between the controllers is necessary, this is termed partial decentralization.

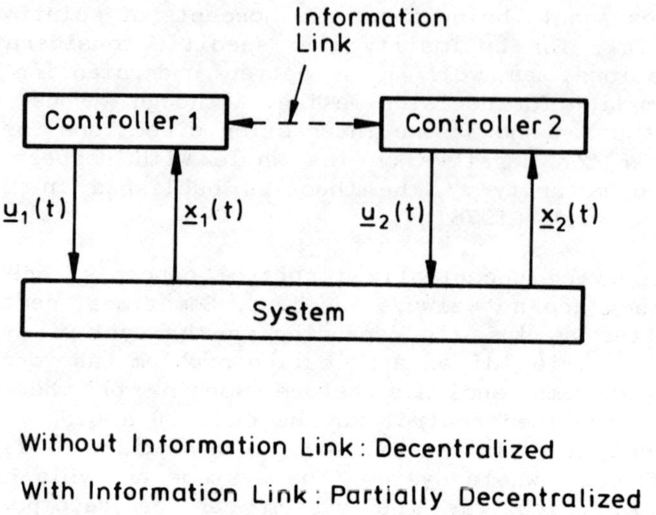

Figure 1. Decentralized and Partially Decentralized Control Structure.

With strong coupling between the subsystems, a perturbation to the left-hand side of the differential equation is assumed:

$$\begin{aligned} \dot{\underline{x}}_1(t) &= \underline{A}_{11}\underline{x}_1(t) + \underline{A}_{12}\underline{x}_2(t) + \underline{B}_1\underline{u}_1(t) \text{ (slow subsystem)} \\ \varepsilon\dot{\underline{x}}_2(t) &= \underline{A}_{21}\underline{x}_1(t) + \underline{A}_{22}\underline{x}_2(t) + \underline{B}_2\underline{u}_2(t) \text{ (fast subsystem).} \end{aligned} \tag{2}$$

In this way, a separation of the systems dynamics into a slowly and a fastly responding part is obtained. This yields a naturally hierarchical structure with different time scales for the controller design. If $\varepsilon = 0$, the fastly responding part can be neglected which results in an order reduction of the system.

There are different methods available for designing control laws for large-scale systems. (Athans, 1978). These include pole-placement design, robust control, and game theoretical approaches. The latter assume that a separate decision maker with its own cost functional exists for each subsystem, (Cruz,

Jr.,1978). It is also possible that the decisions for the control laws of the individual subsystems are based on different model simplifications of the overall system, (Khalil and Kokotovic, 1978). This leads to a combination of game theoretical approaches with singular perturbation methods, (Gardner, Jr., and Cruz, Jr., 1978). The model simplifications depend on the amount of information available to the local decision makers which are, thus, determining control with respect to their specific cost functionals and these specific internal models.

The decision makers are sometimes in a goal conflict with each other if the couplings between individual subsystems are correspondingly strong. Especially, for such decision-making situations, the game theoretical approach seems to be very appropriate. A coordinator or leader on a higher level in the decision structure seeks for a global optimum (see Figure 2). The other lower-level decision makers pursue their individual goals being influenced indirectly, however, by the actions of the coordinator. The basic leader-follower strategy is also called the Stackelberg strategy. It is even possible to design a coordinator with memory, i.e., assuming a higher cognitive level, (Papavassilopoulos and Cruz, Jr., 1980).

If the lower-level decision makers work on a noncooperative basis, they act simultaneously according to a game solution concept. Typical examples are the so-called Nash strategies. If, on the other hand, only one common criterion exists for all lower-level decision makers, then, the solution is termed a team theoretical one.

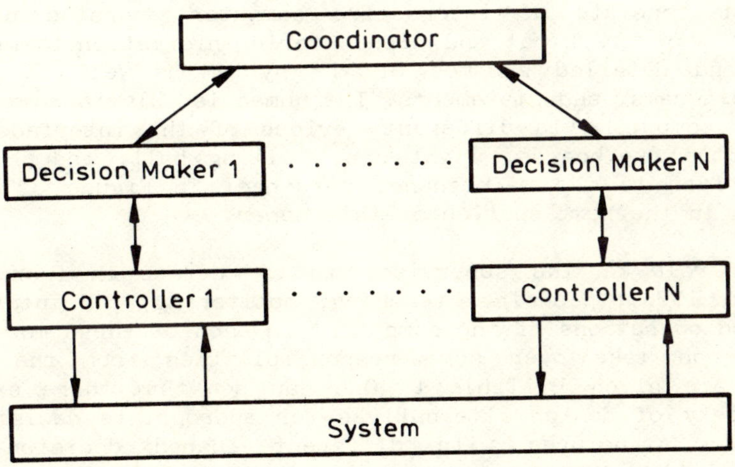

Figure 2. Multilevel Control of Large-Scale Systems.

SUPERVISORY CONTROL AND SHARED DECISION-MAKING RESPONSIBILITY

The before mentioned game theoretical approaches with their specific decision structures so far do not consider human-computer interaction. Principally, this should, however, be possible, especially, as some basic ideas for these theories have been adopted from economy where human decision- making strategies are of similar importance as they are in human-computer interaction.

The notion of supervisory control has been used to characterize "control by a human operator of a computer which, at a lower level is controlling a dynamic system", (Sheridan and Johannsen, 1976). It is based on the philosophy that the final responsibility for goal-setting, system performance, and reliability will be with the human operators in complex man-machine systems, (Schweppe, 1978).

The schematic block diagram of a supervised decentralized control system may look like Figure 3. The lower part of this block diagram corresponds with Figure 2. The process computer plays the role of the coordinator and is itself supervised by the human operator. The interface between the process computer and the lower-level decentralized controllers and decision makers consists normally of the bus system when microcomputers are used on the lower level. The same interface contains analog-digital and digital-analog converters in the case of analog equipment on the lower level.

The interface between the process computer and the human supervisor consists of interactive computer-generated displays which can show graphical and alphanumeric information by means of overall and detailed schemes of the system as well as gauges, scales, diagrams, and statements. The human is able to communicate with the computer via different devices of the interface like, e.g., virtual keyboards, light pen, rolling ball, and joystick. The interface to the maintenance personnel in Figure 3 will be mentioned in the next section of this paper.

The role of the supervisor is to operate in four modes, (Sheridan, 1976): planning, teaching, monitoring, and intervening programmed operations of the computer. In each of these modes, the computer can take over some responsibilities from the human; examples are given in Table 1. One can see that there exists a wide variety of design alternatives for appropriate decision and information structures with different shared decision-making responsibilities between human and computer. In this situation, models of human behaviour are used as a design tool. An overview of these models has recently been published, (Johannsen and Rouse, 1979).

Table 1. Examples of Computer Roles in Man-Supervised System (from T.B. Sheridan, 1976).

	PLAN	TEACH	MONITOR	INTERVENE
SENSING AND PRESENTING RELEVANT INFORMATION ABOUT PRESENT PROCESS STATE				
display data to human operator in given format	+	+	+	+
find and display data which meets given criterion	+	+	+	+
apply given measure (extrapolation, correlation, etc)	+		+	
find best sensory process to meet criterion	+		+	
make diagnosis of measured symptoms			+	+
EVALUATING ALTERNATIVE ACTIONS				
indicate to operator command doesn't meet criterion	+	+		
determine model response to given test input	+			+
determine which control is best by given criterion	+			+
test whether actual response matches model response			+	
suggest an action to human operator	+	+	+	+
request data from operator, process recommendation	+	+	+	+
IMPLEMENTING ACTIONS				
request data from human, process it for action		+	+	
take certain action when operator gives signal		+	+	
take control action unless operator gives signal		+	+	
take control action independently of human operation		+	+	

As an example for shared decision-making responsibility between human and computer, a flight management situation will briefly be mentioned, (Chu and Rouse, 1979). A flight control task and several subtasks have to be accomplished. The subtasks are waiting like customers in a queue for being serviced, either by the human or by the computer. The computer is able to take over the responsibility for subtasks automatically and indicates this to the human if, otherwise, the human operator would be overloaded by the arrival of too many subtasks.

So far, supervisory control and shared decision-making responsibility have mainly been viewed with respect to centralized systems. Basically, the situation is not too different with decentralized systems. This can be seen in Figure 3 where, compared with centralized systems, the lower level of decentralized controllers and decision makers has been inserted with

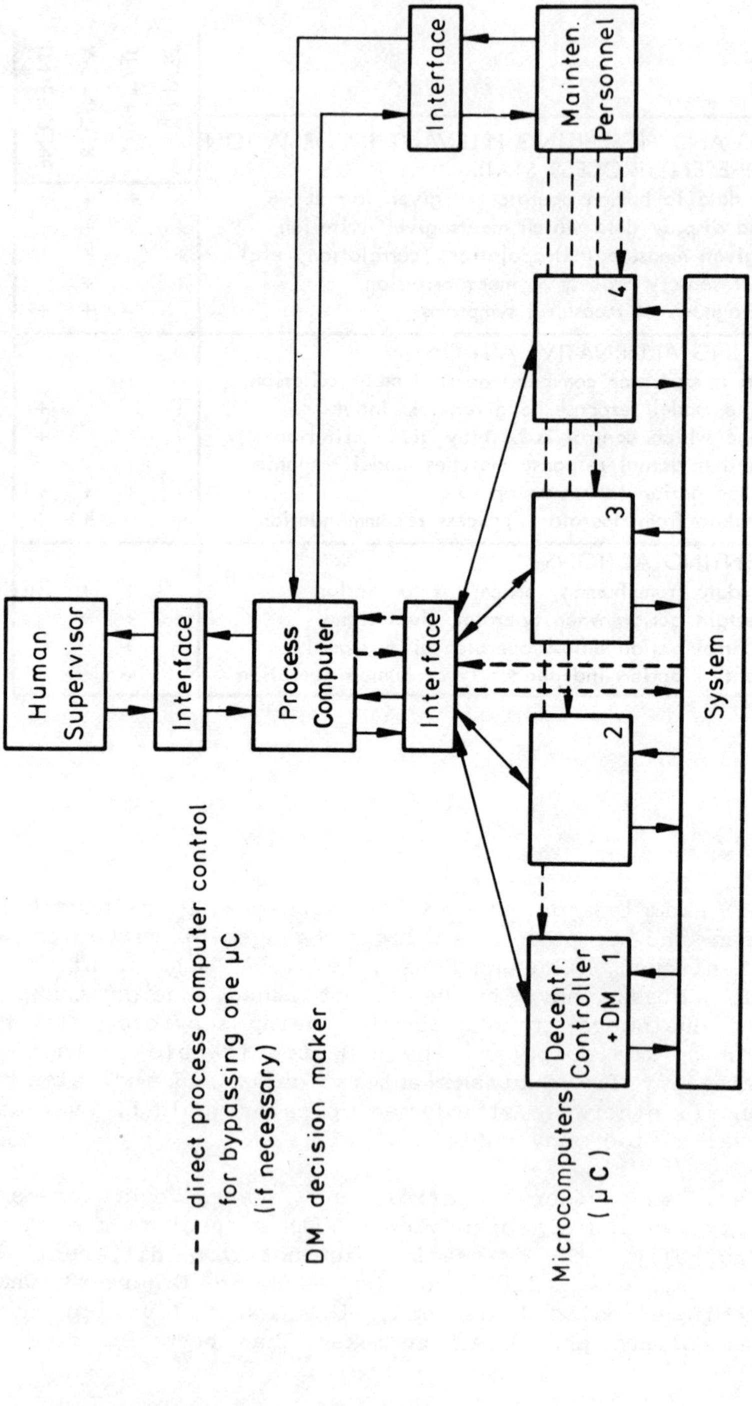

Figure 3. Block Diagram of a Supervised Decentralized Control System.

its own intelligent capabilities. The distribution of computer intelligence to different levels, however, extends the design alternatives for decentralized systems. Some of the new questions which arise are:

1. How much autonomy is appropriate for the lower-level intelligent subsystems?
2. How much information should be transferred from the decentralized intelligent subsystems to the coordinating higher-level process computer?
3. How should the responsibility for supervision and coordination of the intelligent subsystems be allocated between human and computer on the higher level?

The answers to these questions deserve a lot of research effort in the future.

SYSTEM RELIABILITY ASPECTS AND INTERACTIVE FAULT-MANAGEMENT SYSTEMS

The distribution of intelligence between human operators and computers is particularly important in fault-management situations. The possibility that a chain of incidents leads to an accident or plant shut-down has to be reduced, (Kragt, 1978; Rubinstein, 1979). This problem cannot be tackled by only improving the performance and quality of automatic equipment. More often, it is argued today that the role of human operators will increase in importance, (Schweppe, 1978). However, it is necessary to design an appropriate allocation of functions between human operators and computers to assure that the human will be able to perform the tasks allocated to him very reliably, (Kragt, 1978). Also, the human operators should have received intense training for coping with abnormal situations.

Thus, reliability of complex systems has two components, i.e., systems reliability of the technical system and human reliability. Major contributing factors to systems reliability of automated systems are computer hardware and software reliability. A special issue of IEEE on software reliability gives a recent overview, (Lipow, 1979), which shows clearly how important this component is simply because a higher and higher percentage of the cost of computing systems is going into the software development. On the other hand, microprocessors have become so inexpensive that it is possible to increase the hardware reliability, e.g., by using microprocessors in voting blocks of three to five, (Editorial, 1979).

Within a decentralized system, the structure of the intelligent subsystems affects the reliability of the overall system. The individual computers in a distributed process computer system can be linked, e.g., via fiber optics, in such a way that dynamic functional redundancy is achieved in order to reduce the number of system failures, (Syrbe, 1979; Krüger and Nehmer, 1977; Sendler, 1980). During normal operation, this redundancy is used to improve performance. In cases of faults, the respective class of failures is detected and the structure of the computer system changed to allow continued process operation, only with slightly reduced efficiency.

The concepts of coverage of failures and redundancy management are applied in the design of a fault-tolerant engine control system, (Mosca, Rabinowitz, and Kreamer, 1979). Coverage is defined as the conditional probability that, given the existence of a system failure, the system is able to recover and continue operation. The process of redundancy management has the purpose of improving the coverage of failures and making systems fault tolerant. An overview on fault-tolerant digital systems is given in a special issue of IEEE,(Avizienis, 1978).

Under consideration of coverage of failures and redundancy management, the basic reliability goals of automated systems can be formulated as stated by Mosca et al.(1979):

1. The ability to detect (almost) any failure in the system.
2. The ability to limit any system damage caused by a failure.
3. The ability to make failures transparent to system operations.
4. If unable to make failures transparent,
   a) the ability to place the system in a consistent state so that recovery is possible; and
   b) the ability to report the failure to a higher level which can direct an intelligent system recovery, thus, making the failure transparent at higher levels.

On the higher level, a computer or a human operator may be responsible for system recovery. Thus, it is necessary to consider also human reliability when an assessment of the overall system is to be made. An overview on human reliability in complex systems has been published a few years ago, (Embrey, 1976). Some strong attempts have been made to use compatible metrics for systems and human reliability. So far, human reliability metrics have mainly been used as a passive evaluation technique for given or assumed system designs. However, there are also other ways of engaging the topic of human error in man-machine systems as a means of increasing overall reliability, e.g., designing the system appropriately or seeing the human operator in a redundant relationship with the system, (Adams, 1978).

These more active ways of increasing reliability are related to those in systems reliability which have been mentioned above. The result leads to interactive fault-management systems. Fault-tolerant systems that handle also man-made faults are a new challenge to system designers because human interaction faults still remain a major problem area in system operation, (Avizienis, 1978). On the other hand, the value of human insight and intuition is substantial for successful system operation. The main problem is to provide the human with the appropriate information. Underloading as well as overloading the human has to be avoided.

To accomplish this goal, interactive fault-management systems should be designed with a transparent information and decision structure and a high coverage of failures. This is especially important for decentralized systems. Computer and human operator may check each other by means of plausibility testing and fast-time model predictions in order to assure effective supervision of normal operation and reliable fault diagnosis in case of abnormalities. The same means can be used to ease the process of choosing between alternatives in fault correction.

Plausibility testing implies that human operator inputs are screened by the computer for some obvious types of mistakes and vice versa. This will be done on the higher level of supervision between human and process computer, as has been outlined in Figure 3. Both of these systems may have a more or less sketchy or precise internal model of the overall man-machine system. If the system is highly complex, the human's internal model will probably be more sketchy than that of the computer. Therefore, the computer will be responsible for bookkeeping important state information and placing it at the human's disposal on request or if the necessity has been discovered by the computer.

The internal models can be used for fast-time predictions in fault diagnosis and correction by comparing expected and measured outputs of the system. In highly complex systems, it seems to be reasonable to implement some kind of this predictive capability on the lower level of the intelligent subsystems to avoid the transfer of a huge amount of information to the higher level. In that case, the higher-level process computer will only make gross trend analyses.

There exists a variety of alternatives for fault correction in decentralized systems ranging from accepting decreased performance to shifting control responsibility to other levels, either computer or human. The possibility of accepting decreased performance exists mainly when either robust control algorithms have been applied, a predetermined order reduction in strongly coupled systems would result as a consequence of a failure, or functional redundancy has been designed into the system.

An example for shifting control responsibility to other levels has been shown in Figure 3. If one microcomputer working as a lower-level decentralized controller and decision maker has failed, it can be bypassed by direct process computer control. In that case, the process computer may forward the appropriate information via an interface, e.g., a computer terminal, to the maintenance personnel. As the tasks of the microcomputer on the lower level have been taken over by the process computer, online maintenance without interrupting system operation would be possible, e.g., replacing the failed microcomputer and testing a new one in parallel to the ongoing operation.

AN EXAMPLE

The representations of this paper have visualized a multitude of possible design parameters of fault-management and supervisory control aids for decentralized systems. An experimental study has been initiated in order to investigate a restricted number of these parameters in a laboratory simulation.

The decentralized system of this study is simulated on a hybrid computer system. The system structure is the same as that outlined in Figure 3 with weak and strong couplings between the four subsystems and failures occurring in one of these subsystems at a time. Sequences of failures are also possible. The more general systems and experimental designs have been chosen with some hope that the results would be applicable to different real systems ranging from slowly to fastly responding ones.

As the experimental investigation has just been started, the following will cover mainly the outline of the experimental design and preliminary hypotheses. Three experimental variables have been selected for the first of what is supposed to be a series of experiments, namely:

1. the dominant time constant of the system,
2. the information and decision structure, and
3. the mode of computer-aided fault correction.

The first experimental variable, i.e., the dominant time constant of the system, has two levels: it may be (1) relatively low or (2) relatively high. The first level is characterizing a slowly responding system, e.g., a large ship or an industrial process plant, whereas the second level is characterizing a fastly responding system like an aircraft. In the latter case, longitudinal and lateral motion may be viewed as weakly coupled subsystems (see Equation (1)). For many aircraft, both subsystems are approximately decoupled, at least for certain modes of flight control. The longitudinal subsystem can be further subdivided in a slow subsystem using thrust as the control input and a fast subsystem with elevator control signals. As Equation (2) shows, this corresponds to a strong coupling between the two subsystems.

The second experimental variable concerns the information and decision structure. This variable comprises different levels on which the diagnosis of faults may be performed as well as, related to this, different amounts of information presented to the human operator with appropriate display designs. Thus, it may well be that this experimental variable turns out to be too complex in its nature and will be separated into several ones in subsequent experiments.

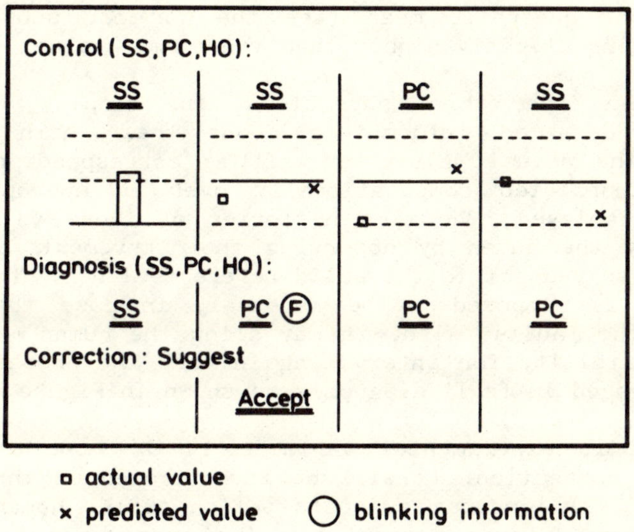

Figure 4. Display Design for Fault Management Experiments (different designs combined).

The fault diagnosis is performed either (1) by the lower-level intelligent subsystems, (2) by the higher-level process computer or (3) by the human supervisor. The amount of information presented to the human in these three cases is illustrated in Figure 4. The three principally possible levels for fault diagnosis are indicated in brackets: subsystem (SS), process computer (PC), and human operator (HO). Status information showing on which level fault diagnosis is being performed is displayed for all four subsystems. A similar indication shows on which level the subsystems are being controlled. This is normally done on the subsystems level (SS) by decentralized controllers and decision makers. In the example, the decentralized controller of subsystem 3 has been bypassed by direct process computer control (PC).

In the example of Figure 4, faults in subsystem 1 are diagnosed on the subsystem level (SS) whereas this is accomplished by the process computer (PC) for the other three subsystems. When a failure, in this case of the lower-level controller, has been detected, a symbol "F" is blinking for the corresponding subsystem. Different types of failures may be indicated by, e.g., "F1" or "F2" (not shown in Figure 4). If this has been diagnosed by the subsystem, it is sufficient to indicate the gross trend for the main controlled variable to the human operator by means of a time-dependent quadratic mean (see the bar for subsystem 1 in Figure 4, being slightly higher than the normal value).

For the other three subsystems, the display of Figure 4 shows more detailed information about their main controlled variables. The actual values as well as corresponding predicted values, extrapolated computationally over an appropriate time span, are displayed. The time histories of these values can be extracted by the human by observing their movements in vertical direction with respect to the solid reference lines. This detailed information is supposed to be necessary in case the human is diagnosing the faults. Also, it may allow the human more insight and the possibility for intervening in case the process computer is being engaged in fault diagnosis as shown in Figure 4.

The third experimental variable is the mode of computer-aided fault correction. It will be investigated on three levels: (1) no aid, (2) computer suggests action to the human operator, and (3) computer takes action when the human operator gives an appropriate signal. In Figure 4, the second level is exemplified: the computer suggests the alternative "Accept" for subsystem 2. There are three alternatives for fault correction among which the human operator may choose: (1) accept decreased performance, (2) shift control responsibility from the lower level to the process computer, and (3) disconnect the failed lower-level subsystem. The human may correct his decisions whenever he wants. On the third experimental level of computer-aided fault correction, the computer is choosing between the three alternatives when the human operator asks for aiding.

The primary hypothesis for the experiments is that best system performance and a high coverage of failures will be obtained when the transparency of the information and decision structure is high. This is assumed to be the case when control is performed on the subsystem level, fault diagnosis and suggestions for fault correction are accomplished on the process computer level, and fault correction itself using these suggestions is handled by the human operator.

The results of the study are to be analyzed with regard to system performance, transparency of information and decision structures, and coverage of failures. This will, of course, require that an appropriate definition of transparency be developed. Further, suitable objective and subjective measures have to be tested for this purpose.

REFERENCES

Adams, J.A., The assessment of human reliability in man-machine systems, European Scientific Notes, ESN 32-3: 116-118 (1978).

Athans, M., (Ed.), Special issue on large-scale systems and decentralized control, IEEE Trans. Automat. Contr., AC-23: 105-371 (1978).

Avizienis, A., (Ed.), Special issue on fault-tolerant digital systems, Proc. IEEE, 66: 1107-1268 (1978).

Chu, Y.-Y. and W. B. Rouse, Adaptive allocation of decision making responsibility between human and computer in multitask situations, IEEE Trans. Syst., Man, Cybern., SMC-9: 769-778 (1979).

Cruz, J.B., Jr., Leader - follower strategies for multilevel systems, IEEE Trans. Automat. Contr., AC-23: 244-255 (1978).

Editorial, Electronics can help prevent more nuclear accidents, Electronics, 52 (8): 24 (April 1979).

Embrey, D.E., Human reliability in complex systems: An overview, National Centre of Systems Reliability, NCSR R 10, United Kingdom Atomic Energy Authority, Warrington (1976).

Gardner, B.F., Jr. and J. B. Cruz, Jr., Well-posedness of singularly perturbed Nash games. J. The Franklin Institute, 306: 355-374 (1978).

Johannsen, G. and W. B. Rouse, Mathematical concepts for modelling human behaviour in complex man-machine systems, Human Factors, 21: 733-747 (1979).

Khalil, H. K. and P. V. Kokotovic, Control strategies for decision makers using different models of the same system, IEEE Trans. Automat. Contr., AC-23: 289-298 (1978).

Kokotovic, P. V., R. E. O'Malley, Jr., and P. Sannuti, Singular perturbations and order reduction in control theory - An overview, Automatica, 12: 123-132 (1976).

Kragt, H., Human reliability engineering, IEEE Trans. Reliability, R-27: 195-201 (1978).

Krüger, G. and J. Nehmer, Methoden zur Steigerung der Zuverlässigkeit von Prozessrechnersystemen, in: "Automatisierungstechnik im Wandel durch Mikroprozessoren", M. Syrbe and B. Will (Eds.), 509-539, Springer-Verlag, Berlin (1977).

Lipow, M., (Ed.), Special issue on software reliability, IEEE Trans. Reliability, R-28: 178-253 (1979).

Litz, L., Ordnungsreduktion linearer Zustandsraummodelle durch Beibehaltung der dominanten Eigenbewegungen, Regelungstechnik, 27: 80-86 (1979).

Mosca, V. G., C. Rabinowitz, and H. Kreamer, Fault-tolerant, high-reliability electronic engine control system, in: Proc. AIAA/SAE/ASME 15th Joint Propulsion Conf., Las Vegas (1979).

Papavassilopoulos, G. P. and J. B. Cruz, Jr., Sufficient conditions for Stackelberg and Nash strategies with memory, J. Opt. Theory Appl., 30 (1980).

Rubinstein, E. (Ed.), Special issue: Three mile island and the future of nuclear power, IEEE Spectrum, 16(11): 30-111 (Nov. 1979).

Sandell, N. R., Jr., P. Varaiya, M. Athans and M. G. Safonov, Survey of decentralized control methods for large scale systems, IEEE Trans. Automat. Contr., AC-23: 108-128 (1978).

Schweppe, F.C., Power Systems "2000": Hierarchical control strategies, IEEE Spectrum, 15 (7): 42-47 (July 1978).

Sendler, W., Eine fehlertolerierende Reglerstation auf der Basis eines busorientierten Multi-Mikrorechner-Systems, Regelungstechnische Praxis, 22: 73-81 (1980).

Sheridan, T. B., Toward a general model of supervisory control, in: "Monitoring Behaviour and Supervisory Control", T. B. Sheridan and G. Johannsen (Eds.), 271-281, Plenum Press, New York (1976).

Sheridan, T. B. and G. Johannsen (Eds.), "Monitoring Behaviour and Supervisory Control", Plenum Press, New York (1976).

Syrbe, M. and B. Will (Eds.), "Automatisierungstechnik im Wandel durch Mikroprozessoren", Fachberichte Messen, Steuern, Regeln 1, Springer-Verlag, Berlin (1977).

Syrbe, M., Basic principles of advanced process control system structures and a realization with distributed microcomputers, in: Proc. IFAC 7th World Congress, A. Niemi (Ed.), 1: 393-401, Pergamon Press, Oxford (1979).

COMPUTER SUPPORT FOR DIAGNOSTIC TASKS IN THE PROCESS INDUSTRIES

F.P. Lees

Department of Chemical Engineering
Loughborough University of Technology
Loughborough, U.K.

INTRODUCTION

Modern process plants are very large and complex and their control involves large flows of information. In fault conditions there is a severe problem of the interpretation of this information in order to diagnose the fault. The task of fault diagnosis is normally assigned, if only by default, to the process operator, but the use of the process computer to assist him in this task appears a natural development. The process computer normally takes in measurements from the plant, compares them with alarm limits and generates alarms. Clearly this task may in principle be extended to that of analysing the alarms to diagnose the fault.

Such alarm analysis has in fact been carried out in the nuclear industry for some years. Computer-based alarm systems, which include alarm analysis, are used on the nuclear reactors at Oldbury and at Wylfa. The former has been described by Kay (1966), by Kay and Heywood (1966), and by Patterson (1968) and the latter by Welbourne (1965 and 1968), by Clarke and Welbourne (1971), and by Jervis and Maddock (1964).

At Wylfa, for example, the system described has some 6000 fuel channels, 2700 mixed analogue inputs and 1900 contacts on each reactor. The alarm analysis is based on the representation of the alarms in the form of an alarm tree. When a fresh alarm occurs, a check is made to determine whether the alarm is part of an existing alarm tree which is developing as the effect of an existing prime cause alarm. If the fresh alarm is not part of the tree, it is classed as another prime cause alarm, while if it is part of the tree, it is classed merely as a new alarm.

The basic alarm tree thus consists of alarmed process variables. In addition, however, there are deduced alarms which are associated with particular process variable alarms and which usually relate to mechanical faults.

Following these early systems work on computer alarm analysis of nuclear reactors has been reported by a number of workers, including the groups at Risø, Denmark, and Garching, West Germany.

An investigation of the use of computer alarm analysis on a chemical plant was described by Barth and Maarleveld (1967), but was not apparently followed up.

Further accounts of computer alarm analysis have been given by the author (Andow and Lees, 1975, Edwards and Lees, 1973, and Lees, 1980), elsewhere.

PROCESS OPERATOR AND ALARM ANALYSIS

Computer alarm analysis is normally considered as an aid to the process operator in his task of diagnosing faults. Therefore before considering the more specifically engineering aspects of alarm analysis it is appropriate to say something about the relative roles of operator and computer in the fault diagnosis task.

On a plant with no automatic protection the process operator has the task of keeping the plant running if he can but shutting it down if he must. His task is to attempt first to prevent the development of the hazardous condition and then, if he is unsuccessful, to shut the plant down safely. On a plant with full automatic protection in the form of trip systems, however, the operator is relieved of the second part of this task. The task is then confined to that of averting the development of the hazardous condition and thus of automatic shutdown. To this extent the function of the operator becomes one of economics rather than safety. In practice, of course, the operator's task normally lies somewhere between these two extremes. Nevertheless, there is no doubt that there is a trend towards the increased provision of protective systems.

The question of whether in fault conditions the function of the operator is one of safety or of economics is of some importance in defining the objectives of computer alarm analysis. In particular, if his function is essentially economic, it becomes less necessary that the system be able to handle fault conditions which occur only very rarely.

A computer alarm analysis facility is an aid to the operator. It is clearly futile, therefore, to create an aid which he will not use. If the facility is to be successful, it must be conformable with his thought processes in fault diagnosis. This is a point which has been emphasised by Rasmussen and coworkers (1968 and 1973). In particular, their work indicates that in a diagnostic task man tends to search the high probability paths first rather than all paths in a logical order regardless of probability.

Computer alarm analysis may have two somewhat different objectives. The first objective is to assist the operator to order and interpret the alarms as they occur in real time. This aim is achieved if the analysis indicates to the operator that several alarms are all part of a sequence caused by the same prime cause alarm, even though the actual fault is not known. The second objective is to diagnose the fault, which most often is a mechanical failure.

FAULT PROPAGATION: GENERIC PROBLEM AND DEVELOPMENTS

The alarm analysis programs used at the nuclear installations described require much time and effort to write. The alarm trees are normally constructed by experienced engineers and take much time to create. It is attractive, therefore, to develop a systematic method. The natural approach to the problem is to devise a method of synthesising and then analysing the fault structure which is general and which is used in the computer in conjunction with data which are specific to the particular plant.

The problem of alarm analysis in real time may be seen as one aspect of a generic problem - that of fault propagation in process plants. Another aspect is the problem of identifying and investigating potential faults at the design stage. Techniques which are currently used for this latter purpose include failure modes and effects analysis, hazard and operability studies, fault trees, event trees and cause-consequence diagrams.

A review of the generic problem of fault propagation has been given by Andow, Lees and Murphy (1980). These authors describe in more detail the various methods just mentioned, discuss the generic features and problems of these methods, indicate some problems of modelling, review current developments in analysis and synthesis of fault propagation structures and outline the desirable characteristics of a general method.

The engineering model which gives the most complete description of a plant is a full unsteady-state model covering both normal and fault conditions. In the investigation of fault propagation some use is made of such models, particularly for critical control systems, but they are used less than some of the other techniques described. The reasons for this have to do not only with the problems of creating such models but also with those of using them. The development of a full unsteady-state model of a plant is a time-consuming task. But this is only part of the problem. Even when it is constructed, there is an infinite number of process and fault conditions for which the model may be interrogated. Thus there is the further problem of selecting the sets of conditions for which the model should be run.

For this reason alternative techniques have been developed. Most of these involve a rather drastic simplification of the inherently complex behaviour of process plants. The fundamental simplification is the use of defined events and states to handle process variables and fault conditions which are inherently continuous.

One of the most widely used of the methods mentioned is the fault tree. The fault tree is a very powerful tool in the investigation of faults in complex systems. But a fault tree does have some fundamental limitations. It is a representation of the state of the system at one instant in time and in principle it is not well adapted to handling systems in which the defined events and states relate to different instants in time, such as a process system.

Andow, Lees, and Murphy (1980) have given the following outline specification for a generalised method for fault propagation in process plants:

1) The method is developed for the study of fault propagation specifically in process plants.
2) The method is systematic, flexible and economical of effort.
3) The method is computer-based with automatic and semiautomatic/interactive features.
4) The basis of the method is the decomposition of the plant into a set of modules, or units, with an associated topography; the use of models of these units; and the creation of a fault propagation structure from these models by application of synthesis features.
5) The method accepts as inputs a variety of types of model and effects interchange between models by application of interchange features.

6) The synthesis features include
   Event/state definition
   Operators (gates and vertices) and associated logic
   Propagation rules
   Methods to handle time aspects.
7) The regular, or canonical, form of the fault propagation structure is that held in the computer and other forms are derivations from or subsets of this form.
8) The models and topography are particular to the plant investigated, but the synthesis features are general.
9) The fault propagation structure can be interrogated to obtain various types of information, including those given by existing methods, e.g., fault trees or minimum cut sets. These types of information are subsets of the information implicit in the fault propagation structure.
10) The interrogation facilities include numerical output, e.g., minimum cut sets; graphical output, e.g., fault tree diagrams; and interactive facilities.
11) The interrogation facilities include facilities for the validation of the models and of the fault propagation structure.

This specification is intended to apply both to design and real-time applications of fault propagation. Computer alarm analysis is then effectively a particular method of interrogating the fault propagation structure.

There is a considerable interest world wide in methods for the creation of the fault propagation structure. In particular, there is currently much work on the synthesis of fault trees. Here mention should be made of the work of Fussell (1973), of Salem and Apostolakis and coworkers (1976 and 1978) and of Powers and coworkers (Powers and Tompkins, 1974a, 1974b, 1976, Powers and Lapp, 1976, 1977, and Lapp and Powers, 1977, 1979).

Fussell (1973) has described a computer code for the synthesis of large fault trees from mini-fault trees for electrical systems. The input data are the system topography and a set of mini-fault trees for each unit. One mini-fault tree is required for each fault output from the unit.

Apostolakis, Salem and Wu (1978) have described a computer code for the synthesis of large fault trees from fault decision tables for process plants. The input data are the system topography and a set of decision tables for each unit. One decision table is required for each fault output from the unit.

Powers and coworkers (Powers and Tompkins, 1974b, Powers and Lapp, 1976, and Lapp and Powers, 1979) have described a method for the synthesis of large fault trees from digraph models for process plants. The input data are the system topography and a single digraph model for each unit. This work has given rise to a considerable discussion in the literature. This discussion is of interest in that it brings out a number of important points in the technology of representing fault propagation.

A method of representing fault propagation which has certain advantages over the fault tree, particularly in handling time aspects, is the cause-consequence diagram, which has been described by Nielsen (1971, 1974) and by Taylor (1974).

Hollo and Taylor (1976) have described a method for the synthesis of large fault trees and cause-consequence diagrams for process plants.

In this context mention should also be made of the work of Berenblut and Whitehouse (1977) on fault monitoring using decision table analysis, although in the work described by these authors the tables are evidently constructed manually.

COMPUTER ALARM ANALYSIS: BACKGROUND AND OVERALL APPROACH

The approach to computer alarm analysis taken by the author and coworkers has been to try to develop a systematic, or algorithmic, method of creating in a process control computer a fault propagation structure which can then be interrogated in real time as alarms occur and thus used to provide an alarm analysis aid to the operator.

The project began about 1971 and is still continuing. The computer used during this period has been a PDP11-20. The languages used have been initially FORTRAN and subsequently RTL/2. The latter has full list processing capability.

The work has resulted in the development of two methods of alarm analysis. In both methods the input data are the system topography and a simple model or set of models for each unit. The first method (Method 1) involves the creation first of a loosely-structured network of the process variables and then a network of the process alarms. This alarm network, which may be created on an off-line computer, constitutes the fault propagation structure and is stored in the process computer. It is interrogated in real time as the alarms occur to give alarm trees.

In the second method (Method 2) the input data are stored in the process computer. When alarms occur a fault tree for the top event alarm is synthesised in real time using these data. There is no prior processing of the input data analogous to that in Method 1.

A laboratory pilot plant in the form of a double effect salt evaporator has been built to test these methods of alarm analysis. The plant is run under the control of the PDP11 computer using the language RTL/2.

Work has also been done to test these methods industrially.

These different aspects of the project are now described in more detail.

COMPUTER ALARM ANALYSIS: PROCESS ALARM NETWORK METHOD (METHOD 1)

In this method, which has been described by Andow (1973) and by Andow and Lees (1975), the input data consist of 1) the plant topography and 2) the unit models. The principle of the method may be illustrated by a simple example taken from Andow and Lees

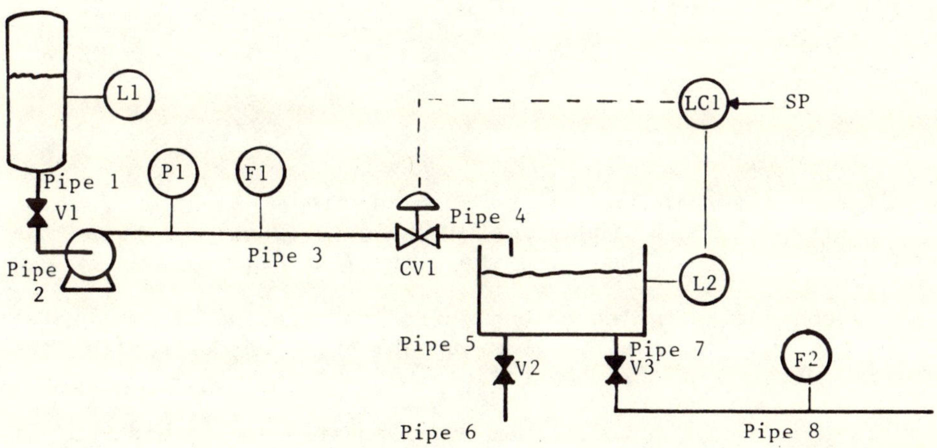

Figure 1. A simple level control system (after Andow and Lees)

(1978). Figure 1 gives the flow diagram of a very simple section of plant. Each of the units in the plant, including pipes, is described by a simple model. The plant topography is then represented as a block diagram of linked units as shown in Figure 2.

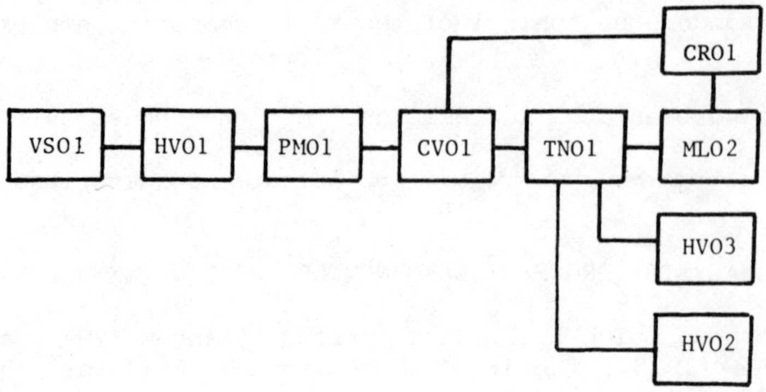

CR	Controller		ML	Level meter		TN	Tank
CV	Control valve		PM	Pump		VS	Vessel
HV	Hand valve						

For simplicity pipe sections and measuring instruments not part of a control loop are omitted.

Figure 2. Block diagram for part of level control system

The unit models are in the form of simple equations, which are referred to as functional models. Figure 3 gives the functional models for the units shown in Figures 1 and 2.

By convention the equations define the pressure at the inlet to the unit and the other process variables at the outlet from the unit. The utility of this convention is described below.

In the form of a functional model equation is considered, it can readily be seen that the equation gives information on which process variables on the right hand side of the equation affect the variable on the left hand side. It is therefore possible to construct directly from these equations a network of process variables which interact with each other. Figure 4 shows such a network for part of the system given in Figures 1 and 2 constructed using the models given in Figure 3.

Most of these process variables, however, are not measured and therefore do not have alarms on them. The process variable network is therefore reduced using a reduction algorithm to the corresponding process alarm network as shown in Figure 5.

Pipe

$$F_B = f\,(P_{A}, - P_B)$$

$$\frac{dP_A}{dt} = f\,(F_{A}, - F_B)$$

Open tank

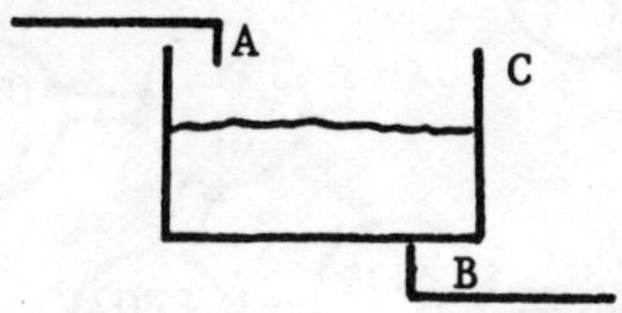

$$\frac{dL_C}{dt} = f\,(F_{A}, - F_B)$$

$$F_B = f\,(L_{C}, - P_B)$$

Control valve

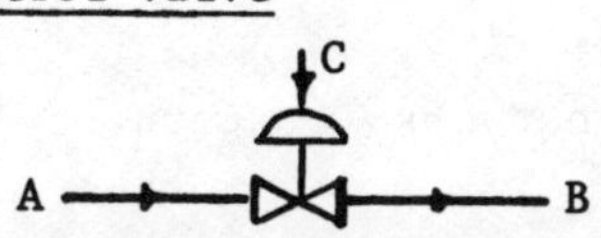

Assumption: air-to-open

$$F_B = f\,(P_{A}, - P_{B}, Z_C)$$

$$\frac{dP_A}{dt} = f\,(F_{A}, - F_B)$$

Controller

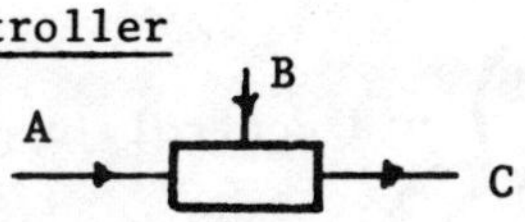

$$Z_C = f\,(Z_{A}, - Z_B)$$

Sensor

$$Z_B = f\,(Z_A)$$

Figure 3. Some functional models

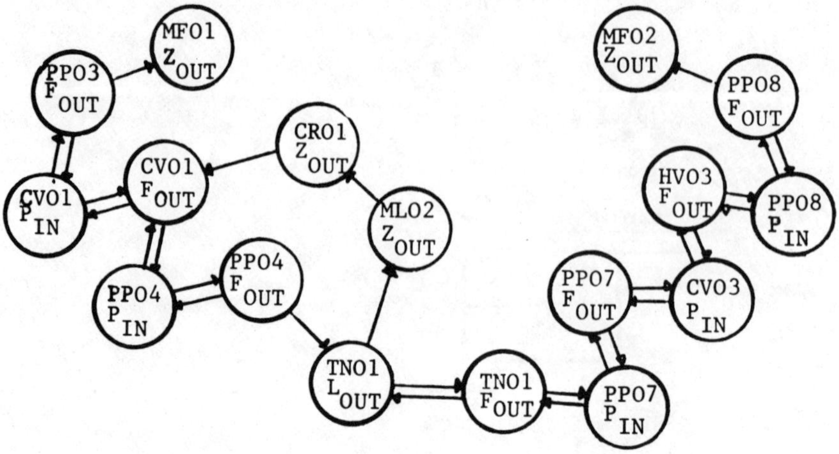

CR Controller
CV Control valve
HV Hand valve
MF Flow measurement
ML Level measurement
PP Pipe
TN Tank

Figure 4. Process variable network for part of level control system

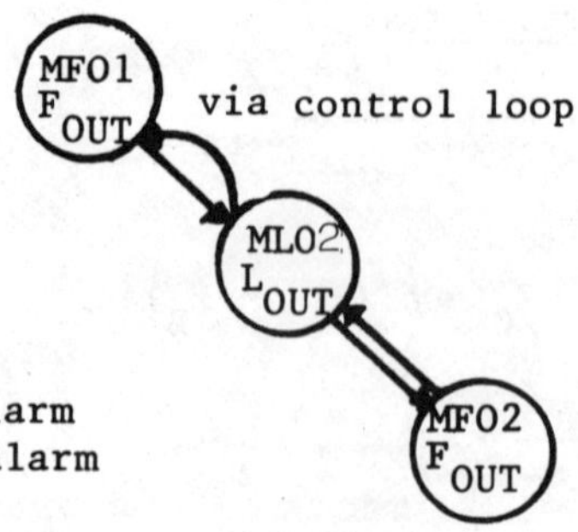

MF Flow alarm
ML Level alarm

Figure 5. Process alarm network for part of level control system

When a set of alarms occurs in real time, the relations between alarms may be derived by interrogating the process alarm network. Figure 6 illustrates the type of alarm information which may be obtained from the alarm network given in Figure 5. It may be noted that in addition to the process alarms there is also a deduced alarm (Drain Valve Open).

*** DRAIN VALVE V3**	**OPEN**
TANK LEVEL L2	**LO**
TANK INLET FLOW F1	**HI**
TANK OUTLET FLOW F2	**LO**

*** signifies deduced alarm**

Figure 6. Possible display of alarms for event 'Drain Valve Open' in level control system

The information given by a process alarm network is rather similar to that given by the alarm trees used in the alarm analysis systems on nuclear reactors described earlier. In both cases the aim is primarily to give the relations between sets of alarms rather than to diagnose a mechanical fault and the structure used is one which is relatively loose but which is economical for real time work.

In applying the method the practice is that the process alarm network is created in an off-line computer program and then transferred to the process computer program. List processing languages are used on both machines. In the project described the languages used in the two machines are respectively ALGOL 68 and RTL/2.

COMPUTER ALARM ANALYSIS: PROCESS FAULT TREE METHOD (METHOD 2)

In this method, which has been described by Martin-Solis (1978) and by Martin-Solis, Andow and Lees (1977), the input data consist of 1) the plant topography and 2) the unit models. The plant topography is again represented as a block diagram of linked units as shown in Figure 2.

The unit models are in the form of mini-fault trees. Figure 7 gives a mini-fault tree for a process variable deviation, F_{OUT}LO, for the pipe model given in Figure 3. A separate mini-fault tree is required for each output deviation on the unit.

It may be noted that the mini-fault tree shown in Figure 7 contains effectively two deduced alarms related to mechanical faults, namely Blockage and Leak to Low Pressure Environment.

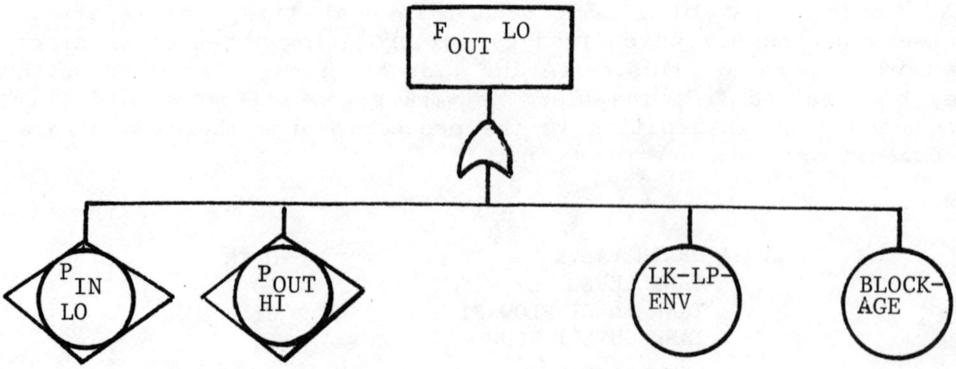

LK-LP-ENV : Leak to Low Pressure Environment

Figure 7. Mini-fault tree for pipe model for event $F_{OUT}LO$

The mini-fault trees can be related to the functional models which are used in Method 1. The deviation in the variable on the left hand side of a functional equation can be taken as the top event in a mini-fault tree. This event is the output of an OR gate, the inputs to which are deviations in the variables on the right hand side of the equation. If deduced alarms are to be included, the functional models must be supplemented by information on these alarms.

These input data are used in conjunction with a set of rules for the construction of the fault tree. These rules are necessary to ensure consistency in the tree.

As already stated the fault trees are not developed and held in the program in advance. When a set of alarms occurs in real time, however, one alarm is selected as the top event and the fault tree is developed.

Figure 8 after Andow and Lees (1978) shows a fault tree for part of the system given in Figures 1 and 2 constructed using mini-fault trees such as that shown in Figure 7.

The fault tree in Figure 8 is also intended to illustrate some other aspects of fault tree construction. The example distinguishes between the event variable L2 LO and the event measurement L2 LO (but variable normal) and shows the effects of these two different events. It also illustrates the point that conflicting events, in this case F1 LO and F1 HI, are liable to occur in the same tree unless rules are included to guard against this.

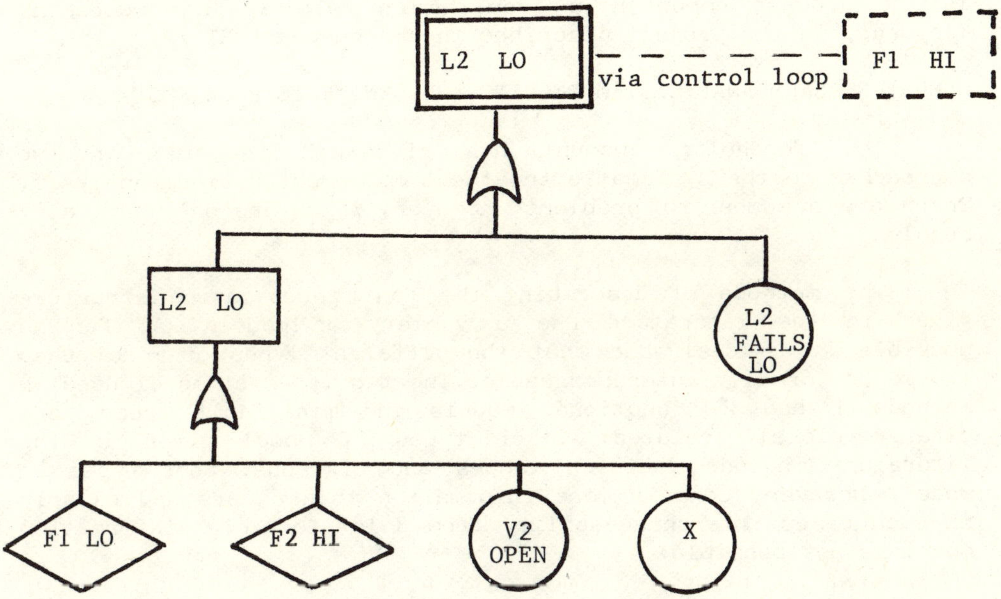

X : Controller disconnected and disturbance too small to trigger alarms

Double rectangle denotes alarm indication; dotted rectangle denotes event incosistent with events further down tree

Figure 8. Fault tree for part of level control system (after Andow and Lees)

The type of information which is given by a process fault tree is somewhat different from that given by a process alarm network or by an alarm tree in that the alarms are related by a more definite structure.

A fault tree constructed for the purpose of alarm analysis in real time differs considerably from one constructed for design purposes. In the real time situation only one, or at most a few, faults have occurred and the development of the tree can be confined to those branches associated with these particular faults.

In applying this method no use is made of an off-line computer. The input data and fault tree synthesis rules are held in the process computer program, which uses a list processing language. In the project described the language is RTL/2.

COMPUTER ALARM ANALYSIS: MODELLING AND SYNTHESIS PROBLEMS

The foregoing account has given the outline of two approaches to the systematic treatment of computer alarm analysis. There are a number of problems, however, which are only partially resolved.

The methods of describing the fault propagation structure given in the literature use a variety of models. As far as possible it is desirable that the different types of model used should be formally interchangeable. The two types of model used in Methods 1 and 2, functional models and mini-fault trees, are clearly closely related. Other types of model given in the literature include decision tables and digraphs. Some types of model, however, contain more information than others and clearly in such cases interchangeability from a low to a high information model is not possible.

Much of the fault propagation process can be modelled using models which describe the normal operation of the units. In most cases, however, it is necessary to include in the model of the unit a description of particular mechanical faults. It may be necessary to include faults even in a very simple model. Thus a unit model for a pipe may need to contain the faults Blockage and Leak to Low Pressure Environment.

In both Methods 1 and 2 the basic models, whether functional equations or mini-fault trees, are used to derive the transmission from an input variable deviation to an output variable deviation. There may not be a unique relation, however, between these two deviations. The relation may depend on the nature and degree of other input deviations which are present. The problem has been considered by Andow (1979) using the example of a control valve.

It is not at present clear how serious this difficulty is. For the purposes of alarm analysis it may be sufficient that a particular input deviation can, not necessarily will, give a particular output deviation.

An important feature of fault propagation in process plants is that propagation occurs in both upstream and downstream directions. This feature is less pronounced in electrical systems. Some of the fault tree synthesis methods described in the literature do not handle two-way flow of fault information but are restricted to downstream flow only. Both Methods 1 and 2 do handle fault propagation in both directions. In the original work on Method 1 full differential equation models were used to describe the process. In these models use was made of high gain differential equations as described by Franks (1966) and of the convention of defining pressure at the unit inlet and the other process variables at the unit outlet. When these full equations were replaced by functional equations, this feature was retained and it is this which gives the capability of handling two-way fault propagation. Further discussions of this feature and examples of fault propagation in both directions have been given by Andow (1973, 1980a, 1980b), by Martin-Solis (1978) and by Martin-Solis, Andow and Lees (1977).

In some cases it is necessary to consider not only two abnormal states of the variable (LO, HI) but multiple states (VLO, LO, ZERO, HI, VHI). Such multiple states can be handled by Method 2. An example has been given by Martin-Solis, Andow and Lees (1980).

The automatic generation of OR gates from the unit models in Method 2 was described above, but a method for fault tree synthesis must be capable of generating AND gates. It is a feature of many fault trees which are given in the literature and which have been obtained by automatic synthesis methods that they contain few or no AND gates. This aspect may be tackled by developing a taxonomy of AND gates. It appears in fact that the number of different types of AND gate used, for example, in such a major study as the Rasmussen Report (Atomic Energy Commission, 1975) is quite limited. Some principal types of AND gate are as follows. One type arises from the requirement for occurrence of a chemical reaction, e.g., flammable mixture and source of ignition. Another type arises from the requirement for failure in a redundant system, e.g., failure of device 1 and failure of device 2. A third type arises from the requirement for transmission of a deviation or fault against which protection is provided, e.g., temperature deviation and failure of temperature trip. And so on.

The automatic generation of AND gates can be handled if the information necessary for their generation is provided in the input data. It turns out that in some cases the information is part of the topography and in others it is part of the unit models.

It is appropriate to emphasise here, however, that Method 1 does not handle the combination of faults which an AND gate represents. The alarm information given by this method is essentially restricted to an indication that a potential fault path exists.

From the discussion in the literature the operation of control loops appears to give rise to a number of problems. In both Methods 1 and 2 no special algorithm is used to handle failure in a control loop. Like other parts of the plant the elements in the control loop are described by unit models. A fault tree representation of a control loop generally includes an AND gate. Method 2 generates an AND gate for a control loop by the normal application of the rules.

Failure of measuring instruments is generally recognised as presenting a problem both in operator fault diagnosis and in computer alarm analysis. Thus, for example, in the work on fault diagnosis by the process operator described by Duncan and Gray (1975) the operator is required to carry out a check on the veracity of the instrument reading before beginning the diagnosis. Both in manual and in automatic systems the diagnosis is much simpler if absolute reliance can be placed on the measuring instruments. Unfortunately, this is not so, since measuring instrument failures are one of the most common types of failure on process plants.

In both Methods 1 and 2 the measuring instrument is described by a unit model just like any other plant unit. The effects of a process variable deviation and of a failure in the corresponding measuring instrument are, of course, different. This is illustrated in the fault tree shown in Figure 8.

Another approach to the measurement problem is to devise methods of instrument malfunction detection. Studies by the author and coworkers on the topic have been described elsewhere (Anyakora, 1971, Anyakora and Lees, 1972, 1973, Bellingham, 1976, Bellingham and Lees, 1977, Lees, 1980).

Another problem area is that of time effects. In a sense this is perhaps more of a problem in Method 2 than in Method 1, since the former is based on fault trees, whereas the latter is based on a network which implies no more than that one alarmed process variable may be affecting, or may have affected, another alarmed process variable. An example in which there is a time effect in control loop failure has been given by Andow (1980a).

Further discussion of these and other problems in computer alarm analysis is given in the work mentioned.

COMPUTER ALARM ANALYSIS: EXPERIMENTAL WORK AND RELATED WORK

Some experimental work has been done to test these methods of computer alarm analysis.

Most of the work has been done on Method 1. The laboratory pilot plant evaporator has been run under the control of the PDP11 computer using RTL/2. Modelling of parts of the plant and creation of the process alarm network has been done off-line using ALGOL

68. The process alarm network has been stored in the process computer using RTL/2. The plant has then been operated so that alarms are generated. The work has been limited in extent, but the alarm analysis program has worked satisfactorily.

On the industrial side, an analysis has been carried out of an industrial fault sequence leading to a serious condition. The fault sequence occurred before the author became involved with the problem. There were available, however, adequate records of the process alarms during the fault sequence. It was possible, therefore, to model the process and derive the process alarm network for Method 1 and to confirm that in the real time situation the alarm analysis program would probably have generated reasonable alarm information.

A second industrial project has been the modelling of a gas distribution system and the derivation of the process alarm network for this system.

None of this experimental work has as yet been published.

Method 2 has been tested by a number of simulations.

Although the problem of computer alarm analysis was the author's point of entry to the general problem area, the work was extended some time ago to cover the generic problem of fault propagation.

This more general investigation of fault propagation consists of simultaneous and iterative work on a synthesis method (trees, networks), on plant topography and on unit models on the one hand and on the interrogation of this basic data structure on the other. The work on computer alarm analysis is now regarded as a subset of this more general work.

ACKNOWLEDGEMENT

The author wishes to thank the Science Research Council for supporting this work.

REFERENCES

Andow, P.K., 1973, "A Method for Process Computer Alarm Analysis", Ph.D. thesis, Loughborough University of Technology.

Andow, P.K., 1979, Private communication.

Andow, P.K., 1980a, "Difficulties in Fault Tree Synthesis for Process Plant", IEEE Trans. Reliab., in press.

Andow, P.K., 1980b, "Real-Time Analysis of Process Plant Alarms Using a Mini-Computer", Computers in Chem. Engng., in press.

Andow, P.K. and Lees, F.P., 1975, "Process Computer Alarm Analysis: Outline of a Method Based on List Processing", Trans. Instr. Chem. Engrs., 54:195.

Andow, P.K. and Lees, F.P., 1978, "Real Time Analysis of Process Plant Alarms", paper presented at NATO Advanced Study Institute on "Synthesis and Analysis Methods for Safety and Reliability Studies", Urbino, Italy.

Andow, P.K., Lees, F.P., and Murphy, C.P., 1980, "The Propagation of Faults in Process Plants: A State of the Art Review", in "Chemical Process Hazards with Special Reference to Plant Design", Vol. 7, Instn.Chem.Engrs., London.

Anyakora, S.N., 1971, "Malfunction of Process Instruments and its Detection Using a Process Computer", Ph.D. thesis, Loughborough University of Technology.

Anyakora, S.N. and Lees, F.P., 1972, "Detection of Instrument Malfunction by the Process Operator", Chem. Engr., London, 264:304.

Anyakora, S.N. and Lees, F.P., 1973, "The Detection of Malfunction Using a Process Control Computer: Simple Noise Power Techniques for Instrument Malfunction", in "The Use of Digital Computers in Measurement", Conf.Pub. 103, Instn. Elec. Engrs., London, p. 35.

Apostolakis, G.E., Salem, S.L., and Wu, J.S., 1978, "CAT - A Computer Code for the Automated Construction of Fault Trees", Rep. EPRI-705.

Atomic Energy Commission, 1975, "Reactor Safety Study. An Assessment of Accident Risks in U.S. Commercial Nuclear Power Plant", Rep. WASH 1400, Washington.

Barth, J. and Maarleveld, J., 1967, "Operational Aspects of a D.d.c. System", in "The Application of Automation in the Process Industries", Instn.Chem.Engrs., London.

Bellingham, B., 1976, "The Detection of Malfunction Using a Process Computer", Ph.D. thesis, Loughborough University of Technology.

Bellingham, B. and Lees, F.P., 1977, "The Detection of Malfunction Using a Process Control Computer: A Simple Filtering Technique for Flow Control Loops", Trans. Instn. Chem. Engrs., 55:1.

Berenblut, B.J. and Whitehouse, H.B., 1977, "A Method for Monitoring Process Plant Based on a Decision Table Analysis", Chem. Engr., London, 318:175.

Clarke, J. and Welbourne, D., 1971, "Display Systems for Use On-Line in Power Stations", in "Displays", Conf.Pub. 8, Instn.Elec.Engrs., London.

Duncan, K.D. and Gray, M.J., 1975, "An Evaluation of a Fault-Finding Training Course for Refinery Process Operators", J. Occup. Psychol., 48:199.

Edwards, E. and Lees, F.P., 1973, "Man and Computer in Process Control", Instn.Chem.Engrs., London.

Franks, R.G.E., 1966, "Mathematical Modeling in Chemical Engineering", Wiley, New York.

Fussell, J.B., 1973, "A Formal Methodology for Fault Tree Construction", Nucl. Sci. Engng., 52:421.

Hollo, E. and Taylor, J.R., 1976, "Algorithm and Program for Consequence Diagram and Fault Tree Construction", Danish Atomic Energy Comm., Res.Est., Risø, Rep. Risø-M-1907.

Jervis, M.W. and Maddock, P.R., 1964, "Electric Power Station Startup and Control", in "Digital Computer Applications to Process Control", (edited by W.E. Miller), Plenum Press, New York.

Kay, P.C.M., 1966, "On-Line Computer Alarm Analysis", Ind. Electron., 4:50.

Kay, P.C.M. and Heywood, P.W., 1966, "Alarm Analysis and Indication at Oldbury Nuclear Power Station", in "Automatic Control in Electricity Supply", Conf.Pub.16, Instn. Elec. Engrs., London, p. 295.

Lapp, S.A. and Powers, G., 1977, "Computer-Assisted Generation and Analysis of Fault Trees", in "Loss Prevention and Safety Promotion in the Process Industries", Vol. 2, DECHEMA, Frankfurt, p. 377.

Lapp, S.A. and Powers, G.J., 1979, "Update of Lapp-Powers Fault Tree Synthesis Algorithm", IEEE Trans. Reliab., R-28:12.

Lees, F.P., 1980, "Loss Prevention in the Process Industries", Butterworths, London.

Martin-Solis, G.A., 1978, "Fault Tree Synthesis for Real Time and Design Applications on Process Plant", Ph.D. thesis, Loughborough University of Technology.

Martin-Solis, G.A., Andow, P.K., and Lees, F.P., 1977, "An Approach to Fault Tree Synthesis for Process Plants", in "Loss Prevention and Safety Promotion in the Process Industries", Vol. 2, DECHEMA, Frankfurt, p. 367.

Martin-Solis, G.A., Andow, P.K., and Lees, F.P., 1980, "Synthesis of Fault Trees Containing Multi-State Variables", paper to be presented at Symp. on "Loss Prevention and Safety Promotion in the Process Industries", Basle, Switzerland.

Nielsen, D.S., 1971, "The Cause-Consequence Diagram Method as a Basis for Quantitative Accident Analysis", Danish Atomic Energy Comm., Res.Est., Risø, Denmark, Rep. Risø-M-1374.

Nielsen, D.S., 1974, "Use of Cause-Consequence Charts in Practical Systems Analysis", Danish Atomic Energy Comm., Res. Est., Risø, Denmark, Rep. Risø-M-1743.

Patterson, D., 1968, "Application of a Computerised Alarm Analysis System to a Nuclear Power Station", Proc. IEE, 115:1858.

Powers, G.J. and Lapp, S.A., 1976, "Computer Aided Fault Tree Synthesis", Chem. Engng. Prog., 72(4):89.

Powers, G.J. and Lapp, S.A., 1977, "Computer-Aided Synthesis of Fault Trees", IEEE Trans. Reliab., R-26:2.

Powers, G.J. and Tompkins, F.C., 1974a, "A Synthesis Strategy for Fault Trees in Chemical Processing Systems", in "Loss Prevention", Vol. 8, Am.Inst.Chem.Engrs., New York, p. 91.

Powers, G.J. and Tompkins, F.C., 1974b, "Fault Tree Synthesis for Chemical Processes", AIChE J., 20:376.

Powers, G.J. and Tompkins, F.C., 1976, "Computer-Aided Synthesis of Fault Trees for Complex Processing Systems", in "Generic Techniques in Systems Reliability Assessment", (edited by E.J. Henley and J.W. Lynn), Noordhoff, Amsterdam, p. 307.

Rasmussen, J., 1968, "On the Communication Between Operators and Instrumentation in Automatic Process Plants", Danish Atomic Energy Comm., Res.Est., Risø, Denmark, Rep. Risø-M-686.

Rasmussen, J. and Jensen, Aa., 1973, "A Study of Mental Procedures in Electronic Trouble-Shooting", Danish Atomic Energy Comm., Res.Est., Risø, Denmark, Rep. Risø-M-1582.

Salem, S.L., Apostolakis, G.E., and Okrent, D.L., 1976, "Computer-Oriented Approach to Fault-Tree Construction", Univ. of Calif., Los Angeles, Rep. UCLA-ENG-7635.

Taylor, J.R., 1974, "A Semi-Automatic Method for Qualitative Failure Mode Analysis", Danish Atomic Energy Comm., Res.Est., Risø, Denmark, Rep. Risø-M-1707.

Welbourne, D., 1965, "Data Processing and Control by a Computer at Wylfa Nuclear Power Station", in "Advances in Automatic Control", Instn. Mech. Engrs., London, p. 92.

Welbourne, D., 1968, "Alarm Analysis and Display at Wylfa Nuclear Power Station", Proc. IEE, 115:1726.

APPLICATION OF PATTERN RECOGNITION TO FAILURE ANALYSIS AND DIAGNOSIS

L. F. Pau

Ecole Nationale Supérieure des Télécommunications, Paris, France; George Washington University, Washington, DC, USA, and French Scientific Mission, 2129 Wyoming Avenue NW, Washington, DC.

INTRODUCTION

In this chapter, we will be dealing with systems, all initially at the same design, manufacturing, quality control and operating standards. For each set of operating conditions, technical specifications describe the expected performance and characteristics of these systems, all of which cannot, in general, be quantified. After having introduced a number of basic concepts in failure analysis and diagnosis, the relation to pattern recognition will be explained.

The Basic Events in Failure Analysis and Diagnosis (Saeks, Liberty, 1977, Pau, 1979).

1) Failure. A failure or defect is a condition (or state) characterized by the inability of a material, structure or system to fulfill its intended purpose (task or mission) and resulting in its retirement from usable service. Due to priorities between tasks, there may be a priority between failures; those failures of highest priority will be dealt with in safety analyses.

2) Degradation. A degradation is the act of impairing or deteriorating the condition, functional or physical property, including the performances for a specified task or mission. This includes improper commands and the effect of the environment.

3) Failure mode. A failure mode is the particular manner in which an omission of expected occurrence or performance of task or mission happens; it is thus a combination of failure, defects, and degradations; for a given task or mission the N possible failure

modes will be noted E_0, E_1, ..., E_{N-1}, where E_0 is the no - failure operating mode fulfilling all technical specifications.

The Basic Troubleshooting Techniques
(Saeks, Liberty, 1977, Pau, 1979).

1) Failure detection. Failure detection is the act of identifying the presence or absence of a non-specified failure mode in a specified system carrying out a given task or mission, or manufactured to a given standard.

2) Failure localization. If the outcome of failure detection is positive, then failure localization designates the material, structures, components, processes, or systems which have had a failure.

3) Failure diagnosis (Pau, 1979). The act or process of identifying a failure mode E upon an evaluation of its signs and symptoms (incl. performance monitoring measurements). The diagnostic process therefore carries out a breakdown of failure detection into individual failure modes.

4) Failure analysis, (Mann, Schafer, Singpurwalla, 1974). The process of retrieving via adequate sensors all possible information, measurements, and non-destructive observations, about the life of the system prior and up to the failure; also, method whereby to correlate these informations.

5) Failure monitoring, (Pau, 1979) The act of observing indicative change of equipment condition or functional measurements, as warnings for possible needed corrections.

The Effects of Failure Analysis and Diagnosis

Whereas system reliability and safety theories are concerned with *a priori* assessments of the probability that the system will perform a required task under specified conditions, without failure, for a specified period of time (Mann, Schafer, Singpurwalla, 1974; Barlow, Proschan, 1965; Tsokos, Shmi, 1977) the field of failure analysis and diagnosis,(Pau, 1979) is essentially focusing on a posteriori processing of all monitoring information (generally in real time) for later decision making prior to disassembly. The decisions to be taken post - failure are:

i: alarms, warnings, and display hereof
ii: maintenance and repair actions
iii: hardware reconfiguration and activation of spare units
iv: software reconfiguration: new sets of measurements,

modified sampling rates and coding, sensor reconfiguration

v: design changes, or modified controls and utilization conditions

vi: recycling or scrapping items detected as bad because of the presence of a failure mode.

In other words, failure analysis and diagnosis have a direct impact (Pau, 1979) on system availability, system survivability and production yield, with all consequences hereof.

Technical Performances of a Diagnostic System
(Pau, 1979; De Dombal, Gremy, 1976; Parkhomenko, 1969; Gross, 1970)

As a decision operator, any diagnostic system can make errors; each of the following errors can be specified either for a specified failure mode, or in the expected sense over the set of all possible failure modes $E_0, E_1, \ldots, E_{N-1}$.

1) Probability of incorrect diagnosis. This is the probability of diagnosing a failure mode different from the actual one, with everything else equal.

2) Probability of reject (or miss). This is the probability of taking no decision (diagnosis or detection) when a failure mode is actually present.

3) Probability of false alarm. This is the probability of diagnosing that a failure mode is present, when in fact none is present (except the normal condition E_o).

4) Probability of correct detection. This is the probability of detecting correctly a failure mode to be present, when it is; when there are only two possible failure modes, it is the complement of the probability of false alarm ($N = 2$).

It is essential to point at the reject option, either because of lack of symptoms and monitoring information, because of improper diagnostic decisions, or because an additional exogeneous confirmation is needed. As it will be realized later, the above performances are directly comparable to those of a pattern recognition system.

Application of Pattern Recognition to Failure Diagnosis and Performance Monitoring
(Pau, 1979; Pau, 1972; Kulikowski, 1970)

Pattern recognition, (Fu, 1976; Fukunaga, 1972) consists in carrying out the processing and taking decisions, whereby the system condition E is classified into one of the N possible failure modes E_0, ... $E_{(N-1)}$, using all available symptoms and monitoring information, and similar learning information.

If we assume the pattern recognition problem to be decomposed into the basic steps as specified classically, then the following correspondences may be established:

1) Pattern measurement, consists of acquiring information about the system condition; this information is either quantitative, visual or verbal, (Pau, 1979):

- Design parameters, operational specifications, nominal signals, tolerances
- Settings, tunings, and controls applied to the system
- Measurements and pictures provided by the on-line monitoring process
- Measurements and pictures provided by off-line non-destructive testing and by automatic test systems
- Maintenance and repair reports, frequencies of overhauls and repairs, component changes
- Observations and measurements related to individual failures, incl. those on the operational environment.

2) Learning (Mendel, Fu, 1970), consists of compiling and retrieving all information about past system failures, also for different however similar equipments, or for common components operating in other equipments.

3) Feature extraction (Fukunaga, 1972), consists of analyzing and combining all available information about the system condition in order to optimize the diagnostic performance of the classification procedure operating on a reduced number of combined symptoms/features.

4) Classification consists of applying a decision procedure to the extracted feature symptoms for the purpose of carrying out one of the actions listed in the previous section on the effects of failure analysis and diagnosis.

5) Inference (Fu, 1974), consists of identifying the chain of events connecting a change of physical condition of one system component or condition with changes in other components or systems which have terminated in system failure; this chain is called failure path.

In the following sections, we will review, for each of the above problem areas 1) - 4), the specific aspects met in applying pattern recognition to failure analysis and diagnosis and next the pattern recognition methods used. Finally, a survey will refer to publications describing some major implementations. It ought to be pointed out, however, that in many practical implementations, all problems areas coexist; as a consequence, reference will only be given to original approaches to each specific area. Also, because of the large diversity of systems to which this methodology applies, the publications are extremely scattered throughout the technical literature. Bibliographies on failure analysis and diagnosis in analog or digital systems, can be found in (Saeks, Liberty, 1977; Pau, 1979; Parkhomenko, 1969).

PATTERN MEASUREMENT

Specific Aspects

1) There are few areas in pattern recognition using such a variety of measurements, (Pau, 1972; Pau, 1974): analog sensor signals, imagery (industrial TV, boroscopy, endoscopy), binary or digital position or level sensors, and last but not least unstructured verbal reports (follow - through cards). Regardless of the level of accuracy achieved in the coding and lay-out of the specifications, of the maintenance or of the test procedures, oral or written reports expressed in plain language will often reveal significant features.

2) Observability is an eventual property of dynamic systems which expresses the ability to infer or estimate the system state or condition at a given past instant in time from quantified records of all measurements made on it at later points in time, (Mendel, Fu, 1970). Whereas most applications of pattern recognition are for observable systems, this does not hold for failure analysis and diagnosis, first because of missing measurements/data, (Pau, 1979; Pau, 1972), and next because of time-dependent changes of the system condition which, in general, cannot be modelled.

   This fundamental remark is related to the non-independence of catastrophic and degradation failures; a catastrophic failure will induce later unpredictable degradations, while degradations may induce a sequence of subsystem or localized catastrophic failures.

3) One of the main limitations to observability is bad accessability of the main test or measurement points because of inadequate design, (Saeks, Liberty, 1977; Pau, 1979; Parkhomenko, 1969). In the case of electrical measurements, impedance and bandwidth mismatch are introduced at the sensor or interface level, resulting in signal distortion features which do not originate in system failures. In the case of human observations, sources to observation errors are many as expected, (Pau, 1979). Another source of

limitations is inadequate selection of the measurement sampling frequency (spatial or temporal), so that fine features revealing incipient failures go unnoticed.

Distortion is a classical problem in measurement acquisition, but added difficulties result from the fact that the sensors themselves cannot be properly modelled outside their limited normal operating bandwidth, whereas it is likely that true measurements on systems which fail will be characterized by extremely large bandwidths. Such large bandwidths also conflict with low noise, and infrequent calibration requirements.

4) Lastly, failure analysis and diagnosis are only possible if the sensors of all kinds survive system failures, (Pau, 1979); this may require sensor redundancy, separate power supplies, and different technologies and computing resources. Besides sensor and processor reliability, short reaction time and good feature extraction are conflicting hardware requirements, all of which contribute to increased costs which in turn limit the extent of possible implementations.

Any subsystem which can be activated separately should be equipped with a meter, unit or cycle counter, eventually replaced by a multiplexed meter which tests periodically the operating status of all subsystems, (Pau, 1979).

In many cases, it is useful to extract and monitor the r. m. s. amplitude in some fixed narrow spectral bands, (Pau, 1979).

Whether the system under surveillance is autonomous or not, analog to digital multiplexing will often be required before data transmission. However, if the data acquisition rate is low under good operating conditions, data transmission may become irrelevant: on - site temporary data storage is then a convenient solution, (Pau, 1979).

Methods

To a very large extent, the specific aspects above will be treated by the sensors and preprocessing hardware.

1) Quantization is frequently adjusted in such a way that acceptable signal fluctuations within each failure mode are only coded on a few levels, while transitions are coded on many more levels.

As a general rule, the maximum quantization level is adjusted to the sensor delivering the poorest quality data, (Pau, 1979).

2) For signals with very large bandwidths and with no stationarity, binary level crossing encoding transforms the signal into a sequence of bits in which binary word frequencies may be estimated, (Becker, 1978).

3) In some more evolved applications, the quantization and coding schemes are controlled by a contextual event tree describing the outside environment and the sequence of controls applied to the equipment. In other words, for each transition of the context or of controls in this event tree, a different quantization and coding scheme is selected.

4) Especially when follow - through cards are used, data are often missing, (Pau, 1979; Pau, 1974). Although the procedure is only justified under very restrictive assumptions, the missing data may be replaced by the product of the marginal probability density estimates derived from the unsupervised learning data, (Pau, 1979; Becker, 1978).

LEARNING

Specific Aspects

1) The fundamental difficulty met at the learning stage can be called the "rare disease problem", (De Dombal, Gremy, 1976; Kulikowski, 1970; Donio, 1972):

i: even for a reduced number $N \geqslant 2$ of possible failure modes, the available number n_j of learning patterns describing the defective mode E_j, $j=1, \ldots, (N-1)$ may be very small.

ii: the number n_0 of learning patterns describing the good operating mode E_0 is very large.

As a consequence, it is essential that the later feature extraction and classification stages take explicity into account n_j, $j = 0, \ldots, (N - 1)$, (Pau, 1979; Pau, 1978).

2) There is also an obvious conflict between the rare disease problem and the requirements for small probabilities of incorrect diagnosis or of false alarm, (Pau, 1978).

3) An additional difficulty is that, while in medical diagnosis, diseases are quite well defined and characterized, this is not at all the case in failure diagnosis: the notion of failure mode is much more fuzzy because of imperfect

observability, variabilities in the physical degradation mechanisms, and sensor inaccuracy, (Pau, 1979; Donio, 1972). The diagnostic process carried out at the repair, maintenance or test stages on a piece of equipment would be more adequately modelled by a fuzzy decision process than by a set relationship.

As a consequence, the learning patterns themselves may be incorrectly assigned: this is the "imperfect teacher" problem, the effect of which is either to require larger learning sample sizes n_j, j = 1, ..., (N - 1) or to reduce the statistical confidence in any classification decision, (Pau, 1978; Kittler, Pau, 1978).

4) Depending on the application, the number (N - 1) of different failure modes with lowered performance values may be large or small:

- i: in the case of visual inspection by a human or an imagery system, such as screening of microelectronic components, N is very large because of many different possible defect locations on the substrate layers, (Muehldorf, 1975; Gupta et al., 1977; Thatte, 1977).
- ii: in the case of equipments built with few rigid mechanical components subject to fracture or rupture, N is usually small, (Pau, 1979).

Considering the present state of the art, the pattern recognition techniques are only applicable for $5 \leqslant N \leqslant 15$, or N = 2. The failure diagnosis performances (see earlier) are too low for larger values of N, whereas simpler approaches or hardware may treat most cases where N = 2, 3, 4, except when the pattern measurements are strongly correlated.

5) Very often, the failure mode definitions are time or temperature dependent because of component drift, physical memory effects or wear, (Saeks, Liberty, 1977; Pau, 1979; Pokrowsky, 1972). In other words, a learning pattern would be assigned as defective by the teacher during the early life of the equipment, although it would be assigned to E_0 at a later point of time, because, e.g., wear is normal up to a certain limit. It is very difficult to account for these changes in the definitions of the pattern classes, especially when no exogeneous model can generally be estimated.

Methods

1) The standard methods by which to account for the "rare disease" problem are:

i: to assume some class conditional probability densities for the classes $E_1, \ldots, E_j, \ldots, E_{N-1}$ and to replace the set of learning patterns from each such class E_j by a n_j - dependent tolerance region (Donio, 1972; Pau, 1978).

ii: to replace the n_j learning patterns of class E_j, by a reduced number of strong patterns in a E_j -kernel, obtained i.e. by the dynamic clusters algorithm, (Fu, 1976; Pau, 1973).

2) The acceptance sampling methodology, (Pau, 1978), inspired by quality control by variables, is essential at the design stage of the learning procedure. Given upper bounds on the error probabilities of 1st and 2nd kind, and assuming the sample size to be equal to one unsupervised pattern, then the acceptance sampling results will tell whether the learning samples of size n_j, $j = 0, \ldots, (N - 1)$ are sufficient or not. Because classical statistical quality control only treated scalar pattern measurements, a methodology for multivariate pattern measurements has been developed, (Pau, 1978; Pau, Chen).

3) The effect of the "imperfect teacher" problem is to increase even more the required learning sample size n_j, (Kittler, Pau, 1978). It can be modelled by specifying a confusion matrix for the teacher, and by studying its effects on the acceptance sampling plans and the probability of correct diagnosis vs. the test statistic, (Pau, 1978).

4) The "drift or wear" problem may be approached by specifying separate learning sub-sets parametrized vs. time or temperature. Another procedure is to operate on the mixture of these sub-sets, that is the full learning set, in which each learning pattern is assigned a weight which decreases with the deviation between the current time and the time at which this pattern was acquired.

5) In some instances, actual learning measurements are replaced by parametric simulation models estimated by identification techniques, (Pau, 1979; Levadi, 1967; Yermachenko, Sdor, 1975).

In the field of syntactic patterns, (Fu, 1974), a similar problem is the use of symbolic layout procedures for integrated circuit mask design, (Larsen, 1978; Van de Wiele, 1977; Bertails, Zirphile, 1977; Vancleemput, 1976; Henriksen, 1977). The layout is not described in these models in terms of its geometry and process requirements, but as sequences of symbols; each of these represents the logical, electrical and geometric functions of a specific piece of chip, and the procedure infers from these sentences of symbols the two - dimensional structure of the circuit and its properties.

FEATURE EXTRACTION

Specific Aspects

1) As expected, feature extraction is a crucial problem, although its criticality depends very much on the subsequent processing:

   i: for data compression purposes only, and display of all learning data in a failure analysis framework;
   ii: for classification purposes also, in an automated failure diagnosis framework without disassembly.

* Failure analysis uses generally statistical features.
* Failure detection (N=2) uses generally syntactic features.
* Failure diagnosis ($N>2$) uses generally statistical and syntactic features.

2) Most features are selected based on solely heuristic or experimental considerations; but the point of view adopted is that the fluctuations of the signals, pictures and the like, when properly filtered and selected, may reveal important information regarding the internal condition of the system, including degradations, and the relations to external causes, (Pau, 1979; Rabiner, Gold, 1975; Braun, 1975).

   The very nature of failure analysis and diagnosis is such that the feature signals or sub-images have very large bandwidths corresponding mostly to random discontinuities, (Friedland, 1978).

3) Another difficulty is to detect such discontinuities in noise:
   - for signals, detection theory does not necessary apply, because the fluctuations and the noise are often not gaussian, (Braun, 1975; Thomas, Wilkins, 1972; Pau, 1975), (except for the condition E_0)
   - for imagery, image substraction techniques can be efficient feature extractors, but only if prototype pictures are available for the various failure modes, which is often not the case (e.g. visual inspection), (Sterling, 1978; Skinner, 1977).

Methods

1) There have only been a few applications of clustering algorithms, whether hierarchical or not, to failure analysis, (Pau, 1979; Solomon, 1970; Jones, 1971; Hughes et al., 1977). So far, emphasis has been on the use of multivariate statistical analysis, first for data compression and feature extraction and next for visual clustering or causality analysis. The most commonly used methods have been principal components analysis, variance analysis. the Karhunen - Loewe expansion (in the time, frequency or pixel domains), canonical analysis, correlation analysis (Fukunaga, 1972).

The pattern features are obtained as coordinates of the original pattern vector in a linear subspace estimated from the learning data after data normalization, (Pau, 1979; Pau, 1972). If a two-dimensional subspace is selected, then all learning patterns may be displayed as points on a map.

Visual clustering is obtained by carrying out a proximity analysis between learning patterns on this map, where each point carries a label which is the diagnosis assigned to it by the teacher, (Pau, 1972; Pau, 1973; Pau, 1974; Pau, 1974).

The goal is to infer causality relations between failure causes, and symptoms or measurements from this visual clustering, (Pau, 1979; Pau, 1972). This requires a symmetric treatment of the pattern vectors (rows) and the measurements (columns), which is not fulfilled by the previous methods. Only contingency table analysis, hypothesis testing in contingency tables and correspondence analysis fulfil this requirement to some extent, (Pau, 1979; Benzecri, 1977; Pau, 1974; Pau, 1974; Hill, 1974). They are therefore widely used, although causality relations will rest on an infinite - sample size assumption rarely met in practice.

Apart from problems in numerical analysis, the following areas of research are still widely open in connection with failure analysis:

i: processing of large learning data amounts, (Pau, 1972), (many patterns and measurements)

ii: theoretical properties of multivariate statistical analysis methods in presence of mixed real and binary pattern measurements, (Pau, 1979)

iii: properties of the extracted features when some measurements are pertubated by non-normal noise or fluctuations.

2) Failure detection usually operates by straightforward threshold comparison or matching of signals, (Saeks, Liberty, 1977), and pattern matching of syntactic features, (Fu, 1974; Burrows, Miles), 1972), with a model of the operating condition E_0 as reference. One theoretical question left is the adjustment of the tolerances on E_0; in the absence of a statistical or syntactic characterization of the fluctuations in E_0, tolerance limits are set as two - sided confidence bands with a specified confidence level, assuming a distribution for the extreme fluctuations, (Gumbel, 1958; Grenander, Rosenblatt, 1957).

A more evolved approach, not yet fully validated, is to let a parsing algorithm, (Fu, 1974) operate on the syntactic description of the failure pattern, and to compute a statistic of the number of primitives in this pattern which accounts for their relative positions in the output of the parser, e.g. the frequencies of binary words, (Becker, 1978).

A simplified version of the previous procedure consists of carrying out a vote based on the outputs of a number of binary feature extractors; the voting rule can be made dependent upon external operating conditions.

3) Failure diagnosis requires a variety of feature extractors; in addition to those described in the previous paragraph , the common requirement is that features are to be based as far as possible on non-parametric statistics, or to be outputs of observers, (Clark, Fobth, Walton, 1975), or filters, (Pau, 1979; Rabiner, Gold, 1975) estimating a state vector, (Martin, Hokins, 1973) characteristic of the internal condition of the system.

Non-parametric statistics, such as rank tests, signed rank tests, Wilcoxon tests or sequential non-parametric statistics, (Pau, 1978; Pau, 1978), have the advantage of being useful regardless of the noise and of the pattern fluctuation distribution.

Observers or filters assume that a state-space (Pau, 1979; Clark, Fobth, Walton, 1975), an autoregressive (Rabiner, Gold, 1975; Markel, Gray, 1973; Pau, 1977), or a transfer function model (Piety, Robinson, 1976) of the system is available, at least for the normal condition E_0. Such a model, combined with the observer or filter, compensates for the lack of observability of the internal condition (see the earlier section on observability).

One of the most successful methods has been the use of recursive Kalman filtering for the on-line joint estimation of time-dependent changes in critical unobservable condition dependent parameters, (Pau, 1979; Piety, Robinson, 1976; Gonzalez, Fry, Kryter, 1974) or of the short-term power spectrum of several measurements' fluctuations (Pau, 1979; Pau, 1977). Critical difficulties are however related to the modelling of failed conditions $E_1, \ldots, E_{(N-1)}$.

CLASSIFICATION

Specific Aspects

1) The importance of sequential classification methods, (Mendel, Fu, 1970; Fu, 1968) in failure diagnosis comes from the fact that they best can be represented by a fault tree, with a decision taken at each node based on changing sets of feature measurements, thus allowing for the integration of a wide variety of sensors, (Spire, 1970; Cardillo, Fu, 1968; Slagle, Lee, 1971). It should be pointed out that failure localization is in turn also a sequential process, since it conducts the search through the set of possibly failed parts, components or processes, (Pashkovskiy, 1971; Peron, 1970; Cohn, Ott, 1971; Chu, 1968).

2) The major difficulty in applying sequential classification comes from the possibly large number N of not well defined failure modes (see the earlier comments on medical diagnosis), and from the non-independence of failure events because of physical connections, proximity or memory effects, (Persoon, 1971).

3) The major difficulty in applying Bayesian decision theory (respectively syntactic pattern recognition) is that, whereas the normal operating condition E_0 is generally characterized by normally distributed features (respectively a regular context-free grammar of primitives), all degraded conditions $E_1, \ldots, E_{(N-1)}$, ought to be treated as if they had non-normally distributed features (respectively non--regular context-sensitive grammars of primitives). The explanation is partly physical, partly due to design because of specifications on the performances in the condition E_0: feasible feature measurements will belong to a bounded convex subset of the feature space; as a consequence, the subsets which can be reached from E_0 because of a transition/failure will in general be non-convex.

4) Finally, the small sample properties of most classification rules are difficult to analyse, (de Dombal, Gremy, 1976;

Pau, 1978); the same remark holds for finite sample estimators of the error probabilities and of the risks of 1st and 2nd kind.

Methods

1) Sequential pattern recognition, (Fu, 1968) has been applied in various forms in a number of cases, either using dynamic programming, (Bellman, 1966) with sequential feature selection, or sub-optimal questionnaire theory, (Cardillo, Fu, 1968; Slagle, Lee, 1971; Pashkovskiy, 1971; Persoon, 1971), or sequential probability ratio tests, (parametric or non-parametric), (Pau, 1979; Wald, Wolfowitch, 1948).

2) Discriminant analysis, (Fukunaga, 1972) and piecewise discriminant analysis, have only been used when problems with detecting discontinuities in noise vanish, and when the learning sample sizes are sufficiently large (de Dombal, Gremy, 1976). The usual statistical classification rule is Bayesian decision theory (with or without reject option) combined with non-parametric probability density estimators, (Fukunaga, 1972), (nearest - neighbour rule, Parzen kernels, etc. ...); this technique is applied as well to continuous feature measurements and to binary feature vectors, (Waidelich, 1977), using the L^1 metric. The a-priori class conditional probabilities may sometimes be specified using time and/or temperature dependent reliability functions, (Tsokos, Shimi, 1977; Grellin, 1972).

3) Syntactic pattern recognition, (Fu, 1974), whether deterministic or stochastic, uses a battery of parallel class conditional parsers and acceptance automata. When such features are combined with statistical features, a final decision tree is introduced, and this can be made context-dependent, namely with respect to threshold selection.

4) When feature fluctuations cannot be modelled, these will be characterized by class conditional histograms with confidence bands; failure diagnosis results from matching an observed feature signature histogram with the N learning histograms, (Pau, 1979; Pau, 1977).

FIELDS OF APPLICATION OF PATTERN RECOGNITION TO FAILURE ANALYSIS AND DIAGNOSIS

Although the pattern analysis and classification methodologies are inherent to all failure analysis and diagnosis problems as defined in the introduction, the earliest papers making reference to it are approximately 10 years old. Since then, an increasing number of actual implementations has been reported in very diversified areas, (Pau, 1979); however, mostly in relation to advanced equipments. Any specific implementation is a unique blend of the specific aspects and methods surveyed in the previous sections.

Main Fields of Application

They are, at the functional level:

1. analysis of reliability and maintenance data banks, (Pau, 1979; Pau, 1972; Pau, 1974)
2. control of a maintenance policy, (Pau, 1979; Pau, 1974)
3. component or part selection, (Pau, 1979; Hughes, 1977)
4. individual or lot acceptance sampling and screening for quality control (discrete parts or processes), (Pau, 1978; Pau, Chen; Pau, 1977; D'Angelo, 1978)
5. failure detection and diagnosis, with or without failure localization, (Burrows, Miles, 1972; Hsiung, Cox, 1972)
6. performance monitoring for alarms or warnings, and error analysis (Pau, 1979; Hankley, Merrill, 1971)
7. sensor redundancy management, (Pau, 1979; Yermachenko, Sdor, 1975; Clark, Fobth, Walton, 1975)
8. automated parts manufacturing and control, (Pau, 1979; Sterling, 1978; Skinner, 1977; Jones, 1971)
9. accelerated life testing and prototype design reviews, (Pau, 1979; Pokrowsky, 1972)

Main Classes of Systems

From earlier considerations it is clear that the pattern recognition methodology does not apply to digital systems such as computers, for which the signals are essentially binary with a considerable number of combinatorial states.

Thus, those equipments on which implementations have been made so far are generally:

1. mechanical assemblies, tools, structures or parts, (Pau, 1979, Solomon, 1970; Piety, Robinson, 1976; Hoffman, Fukunaga, 1969; Cohn, Ott, 1971; Frarey, 1970; Hsiung, Cox, 1972; Fang, Pavlidis, 1972)
2. electrical or electrotechnical (discrete components), (Saeks, Liberty, 1977; Pau, 1973; Sterling, 1978)
3. electronic (discrete, components or structures such as integrated circuits masks), (Saeks, Liberty, 1977; Muehldorf, 1975; Thatte, 1977; Waidelich, 1977; D'Angelo, 1978)

4. electromechanical assemblies or parts, (Pau, 1979; Pau, 1972; Pau, 1974; Braun, 1975; Frarey, 1970)
5. thermal
6. nuclear, (Pau, 1979; Piety, Robinson, 1976; Gonzalez, Fry, Kryter, 1974; Joksimovic, 1969; Gonzalez, Howington, 1977)
7. engines (cars, trucks, aircraft), (Pau, 1979; Becker, 1978; Braun, 1975; Thomas, Wilkins, 1972; Pau, 1975; Solomon, 1970; Hughes, et al, 1977; Pau, 1977; Cortina, Engel, Scott, 1970; Cortina, 1971)
8. microwave components, (Pau, 1979)
9. flow control, (Pau, 1979)
10. control systems, (Pau, 1979; Pau, 1975; Pau, 1977; Piety, Robinson, 1976)
11. metallurgical products, (Pau, 1979; Jones, 1971)

REFERENCES

Barlow, R.E., F. Proschan, "Mathematical theory of reliability" (Wiley, New York 1965).

Becker, P.W., "Recognition of patterns using the frequencies of occurrence of binary words", (Springer Verlag, New York 1978) 3rd ed.

Bellman, R., "Dynamic programming, pattern recognition and location of faults in complex systems", J. of Applied Probability, Vol. 3, 1966, 268-271.

Bebzercri, J.P., "L'analyse des données", Vol. 1 & 2 (Dunod, Paris 1977).

Bertails, J.C., J. Zirphile, "Une méthode de conception rapide et fiable des circuits intégrés avec controle automatique de l'implantation", Revue Technique Thomson-CSF, Vol. 9, No. 4 December 1977, 717-735.

Braun, S., "Signal analysis for rotating machinery vibrations", Pattern Recognition J., Vol. 7, 1975, 81-86.

Burrows, A.A., W.L. Miles, "Aircraft fault isolation based on patterns of cockpit indications", The Aeronautical J., September 1972.

Cardillo, G., K.S. Fu, "On suboptimal sequential pattern recognition", IEEE Trans., Vol. EC-17, No. 8, August 1968, 565-588.

Chu, W.W., "Adaptive diagnosis of faulty systems", Operations Research, Vol. 16, 1968, 915-927.

Clark, R.N., D.C. Fobth, W.M. Walton, "Detecting instrument malfunctions in control systems", IEEE Trans., Vol. AES-11, No. 4, July 1975.

Cohn, M., G. Ott, "Design of adaptive procedures for fault detection and isolation", IEEE Trans., Vol. R-20, No. 1, February 1971, 7-10.

Cortina, E., H.L. Engel, W.K. Scott, "Pattern recognition techniques applied to diagnostics", Soc. Automotive eng., report 700497, 1970.

Cortina, E. "Automatic diagnostic equipment studies", USATACOM TR-11289, 29 January 1971.

D'Angelo, H., "Testing networks and the computer-aided design of multi-stage screening processes", Proc. Southeastcon' 1978, IEEE, N.Y. 1978, 213-216.

De Dombal, F.T., F. Gremy, (ed.), "Decision making and medical care", (North Holland, Amsterdam 1976).

Donio, J., "Problémes de diagnostic par construction d'espaces mesurables", Metra, Vol. 11, No. 2, 1972, 315-331.

Fang, G.S., T. Pavlidis, "Signal classification through quasi--singular detection with application in mechanical fault diagnosis", IEEE Trans., Vol. IT-18, No. 5, September 1972, 631-636.

Frarey, J.L., "Mechanical basis for pattern analysis", Soc. Automotive eng. report 700496, 1970.

Friedland, B. "Maximum-likelihood estimation of a process with random transitions (failures)", Proc 1978 IEEE Conf. on decision and control, IEEE Catalog 78, CH1392-OCS, 427-432.

Fu, K.S., "Sequential methods in pattern recognition and machine learning", (Adacemic Press, New York 1968).

Fu, K.S., "Syntatic methods in pattern recognition" (Academic Press, New York 1974).

Fu, K.S., (ed.), "Digital pattern recognition", Communications and Cybernetics, Vol. 10 (Springer, Berlin 1976).

Fukunaga, K., "Introduction to statistical pattern recognition", (Academic Press, New York, 1972).

Gonzalez, R.C., D.N. Fry, R.C. Kryter, "Results in the application of pattern recognition methods to nuclear reactor core component surveillance", IEEE Trans Nuclear Science, 1974, No. 21, 750-756.

Gonzalez, R.C., L.C. Howington, "Machine recognition of abnormal behaviour in nuclear reactors", IEEE Trans., Vol. SMC-7, No. 10, 1977, 717-728.

Goto, N., et al., "An automatic inspection system for mask patterns", in Proc. 4th Int. J. Conf. pattern recognition (Kyoto), IEEE, N.Y., November 1978.

Grellin, G.L., "Special issue on Bayesian reliability techniques", IEEE Trans. Vol. R-21, No. 3, August 1972.

Grenander, U., M. Rosenblatt, "Statistical analysis of stationary time series", (Wiley, New York, 1957).

Gross, A.J., "An approach to the minimization of misclassification in the repair of equipments", IEEE Trans., Vol. R-19, No. 1, February 1970, 10-13.

Gumbel, E.J., "Statistical theory of extremes (Columbia University Press, New York 1958).

Gupta, A. et al., "Defect analysis and yield degradation of integrated circuits", IEEE J. Solid State Circuits, Vol. SC-9, No. 3, 1974, 96-103.

Hankley, W.J., H.M. Merrill, "A pattern recognition technique for system error analysis", IEEE Trans., Vol. R-30, No. 3, August 1971, 148-153.

Henriksen, G.M., "Reticles by automatic pattern generation", in Semiconductor Microlitography II, SPIE Proc., Vol. 100, 1977, 86-95.

Hill, M.O., "Correspondence analysis: a neglected multivariate method", Applied Statistics, Vol. 23, No. 3, 1974 (Series C), 340-354.

Hoffman, R.L., K. Fukunaga, "Pattern recognition signal processing for mechanical diagnostics signature analysis", IEEE Trans., Vol. C-20, No. 9, September 1969, 1095-1100.

Hsiung, C.Y., C.W. Cox, "Pattern classification in scan-type non-destructive tests", Int. J. Non-destructive testing, Vol. 4, 1972, 231-247.

Hughes, R.A., et al., "Using pattern recognition in product assurance", Proc. Ann. Reliability and Maintainability Symp., Philadelphia, 18-20 January 1977.

Joksimovic, V., "Statistical fault analysis method applied to advanced gas cooled reactors", J. British Nuclear Energy Society. Vol. 9, No. 4, October 1969, 275-302.

Jones, J.A., "The analysis of metallurgical data using pattern recognition techniques", Proc. NAECON, IEEE Catalog 71-C--24-AES-1971.

Kittler, J., L.F. Pau, "Small sample properties of a pattern recognition system in lot acceptance sampling", In Proc. 4th Int. J. Conf. on pattern recognition (Kyoto), (IEEE, New York, 1978).

Kulikowski, C.A., "Pattern recognition approach to medical diagnosis", IEEE Trans., Vol. SSC-6, No. 3 July 1970.

Larsen, P., "Symbolic layout system speeds mask design for IC's", Electronics, Vol. 51, No. 15, 20 July 1978, 125-128.

Levadi, V.S., "Automated learning applied to fault diagnosis", IEEE Trans., Vol. IES-3, No. 6, November 1967, 941-946.

Mann, N.R., N.D. Schafer, R. Singpurwalla, "Methods for statistical analysis of reliability and life data", (Wiley, New York, 1974).

Markel, J., A. Gray, "On autocorrelation equations as applied to speech analysis", IEEE Trans., Vol. AU-21, No. 2, 1973, 69-79.

Martin, W.C., W.D. Hokins, "A state space basis for sequential pattern classification", in Proc. 1st Int. J. Conf. on pattern recognition, IEEE Catalog 73 CHO-821-9c, 1973.

Mendel, J., K.S. Fu, (ed.), "Adaptive learning and pattern recognition systems", (Academic Press, New York 1970).

Muehldorf, E.I., "Fault clustering: modelling and observation on experimental LSI chips", IEEE J. Solid State Circuits, Vol. SC-10, No. 4, 1975, 237-244.

Parkhomenko, P.P., O., "Tekhnickeske-diagnostike", (Znaniye Press Moscow, 1969).

Pashkovskiy, G.S., "Optimization of sequential fault detection procedures", Engineering cybernetics, Vol. 9, No. 2, March 1971, 259-270.

Pau, L.F., "Topics in pattern recognition, IMSOR", (Technical University of Denmark, Lyngby 1973).

Pau, L.F., "Applications of pattern recognition to the diagnosis of equipment failures", Pattern Recognition J., Vol. 6, No. 3, August 1974, 3-11.

Pau, L.F., "Diagnosis of equipment failures by pattern recognition", IEEE Trans. Vol. R-23, No. 3, August 1974, 202-208.

Pau, L.F., "Diagnostic Statistique", Onde Electrique, Vol. 54, No. 10, Decembre 1974, 529-537.

Pau, L.F., "Adaptive failure mode diagnosis based on pattern recognition of acoustical spectral measurements", in Proc. Int. Conf. Monitoring Diagnostics in Industry, (House of Technology, Prague, 1975).

Pau, L.F., "Diagnostic statistique: Synthese des informations relatives a la fiabilité et a la maintenance d'un material aeronautique", L'Aeronautique et l'Astronautique, No. 34, 1972 - 2, 69-76.

Pau, L.F., "Analyse continue des fluctuations d'un parametre de fonctionnement: Application au diagnostic automatique des changements de regime d'une machine", R.A.I.R.O., Vol. Automatique-11, No. 1, March 1977, 5-15.

Pau, L.F., "An adaptive signal classification procedure, application to aircraft engine monitoring", Pattern Recognition J., Vol. 9, No. 3, October 1977, 121-130.

Pau, L.F., et al., "Controle statistique de qualite pour l'instrumentation", (Statistical quality control by variables: application to instrumentation), (Editions Chiron, Paris 1978).

Pau, L.F., "Finite learning sample size problems in pattern recognition", in C.H. Chen (ed.), Pattern recognition and signal processing, NATO ASI Series E, No. 29, (Sijthoff & Noordhoff, Alphen aan den Rijn (The Netherlands, 1978), 83-116.

Pau, L.F., "Signal Classification by non-parametric sequential tests", in Proc. AFCET/IRIA Congress on pattern recognition and picture processing, (IRIA, Rocquencourt, 1978; or T.R. ENST-C-78020, ENS Télécommunications, Paris 1978.

Pau, L.F., "Failure diagnosis and performance monitoring", (Marcel Dekker Inc. New York, 1979)

Pau, L.F., C.H. Chen, "Multivariate classification rule subject to small learning samples: application to quality control and signal classification", to appear.

Peron, V.I., "Majority gradient method for the optimization of a sequential procedure for checking operatibility, automation and remote control", February, 1970, No. 2, 282-288.

Persoon, E., "Dynamic sequential pattern recognition applied to medical diagnosis", (T.R. Purdue University, AD-734292, July 1971).

Piety, K.R., J.C. Robinson, "An on-line reactor surveillance algorithm based on multivariate analysis of noise", Nucl. Sci. Eng., Vol. 59, No. 4, 1976, 369-380.

Pokrowsky, F.N., "On reliability prediction by pattern classification", Proc. Ann. Reliability and maintainability Symp., IEEE Catalog 72-CH-0577-7R, 367-375.

Rabiner, L.R., B. Gold, "Theory and application of digital signal processing", (Prentice Hall, Englewood Cliffs 1975).

Saeks, R., S. R. Liberty, "Rational fault analysis", (Marcel Dekker Inc., New York, 1977).

Saridis, G.N., R.F. Hofstadter, "A pattern recognition approach to the classification of non-linear systems", IEEE Trans., Vol. SMC-4, No. 4, July 1974, 362-371.

Skinner, J.G., "The use of an automatic mask inspection system in photomask fabrication", in: Semiconductor Microlitography SPIE Proc. Vol. 100, 1977, 20-36.

Slagle, J.R., R.C.T. Lee, "Application of game tree searching techniques to sequential pattern recognition", Comm. of the ACM, Vol. 14, No. 2, 1971, 103-110.

Solomon, H., "A first application of clustering techniques to fleet material condition measurements", T.R. T-238, (George Washington University, Washington, D.C. June 1970).

Spire, O., "Determination de l'importance relative de differents parameters descriptifs pour l'obtention d'un diagnostic", R.A.I.R.O., Vol. 4, No. 1, 1970, 85-99.

Sterling, W.M., "Automatic non-reference optical inspection of printed wiring boards", Proc. OSA/IEEE Conf. on laser and electrooptical systems, San Diego, 7-8 February, 1978, IEEE (New York 1978), 66.

Thatte, S.M., "Fault diagnosis of semi-conductor random access memories", AD-A-044281, 1977.

Thomas, D.W., B.R. Wilkins, "The analysis of vehicle sounds for recognition", Pattern Recognition J., Vol. 4, 1972, 379-389.

Tsokos, C.P., I.N. Shimi, "The theory and applications of reliability with emphasis on Bayesian and non-parametric methods", Vol. 1 & 2 (Academic Press, New York 1977).

Vancleemput, V.M., "Topological circuit layout, AD-A-048050, 1976.

Van de Wiele (ed.), "NATO ASI on process and device modelling for integrated circuit design", (Noordhoff, Groningen, The Netherlands, 1977).

Waidelich, J., "Methoden der Mustererkennung zur Fehlerdiagnose von Digital rechnern", T.R. KFK-PDV-127, Kernforschungszentrum Karlsruhe, Karlsruhe, November 1977.

Wald, A., J. Wolfowitz, "Optimum character of sequential probability ratio test", Ann. Math. Stat., Vol. 19, 1948, 326-329.

Watanabe, Y. et al., "A fundamental experiment on automatic LSI mask pattern drawing reading", in Proc. 4th Int. J. Conf. pattern recognition (Kyoto), IEEE, N.Y. November 1978.

Yermachenko, A.I., V.V. Sdor, "Automatic classification and pattern recognition diagnostics and realiability", Soviet Automatic Control, Vol. 8, No. 4, July 1975, 1-5.

THE USE OF FLOW MODELS FOR AUTOMATED PLANT DIAGNOSIS

Morten Lind

Risø National Laboratory
DK-4000 Roskilde
Denmark

INTRODUCTION

Automatic, computerized diagnosis can be based on several different search strategies, e.g. a search for a match between a pattern of measured data and some stored symptom patterns, or a search to locate a change in the system's functional state with reference to a stored model of normal or specified plant state. The latter strategy has a number of basic advantages: it is independent of the prediction and analysis of specific faults and events; the reference for search, the normal state, can be derived from actual plant operation by the computer; the strategy can be based on invariate relations such as conservation laws; etc.

In the paper we will explore the use of conservation laws for mass and energy in diagnosis of plant malfunction. An advantage of this approach is that it is possible to diagnose unforeseen plant disturbances; this is due to the general nature of the conservation laws. Furthermore, accumulation of mass and energy is a potential source of risk in plant operation, and the identification, counteraction and location of unbalances is thus an important aspect of process plant diagnosis.

We will discuss how a category of plant models, developed by the writer, can be used as a formal basis for diagnosis using conservation laws. A plant flow model describes the topology of the pattern of mass and energy flows and represents qualitative aspects of plant function in a given operational regime. A short introduction to flow models is given in appendix A.

Diagnostic strategies based on flow models can be used in the design of automatic disturbance analysis and control, but they can also be used for organizing the available measured plant data in a meaningful way for an operator and for supporting him in the need for rapid "zooming in" on the relevant details in a complex situation.

TASKS IN PLANT DIAGNOSIS

As discussed in (Rasmussen, 1978) we can divide the diagnosis of plant malfunction into three basic subtasks

- plant state identification
- disturbance compensation (control)
- fault localization

These subtasks constitute the basic elements of every diagnostic task but their sequencing, i.e., their order of occurrence in time depend on the specific situation considered. However, plant state identification plays an important role as knowledge about the plant state is a necessary basis for the planning of control actions in disturbance situations and for fault localization. In fact the problem of plant diagnosis is largely a question about state identification because appropriate control actions usually can be derived directly from knowledge about the state. In the following we will formulate the problems of state identification, disturbance compensation and fault localization within the framework of flow models (see appendix A). This will lead to a decomposition of the diagnosis problem into some simpler subproblems. The solution of these subproblems may provide the basis for the construction of computerized diagnostic tools for the plant operator.

PLANT STATE IDENTIFICATION

Before we formulate the state identification problem it may be necessary to emphasize the relative nature of the concept of state. The state of a system can be defined as the information we need about the system history in order to predict its future behaviour. Accordingly, the nature of the information needed depends on the kind of prediction performed, the behavioural characteristics of the system which should be predicted and the level of detail required. This means, that the state concept is an integrated part of the modelling framework chosen and that state information refer to a given level of detail used in system description.

Considering plant diagnosis using flow models we must then choose our state concept in accordance with the requirements of diagnosis. This means that the system state should include information about extensive plant variables (content properties) and through variables (flow properties). The behaviour of extensive variables (accumulated mass or energy) is closely related to the risk potential in the plant. Through variables are important for characterizing the production state of the plant and for prewarning (disturbances in contents are caused by flow disturbances and prewarning based on the flow state could thus be used to avoid the occurrence of serious disturbances).

The availability of plant information determines to what level of detail the plant state can be identified and the identification problem can be divided into two main problems

- what kind of information is available for plant diagnosis and how should it be related to a description in terms of contents and flows of mass and energy. This is basicly a question of the thermodynamic degrees of freedom of the processes considered and the use of data transformations to infer new plant information.

- for what plant subsystems is there sufficient information available in order to characterize the state (in the sense defined above). This leads to the definition of the concept of a completely observable subsystem.

These two problems will be discussed below and their relation to the structure of the instrumentation system will be indicated. We will furthermore consider a state classification scheme which can be used to distinguish between normal and abnormal operating situations for a completely observable subsystem. The classification also allows for detection of information inconsistenties which may be caused by instrument error or by certain unbalance situations.

Information Sources in Plant Diagnosis

Measurements of plant variables constitute the basic information for diagnosis. However, when doing diagnosis on an overall level (as e.g. by using flow models), we must take into account other types of information. This is because measurements do not always relate directly to the level of plant representation used. We have the following three sources of information:

- Measurements

  This includes the direct measurement of properties of flows and contents. Plant variables may be related to the measured variables by invariant scaling or conversion factors. Properties of the measurement system will determine the conversion factors to be used. In this class is included the measurement of mass flow rate, temperatures and pressures. The reliability of the information given by measurements depends only on the quality of the instrumentation used.

- Inferences

  New plant information is in this case obtained by transformation of known information. It is convenient to distinguish between inferences based on plant specific properties (as e.g., the nature of the materials processed) and inferences based on the use of the general conservation laws for mass and energy. In the first category we find the computation of energy flow from measured values of mass flow rate, temperature and pressure and more complex types of data transformations as e.g., observers (as e.g., Kalman filters used to estimate nonmeasurable variables in linear dynamic systems). The use of mass and energy balances in plant diagnosis for inferring new information is discussed in appendix B. The data transformations in the first case rely on knowledge of a model of the functional structure of the system considered and in the second case on a model of the abstract function (the model categories used here are described in (Rasmussen, 1979)).The reliability of the plant variable values obtained depends in both cases on the validity of the models applied (i.e. the assumptions made) and the quality of the primary information.

- Conditions

  Plant variable values may be given as conditions i.e., they can be given as part of the primary data for the diagnostic system or they may be provided by the operator as part of his diagnostic activities or by the status of the control system. The furnishing of plant variables by conditions is one of the key factors in the cooperation between the operator and the plant computer in diagnosis. Furthermore conditioning may be used in diagnosis as part of a strategy employing hypothesis generation (e.g., doing plant state identification under the assumption that an unknown variable has a given value).

Information from all the types of sources mentioned above is used in diagnosis using flow models. This is necessary because energy quantities cannot be measured directly and because sufficient information is in most cases not available in the form of measurements.

Finally, to widen the perspective, plant variables can be obtained at several levels of inference i.e., based on successive transformation of measured and inferred variables and conditions.

In such cases it is important to keep track of the assumptions inherent to the inferences made. For this purpose an "assumption tree" can be constructed as shown for example in figure 1. It should be noted that several inferences may be available for the same plant variable allowing a check of the validity of assumptions made.

Completely Observable Plant Subsystems

By the state concept defined above and the nature of the diagnosis problem in process plant it is natural to define the concept of a completely observable plant subsystem:

A plant subsystem is completely observable if extensive properties of all its associated contents and the through properties of all the flows connecting the subsystem to the environment are known.

This kind of plant subsystem constitutes an ideal as sufficient information is available to determine its state. But at the same time it defines the level to which the plant can be decomposed on the basis of the available information. The size of the subsystem and its internal complexity (as e.g., cycles in the flow structure) contribute to the uncertainty by which the internal behaviour can be inferred from the mass and energy balance condition of the subsystem boundary. This means that the quality of the diagnosis which can be obtained using flow structures is determined by the level of decomposition and the internal structure of the completely observable subsystems.

It should be strongly emphasized that content and flow properties need not necessarily be measured. They could also be inferred information or conditions. This implies that the plant decomposition should be changed when new information is generated as part of the diagnostic activity. The validity of the information for each subsystem can be judged on the basis of the assumption trees for the associated inferred subsystem variables.

The subsystems can be derived from the plant flow structure and information about the known plant variables and their localization in the flow structure. Algorithms for decomposing the flow structure into aggregates representing completely observable subsystems can be formulated on the basis of fundamental graph operations. This will be discussed in (Lind et al. 1980).

By identifying completely observable subsystems we have divided the plant into two parts, one for which the state can be identified and one for which we have insufficient information. For these latter parts either content or flow information may be missing. The network created by the completely observable

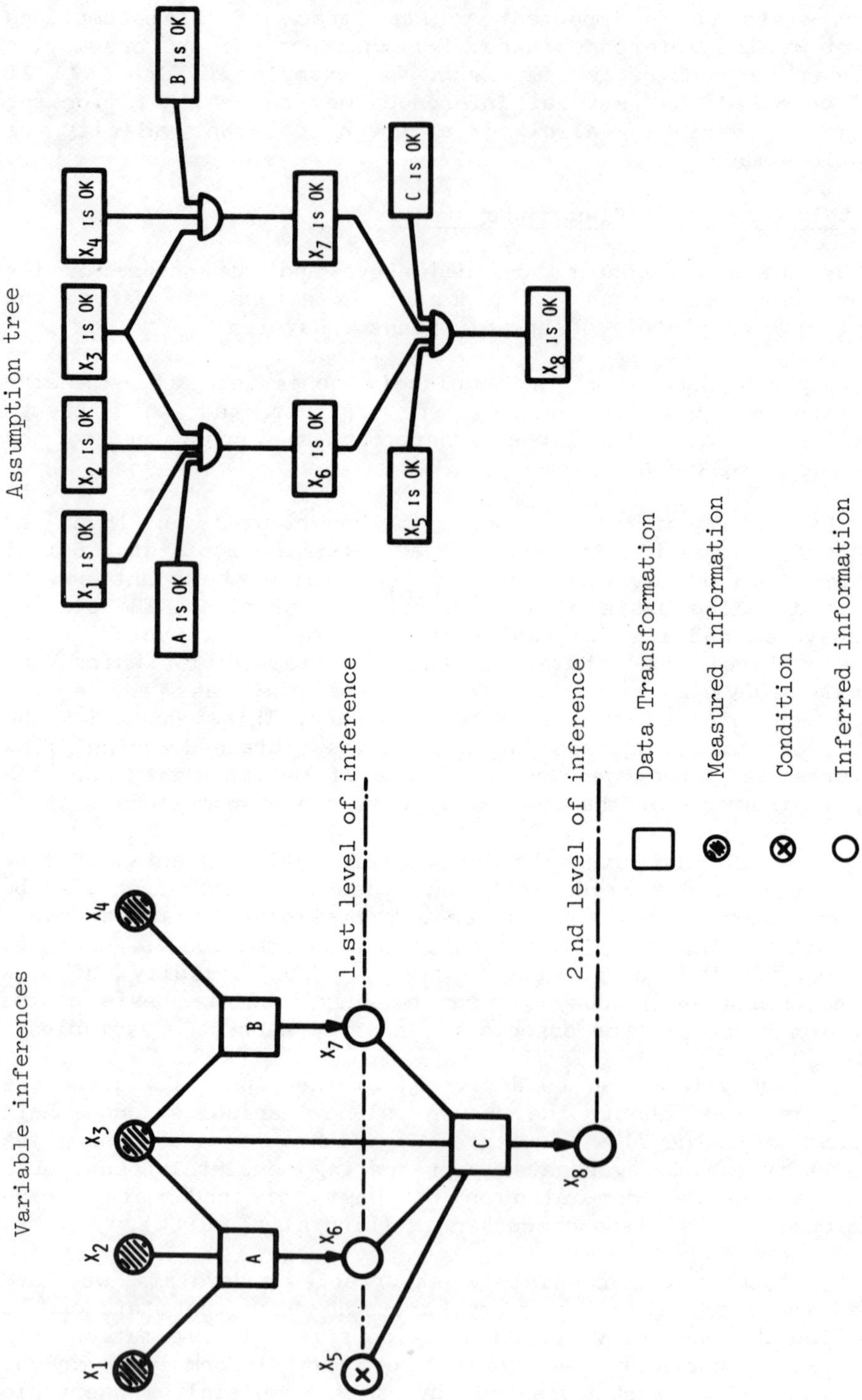

Fig. 1. Plant variable inferences and corresponding assumption tree, an example.

subsystems and their interconnections is important as it defines the overall topology which must be followed if a diagnosis based on a consistent use of flow variables and changes in mass and energy balances is to be obtained. Furthermore, the decomposition determines the level of detail on which we have sufficient information to support topographic strategies (see Rasmussen, 1978) based on a separate use of the material and energy flow structures.

The division into observable and not observable subsystems is a problem of design of the instrumentation system. The criteria which must be used to structure the instrumentation system are closely related to the location and nature of the risk potential in the plant. Thus it must be a design requirement that the plant parts which are potentially dangerous should be completely observable on the basis of measurements or on inferred information which is reliable. Furthermore cycles internal to subsystems should be avoided as they contribute to the uncertainty about the plant state. Cycles can be avoided by a suitable layout of the instrumentation system. These requirements ensure that the "level of risk" and its tendency to change can be continuously supervised. Problems of measurement technology may in some cases prevent an adequate subdivision.

Criteria for choosing a convenient plant decomposition may also be derived from the requirement of effective cooperation between the plant operator and the computer. During computer aided diagnosis it is necessary that the results of the computer can be communicated to the operator. If an inadequate decomposition into subsystems has been used by the computer it may be difficult for the operator to relate his overall conception of the plant function to the results obtained by the computer. This is important if the operator is to have a possibility of taking over in situations where the computer cannot proceed due to the lack of sufficient information.

State Categories

In plant diagnosis it is important to be able to distinguish between normal and abnormal behaviour of plant subsystems. In the following we will develop a set of state categories that allows for such a distinction and that is suitable for characterizing the state of a system governed by conservation laws. We will implicitly assume that the subsystem considered is completely observable as this is the only case where the state can be uniquely classified. The state classification will be based on the values of content (extensive) and flow (through) variables and their interrelations, and the categories constitute a discretization of the space spanned by these variables.

The distinction between normal and abnormal behaviour is given by the functional specifications for the subsystem considered. Accordingly, we need to consider the nature of functional specifications in order to set up relevant state categories. Functional specifications for the plant are provided by the plant designer or by operating experience and are given within each operating regime as plant variable values with some prescribed limits of acceptable deviation. Specifications may be obtained from the operational experience in the form of mean values and variances of plant variables generated by on line processing of collected plant operating data. In this way mean values and variances constitute a standard of acceptable (normal) plant behaviour. It is a primary requirement of specifications for plant operations that they are consistent, i.e. parts of the specifications must not be in contradiction. This is necessary in order to obtain distinct categories of normal and abnormal behaviour.

In the following we will consider the categorization of flows and contents separately. Finally, we will combine flow and content categories into a complete set of states.

For the characterization of a set of through variables associated with the flows over a subsystem boundary we could compare with their individual expected or prescribed values. In this way, we can classify a set of through variables as disturbed if the corresponding point $(\tau_1, \tau_2, \ldots, \tau_N)$ in a N-dimensional space is outside a certain volume of acceptable (normal) behaviour. But in addition to this we can also characterize the set of through variables according to their interrelation. In the present context it is particularly useful to distinguish between balanced and unbalanced flows i.e.

$$|\tau_1 + \tau_2 + \ldots\ldots \tau_N| \leq \varepsilon\text{, balance}$$

$$|\tau_1 + \tau_2 + \ldots\ldots \tau_N| > \varepsilon\text{, unbalance}$$

(ε is tolerance)

The balance or unbalance condition may be normal or disturbed. Accordingly, we will distinguish between normal and disturbed balance and normal and disturbed unbalance. By combining the two set of categories we get eight classes as shown in fig. 2. However, two of these are inconsistent and must be ignored (a disturbed balance and a disturbed unbalance are not consistent with a normal flow). The two types of categories are illustrated in fig. 3 in the case of N = 2.

Considering the contents of a plant subsystem they can be characterized as being steady or transient and each of these can again correspond to normal or disturbed behaviour. This means that we get four content categories as shown in fig. 2.

Flow categories		FN (Flow normal)	FD (Flow disturbed)
Balance normal	BN	1	1
Unbalance normal	UN	1	1
Balance disturbed	BD	1 (shaded: Rejected)	1
Unbalance disturbed	UD	1 (shaded: Rejected)	1

Content categories		N (Normal)	D (Disturbed)
Steady	S	1	1
Transient	T	1	1

Fig. 2. Flow and content categories

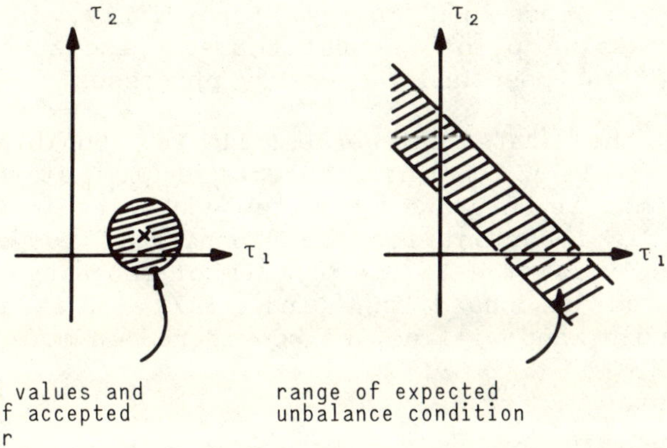

Fig. 3. Illustration of flow categories

State categories for a completely observable subsystem can now be obtained by combining flow and content categories. This is done in fig. 4. However, not all combinations correspond to proper plant information. This is because we have considered the balance/unbalance condition for the flows as independent of content behaviour. This is clearly not possible as an unbalance is associated with transient behaviour and balance is associated with steady behaviour of the contents. This means that some of the combinations in fig. 4 can be interpreted as corresponding to cases with detectable instrument error or leakages. Furthermore, some of the combinations correspond to inconsistent specifications and are ignored (shaded in fig. 4).

	BN,FN	BN,FD	BD,FD	UN,FN	UN,FD	UD,FD
SN	OK	OK	IE or L			
SD	OK	OK	IE or L	IE or L	IE or L	OK*
TN				OK	OK	IE or L
TD	IE or L	IE or L	OK	OK	OK	OK

Abbreviations: OK :operating values consistent
IE :instrument error or bad inference
L :leak
OK*:may be due to error, but information is consistent

Fig. 4. State categories

The categories above are valid for systems which process a substance governed by conservation laws. This means that they apply to both mass and energy. Accordingly, in a material and energy processing plant we must characterize subsystem states by the state of the mass and the energy processes.

From the state categories it is possible to set up a transition table which interconnects system states which can be reached from a given state by suitable changes in flow conditions. This table is important for the planning of compensating actions and is closely related to the problem of choosing suitable control heuristics as discussed in (Lind, 1979). This is a subject of further studies and will not be considered in more detail here.

The plant state

The content of the previous section leads in a natural way to an overall characterization of the plant state.

In order to define the plant state we create a simplified flow structure by aggregating completely observable subsystems into simple nodes. This graph shows the pattern of flow with a level of detail given by the available plant information.

If we now consider the network created by the observable and non-observable subsystems we can divide the plant into an observable and non-observable part. These parts are connected by arcs where the flow state is known. This implies that we can separate the two parts to analyse the observable part in isolation. The observable part is the part of the plant where we have sufficient information to do mass and energy balance calculations on the boundary and we have information about the substance (mass or energy) accumulated internally in the subsystem.

The subgraph of the flow structure consisting of observable parts is important as it shows the paths which should be followed in the plant if a diagnosis consistent with the given data is to be accomplished.

By classifying the state of each completely observable subsystem using the categories shown in fig. 4 we can obtain an overall state description which will be called the plant state. This is obtained by collecting subsystems which belong to the same state category into state aggregates. By comparing the states of a given aggregate with the states of its environment aggregates we can determine whether a given disturbance is internal to the aggregate considered or is imposed from the environment (a balanced aggregate with transient extensive variables is clearly disturbed from an inside source of flow whereas a disturbed but balanced flow indicates that the aggregate is disturbed from the outside). These considerations should be further developed and will lead to criteria for localizing a disturbance to a given plant subsystem. It should be emphasized that the aggregation into state aggregates changes with the operating condition and must accordingly be done continuously in a supervisory system.

DISTURBANCE COMPENSATION

Disturbance compensation in plant diagnosis is a control problem. As this has been discussed by the writer within the framework of flow models elsewhere (Lind, 1979) we will only give a short account of this subject here. In op-cit. it is shown how flow models may be used for decomposing a control task into a set of coordinated basic tasks. In addition to the plant flow structure for the operational regime considered, the decomposition also requires a set of control heuristics. As mentioned earlier these heuristics can be based on state transition tables, but in addition to this it is also necessary to consider the coordination of the different subsystems during a change of states and to consider the potential for control of the different subsystems (position of conditioned transport nodes in relation to subsystem boundary).

The decomposition considered in op.cit. leads to a representation of the task structure in a state-activity diagram which can be used for generating operating procedures or sequential controls. The procedures are generated by the use of project planning techniques as e.g. CPM or PERT.

SEARCH STRATEGIES

Search strategies in plant diagnosis is discussed in (Rasmussen 1978). Here we will limit ourselves to some comments which relate search strategies to the material in the present paper.

The major task in plant diagnosis is to identify plant state at the level of detail where it is possible to localize disturbances or provide adequate compensating actions. Due to reasons of effectivity the level of detail must not be too high but on the other hand a certain level is necessary in order to reduce the uncertainty of detail about the plant state.

Topographic search in plant diagnosis can be considered as a procedure to increase the level of detail in plant state information. An effective strategy minimizes the number of steps which are necessary to obtain an information level which is sufficient to support a symptomatic search (i.e. where symptom patterns are available). The state categories presented previously in the paper provide a set of patterns which can be used for a completely observable subsystem on each level of detail. (It may be necessary to consider combinations of states in order to obtain a set of symptoms for a network of subsystems.

COMPUTER AIDED PLANT DIAGNOSIS

Generation of the plant information which is necessary to support diagnosis in the flow structure requires complex transformations of primary plant data into variables describing the state of plant subsystems in terms of mass and energy. This computational task can be allocated to the computer in order to support a presentation of the plant situation for the operator in terms of flows and contents of mass and energy.

As discussed earlier it is possible to infer plant information on many levels of detail by using variable interrelations which are specific to the plant and by using balance equations. This means that the computer could be used as a tool for the operator in zooming in on plant details. However, it is necessary to indicate for the operator the degree of uncertainty of the information which is generated by variable inferences. This implies that the assumption tree corresponding to the inferences made (see fig. 5) must be stored in the computer and used as an aid for the operator to choose between plant representations with different degrees of detail and uncertainty. The computer can thus be used as an advanced data transformation facility. Different automated aids for reducing or increasing the degree of detail and for search in the flow structure may be made available to the operator as e.g. zooming-in and path-following. These operations are based on inferences made inside a completely observable subsystem using boundary information and information which might be available internally (as e.g. isolated measurements).

Above we have considered the computer as a diagnostic tool for the operator. However, the computer could also be used to perform an automated diagnosis. In this case the data transformations necessary for inference would be performed as an integrated part of the topographic search. In this way the amount of computation may be reduced as more effective use can be made of the information generated. Furthermore, it becomes easier to keep track of the validity of assumptions made in variable inference. The automation of diagnosis requires the formulation of a set of heuristics for choosing suitable decompositions of a completely observable subsystem. This will be considered in (Lind et al., 1980). The heuristics should take into account information about the sequences of changes in measured plant variables as this information is important for the localization of disturbances in the flow structure. Consideration should also be given to the nature of disturbance propagation for mechanical (fluid flow) and thermodynamic transport processes involved in the plant. This involves the use of plant information related to a model of the functional structure (see Rasmussen, 1979). Thus, the use of heuristics which take into account this kind of information implies that we consider the diagnosis problem within a multilevel modelling framework.

ACKNOWLEDGEMENT

This work was carried on in connection with the Scandinavian project on Control Room Design and Operator Reliability which is supported in part by the Nordic Council of Ministers.

REFERENCES

Frank, O., "Statistical Inference in Graphs", FOA 1971.

Lind, M., "The Use of Flow Models for Design of Plant Operating Procedures", paper presented at: IWG/NPPCI Specialists Meeting on Procedures and Systems for Assisting an Operator During Normal and Anomalous Nuclear Power Plant Operation Situations, December 5-7, 1979, Garching FRG.

Lind, M., "Flow Models of Material and Energy Processing Systems", Risø-M-2201, Risø 1980, (in preparation).

Lind, M. and Talmon, H., "The Use of Mass and Energy Balances for Observation in Process Plant Diagnosis (in preparation), 1980.

Rasmussen, J., "Notes on Diagnostic Strategies in Process Plant Environment", Risø-M-1983. Risø 1978.

Rasmussen, J., "On the Structure of Knowledge - a Morphology of Mental Models in a Man-Machine System Context, Risø-M-2192, Risø 1979.

Rasmussen, J., "Models of Mental Strategies in Process Plant Diagnosis, (this volume), 1980.

APPENDIX A

DEFINITION OF FLOW MODELS

In the following we will give a short introduction to the concepts used in flow modelling of process plant. For a more detailled discussion the reader is referred to (Lind, 1980).

In flow modelling, the basic assumption is that every material and energy process can be described as an interaction between two fundamental types of processes. These are storage and transport processes. Storage processes represent phenomena which occur in a volume fixed in space whereas transport processes represent the phenomena associated with transfer of substance (mass or energy) between two locations in space.

Storage Processes

Storage processes include simple accumulation phenomena, i.e. pile-up of material or energy in a volume. But in addition to accumulation phenomena, storage processes may also include chemical processes, i.e. changes of material composition and changes of phase. To a storage process is associated a set of variables describing the content of substances established by the process. These variables divide into two groups, one describing quantity of substance stored; these are the so called extensive variables. The other group describes the qualities of the substance stored and comprises the intensive variables. Storage processes are governed by the conservation laws for mass and energy, chemical reaction kinetics and state-equations for the material processed. The extensive variables are controlled by the conservation laws and the chemical kinetics and the relations between extensive and intensive variables are given by the state equations. A description of the storage process in terms of all these variables and their interactions constitutes a model of the functional structure (following the categories given by Rasmussen 1979) of the storage process. But in a flow model we only consider the aspect of the storage function which is related to accumulation of mass and energy. This means that a flow model of a storage process is given in terms of the extensive variables and the conservation laws.

Transport Processes

Transport processes include the transfer of material and energy between two locations in space by convection, conduction and diffusion phenomena. Transport processes are the origins of irreversible behaviour as they lead to entropy production by tending to reduce gradients in temperature, concentrations and pressure. To a transport process is associated a set of variables which describe the flow phenomenon generated by the process. Again we can divide these variables into two groups, one representing generalized forces and these are called across variables. The other variables represent generalized fluxes and will be called through variables. Transport processes are governed by the laws of fluid mechanics (momentum balance) and by the generalized diffusion equation. These laws interrelate across and through variables. As for the storage process we can describe the transport process by a model of its functional structure (across and through variables and their interrelations) but in flow modelling only through variables are of interest as these variables relate to the transport of mass and energy.

Boundaries

From these basic definitions we can then describe a process plant as an interconnection of material and energy storage and transport processes. The interconnection between two processes is denoted a boundary and is associated with a set of variables which represent the variables which are shared by the two processes. We distinguish between two groups of variables, potentials and fluxes. As the function of a boundary is to interface two processes rules are given which interrelate boundary variables to the variables for the two processes in question. This means that in a flow model boundary fluxes provide the link between through variables and extensive variables.

Conditioned Processes

A boundary represents a potential for interaction between two processes. The interaction is created by interchange of material and energy and is denoted process interaction. But a process can also be influenced by conditioning. As an example can be mentioned the change of stem position in a control valve which is a conditioning of the transport process associated with the valve. A tank open to the air is an example of a conditioned storage process because the pressure is not a free variable (it is determined by the environment). The major difference between process interaction and conditioning interaction is that the former is bidirectional whereas the latter is unidirectional. This means that a conditioning defines a unique direction of control whereas the process interaction implies that the two interacting processes are functionally integrated.

Flow Structures

The concepts underlined above constitute the basic vocabulary of flow modelling. The concepts are summarized in fig. 5 and illustrated by an example in fig. 6. In fig. 5 we have introduced symbols to represent transport and storage processes and boundaries. These symbols are used for the construction of graphs called the <u>flow structures</u>.. In these graphs processes are represented by nodes and boundaries as arcs. The flow structure describe the topology of the pattern of material and energy flow in the process plant modelled. As can be seen from fig. 5 we use different symbolisms for pure energy processes and for material processes.

		PROPERTIES	VARIABLES	SYMBOL IN FLOW STRUCTURE MATERIAL	PURE ENERGY
STORAGE	CONTENT	INTENSIVE EXTENSIVE	θ ε		
TRANSPORT	FLOW	ACROSS THROUGH	α τ		
BOUNDARY	INTERFACE	POTENTIAL FLUX	ψ ϕ	(condition)	

Fig. 5. Flow modelling concepts.

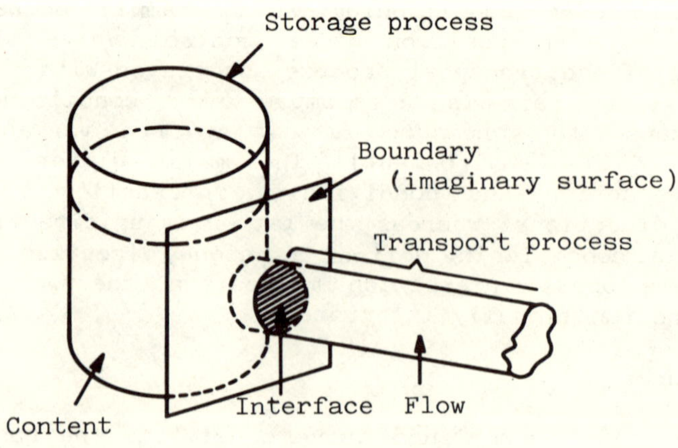

Fig. 6. Illustration of modelling concepts.

In addition to the basic concepts defined above, the concept of an aggregate is also used in flow modelling. An aggregate is a collection of interrelated transport and storage processes. Aggregates are used for representing plant subsystems for which the internal structure is ignored and the symbolism used in modelling is shown in fig. 7.

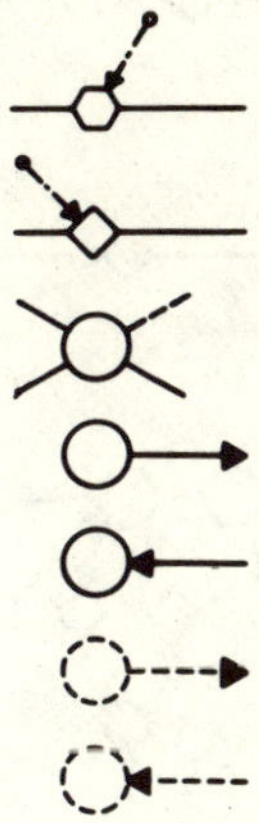

Fig. 7. Additional definitions.

An example of a flow structure for a conventional power plant is shown in fig. 8. The flow structure describes plant function in an intermediate operating regime during boiler start-up (boiler is filled with water and heating is initiated, steam produced is absorbed in the start-up system).

Two graphs can be derived from a plant flow structure, one by ignoring all dotted lines (pure energy boundaries) and the other by replacing all solid arcs (material boundaries) with dotted lines. The first graph represents the material flow structure and the other the energy flow structure. These graphs define two networks of process interaction which can be used for process analysis. For processes involving multicomponent flows we can in addition define component flow structures from the material flow structure (one graph for each component). The flow structure for component no. j is constructed from the material flow structure by ignoring all processes (and associated boundaries) where the component j does not participate.

After the definition of flow models presented previously we will now summarize the characteristics of flow models. Flow modelling of process plant can be considered as a two level activity. It is a description of the plant in terms of basic

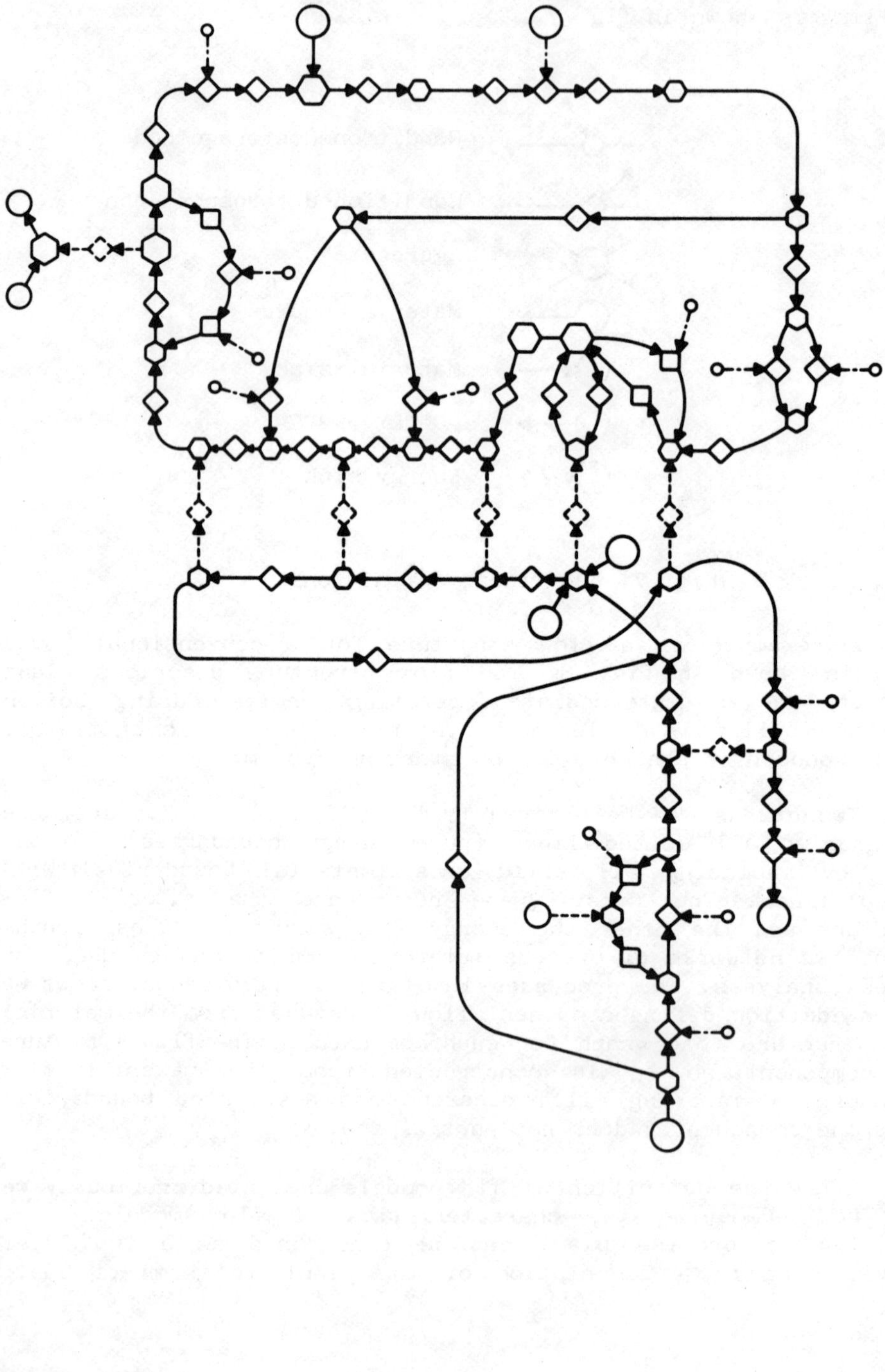

Fig. 8. Flow structure for conventional power plant.

processes and their interrelations as represented by the flow structure (model of abstract function) and it is a representation of the individual processes by models of their functional structure. Both types of models are used in diagnosis as described in the main part of this paper. The functional structure of storage and transport processes can be represented by general variable bigraphs as shown in (Lind, 1980). These bigraphs represent the logic structure of irreversible thermodynamics.

A flow model is basicly a description of plant function and accordingly, the subdivision of the plant into processes will, as a rule, not follow the physical structure of the plant. Each process represents a phenomenon which, with the chosen level of detail, can be considered to be homogeneous in time and space. The description of plant function provided by a flow model is usually valid only within a range of plant operation. The structure of a flow model may change because processes which are important in a certain range of operation may not contribute when the plant is outside this operational regime. This change in structure is basicly a consequence of the approximations made in a flow model. Furthermore, it is not uncommon that the functional structure of a process plant is changed as part of a change in operation (e.g., start-up and shut-down). This is accomplished by e.g., opening and closing valves which in the flow structure is represented by removing and including transport nodes.

APPENDIX B

The use of conservation laws to compute unknown plant information

The laws of mass and energy conservation can be used in plant diagnosis for derivation of unknown plant variables. These laws apply to any thermodynamic system and can be used to set up data transformations which have application for all process plants independently of the nature of the processes carried out.

For a system (an aggregate in the flow structure) in internal equilibrium which interacts with the environment by N material flows and M pure energy flows we have the following formulation of the conservation laws (balance equations)

$$\sum_{n=1}^{N} m_n = 0 \qquad (1)$$

$$\sum_{n=1}^{N} u_n m_n + \sum_{i=1}^{M} e_i = 0 \qquad (2)$$

where

m_n is the mass flow rate of material flow no. N

$u_n m_n$ is the energy flow rate associated with material flow n (u_n is the energy per mass unit)

e_i is the flow rate of pure energy flow i.

Flow rates are through properties of the corresponding flows (see appendix A). It should be noted that it is assumed that the system considered is in internal equilibrium. This means that we either consider a static situation or that changes in content are represented as flows to the environment (storage flows) and are included in the N material flows and the M pure energy flows.

The mass and energy balance (1 and 2) represent two equations from which we can determine at most two unknown variables. We have the following situations where missing information can be determined by the use of balance equations:

1) m_j unknown (N-1 variables known)

$$m_j = -\sum_{n \neq j}^{N} m_n$$

2) e_k or u_r unknown (2N-1+M variables known)

$$e_k = -\left[\sum_{n=1}^{N} m_n u_n + \sum_{i \neq k}^{M} e_i\right]$$

or

$$u_r = -\frac{1}{m_r}\left[\sum_{n \neq r}^{N} m_n u_n + \sum_{i=1}^{M} e_i\right]$$

3) m_j and u_r or m_j and e_k unknown (2(n-1)+M variables known)

$$m_j = -\sum_{n \neq j}^{N} m_n$$

$$u_r = -\frac{1}{m_r}\left[\sum_{n \neq r}^{N} m_n u_n + \sum_{i=1}^{M} e_i\right]$$

or

$$e_k = -\left[\sum_{n=1}^{N} m_n u_n + \sum_{i \neq k}^{M} e_k\right]$$

4) m_j and m_r (2N-2+M variables known)

$$m_j + m_r = -\sum_{n \neq j,r}^{M} m_n$$

$$u_j m_j + u_r m_r = -\left[\sum_{n \neq j,r}^{N} m_n u_n + \sum_{i=1}^{M} e_i\right]$$

(the equations are singular for $u_r = u_j$)

By the equations above we can use known plant information (through variables and the rates of extensive variables) to compute unknown flow properties. However, in general there will be many ways to combine the available information for different inferences.

In (Lind et al. 1980) consideration is given to the problem of decomposing the plant flow structure into aggregates corresponding to the different types of inferences discussed above. These aggregates are called observation aggregates as their associated balance equation allows for observation of nonmeasured variables.

It should be noted that we have implicitly assumed that the system considered is a single component system (only one chemical species is processed). In the multicomponent case we must take into account partial mass balances and the information given by component concentrations. This is important for inference in the diagnosis of chemical process plant and should be considered in future work. The case discussed above is a special case of the multicomponent case.

Above we have considered inference by using balance equations as a deterministic problem because we have ignored the influence of noise. This should be considered in future work in relation to the work done by Frank (Frank, 1971) on statistical inference in graphs.

DISCRIMINATIVE DISPLAY SUPPORT FOR PROCESS OPERATORS

L.P. Goodstein

Risø National Laboratory
DK-4000 Roskilde, Denmark

INTRODUCTION

The advent of computers has had a rather haphazard influence on information presentation in industrial control rooms. Some areas, as exemplified by the American nuclear field, have almost completely ignored their possibilities while others such as chemical installations, el-distribution networks, etc. have taken great strides in incorporating computers with visual display units such as CRT's (with colour and/or graphics) to serve as the main vehicle for information presentation for the operators. However, it seems safe to state that the availability of this new technology has not brought with it any evident advances in the design philosophy employed so as to better enable the operator to cope with the vageries of the complex and potentially risky processes which technology also has made possible. Thus, current presentation techniques essentially preserve the <u>one measurement - one indication</u> approach from conventional installations and use the relatively modest area on the VDU to display information in ways which reflect both pre-computer practises as well as the influence of digital computer inspired alphanumeric presentations as the basis for the display repertoire which is utilized.

This is indeed unfortunate on several counts:

- A <u>one measurement</u> - <u>one indication</u> approach forces the operator to use sequential observations and data processing in unfamiliar situations. This is a very demanding activity - especially with respect to loading of short term memory and usually results in operators selecting a relatively restricted set of indicators as signs of total system state. This can be risky.

- The computer offers enormous (and virtually untapped) capabilities for information transformation and display which can be tailored to the needs of any situation.

- There exists a knowledge base (although it certainly is incomplete) on human functioning which can support to a certain degree the design of interface systems and thus reduce eventual mismatches between system demands and human capabilities and limitations (fx, in stressed situations).

This paper will consider these points by first discussing the spectrum of tasks encountered on the control room - especially in terms of the operator's identification of the situation underlying the task. This will then be further discussed in terms of a three-level representation of human behaviour which, in turn, will be used to highlight the important concept of discrimination and its significance - particularly for display design.

TASK SPECTRUM

Tasks depend directly on the operator's identification of the situation and, for our purposes, can fall into one of four main categories:

- Routine.
- Familiar.
- Preplanned.
- New.

In a well-balanced plant design, one would expect the frequency of these categories of tasks to vary somewhat as shown in fig. 1. Task frequency and risk potential are inversely related so that, fx, new, unforeseen (and potentially risky) situations would hopefully belong to the relatively rare variety.

With regard to identification of task, it is important to note that this occurs in different ways for the various categories as the result of the operator's experience in the control room as well as his personal preferences and criteria for performance.

Routine task identification is characterized by its subconscious nature and amounts to a kind of automatic recognition - such as, fx, when the kettle whistles at breakfast time, make tea. Familiar situations lead to more of a conscious identification based on a limited set of descriptors of the world and result in a recognition - either of the name of the situation, the associated event or the task to be carried out. When the kettle whistles in the evening and there are guests, make coffee.

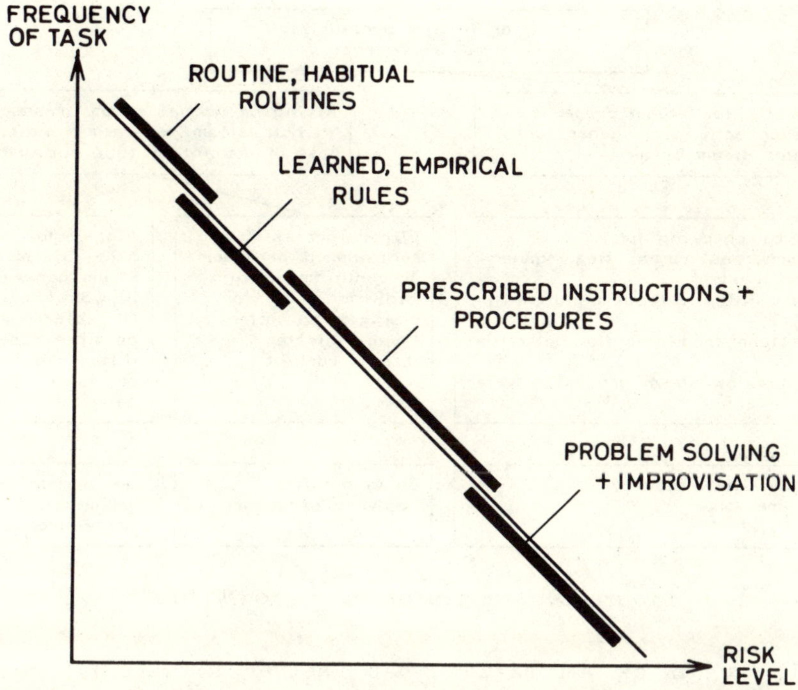

Figure 1. Task Spectrum

Pre-planned situations are generally more complex and are often subject to different kinds of prerequisites, conditions, etc. They are also usually pre-planned by some one other than the operator. An identification can require the utilisation of decision tables of some kind to link system descriptor state (sometimes perceived as patterns) with the applicable rules and procedures which have to be followed. Fig. 2 is an illustration of such a high level "decision table" or, in this case, flow chart.

New situations fall into the category as far as the operator is concerned where the state of the system as reflected in the available set of descriptors is not immediately identifiable. Thus more of a problem solving activity aimed at removing this doubt is required and consists of a search through the system (see Rasmussen 1980) to narrow down the problem area to a satisfactory minimum in accordance with the situational requirements. This will involve using information from the system to "check that -" and "see whether -" in accordance with the particular search strategy being used and the associated reference model of the system.

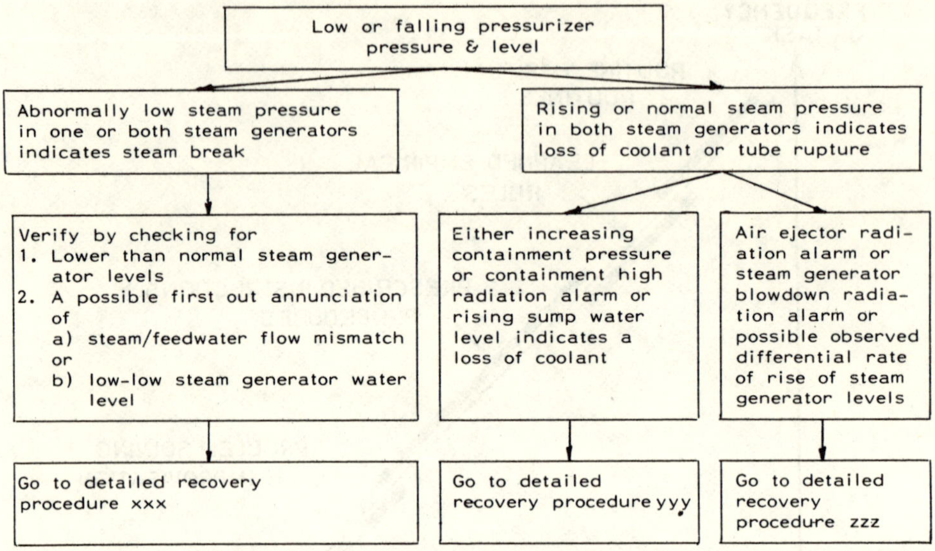

Figure 2. Example of "Decision Table"

It is to be emphasized that whether situations, fx, are considered to result in "routine" or "familiar" tasks is quite operator dependent. It is also clear that these situations can be "normal", "expected", "abnormal", "unexpected". Thus, although the term diagnosis is usually linked with the response to non-normal states, it seems evident that the operator in reality is constantly engaged in a diagnostic activity when interacting with the plant.

A SUITABLE MODEL

Operator responses to the spectrum of situations described previously and illustrated on Fig. 1 correspond very well with a useful three-level model of human behaviour (see also Goodstein and Rasmussen 1980 and Rasmussen 1980) the main components of which can be identified and located by means of the overlay shown on Fig. 3. Thus routine and familiar tasks will be executed by skill-based behaviour. With some overlap, other familiar and preplanned tasks will fall into the category of rule-based behaviour while the response to new situations requiring problem solving and improvisation will call for knowledge-based behaviour.

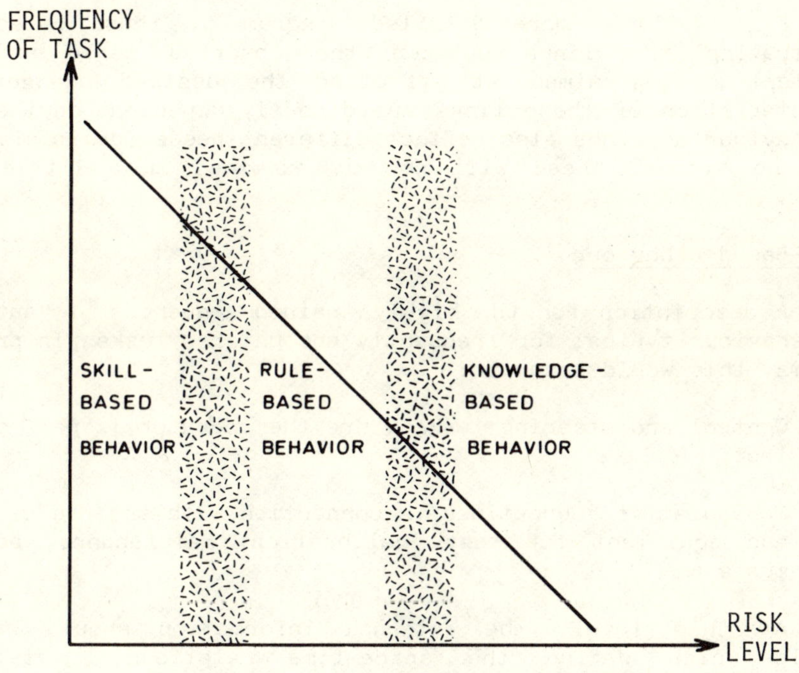

Figure 3. Overlay with Behavioral Categories

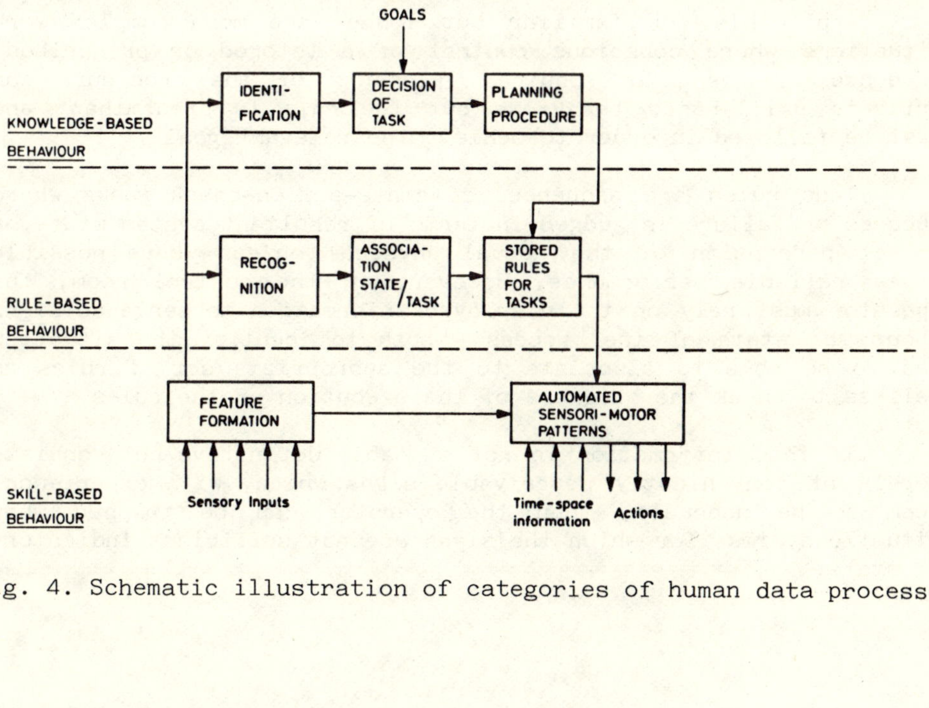

Fig. 4. Schematic illustration of categories of human data processes.

Fig. 4 is a more detailed diagram of these categories illustrating the links between the observed world and the resultant action aimed at effecting the desired change. The characteristics of these links are directly dependent on the mode of behaviour and thus also reflect different needs for information from the system. These will now be examined in a little more detail.

Skill-Based Behaviour

A description for the highly trained sequences of "automated" behaviour typical for frequently encountered tasks. In process systems, this would occur in:

- Control and steering tasks where the operator is part of the loop.

- Manipulative subroutines in connection with the use of tools and equipment for test, calibration, maintenance, adjustments.

At this level, the external information should act as signals which define the space-time relations, deviations, variations, margins between the controlled object - be it the thumb and forefinger in a manipulative task or a graphical object on a display screen in a tracking task - and the environment.

Rule-Based Behaviour

Applicable for familiar but longer and more complex work situations where conscious control of a (stored or prescribed) sequence of tasks is required. Elements of the procedure can activate skill-based behaviour but the rule is predominant and must be followed in order to achieve the relevant goal.

Thus rules are sequences of state-action-check tasks where success or failure is judged in terms of resultant system state. A direct perception of the actual physical object where possible gives reliable performance. However, in the control room, the operator must rely on the displayed information to serve as signs about the state of the process - both to identify the situation and to be able to associate to the appropriate set of rules as well as to check the progress of the execution of the rules.

If this information is not suitably definitive but consists merely of conveniently perceivable signs which, with experience, seem to be adequate, then the operator can be trapped when situations arise for which the signs are not sufficient indicators of state.

Knowledge-Based Behaviour

Necessary where skills or rules are either unavailable or inadequate so that conscious problem solving and planning are called for in order to meet the demands of the unfamiliar situation which has arisen.

In this mode, information needs to be treated as symbols which can be directly utilized and manipulated within the particular system model and structure which form the reference frame for thinking about the system. Proper symbols will avoid the need for resource-demanding mental transformations. Examples from everyday include playing cards with their suits and numbers, chessmen, algebraic equations. Finding suitable process symbols is a more complex problem since representations of physical variables and relations at various levels are required. This is discussed later.

DISCRIMINATION

A concept associated with the foregoing model formulation is related to the importance in the initial phases of the operator's response to a change in the process of "alerting" him to the full significance of the change so as to suppress normal tendencies to make superficial analyses, premature judgements and take hasty and inappropriate actions. This has great significance for information presentation. It also makes clear the inadequacy of the concept of detection as a viable stand-alone function and the need for its replacement or, more properly, inclusion in a more suitable three-modal discrimination function which is compatible with the behavioural categories described previously. In its simplest formulation, discrimination occurs optimally when the displays present information by means of the necessary and sufficient set of signals, signs and symbols so that:

- Skill-based behaviour is supported - but only "until ...".
- Rule-based behaviour is supported but potential traps are warned against.
- Knowledge-based behaviour is always supported as the "last resort".

Thus discrimination incorporates detection, an initial identification and an indication of the applicability of the available behavioural modes to cope with the immediate situation.

HOW NOT TO IMPLEMENT A DISCRIMINATION FUNCTION

In process plant today, changes in operating conditions are, after the designer's intention, made apparent to the operator by the instrument and control system's alarming and annunciating function. In practise, one hopes also for (and/or optimistically relies on) the operator himself often being able to "feel" an imminent change in advance of any alarm by means of his continuous interfacing with the process through the control room displays as well as his knowledge of the activities of the rest of the staff. Modern control rooms are becoming very isolated and somewhat sterile installations and thus the implementation of suitable means for information presentation is especially important if this "feel" generation is to be supported.

Conventional alarm systems (see, fx, Andow and Lees 1974), with associated annunciation are based on (very) large sets of one measurement - one indicator elements which, in the control room, are built up in the form of arrays of fascia with identifying text about the source of the alarm. The operator's attention is usually attracted by a combination of audible and blinking signals. Identification on the basis of geographic location is possible as long as the number of alarms is restricted and the operator is not busy with other duties. However, complex scanning and serial processing operations are required when more than one alarm occurs - as is usually the case.

It is safe to say that current alarm systems are quite "contaminated" because of the non-use of any method which reflects a systematic modelling of the plant by the designer as the basis for alarm selection. Instead, other less objective criteria seem to apply so that, fx, in modern nuclear plants, there exist literally thousands of "alarms". In response to a recognition that this indeed is intolerable, considerable efforts are being devoted to alarm filtering, conditioning and analysis while, however, retaining the basic approach of attempting to capture and display in essentially raw form the behaviour of thousands of bits of information reflecting system state at many levels of importance and urgency. It seems clear to us that the discriminating capabilities of these types of alarm systems are very limited. As a practical illustration, consider this report from the TMI control room.

> Mr. FREDERICK. The alarms—this is a big problem. There is only one audible alarm in the control room for the 1,600 alarm windows that we have, in other words, the ones that are displayed on the front of the console along with the ones on the reverse panel. So that during the emergency, I made a point of announcing that I did not want anybody to acknowledge the alarm, that is, push the acknowledgment to silence the alarm, because that would make all the windows stop flashing, and

I wanted to read them all to see what was happening. As we began to run out of ideas, I wanted to review all of the alarms that we received to see if anything was happening that we could not see.

So the alarms that came in on the drain tank were not displayed as being different from any other alarms.

Dr. MYERS. So if you had turned off the audible alarm, that meant that would have turned off all of the flashing?

Mr. FAUST. They would be in and you would not be able to determine which ones were there first and which ones were normally in.

Dr. MYERS. OK.

Mr. FREDERICK. There are several steps in the alarm process. As the alarm comes in, it sounds an alarm and a flashing light. As long as the alarm stays in—and you push the button, and the light will go solid. If in the meantime the alarming condition clears itself or goes away and you push the button, the light will go out and you will not be able to tell that it ever came in.

If you have three or four alarms at the same time, only one may stay lit out of the three or four. We had probably 100 or 200 alarms flashing within the first few minutes.

Mr. CHENEY. But only one audible?

Mr. FREDERICK. Yes.

Mr. VENTO. If they correct themselves, the alarm turns off and the light quits flashing?

Mr. FREDERICK. Yes.

Mr. VENTO. Well, if you were not pushing the buttons and you just told us you wanted to have a full history of this before you—if you do not push the button and the problem clears——

Mr. FREDERICK. Then there is a difference in the way it flashes. It would flash brightly if they are coming in, and they would flash somewhat dimmer.

Mr. VENTO. But you wanted a chronology of that?

Mr. FREDERICK. I would not be able to establish the chronology. I just wanted to see what systems were affected by the transient and if we could see something. There are some alarms you expect to get. If you read over them, you just discount them as being normal.

But there may be a few that come in that you had not expected, and those are the ones I was looking for.

From Oversight Hearings 1979

Thus, detection is often muddled by difficulties in "seeing the forest for the trees", identification requires complex data processing if at all possible and therefore the category of behavioural response will either be controlled by an arbitrary subset of alarms or by other completely separate and individual elements of the display interface (indicators, recorders, lamps ...).

A START TOWARDS DISCRIMINATION

As an eventual substitute for or, more realistically speaking, as a supplement to present alarm systems, the possibilities for aided discrimination from VDU-presentations backed up by the necessary computer processing need to be considered. Since pre-alarm sensing by the operator amounts to a kind of "screening" of the plant to find deviations from normal[x)], an immediate possibility is to display updated sets of information as patterns which are amenable to reliable perception by the operator as a prerequisite to activating a conscious response.

Fig. 5 illustrates an approach which can be applied at almost any level within the system hierarchy. It is based on deviations from a circle which, in its undisturbed state, is the epitomy of stability and normality. Its use in displays in our sense of the word dates back to at least 1967 and Coekin 1969 has described some limited evaluative experience with a non-computerized version which, on the whole, indicated quite good performance in detecting and characterizing deviations from normal. The examples given in the figures are suggestive of an application in a nuclear reactor where twelve critical parameters which reflect safety-related aspects of total system state have been selected for inclusion in the integrated display. Each parameter's magnitude is plotted along an assigned polar axis and it is important to point out that the transformations and scaling for each parameter (and thus the weight assigned to changes) can be completely different - provided, of course, that their locus forms a circle when they are normal. In fig. 5, primary system parameters are grouped on the left and secondary parameters on the right. This gives rise to the distinctive patterns illustrated for the postulated abnormalities. Since Coekin could note consistent detection of deviations as low as 10% of the displayed full-scale values, it is reasonable to suppose that a wide range of variations of these situations would also be discernable.

SUPPORTING DIAGNOSTIC SEARCH

A good discriminative function will result in the operator reverting to knowledge-based behaviour when he realizes that his skills and/or available rules are inadequate or unavailable. A suitable display support system must thus be able to assist him in his diagnostic efforts at this level with appropriate information in symbolic form.

The operator's search through the system in response to a deviation from normal is aimed at narrowing down the problem to

---

[x)] "Normal" is often relatively undefined - compared with the degree of refinement (and effort) which goes into defining abnormalities. It seems reasonable to expect that a combination of design intentions coloured with a good portion of operational experience would permit a usable clarification to be made.

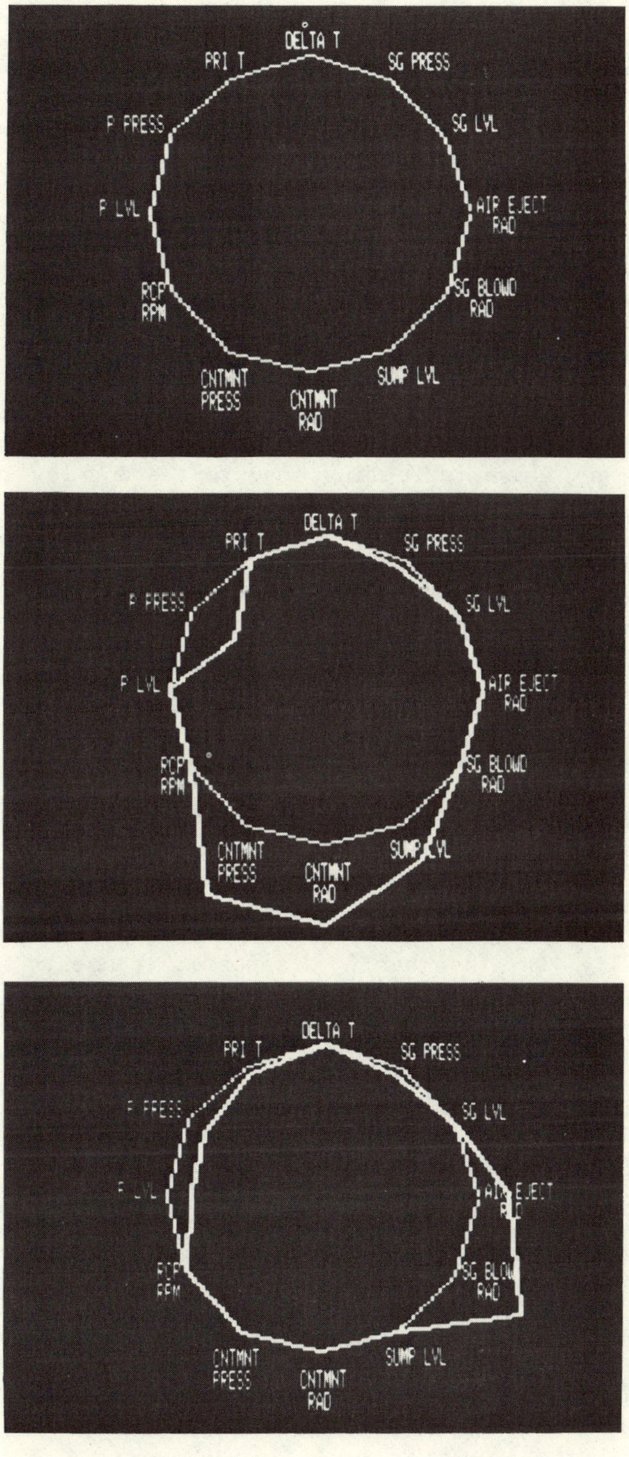

NORMAL

LOSS-OF-COOLING ACCIDENT

PRIMARY TO SECONDARY LEAK

Figure 5

the point where an identification having an acceptable solution can be found. As was stated in Rasmussen 1980, this "point" will depend strongly on the particular situation. In general, however, searching can be characterized by two complementary concepts which have significance for display design - "field of attention" and "level of abstraction". The first corresponds to "windowing and zooming" and calls for display support which follows the operator's needs for higher amounts of detail about more restricted portions of the system as attention becomes more concentrated. The second is a more subtle concept which reflects human ability and tendencies to speculate consciously about the world (process plant) in different ways depending on needs and abilities. Thus at the earlier stages of a diagnosis where attention has to be paid to the entire system to evaluate the propagation effects of changes, deviations, possible counteractions, the operator will/should use abstraction levels which involve fundamental mass flows (affecting inventories and levels) and energy flows (affecting power control and distribution). When attention at a later point is directed to a particular sub-system, the level of abstraction will probably move "down" to considerations of physical-function relating to parts, components and their interaction and/or physical variables and their relations. Thus a "high" abstraction level usually demands a large field of attention with limited detail and vice versa. This creates the need for a large set of displays to enable different portions of the system to be presented in support of thinking at the various levels of abstraction.

One of the points of this paper is that most attention to date in designing VDU displays for search support has been given to lower abstraction level aids in the form of mimic diagrams of plant (sub)systems which in reality are structured around the more physical aspects of the system such as components. However, there is a need, especially in the early diagnostic phases, for depicting system structure and behaviour at a higher device-independent level of abstraction and for supporting operator thinking at this level by incorporating suitable transformations of the basic process data. Thus, at these higher levels, it is our feeling that operator attention should be centered around fundamental process behaviour relating to the mass and energy in the system and their status with respect to normal. This requires that pressures, temperature, and valve positions which on typical mimic diagrams are visualized as pressures, temperatures and valve positions have to be converted to flows and inventories and suitably represented within a suitable system flow structure.

The remainder of this section will be devoted to some examples of a possible implementation which draws on some of the work described by Lind 1980 on the use of flow models for automated plant diagnosis.

Thus we suggest creating displays using a limited set of symbols (see fig. 6) similar to those described by Lind in a high level representation of the system flow structure to which state information can be appended in various ways - either as the result of computer analyses or in response to operator requests in the course of his search process. Figs. 7 and 8 are examples of some preliminary ideas. The accompanying text should be reasonably self-explanatory. The first is a relatively simple representation at the overall system level where computer-based detection of mass and energy balance disturbances is reflected directly on the structure - fx in the steam generator, - as the first indication to the operator who is then pointed in the relevant direction. The second example illustrates the presentation of flow information - in this case mass flow through a subsystem further down in the hierarchy as well as energy distribution in the main process. These are preliminary illustrations of the ultimate possibilities for display support in connection with:

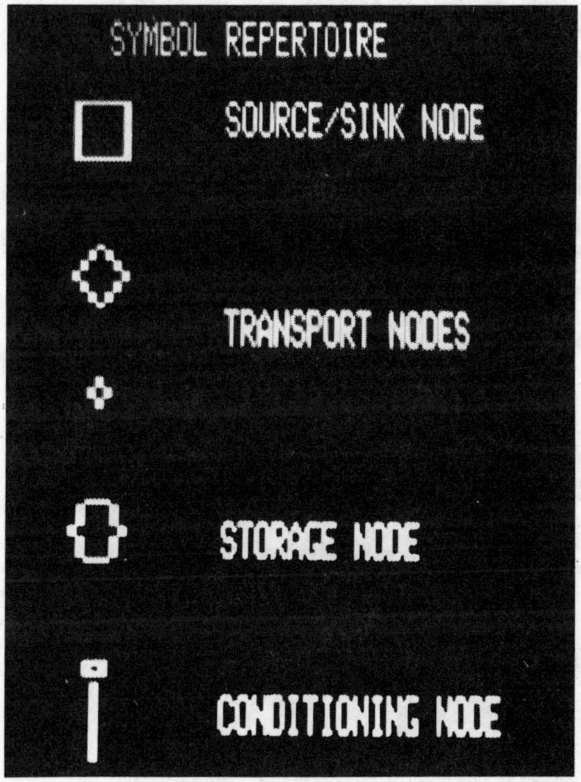

Figure 6

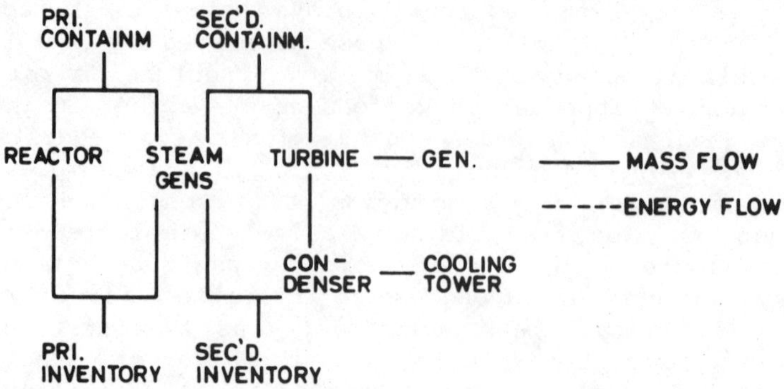

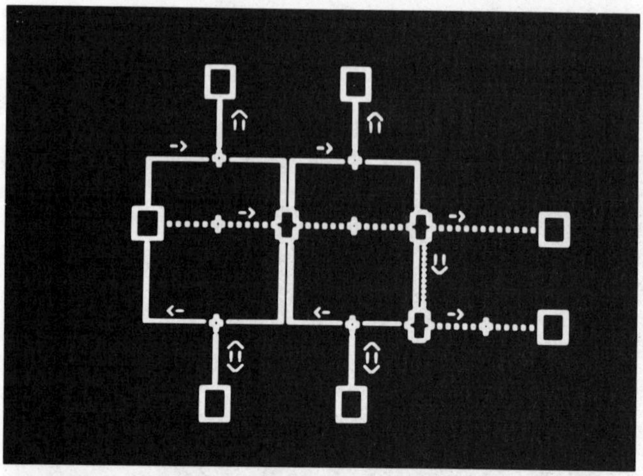

FLOW STRUCTURE

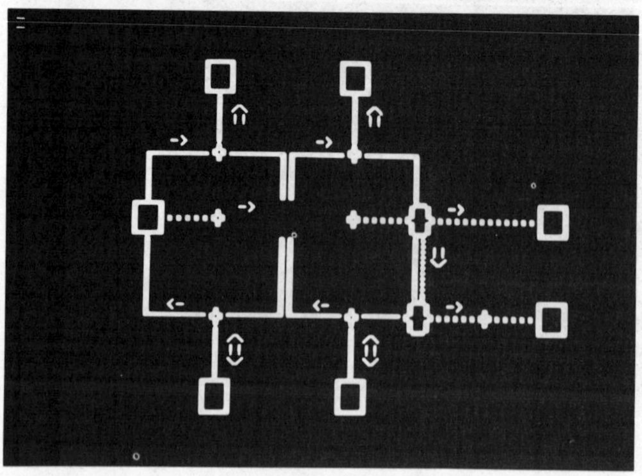

WITH DISTURBANCE
IN STEAM GENERATOR
ENERGY BALANCE
(COLOUR CHANGE
+ BLINK)

Figure 7

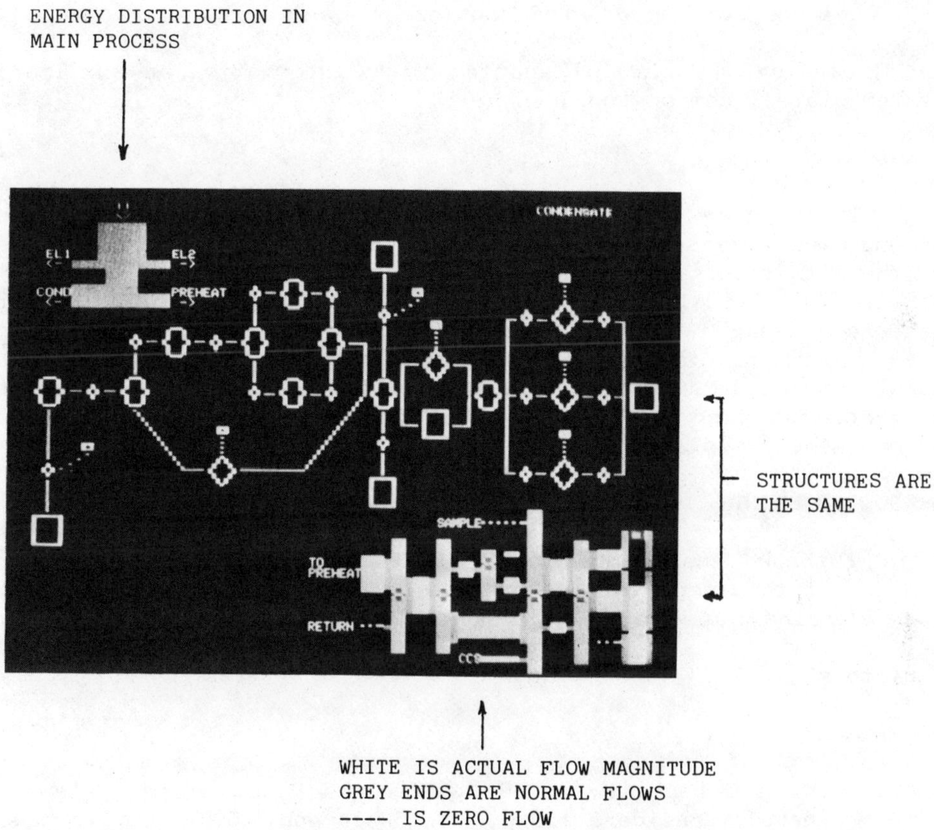

Figure 8. Condensate system flow structure

- Power control
  - . Branchings, feedback of energy
  - . Levels of energy "accumulation" as reflected in state of critical variables
  - . Means for control and routing

- Inventory control
  - . Supply and loss
  - . Levels of accumulation
  - . Means for control and routing

and including the host of supplementary information on auxiliary system states, common supplies, etc.

CONCLUSION

This paper has treated display-related aspects of the man-machine interaction in the control room - first on the basis of the situation/task spectrum encountered and then from the point of view of a model of the operator the significance of which for interface design was pointed out. This led to the introduction of the concept of discrimination and a proposal for its use as a substitute for detection which was considered to be too limited a concept. The importance of adequate computer support was pointed at and some examples given.

ACKNOWLEDGEMENTS

I thank Jens Rasmussen for useful comments on the form and content of this paper. This work was carried on in connection with the Scandinavian project on Control Room Design and Operator Reliability which is supported in part by the Nordic Council of Ministers.

REFERENCES

Andow, R.K. and Lees, F.P., 1974, "Process Plant Alarm Systems - General Considerations", in Buschmann, C.H. (ed.), Loss Prevention and Safety Promotion in the Process Industries, Amsterdam (Elsevier).

Coekin, J.A., 1969, "A Versatile Presentation of Parameters for Rapid Recognition of Total State", International Symposium on Man-Machine Systems, Sept. 8-12, Cambridge, IEEE Conf. Record 69 C58-MMS.

Goodstein, L.P. and Rasmussen, J., 1980, "Man-Machine System Design Criteria in Computerized Control Rooms", in "ASSOPO 80", IFIP/IFAC Symposium in Trondheim, June 16-18, to be published by North Holland.

Lind, M., 1980, "The Use of Flow Models for Automated Plant Diagnosis" (this volume).

Oversight Hearings, 1979, "Accident at the Three Mile Island Nuclear Power Plant", Washington DC, May 9, 10, 11, 15, Document Serial No. 96-8 Part I, US Government Printing Office.

Rasmussen, J., 1980, "Models of Mental Strategies in Process Plant Diagnosis" (this volume).

Rasmussen, J., 1980, "Some Trends in Man-Machine Interface Design for Industrial Process Plants, in "ASSOPO 80", IFIP/IFAC Symposium in Trondheim, June 16-18, to be published by North Holland, also Risø-M-2228.

DISTURBANCE ANALYSIS SYSTEMS

W. Bastl and L. Felkel

Gesellschaft für Reaktorsicherheit (GRS) mbh
8046 Garching, Germany F.R.

GENERAL

The safety requirements of nuclear power plants are completely different from those of conventional power plants. Safety regulations have been established to account for these requirements. This has led to an increase of the amount of instrumentation installed in nuclear power plants, which as a further consequence enlarged the amount of information available for the operators in the control room to such an extent that in case of disturbances the operators may be overloaded, a problem of man-machine-communication.

First attempts to improve the man-machine communication with the aid of computers were made in the early sixties in the United Kingdom but on account of equipment problems the project has been abandoned. However, the idea of alarm suppression, alarm filtering, alarm analysis still persists and had a strong impact on the system which will be described in the subsequent chapters, (Welbourne, 1968; Jervis, 1972; Potts & Tabernacle, 1966; Filby & Glew, 1971; Patterson, 1968; Kay & Heywood, 1966).

Another problem is that the operator is not really informed about the effects of some events in the plant itself from the annunciators in the control room.

Notice:
This study was prepared by Gesellschaft für Reaktorsicherheit sponsored in part by the Commission of the European Communities Joint Research Center, establishment of ISPRA.

Thus the operator loses a feeling (Holmgren, 1980) for the process itself, a situation which is completely different for example from the awareness of an aircraft pilot about the status of his aircraft. Whenever the pilot changes some variables he will get the effect almost immediately. The situation in a nuclear power plant is quite different. There may be significant time delays between an action and its effect. A typical example here is the changing of the boron concentration. This causes the necessity for predicting the behaviour of the process. The only way to do the prediction is by modelling the process and simulating it thereby. One of the problems, however, is that simulation usually requires extensive computation time if the model is sufficiently detailed which unfortunately prevents the desired data to be available in due time.

Nowadays computers have become fast and computer memory has become cheap. However, they are still not fast enough to simulate a complex process like that of a nuclear power plant faster than real-time. In order to speed up simulation there is today only one solution: to simplify the process model to such an extent that the simulation does no longer require that much time.

The complexity of today's nuclear processes makes it necessary to supervise the process thoroughly and to detect disturbances at their very beginning. Only recently the Three Mile Island - 2 accident showed how a very simple event (loss of a main condensate pump) could develop into a severe accident. It should also be noted that this accident was based on insufficient information to the operators and that there was plenty of time to take appropriate corrective action (TMI-2 Lessons, 1979).

OBJECTIVES

The need to detect, for example, sequences of alarms was already discovered in the early sixties by Central Electricity Generating Board (CEGB) in the United Kingdom. They developed an alarm analysis system for Oldbury and Wylfa power stations. Owing to shortcomings in the computer equipment and inflexibility in constructing the alarm trees, the approach fall below expectations (Filby & Glew, 1971).

The principal idea, however, has been retained in the present day disturbance analysis systems. The main objectives and requirements are listed below (Long, 1980):

plant interface:	The DAS shall obtain plant status information directly from the plant instrumentation and control systems or through the process computer.
operator interface:	The method for displaying the results from the DAS analysis shall be integrated into the control room design.
timely analysis:	The DAS shall operate in real time and present results within the time frame of the disturbance so that the operator can take corrective action.
information enhancement:	The DAS shall enhance the quality and content of the information being displayed to the operator and reduce the number of secondary alarms depending on the current mode of plant operation.
disturbance analysis:	The DAS shall be capable of analysing disturbances based on a preestablished plant model stored in a data base. The DAS should be able to determine the nature, cause, consequence, and possible corrective actions of disturbances.
scrutability:	The DAS shall be able to tell the operator on demand how it arrived at any specific conclusion.
plant model:	The DAS shall comprise software tools for developing and modifying the plant model used by the DAS.

Disturbance analysis systems usually have the structure outlined in Fig. 6. The plant data gathered from the instrumentation are preprocessed so as to form the plant data base. Preprocessing comprises limit checking, data validation, filtering and deriving variables from other plant signals. Dedicated preprocessor modules for noise analysis and loose-parts monitoring also contribute to the plant data base.

To detect disturbances, so-called disturbance models (represented by cause-consequence diagrams (CCD)) are used. These are stored in a background data base and are readily accessible by the disturbance analysis routines. The models contain the anticipated flow of events during disturbances. The disturbance models are overlaid by the actual plant data available from the plant data base. After association of pre-defined disturbances models with actual data the models are considered activated.

The disturbance analysis routine scans the activated models and detects disturbances, if there are any, at their very beginning. After determining the status of the process, the further possible consequences will be evaluated and if possible and feasible, corrective actions suggested as well as primary causes detected. Since the disturbance analysis module is a computer program, results delivered by it have to be transformed into a readable form according to ergonomic standards. This task is performed by an operator-communication system. The communication system also provides means for retrieving and supplying information from and to the disturbance analysis system.

The models (CCD's), (Nielsen, 1974) used by the disturbance analysis routine are mainly produced by systems analysis. The information required stems from engineering judgment as well as from plant design models. There is a significant amount of experience though, gained during operation of the DAS, by which most likely new insights evolve and may result in modifications, adaptations and extensions of the existing models. The DAS data background may thus be enhanced continually.

DASs UNDER DEVELOPMENT

There are two major efforts on developing a disturbance analysis system underway. One under sponsorship of the Electric Power Research Institute (EPRI), (Frogner & Meijer, 1980), the other in a cooperative project by Gesellschaft für Reaktorsicherheit and Institutt for Atomenergi, Kraftwerk Union and the Bayernwerk utility sponsored by the German Federal Ministry for Research and Technology and a German Utility Group, (Felkel et al., 1978; Felkel et al., 1979).

Besides these major projects there are developments in Hungary by the Central Research Institute for Physics of the Hungarian Academy of Sciences, where the disturbance analysis is an integral part of a computerized reactor control system for the WWR-SM research reactor, (Bürger & Zobor, 1978).

Common to all developments is the use of one or more colour cathode ray tubes (CRT) for display of analysis results and status information to the operators.

Since the first attempt to analyse, filter and suppress alarms has been made in the U.K. and all DAS systems in a way have adopted the principal idea therein, this system is described first.

CEGB Approach

The prime objective of the system installed at Oldbury and Wylfa in the United Kingdom in 1966 was to reduce the amount of alarms in the nuclear power stations, (Welbourne, 1968; Jervis, 1972; Potts & Tabernacle, 1966; Filby & Glew, 1971; Patterson, 1968; Kay & Heywood, 1966). To achieve this, so-called alarm-trees were established. Alarm-trees are essentially a set of alarms which are assumed to be in a cause-effect relationship denoted by an edge between the two alarm nodes (Fig. 1). The nodes are connected on-line to the alarm generation hardware. In Fig. 1, for example, if both alarms A and B, where B is caused by A, are active, both alarms would be displayed on a CRT.

Provisions are made to prevent suppression of alarms which are considered significant. Since the cross in the node denotes S (Fig. 2) to be a non-inhibited alarm, it will always be displayed when it is active.

Messages can be associated with alarms indicated by a small triangle connected to an alarm node. In Fig. 2 if Q, S, U and W are active the alarms Q, U, S and W together with the message w are displayed.

In preparation of the alarm trees about 3000 alarms were considered; however, a reasonable number bearing little significance in a context of other alarms have been chosen to be detected on-line, analysed on an alarm-tree basis and then treated accordingly. The alarm scanning rate was one fourth of a second for 3000 alarms.

The alarms displayed as the result of any analysis were always in the following order:

- last active alarm
- fault messages associated with last active alarm
- non-inhibited alarms
- fault messages associated with non-inhibited alarms
- highest active alarm
- fault messages associated with highest active alarm

- deduced alarm
- fault messages associated with deduced alarms.

The alarm trees have been constructed by systems analysts; although the effort of producing the alarm trees amounts to ten man years, the benefits of the analysis system have fallen below expectations. This was due to severe restrictions in computer power and computer storage, primitive display technology and the inflexibility of changing the alarm trees after some of them have been found to be inadequate.

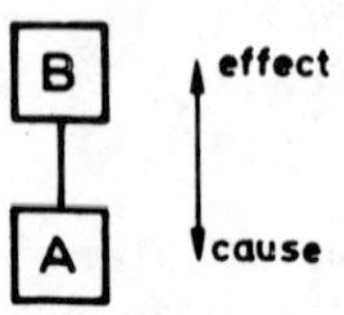

Fig. 1. Simple alarm tree

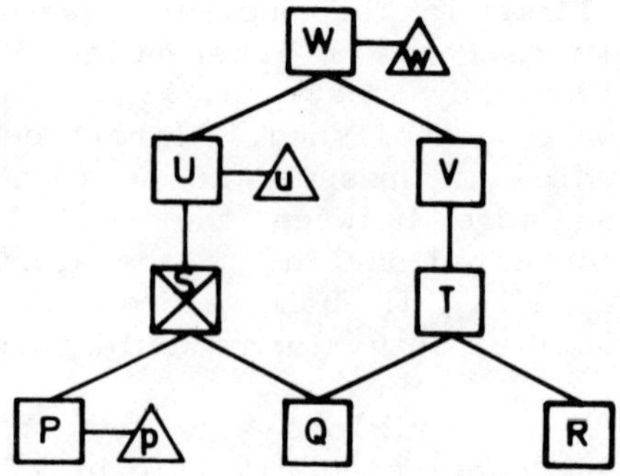

Fig. 2. More complex alarm tree

EPRI Approach

The analysis of disturbances in the EPRI-DAS, (Long, 1980; Frogner & Meijer, 1980), is subdivided into three levels. The first level of analysis is a simple table look-up to identify commonly occurring plant disturbances. It is also intended to reduce the number of disturbances subjected to the more detailed and time consuming analysis of levels two and three. This can be done for very simple disturbances and is meant to be a computerised implementation of the conventional annunciator system.

To analyse more complex disturbances the logical relationships and the sequence of events are taken into account. The underlying methodology for performing this second level of analysis are the cause-consequence trees.

On the third level analysis the DAS is supplied with dynamic models to perform quantitative calculations. However, for this feature only an appropriate interface has been provided and it is not implemented in the present version of the DAS.

The general outline of the EPRI-DAS is shown in Fig. 3. The system can be subdivided into three main parts, the preprocessor, the disturbance-analyser and the operator-interface system.

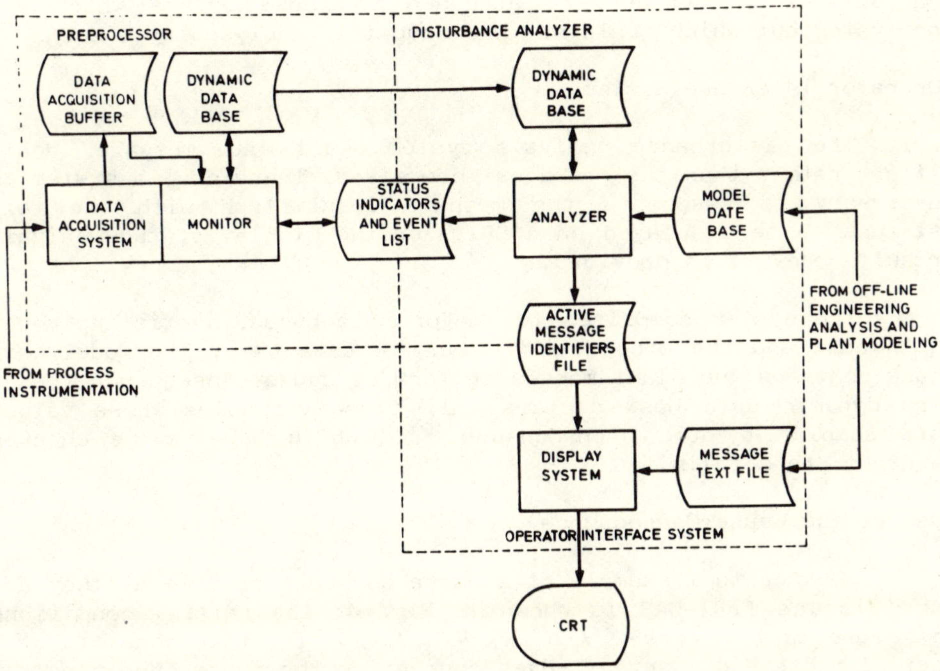

Fig. 3. Modules of the EPRI-DAS

Preprocessor

The preprocessor contains the data acquisition system and a module called the DAS monitor. The data acquisition system is the link between the process instrumentation and the DAS. Input signals are sampled every one second; however, this is adjustable. The input signals will be checked for consistency, the sensor validity will be examined, low pass filtering performed and the signals are converted to engineering units.

The DAS monitor is to activate and schedule the execution of the disturbance analysis. Typical tasks of the monitor are to update status indicators, to update derived variables and to update variable limits. In case an event occurs which has higher priority than those being analysed, the previous analysis is disrupted and the analysis for the new events started.

Disturbance Analyser

The most important and the most commonly used level seems to be the level two (Cause-consequence analysis). After completion of the multi-level analysis some clean-up is performed and so-called "second-best" messages are attached to those events which are activated but which could not be adequately analysed by the DAS.

Operator Interface System

The disturbance analysis system deals with message identifiers rather than the messages themselves. Therefore, a system is necessary to associate the message identifiers with the text strings to be displayed on a CRT. For the display of the analysis results one CRT is provided.

The system comprises two major data bases, one is the model data base and the other is the dynamic data base. The model data base contains the plant models in form of cause-consequence trees. The dynamic data base contains all those variables whose values are sampled by the preprocessor, and which have to be checked against pre-set limits.

Use of the Cause Consequence Tree

A conceptual example of a cause-consequence tree as they are used in the EPRI-DAS is shown in Fig. 4. The initial conditions here are that V 1, V 2 $\neq$ high, V 4 $\neq$ low. The tree consists of a set of nodes and a set of edges connecting the nodes. Nodes on the same level may be combined by logical units, these may be AND, OR and NOT gates. Each node denotes an event whose value for the most cases is derived from process signals. Nodes as well as edges may be associated with messages denoted by the message identifier in a triangle. Edges may be labeled with time delays denoted by squares containing the amount of the delay.

Fig. 5 shows the status of a hypothetical disturbance at specific time points. The third column shows which messages are displayed to the operators at each time point. Active and potentially active messages are discriminated by different colours.

The EPRI-DAS comprises 25 types of disturbances which were modeled in detail for the feedwater control system and the component cooling water system. Nine of these were subjected to extensive testing at the C-E training simulator.

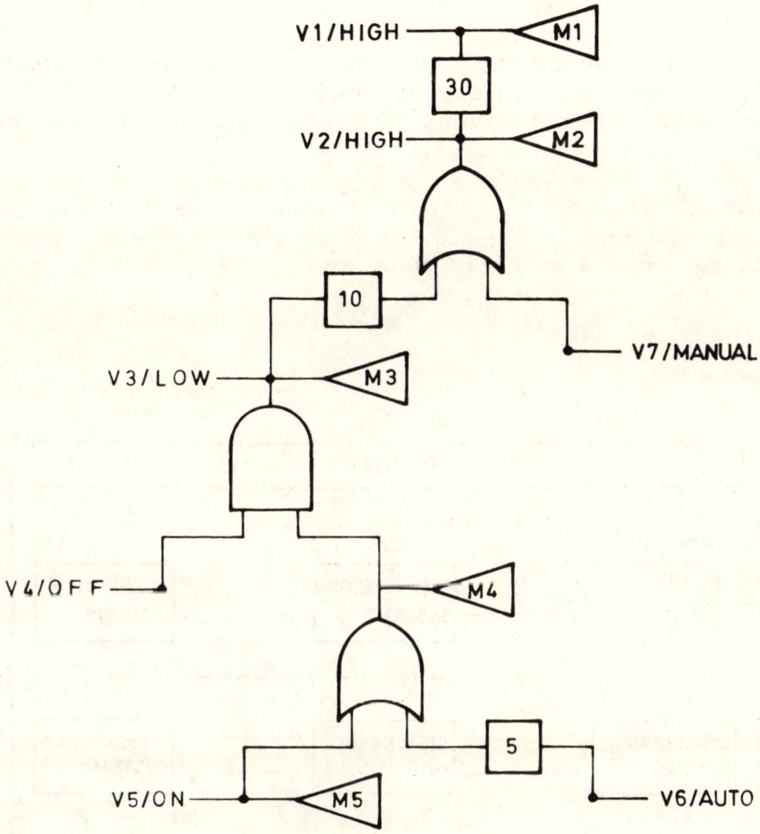

Fig. 4. Cause-Consequence Tree.

TIME IN SEC	EVENT	RESULT OF THE CAUSE CONSEQUENCE ANALYSIS
0	V4 = OFF AND V5 = ON	M4 AND M5 ACTIVE M3 POTENTIALLY ACTIVE
1	V3 = LOW	M3, M4, AND M5 ACTIVE M2 POTENTIALLY ACTIVE
11	V2 = HIGH	M2, ..., M5 ACTIVE M1 POTENTIALLY ACTIVE
41	V1 = HIGH	M1, ..., M5 ACTIVE

Fig. 5. States of a disturbance.

GRS-Approach

The objective of this approach was to extend the DAS version (Grumbach & Hoermann, 1976), so that it would be applicable to a large nuclear power station. The Grafenrheinfeld nuclear power plant was chosen as test site.

The STAR-system [x)], (Øwre & Felkel, 1978; Felkel et al., 1979), is outlined in Fig. 6. It consists of several modules each of which performs a specific function.

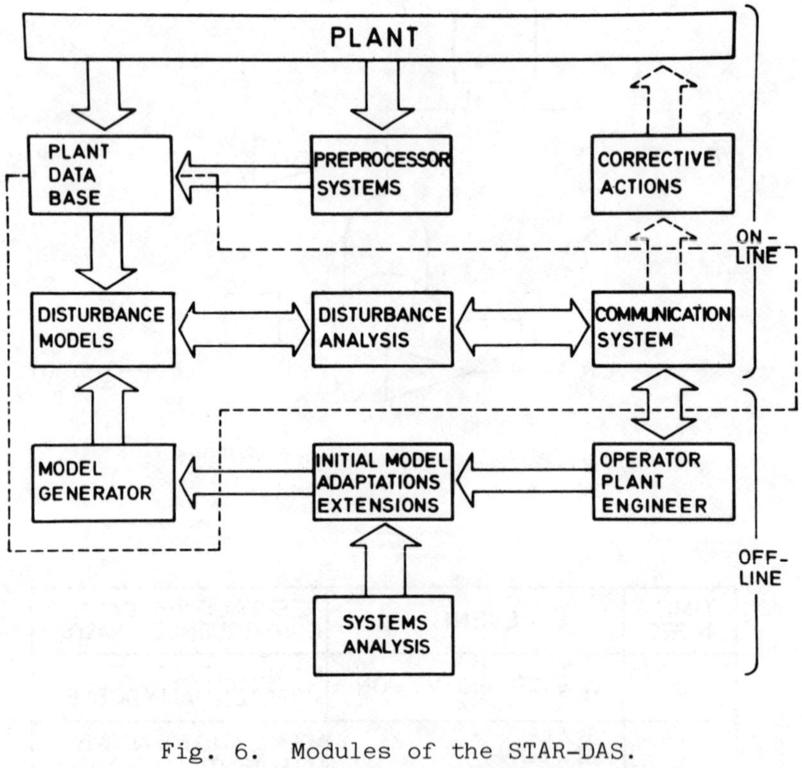

Fig. 6. Modules of the STAR-DAS.

[x)] abbrevation from Störungsanalyserechner (disturbance analysis computer).

An already existing plant data base was used to access plant data located on one of the plant supervisory computers. Therefore analog and binary values are updated only between five seconds and sixty seconds according to the appropriate transient behaviour.

The disturbance models that are used for disturbance analysis are cause-consequence diagrams. These are similar to alarm trees and cause-consequence trees but are more elaborate and usually contain more information than the latter ones. During one update cycle which can be adjusted between five seconds and 60 seconds the on-line data in the plant data base are superimposed on the disturbance model.

The next step is the analysis of this "snap-shot". To optimise the speed of analysis the disturbance analysis module produces only text identifiers to be transmitted to the operator-communication system.

The operator-communication system consists of two colour CRT's and two keyboards comprising a function push-button part and an alphanumeric keyboard. This hardware equipment is controlled by the communication program which for example handles the paging in case there is more information available for the operator than fits on one screen image, and associates test identifiers with actual text. Also the information will be structured by the communication system in such a way as to satisfy ergonomic constraints.

According to the results of the disturbance analysis the operators may be able to take corrective actions much earlier than in conventional systems to prevent or mitigate disturbances.

The STAR-DAS comprises an efficient tool, the model generator MOGEN, for the generation, adaptation and extension of the cause-consequence diagrams. This is one of the most important features in the STAR-system since from prior experience it is reported that the modification of the disturbance model data base is a time-consuming and error-prone task.

Additionally, these tools allow the operators, plant engineers and systems analysts to express their knowledge about the behaviour of the process in their own terms rather than in terms of computer structures.

The core of the STAR-system is the disturbance analysis module. This module is executed each update cycle in the sequence outlined in Fig. 7. The data preprocessing module pre-selects the data used for disturbance analysis, it checks whether or not limits have been violated and it sorts out those data which have been invalidated for sensor failures etc. Finally, the preprocessing module also stores all the data collected for disturbance analysis on a magnetic tape to allow replay and off-line evaluation of the disturbance analysis results.

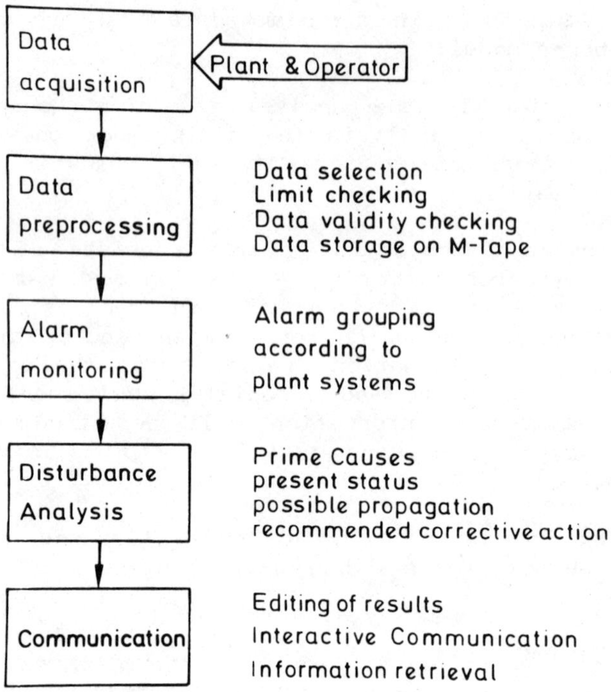

Fig. 7. Tasks of the STAR-DAS performed during one update cycle.

In the next step those events which have been classified as alarms are grouped together according to plant subsystems. For the display of alarms a dedicated screen exists (one of the two screens utilized) to impose a hierarchical order on the information available to the operators to aid them to retrieve the relevant information.

After alarm monitoring disturbance analysis is invoked. This module is to detect disturbances, determine the present status and the prime causes of the disturbance and evaluate the possible propagations of the disturbance. Where it is unambigously possible corrective actions will be recommended to the operators. A detailed example of the analysis of a disturbance is given in the sequel.

Sample Cause-Consequence Diagram

Fig. 8 is a schematic diagram of the feedwater and condensate circuit of the Grafenrheinfeld nuclear power plant. On the right hand side the three main condensate pumps from where the condensed water is transported via the low pressure feedheater into the feedheater tank. From there the main feedwater pumps push the feedwater via the high pressure feedheaters to the steam generators. Fig. 9 shows a small part of a cause-consequence diagram for a disturbance concerning the main condensate pumps.

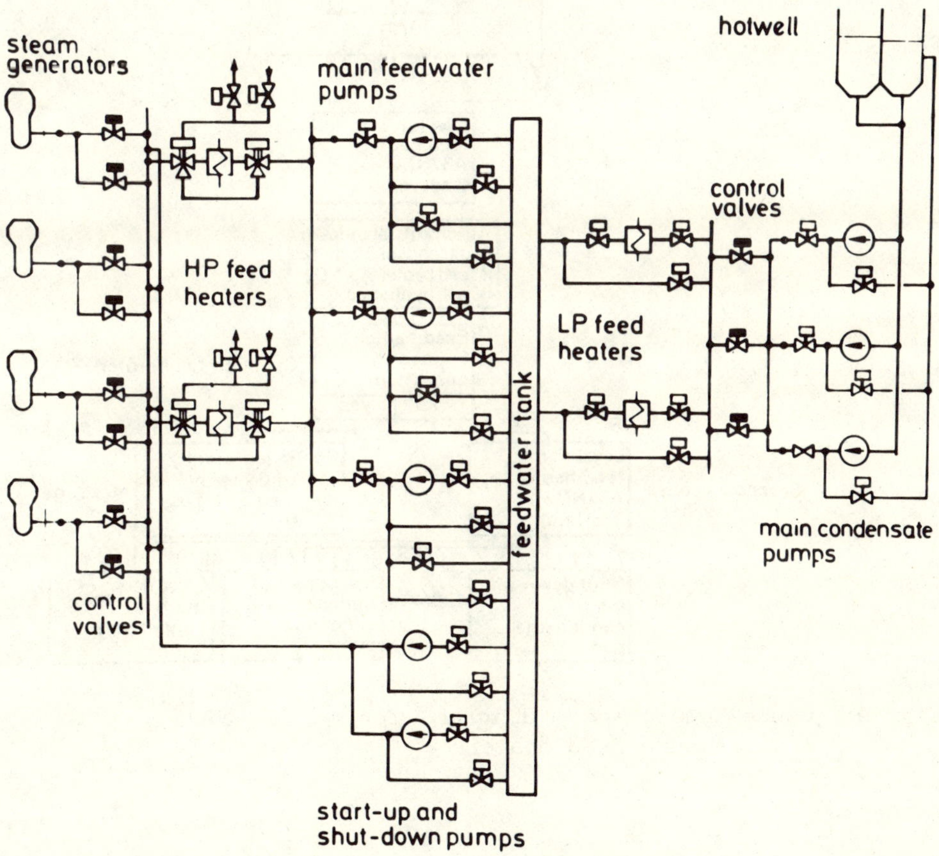

Fig. 8. Feedwater circuit of the Grafenrheinfeld 1300 MWe PWR.

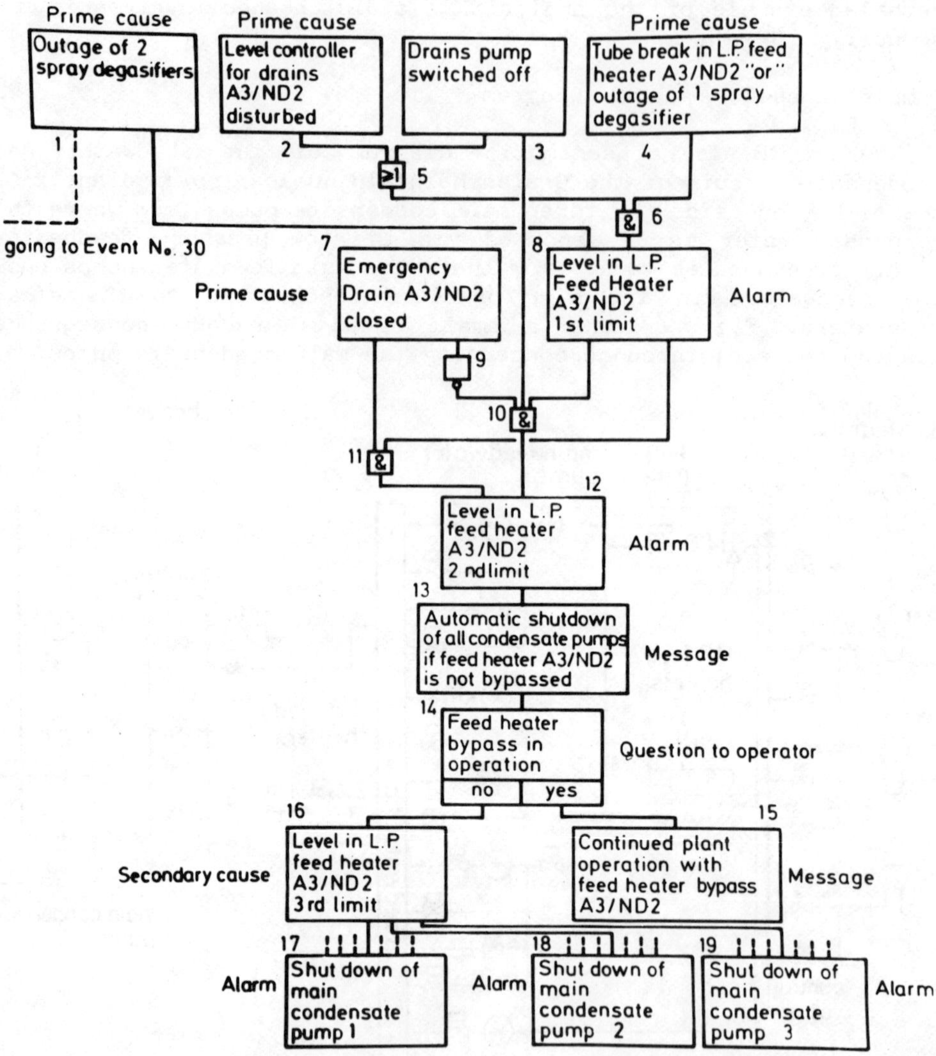

Fig. 9. Cause-Consequence diagram for a disturbance in the low-pressure feed heater.

The basic assumption is that the drains pump for the low pressure feedwater is either switched-off or in repair (event 3). There may be a tube break in the low pressure feedheater or a spray degasifier in the feedwater tank is lost. Subsequently, there will be an increase in the water level of the feedheater. This increase will continue until the first limit is reached (event 8). This is the first observable event from which the disturbance analysis system will conclude that the primary cause can only be the tube break in the low pressure feedheater (event 4). At this point appropriate corrective actions could already be taken. In case the corrective action is not taken, the water level will continue to rise and will exceed a second limit (event 12). The control system automatically isolates the defective low pressure feedheater after the second limit has been exceeded and the appropriate pre-heater will be by-passed. This may not be successful, however, in which case an emergency shutdown of all main condensate pumpts is imminent. The operator will be notified about this situation by appropriate messages. In the Grafenrheinfeld nuclear power plant the status of the valves for the pre-heater by-pass is not available on the process computer and there is only the operator to know whether or not the pre-heater by-pass was successful. The disturbance analysis system therefore asks the operator an appropriate question (event 14). The question is answered by means of a special keyboard in the disturbance analysis system. If the pre-heater by-pass was unsuccessful there will still be continued increase of the water level in the low pressure feedheater which after the excess of a third limit causes the automatic shutdown of the two operational main condensate pumps and prevents the stand-by condensate pump from being started. As a consequence the feedwater tank will be emptied by and by which eventually leads the shutdown of the main feedpumps and reactor scram.

The Hungarian On-line Alarm Analysis System

For the WWR-SM research reactor there exists a computer based hierarchical control system, (Bürger & Zobor, 1978), which has been supplemented with an on-line alarm analysis program. It is reported that when the reactor operates under normal conditions (i.e. all process variables are within their permitted ranges) an optimal control strategy is applied, executed in direct digital control on the basis of a fairly complex model. It is felt, however, that when certain variables exceed their specified limits alarm analysis may be feasible.

The layout of the system which is called PROCESS 24K is shown in Fig. 10.

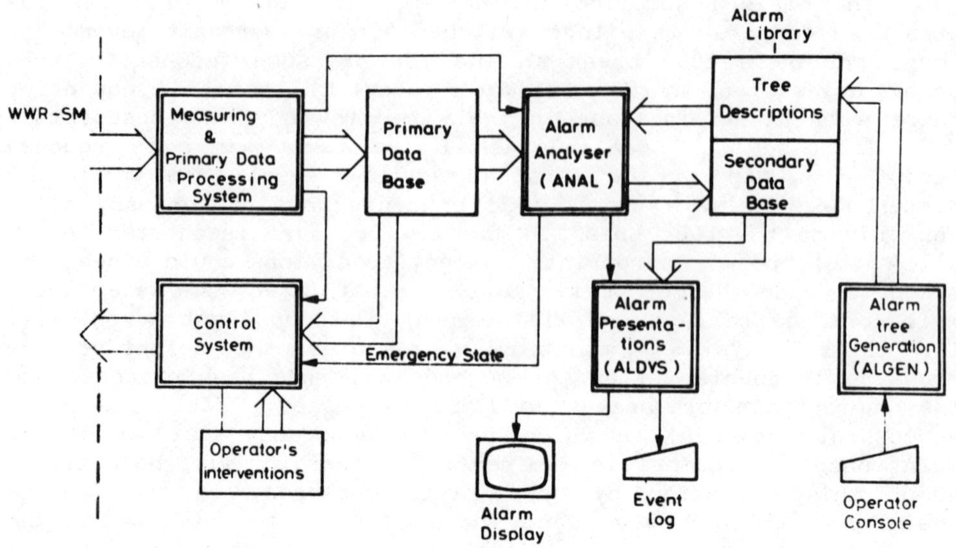

Fig. 10. Modules of the Hungarian alarm analysis system.

The process data are obtained by a measuring and primary data processing system and are then collected in a primary data base. Some of the data are directly fed on to the control system to which the operator also has access to take corrective actions. The data base as well as the data acquisition system are able to handle 128 analog inputs and about 200 digital inputs and outputs.

These data are used by the alarm analyser program which is invoked each time a limit has been violated. In all other cases the alarm analysis program is activated each second. An alarm event is defined here as the changing of a state of a digital input, or whenever an analog variable goes past its trip level or returns back within its limits. The analysis is based on the alarm-tree methodology where the tree descriptions are contained in the secondary data base. These data are associated with the on-line data in the primary data base, and the actual status of the plant is determined.

After analysis has been performed the results are displayed to the operators in a CRT by means of an alarm presentation program. The presentation is subdivided into three categories:

- messages which have to be written only on the event log
- messages to be acknowledged by the operators are written on the CRT <u>and</u> on the event log

- alarm-trees are presented on the CRT only.

For the generation and maintainance of the alarm library (the secondary data base) a utility program ALGEN (alarm generator) has been provided to facilitate the establishment of the alarm-trees without any programming knowledge necessary. This program can be run also during the operation of the whole process control system.

Fig. 11 shows a sample alarm tree as used by the Hungarian system. The alarm tree consists of several alarm nodes logically related by AND, OR and NOT functions. By relating several alarms logically, new alarms are generated which are called deduced alarms. The alarm tree is the result of detailed systems analysis and failure mode and effect analysis usually used for reliability evaluations of complex systems. The alarm-tree may also contain time-delays, but these are associated with the alarms themselves denoting the time delay to every subsequent node. To save storage space parts of the trees are used multiplely.

The alarm analysis program searches for alarms associated with other alarms in the alarm-tree. If an alarm occurs that is not associated with any other alarm it will simply be written on the event log. If there is an alarm associated with others in the alarm-tree the "activated alarm tree" will be displayed to the operator. No prediction or special search for causes and consequences is performed.

COMPARISON OF THE DASs

Keypoints of the general layout of a DAS are taken and the systems under consideration are examined for similarities and differences.

Hardware requirements

All systems described use mini-computer systems. Storage requirements do not differ very much and 256 kbytes of core storage seems to be sufficient for DAS applications. The CEGB system uses much less storage, but this is due to the time (1966) when it was first installed. According to the different methods of data acquisition, though, depending on the type of installation there may be significant differences in peripheral equipment. In Germany, for example data can be acquired either via an existing plant computer or via special buffer amplifiers from the central connection rack.

Software

The EPRI-DAS as well as the STAR-DAS have been programmed for the most part in FORTRAN. The Hungarian approach seems to use

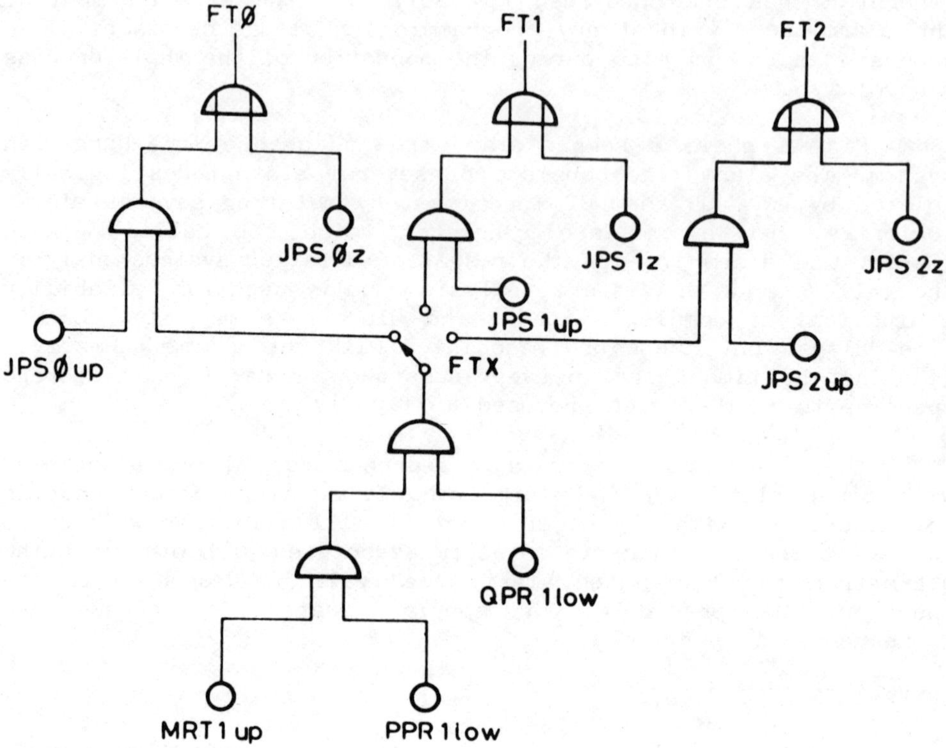

MRT1 - Water level in the reactor tank
PPR1 - Pressure in the primary coolant cct.
QPR1 - Primary water flow
JPSi - Current of the i pump
JPSiz - Current of the i pump = Ø
FTX - Short circuit at the pumps
FTi - Short circuit at the i pump

Fig. 11. Alarm tree used in the Hungarian system.

assembly language although this is not documented explicitly. There are, however, parts in all systems where assembly language has to be used to meet time constraints in data acquisition or where special hardware (keyboards etc.) is connected. By the time the CEGB system was designed no high level languages were available which was a significant factor contributing to the discussed shortcomings since it was almost impossible to alter the data bases and programs after they have been found inadequate. The German application, however, had the DAS programs rewritten in the language PEARL (process and experiment automation real-time language) to meet software reliability requirements and to facilitate software verification, (Basic Pearl Language Description, 1977).

Data Base Construction

Besides other methodologies all systems refer to cause-consequence diagrams or alarm trees. These are stored on disk. Programs to support the construction of the data base library are reported by the Hungarian system as well as the STAR system. However, little is known about the Hungarian program ALGEN whereas the STAR system comprises a powerful Model Generator MOGEN, (Felkel, Grumbach & Hoermann,1978), to aid systems analysts to construct the CCD's and related information needed by the DAS. The goal of this approach is to relieve the systems analysts from any programming. Furthermore, the model generator should support the testing and validation, checking for consistency, completeness and correctness, (Taylor, Hollo, 1977); an interactive version is being developed.

The EPRI system does not comprise such comfortable facilities, it is, however, felt that such software tools are an impelling need to reduce the manpower required to design and test the CCD's.

Methodologies used

All systems comprise as basic methodology some sort of logically combined sets of events denoted by tree structures where the nodes are connected on-line to process data.

However, the algorithms processing the structure vary. In the CEGB system the structure is used to suppress alarms ("last-up alarm" concept). In the GRS as well as EPRI DAS, look-ahead is performed which adds predictive capabilities to the DAS. In the Hungarian system it seems that very little analysis is done other than verifying that there exists a combination of alarms and displaying their logical structure to the operator.

Present Installation

The EPRI-DAS has been connected and tested at the Combustion Engineering training simulator. The main goal was to gain insights in operator performance evaluation exposing the trainees with disturbances and observing their behaviour in such situations with and without a DAS.

The STAR application aimed at installing a DAS in the 1300 MWe Grafenrheinfeld PWR after the feasibility had been verified, (Grumbach & Hoermann, 1976), in tests at the Halden Boiling Water Reactor.

The Hungarian system, (Bürger & Zobor, 1978) has been operating since 1976 in a phased application. In the second phase since 1978 the computer system controls in DDC the neutron level and the outlet temperature of the primary cooling circuit.

JUSTIFICATION OF DAS DEVELOPMENT

Prior to the initiation of work on DASs careful investigations have been performed in the EPRI as well as STAR projects to establish the benefits of a DAS in terms of improved plant safety and availability. A number of plants in the U.S. and in Germany have been examined in view of the total number of outages versus those that could have been prevented by having a DAS at the operator's disposal. By means of early detection of disturbances and notification of the operators in due time (usually before the conventional annunciation system does so), many disturbances can be cancelled before any protection system has to take over. As far as availability is concerned, an increase of as much as 2.5% is suggested. The impact on plant safety, however, is more difficult to quantify. On the other hand the TMI-2 accident according to the Kemeny report, (Kemeny et al, 1979):

> "In conclusion, while the major factor that turned this incident into a serious accident was inappropriate operator action, many factors contributed to the action of the operators, such as deficiencies in their training, lack of clarity in their operating procedures, failure of organizations to learn the proper lessons from previous incidents, and deficiencies in the design of the control room."

suggests that a DAS may be necessary to improve the quality of information available to the operators and thus could considerably contribute to the overall plant safety.

CONCLUSIONS AND FURTHER DEVELOPMENT

As a consequence of the TMI-2 incident the U.S. have initiated two research projects in the field of process surveillance, i.e. a DAS with a wider scope. These projects are being carried out under sponsorship of EPRI and the U.S. Department of Energy (DOE) and the main contractors are Westinghouse and Babcock & Wilcox resp.

The STAR system will be developed further and an extended version is planned to be implemented in the Biblis Block B plant.

An upgraded version of the original CEGB alarm analysis system installed at Oldbury and Wylfa has been offered by Nuclear Power Company and is being implemented in the Heysham AGR power plant in the U.K.

The experience with the STAR-system as well as the EPRI-DAS indicate that a DAS may significantly improve availability and safety of nuclear power plant. However, all disturbances detectable by a DAS have to be entered (have to be considered) before they occur. It is therefore felt, that an enhanced DASS (disturbance analysis and surveillance system) should also include possibilities for the operators to obtain as much high quality information as he needs to determine the process status even in those situations which have not been anticipated by design engineers. However, the disturbance data base containing the CCD's is supposed to be flexible enough to be enhanced continually.

Elaborate mass and energy balance and flow diagrams should inform the operator about the main coolant inventory. Since all systems combine process signals logically by elaborate tree structures multiple faults may be detected. Further use of CRT's is suggested; however, ergonomic considerations have to be taken into account more than has been done previously. Specific systems for process surveillance based on process computers have to be developed, the number of sensors in nuclear power plants increased and measures for the selection of information, (Kiguchi & Sheridan, 1979) determined.

The integration of process computer-based systems for plant surveillance into an advanced control room concept and the redefinition of the operator's role is a major issue to be resolved.

REFERENCES

Basic Pearl Language Description, Gesellschaft f. Kernforschung, KfK-PDV 120, Karlsruhe, Germany, 1977

Bürger, L., E. Zobor, On-line Alarm Analysis in the Hierarchical Control System of the WWR-SM Research Reactor, Symp. on NPPCI, IAEA-SM-226/17, Cannes, France, Apr. 1978.

Büttner, W.E., L. Felkel, R. Grumbach, F. Øwre, B. Thomassen, Functions and Design Characteristics of the STAR Disturbance Analysis System, IAEA Specialists' Meeting on NPPCI, December 5-7, Munich, Germany, 1979.

Felkel, L., R. Grumbach, A. Zapp, F. Øwre, J.K. Trengereid, Analytical Methods and Performance Evaluation of the STAR Application in the Grafenrheinfeld Nuclear Power Plant, idib.

Felkel, L., R. Grumbach, H. Hoermann, Automatic Generation and Application of Disturbance Analysis Models, Halden Project Report, HPR 214, Jan. 1978.

Felkel, L., R. Grumbach, E. Saedtler, D. Wach, Treatment, Analysis and Presentation of Information about Component Faults and Plant Disturbances, Symp. on NPPCI, IAEA-SM-226/76, Cannes, France, Apr. 1978.

Filby, R.J., D.A. Glew, Review of Oldbury Alarm Analysis System, CEGB/Oldbury Internal Report, Aug. 1971.

Frogner, B. Ch. Meijer, On-line Power Plant Alarm and Analysis System, Project NP-891, Final Report, EPRI, Palo Alto, Ca., Feb. 1980.

Grumbach, R., H. Hoermann, Plant Disturbance Analysis by Process Computer-Basic Development and Experimental Tests. Procs. IAEA Specialists' Meeting on NPPCI, MRR 160, Technical University Munich, 1976.

Holmgren, M., The development of "process feeling" and problem solving behaviour in computer based control rooms. EHPG Lillehammer, June 1-6, 1980.

Jervis, M.W., On-line Computers in Power Stations, IEE Rev., 119(8R), Aug. 1972.

Kay, P.C.M., P.W. Heywood, Alarm Analysis and Indication at Oldbury Nuclear Power Station, IEE Conf. Publ. 16, Part I, 1966.

Kemeny, J.G. et al., Report of the President's Commission: The accident at Three Mile Island, Oct. 1979.

Kiguchi, T., T.B. Sheridan, Criteria for Selecting Measures of Plant Information with Application to Nuclear Reactors, IEEE Transactions on Systems, Man and Cybernetics, Vol. SMC-9, No. 4, April 1979.

Long, A.B., Technical Assessment of Disturbance Analysis Systems, Nuclear Safety Journal, Vol. 21, Nr. 1, Jan. Febr. 1980.

Nielsen, D.S., Use of Cause-Consequence Charts in Practical Systems Analysis, Procs. Conf. on Reliability and Fault Tree Analysis, R.E. Barlow Ed., University of California, Berkeley, 1974.

Patterson, D., Application of a Computerised Alarm Analysis System to a Nuclear Power Station, Procs. IEE, Vol. 115, No. 12, Dec. 1968.

Potts, K.H., R. Tabernacle, Alarm Analysis by Computer Methods, CEGB Internal Report, May 1966.

Taylor, J.R., E. Hollo, Experience with Algorithms for Automatic Failure Analysis, International Conference on Nuclear Systems Reliability Engineering and Risk Assessment, Gatlinburg TN, June 20-24, 1977.

TMI-2 Lessons learned task force: "Status report and short-term recommendations", NUREG-0578, July 1979.

Welbourne, D., Alarm Analysis and Display at Wylfa Nuclear Power Station, Procs. Inst. Electr. Eng., 115(11), Nov. 1968.

Øwre, F., L. Felkel, Functional Description of the Disturbance Analysis System for the Grafenrheinfeld Nuclear Power Plant, Halden Project Report, HPR 221.14, Nov. 1978.

AUTOMATIC ERROR DETECTION AND ERROR RECORDING OF A DISTRIBUTED, FAULT-TOLERANT PROCESS COMPUTER SYSTEM

Max Syrbe

Fraunhofer-Institut für Informations- und Datenverarbeitung
D 7500 Karlsruhe

INTRODUCTION

In technical systems, especially in process computer systems, errors occur because of:

(1) physical defects,
(2) design faults,
(3) operating mistakes.

The usual way to counteract these errors is to use better components, more exact design methods or more improved man-machine interfaces combined with operator training. Today, progress in these areas is no longer sufficient to meet the growing number of errors due to growing complexity. The latter is caused by an increasing degree of automation combined with an increasing coupling of functions and/or systems (Syrbe, 1979) which have become necessary because of growing demands for productivity, quality, conservation of raw materials and energy and for security and environment protection. Fig. 1 shows the complexity of a system increasing with the degree of automation, which, in turn, is approximately equal to the number of the relevant process signals. An upper and a lower boundary are given for the complexity. For the upper boundary, a coupling was assumed between each process signal and every other process signal; for the lower boundary, each signal was coupled with five other signals. It should be clear that, even in the case of complexity growing near to the lower boundary, error prevention by the "principle of perfection" must be supplemented by an additional method. This utilizes the principle of "fault-tolerance" which uses system redundancies.

The trend to complex systems has led to a modularization of the systems to improve their design (fig. 2). A complex system is built with highly independent working aggregates and these are again built with the utmost in compatible modules. Thus, introducing system redundancy is relatively straightforward and also makes possible the introduction of fault tolerance.

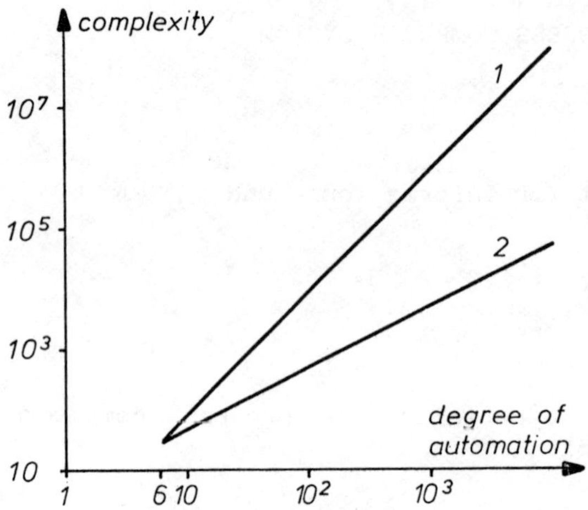

Fig. 1. System complexity as a function of the degree of automation (process signal coupling)

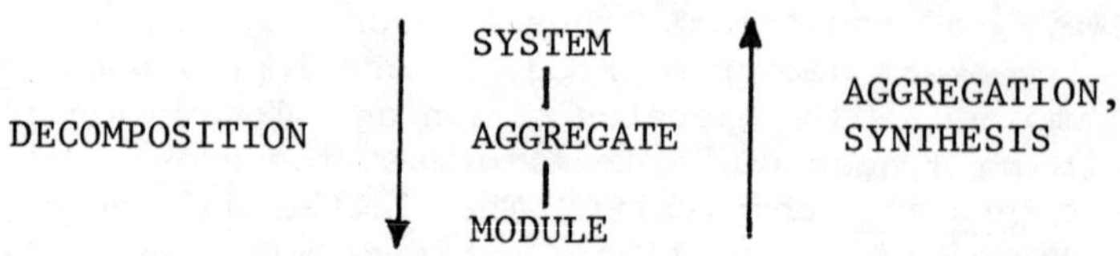

Fig. 2. Trends toward system modularization

FUNDAMENTALS OF FAULT-TOLERANT SYSTEMS

Fault-tolerant systems can be built either with a fixed ("static") structure, or with a variable ("dynamic") structure. The left hand side of Fig. 3 illustrates an example of static redundancy in a 2-out-of-3 circuit with 3 modules for the same processing task and a comparator. Faults (errors) in a module will be detected and masked. The right hand side of this figure is an example of dynamic redundancy in the form of two modules, each one equipped with a checker (fault detector) and with switches to direct the processing tasks. In the case of normal conditions, each module processes its own task. In case of a fault in a module detected by its own checker, the faulty module will be disconnected and both tasks will be switched to the perfect module for further processing within the reconfiguration including a recovery mechanism in a function-sharing manner. For this it is necessary to make processing capacity available, as shown in Fig. 4.

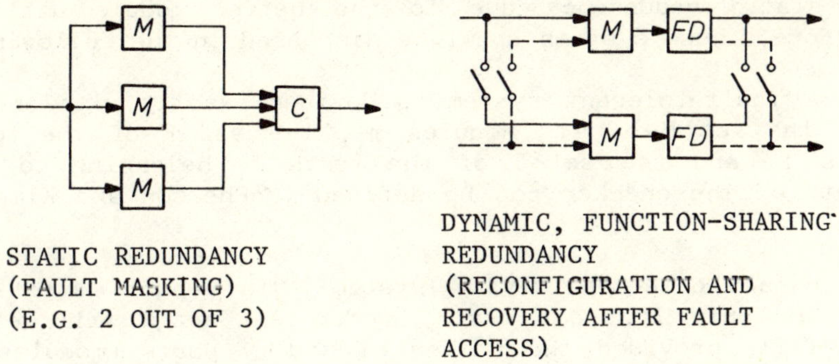

Fig. 3. Fault-tolerant system structures

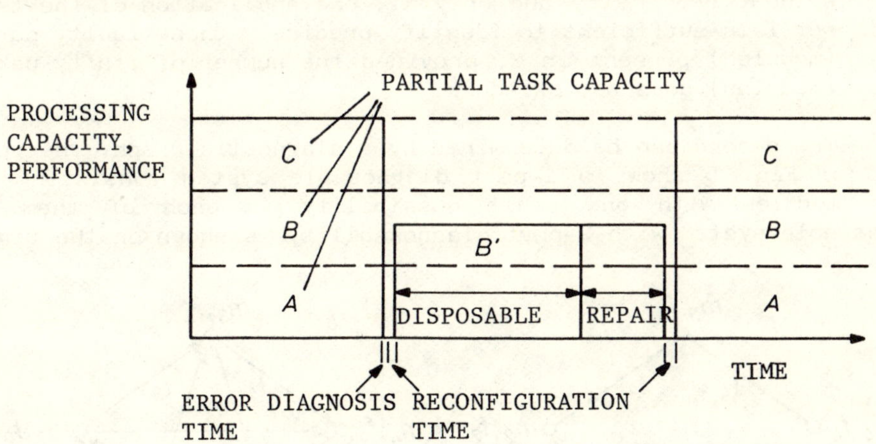

Fig. 4. Processing capacity and partial task control

The tasks are divided into partial tasks belonging to priorities A to C. In the case of a faulty module, the partial tasks with the priority B are set to a half demand of their processing capacity via parameter changing (for example, the sample rate) and the partial tasks with the priority C are suspended.

In the case of static redundance, the degree of fault tolerance is dependent upon the quality of the comparator. In the case of dynamic redundancy, this depends on the checker. Systems with dynamic redundancy are definitely more economic than those with static redundancy due to the better module utilization. Therefore, these systems shall be discussed in the following.

A fault-tolerant system is defined exactly with only three sets: the set M of its modules m_i, the set F of the tolerable faults f_i and the set T of the tests t_i belonging to f_i. The quality of the checker can be defined (McPherson and Kime, 1976) by the

- n-part detectability: A system M is n-part detectable if, and only if, some tests in the test set T will definitely fail, provided the number of faulty parts (modules) is at least one, but does not exceed n;

or by the

- k-part diagnosability: A system M is k-part diagnosable without repair if, and only if, one application of the test set T is sufficient to identify precisely those faulty parts (modules) present in M, provided the number of faulty parts (modules) does not exceed k.

The second can be determined by a diagnostic graph. The left hand of Fig. 5 shows a 1-part diagnosable system consisting of five modules with one fault possibility in each of them. A comparable system with 2-part diagnosability is shown on the right hand.

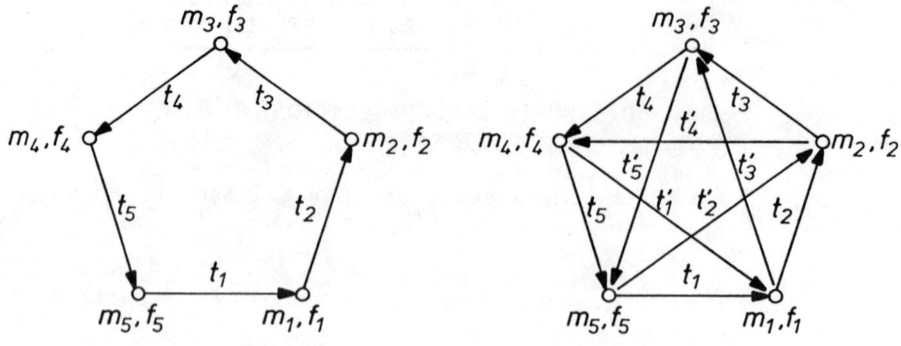

Fig. 5. Diagnostic graphs

In the following, a practical application is described in the form of a newly developed Really Distributed Control Computer System (RDC-System, Syrbe, 1978; Heger et al., 1979). While now controlling four pit furnaces in a steel plant, the system will be expanded in the coming months to control 28. This system has been in operation since the beginning of 1979.

A DISTRIBUTED, FAULT-TOLERANT PROCESS COMPUTER SYSTEM

The hardware structure of the installed RDC-system is presented in Fig. 6. In the lower part of the figure, an input/output-colour-screen-system (EAF-System; Grimm, 1976; Grimm et al., 1978) is shown which works as a complete control panel intended for the central operator control of the entire system including plant and automation system. Its structure is built fault-tolerantly with checkers (error detectors) and electronic switches ( ). In the upper part of the figure are the microprocessor stations (microcomputers) which are allocated to the pit furnaces. Their structure is also fault-tolerant, as described later. All stations are in connection with each other by means of a fiber-optic ringbus.

All microcomputer stations have the same structure (Fig. 7). They contain

- one microcomputer (PμP, based on INTEL's 3000, 16 bit working length with working storage) performing the total direct digital process control programmed in the real-time language PEARL

- a second microcomputer (LμP with a fast working storage) providing the message control at the coupling of the station to the ringbus. It also performs error diagnosis on the serial ringbus and its reconfiguration. In order to achieve a higher performance, the total procedure is micro-programmed

- a light-transmitter/receiver module, providing the conversion of electrical signals into light-signals, vice versa

- the I/O parallel bus where the various process I/O modules (digital I/O, analogue I/O) as well as an operator control panel for the station (CPL) are located. Here actual values of the process signals, set points, signal limits or other parameters can be put into the data base of the station. Thus a simplified operation of the station can be guaranteed in spite of a breakdown of the communication

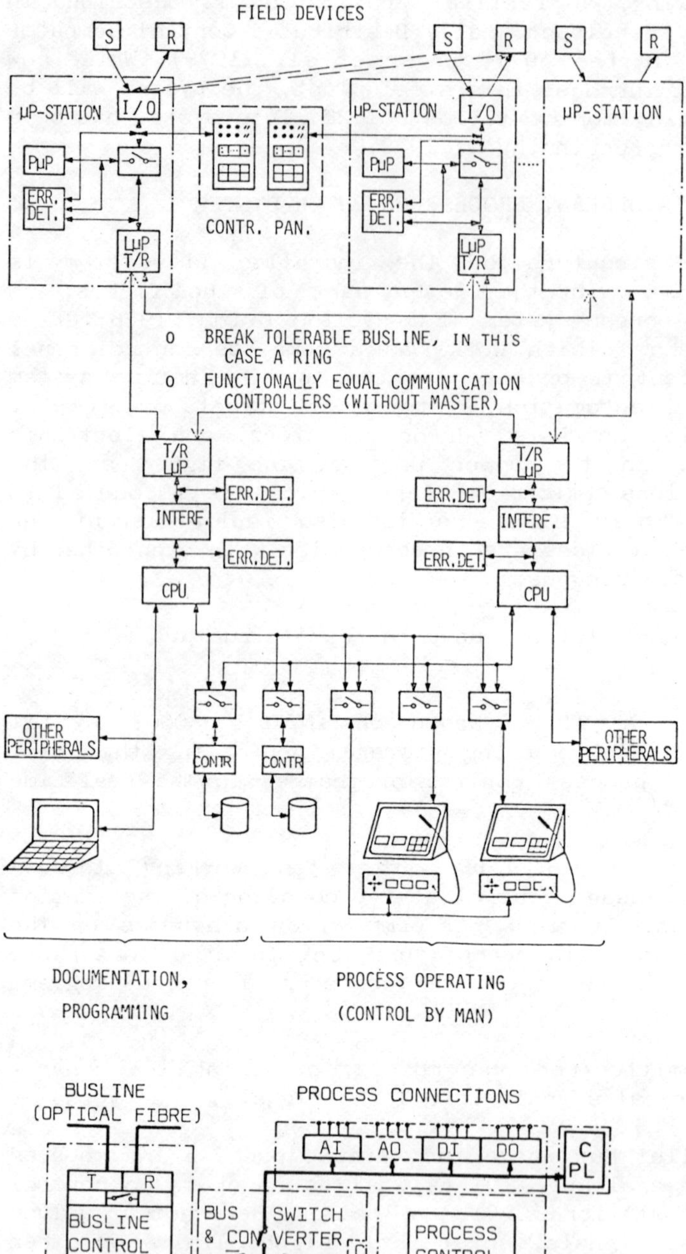

Fig. 6.
The distributed, tault-tolerant RDC-system

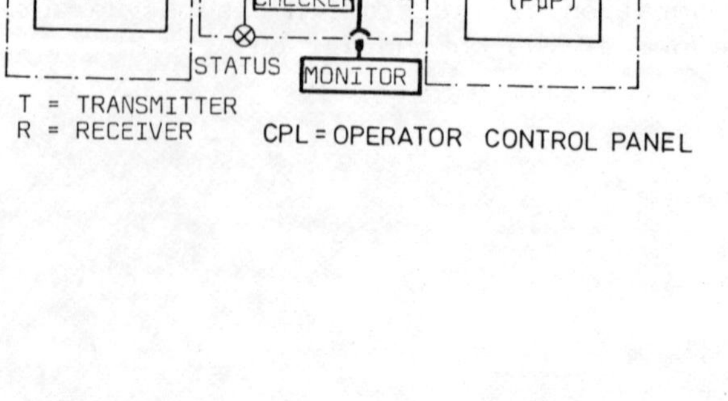

Fig. 7.
Fault-tolerant RDC-microcomputer station

- a bus switch and converter (BS) supervising the functions of the station and performing the reconfiguration combined with a fault-checker. The check (test) results are put into status registers (SR).

The diagnostic graph belonging to the RDC-microcomputer station is shown in Fig. 8. In this case the design of the tests does not achieve a complete 1- or 2-part-diagnosability. The part BS/SR must be perfect as a hard core.

A special problem in the design of fault-tolerant systems is the design of the tests. A highly systematical design method was attempted. Fig. 9 shows the results. The hardware and software components of the distributed computer system are divided into shells (planes). The upper shell forms the connections to the technical process; the lower shell forms the bus interconnection. Using the design method requires that a context be given for a sufficient number of attributes of the shells (left side of Fig. 9). Increasing context brings more efficient and more numerous tests. By means of context interpretation, the tests can be designed.

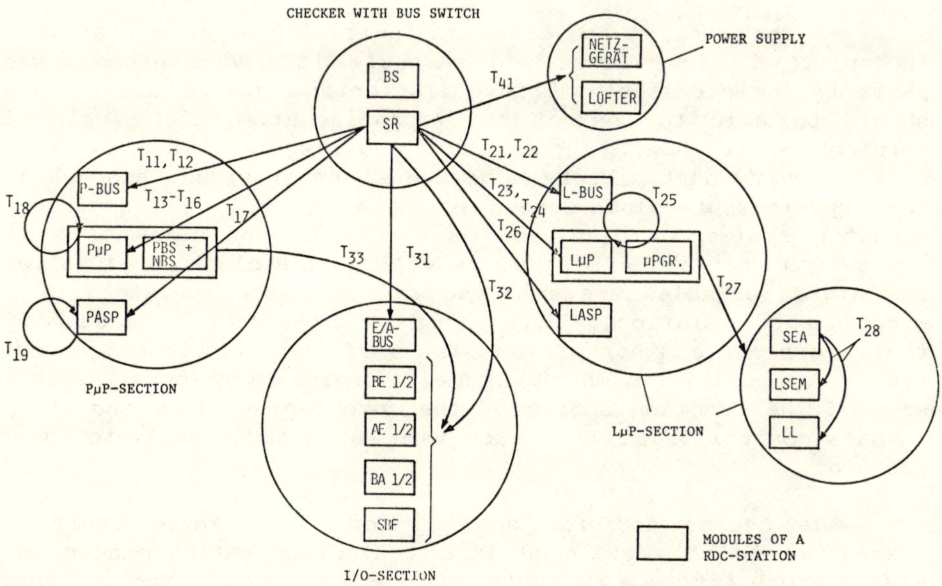

Fig. 8. The diagnostic graph of a RDC-microcomputer station

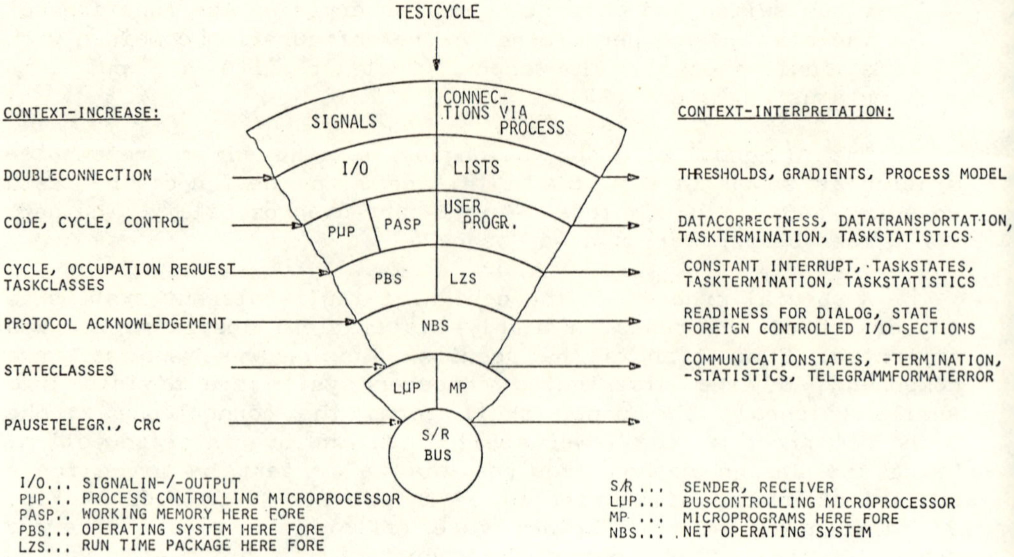

Fig. 9. A design method for tests

HUMAN TASKS IN A FAULT-TOLERANT SYSTEM

As the section above shows, fault-diagnosis in practise is incomplete in some cases and secondly it covers often several parts in their entirety. This calls for an operator aid. This aid should be able to control the reconfiguration and to plan the maintenance (repair). The latter requires the diagnosis of the single faulty part. Hereby some means are provided, supported by the input/output-colour-screen-system (Fig. 10). The distributed computer system is displayed on the screen by means of a flow chart with the system states, as well as the plant. In this case, the state variables are measured by the state registers of the microcomputer stations. Thus enough transparency is acquired so that a central control of the states of the distributed computer system is possible. A matrix-shaped overview image permits a quick scan of the function states of the computer sections and the pit furnace control (Fig. 11) to be made as an additional aid for the operator.

The second aid for an operator, the single fault part diagnosis, is not developed in a central operating manner up to now. At present, single part diagnosis can be carried out at microcomputer station's control panel (Fig. 12). In this way more registers can be displayed. An evaluation of the register data can be supported with an adjoined matrix (Fig. 13). The context of register data allows decisions about the single fault part to be made.

Fig. 10. EAF-system displaying the computer system as flow-chart; reconfiguration can be controlled via light-pen or teletype

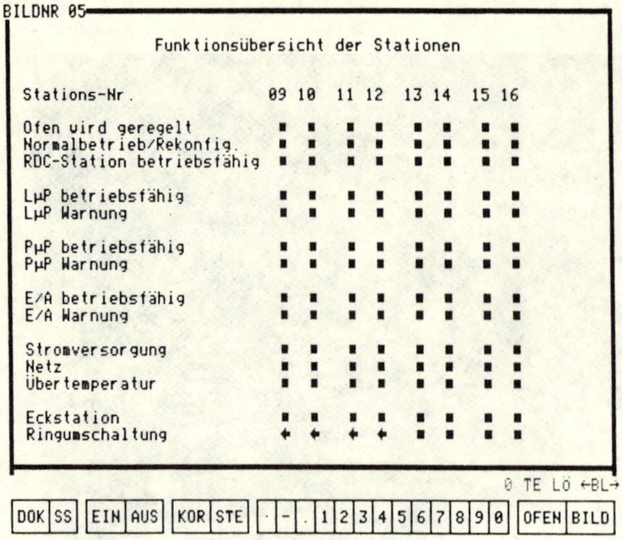

Fig. 11.
Summary display of the function-states of the computer sections and the pit furnace control by the EAF-system

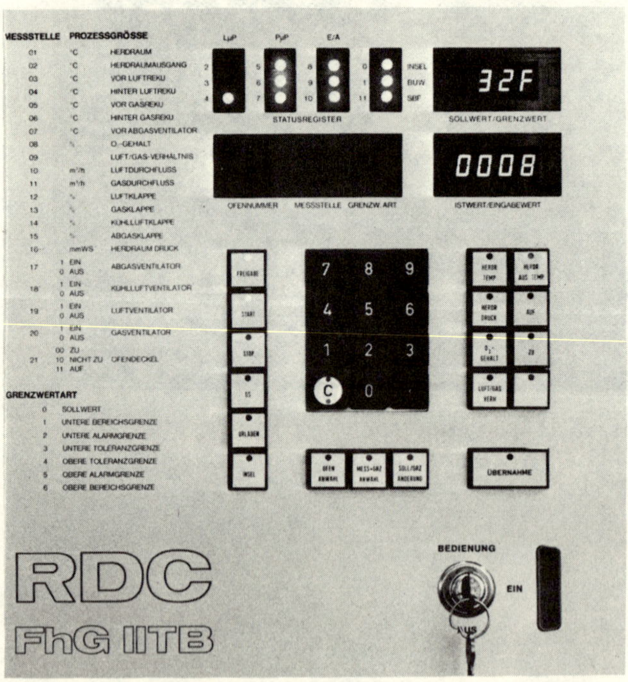

Fig. 12.
Microcomputer station's control panel

PART / DATE	PμP	PASP	BS	SR	AE 1	and so on
C 100 / 0001	X			X		
C 100 / 0002		X				
C 100 / F 013			X			
C 101 / 1111	X					
C 101 / F 011				X	X	
and so on						

Fig. 13.
Adjoined matrix for determination of single fault parts

ACKNOWLEDGEMENT

This paper publishes results from research projects supported by the German Federal Ministries of Defense and of Research and Technology with the data processing program of the German Federal Government, projects "DV-Systeme" and "PDV". The responsibility for the paper's content lies only with the author.

REFERENCES

Grimm, R., 1976. Autonomous I/O-colour-screen-system for process control with virtual keyboards adapted to the actual task. NATO Conference Series "Monitoring Behaviour and Supervisory Control". Plenum Press, New York, pp. 445-457.

Grimm, R., Hellriegel, W., Laubsch, H., Rudolf, M. Sassenhof, A., and Syrbe, M., 1978. Bildprograaierbares Ein-/Ausgabe-Farbbildschirmsystem (EAF) als Warte - Grundprinzipien, Realisierung, Erprobung. Report No. Kfk-PDV 134, Kernforschungszentrum Karlsruhe, GmbH.

Heger, D., Steusloff, H., Syrbe, M., 1979. Echtzeitrechnersystem mit verteilten Mikroprozessoren. BMFT-Report No, BMFT-FB DV 79-01, Kernforschungszentrum Karlsruhe GmbH.

McPherson, J.A., Kime, C.R., 1976. A two-level approach to modelling system diagnosability. International Symposium on fault-tolerant computing, FTCS-6, IEEE, Pittsburgh, pp.33--38.

Syrbe, M., 1978. Basic principles of advanced process control system structures and a realization with distributed microcomputers. IFAC 7th World Congress, Helsinki, Pergamon Press, pp. 393-401.

Syrbe, M., 1979. Regelungstechnik auf dem Wege. Regelungstechnik, Oldenbourg Verlag, pp. 130-134.

THE USER'S ROLE IN AUTOMATED FAULT DETECTION AND SYSTEM RECOVERY

W. J. Dellner

Bell Telephone Laboratories
6 Corporate Place 1K-261
Piscataway, N.J. 08854

INTRODUCTION

Systems whose failure can cause loss of life or large economic loss need to be tolerant to faults (i.e. faults in system hardware, software, and procedures). Examples of such systems include airplane autopilots in the automatic landing mode, electricity utility power generation plants, and telephone electronic switching systems (ESS). Such systems are characterized by high reliability; they fail infrequently and recover quickly when a fault does occur. The user usually cannot respond fast enough if and when a fault is detected. Even if he could respond, his proficiency would not be high because the fault occurs infrequently.

These systems need a mechanized fault monitoring, detection, and recovery capability accompanied by displays and controls for human monitoring. A need exists to better understand and better design the user's role in fault tolerant systems.

APPROACH

Better understanding of the users role will be obtained by (1) understanding existing fault tolerant systems, and (2) reviewing the displays, controls, and responses available in existing systems. With knowledge of existing systems I will build a generalized model of man's role in fault tolerant systems. This model will simplify description and understanding of the three system types previously mentioned. Also, the model will be used to integrate and relate real world data from fault tolerant systems in general (i.e. the above three and those involving other technologies). In essence, I will build a task taxonomy, as defined by Companion and Corso (1977), for the user's role in fault tolerant systems.

After doing the above, the foundation will have been laid for designing the user's role for a specific fault tolerant system. Design factors which will be reviewed include: research on displays, needs for system testing and needs for gaining and maintaining expertise.

EXAMPLES OF FAULT TOLERANT SYSTEMS

Examples of fault tolerant systems that will be reviewed include: the electric utility power generation plant, electronic switching systems, and the autopilot in the automatic landing mode.

Electricity Utility Generation Plant

An integral part of power generation is the steam turbine. Heated steam generates power by passing through a turbine thereby rotating a generator. If the electrical load is suddenly dropped from the turbine, it will overspeed and possibly explode. Within five seconds, a 600 MW turbine at rated speed of 3600 R.P.M. will reach overspeed if the electrical load drops off. To prevent this, two speed sensors, each of a different speed measuring technique, are connected to a processor which independently monitors abnormal rotational acceleration. If abnormal acceleration is found, then the steam to the turbine is automatically cut off. The human monitor is told of the system shutdown via a general warning light/horn and a red turbine overspeed trip light. Further descriptions of turbine control systems can be found in Eggenberger (1965). As a backup process, power is obtained from another generating plant.

Electronic Switching Systems (ESS)

The ESS switches a customer's call from caller to called. The ESS switching machine is a stored program computer. This is in contrast to earlier switching machines which were electro-mechanical. The large ESS machines handle over one half million calls per hour. Today about 40% of all calls in the USA are handled by ESS machines which number a few thousand. ESS are designed so as to be out of service no more than a few minutes per year.

The ESS machine uses its processor 1 to control call switching and routing. Processor 2 is running synchronously and simultaneously processing the same data. Key outputs are compared for error detection. If a mismatch occurs, then an interrupt is generated which causes the fault recognition program to run. In essence, the fault recognition program determines which portion of the system is faulty. The following occurs: (1) the suspected module is removed from service, (2) the system reconfigures itself using the good processor or module,and (3) the new configuration is checked for adequacy. A diagnostic program is run to pinpoint the defective circuit pack. Further descriptions of ESS fault tolerant tures are found in Averjard and Haverty (1977) and Toy (1978).

The ESS system designer has specified the criteria and means for fault detection and recovery to the good processsor. The ESS indicates, via a display, the processor which is in control. From the display, one can infer if a fault detection module is "insane", since it shows ping-pong of control from processor 1 to processor 2 and back and forth. This latter, infrequent state requires manual override to lock one processor into control (LaCava, 1978).

Autopilot in Automatic Landing Mode

The autopilot accepts redundant air data and radio signals and processes them through redundant processors. The controlling processor 1 produces pitch, roll and yaw output signals which are compared with processor 2's outputs. These outputs are assessed. If deviance from criteria are found then the backup processor takes over. If only dual processors are used, then complete control is usually given back to the pilot (passive recovery). If triple redundant processors are used then the control reverts to the remaining two processors (operative recovery). Displays should tell the pilot what is happening (i.e. autopilot in complete control or complete control given back to pilot). Whereas the ESS recovers by reconfiguring its mechanized functions, the autopilot may reconfigure task allocations by asking the pilot to assume some functions.

Other examples of fault tolerant systems include: the Air Traffic Control System, Space Systems, and Military Command and Control Systems.

GENERIC MODEL

Having reviewed three examples of fault tolerant systems, I now proceed to develop generic models of the machine subsystem and the Personnel Subsystem of the user.

Machine Subsystem

The machine subsystem usually requires redundant components in order to achieve high reliability (i.e. redundant sensors, processors or modules thereof, and software).

Redundant sensors of the real world environment are needed in case one sensor path fails (i.e. the turbine speed sensor, and the autopilot's radio inputs).

Redundant processors of input data are used to compensate for hardware faults. Systems that can dynamically reconfigure may not need to duplicate whole processors but only certain modules therein.

Redundant but different software algorithms are used to compensate for incomplete specifications or problem understanding. Usually the redundant software is running on a redundant processor. This is called "N VERSION PROGRAMMING" by Hecht (1979). Another method, called Recovery Block Technique, allows a single program to have redundant criteria tests. This allows one processor to be used but considers incomplete specifications or understanding.

Given the above, both processors are set in motion. A comparator assesses the output of both processors to detect faults. Fault recognition logic assesses which processor is bad. The comparator is a critical part of fault tolerant systems. Toy (1978) relates that 35 percent of all ESS downtime is related to fault recognition and recovery deficiences.

Recovery from a fault occurs when the system or human switches to the good processor or module. This switch of control should be indicated to the human monitor.

Personal Subsystem - Operations

The personnel subsystem is characterized by: displays, controls, level of human intervention, skill and knowledge required, and the interaction of each. Each is explored below:

1. Display should be made to the human monitor as to which processor or module thereof is in control. If the human must assume control, then appropriate control displays must also exist.

2. The level of human intervention required is determined during the task allocation phase of design and ranges from monitor only to full manual control. The various degrees of intervention include:

   A. Lowest: System automatically recovers, no possibility for maintenance (e.g. space systems).

B. System automatically recovers to backup; dispatch maintenance person to fix the defective processor or module (e.g. ESS and power plant).

C. User must assist the system to reconfigure itself (e.g. ESS).

D. User assumes control over part of disabled processor function (e.g. autopilot and traffic air control system).

E. Highest: User manually overrides system and performs all processor tasks (e.g. autopilot).

3. Controls are needed to implement the various degrees of manual intervention required. For example, the human ESS monitor must have a means to lock one processor into control (La Cava, 1978).

4. Skills and Knowledge are needed to support the level of human intervention required. The user must be initially trained and somehow be kept proficient so that when the rare fault occurs, the user will respond properly. This requires consideration during the task allocation phase; the level of human intervention required must be consistent with the level of proficiency available when the fault occurs.

5. The proper interaction must exist between the displays, controls, level of intervention required, and skills and knowledge. In essence, the user must know that a successful recovery has/is occurring and what action to take next (i.e. what to do as well as what not to do). With respect to the latter we should pay particular attention to Pascal, who said that most of the evils in life come from "man being unable to sit still in a room". In essence, many system recoveries fail because a well meaning user performed an inappropriate action. In addition, the user must accept the fault tolerant system as being capable of doing its recovery job.

Personal Subsystem - Maintenance

During the successful recovery, the system should record for later display the processor and/or module thereof which was suspected to be at fault. Some systems will be able to mechanically pinpoint a fault to a module whereas others will list diagnostic tests that failed and suggest, via a fault dictionary, which of several modules may be faulty (Hecht, 1979).

USER'S ROLE - DESIGN, TESTING, TRAINING

Design of User's Role

The user's role and the supporting environment for normal and fault conditions must be designed and then tested. The task allocation phase of system design determined the level(s) of human intervention required when a fault occurs. The decisions made during task allocation determine the requirements for the design of displays, controls and training strategies. For example, the overspeeding turbine is stopped automatically and hence requires a simple display, no manual controls and minimal training. More complex recovery situations, which have multiple levels of human intervention possible, require more consideration as to the design of displays, controls, and training techniques. A review of the design of displays, controls, and training techniques is now conducted.

Design of Displays and Controls

The displays provided should be in accord with the level of human intervention required (i.e. no human intervention requires a simple display while manual intervention may require many displays). In addition, for all levels of human intervention a single master display of overall system status is needed.

The master status display indicates whether the system is operating correctly, whether it has failed and recovered, and whether the recovery is stable (i.e. the user need not intervene). Research on stress, vigilance, and cockpit design supports the idea of a master status display. When a fault recovery occurs, stress is induced in the human monitor since the situation is drastically different than normal (i.e., users of these systems classify their activities as long periods of boredom broken by minutes of frantic activity). Under stress conditions the human loses information processing capability (Broadbent, 1971; Selye, 1956). The human monitor may not be able to quickly integrate the meaning of many separate displays. A master status display would do the integration for the monitor.

Vigilance studies (Teichner, 1974) also support the master status display concept, since the systems we are dealing with often go for long periods of time without a fault occurring. Thus, even though temporal uncertainty exists relative to fault occurrence, spatial uncertainty of the overall status would not exist. The role of spatial certainty is illustrated by cockpit design research (Bunker Ramo, 1967). In this study, fault indicators on individual cockpit instruments had delays in detection greater than that of a master status display accompanied by the individual fault indicators. The study found that second faults were often completely missed when individual fault indicators were used without an accompanying master status display.

Other displays should state the exact nature of the failure to aid manual intervention if necessary and for future trouble shooting. These displays should all be in one location to reduce location uncertainty (Beattie and Blum, 1979). Faults are usually accompanied by horns and flashing lights. These may not be needed since they are stressful into themselves and may induce unwanted action.

Displays of important parameters and time events are needed both as a means of human assessment of the overall situation and as a means for human control. The fault and normal displays complement each other. Spady (1976) related that pilots scan many displays when monitoring an autopilot approach whereas they concentrate on only a few when actually flying the landing approach. Some European users of autopilots believe the constant use of the automated system is necessary to maintain proficiency and acceptance of fault tolerant features (Ropelewski, 1979). These airlines use the autopilot on all landing approaches to maintain pilot knowledge of normal display patterns and cues. Hopefully, their familiarity with normality will enable them to detect abnormality as indicated by a master status light and the integration of the data on the numerous individual displays. For example, normal time events are important for the pilot to learn (i.e. localizer capture, glide slope capture, outer marker, etc.). Controls must be available for the various types of human intervention identified during the task allocation.

Testing of the User's Role

One the system is built, the user's role must be tested (i.e., can the user perform the role expected with the displays and controls provided?) If not, then the system or the user's role must be redesigned before the system is introduced to the field. The best place to test user capabilities is in a controlled environment such as a simulator.

It is important that the user's role be clearly specified during the specification and task allocation phases. If it is, then it can be tested during the system test phase. If the role were not specified earlier, then the later test is ill-defined.

An example of a positive use of testing is the detection of the lack of a manual control to remotely lock one ESS processor into control (La Cava, 1978). Also, an experimental ESS display provided too much information and caused the ESS human monitors to introduce a system failure.

Training of the User

Training is critical so that the user will know what to do in a critical fault situation. Training will: (1) give the user knowledge of what the system will do to recover and/or what he must do, (2) give the user knowledge of what the system will do if he gives unwarranted inputs, (3) reduce the user's stress when fault displays are seen, since they have been seen before. Stress may cause unwarranted actions. Duncan and Shepherd (1975) relate that the user develops competence as well as confidence in his ability to successfully perform due to fault diagnosis training.

How is this training to be achieved? On the job training is impractical because of the infrequency and dangerous nature of the faults. Simulators seem to be the best way of providing training without the risks of the real environment. Power plant, aircraft, and ESS simulators are in use and their use is increasing for both initial and refresher training.

Learning benefits of simulation include: (1) more scenarios per hour than in the real system, (2) immediate knowledge of results (i.e. the user's performance may be compared to the ideal as soon as it occurs), and (3) user restraint (i.e. simulation easily provides the user with stimuli to which the appropriate response is to do nothing). Simulators are also more economical than using the real system, risk aside. For example, a large aircraft, like the BOEING 747, costs $ 8000 per hour to use for training purposes. Simulation costs less. System fidelity may not be critical for some simulations (Adams, 1979; Fenwick, 1966). In some simulations, part task learning is adequate since only the critical stimuli need be shown and responses or no responses practiced.

Transfer of learning studies cannot be done to justify simulator use because the events being prevented happen infrequently. Justification must be based on the benefits that result from not having serious events.

The importance of simulation training cannot be overemphasized. However, simulation does cost and in some cases operations-management has elected to bypass it.

SUMMARY

The user's role must be analyzed and specified as a part of the task allocation processs. Only by explicitly stating what the level(s) of human intervention is to be can we attempt to design displays and controls which will allow the user to perform that role(s). This process becomes complicated when the user can perform many levels of intervention (i.e. no intervention, some intervention or complete manual control).

A master status display is desirable regardless of the level of human intervention required. If manual control is envisioned then appropriate displays and controls are needed.

Once a prototype system is built, the human roles can be tested. Again, the roles are specified it is easy to set up tests. If the roles have not been specified then design effort (i.e. design of the human role) must be conducted during the test phase. Lack of specificity of function of the personnel subsystem is not unique. The machine subsystem be the computer or mechanical, also suffers during testing if specifications do not relate exactly what the subsystem should do.

The user must be taught exactly what to do or not to do given a fault recovery situation. This is best achieved through simulation for initial and refresher training.

REFERENCES

Adams, J.A., 1979, On the Evaluation of Training Devices, Human Factors, Vol. 21(6), 711.

Averyard, R.L. and Haverty, M.B., 1977, No. 4ESS - through a wide-angle lens, Bell Laboratories Record, Vol. 55(11), 290.

Beattie, J.D. and Blun, L.S., 1979, Design Goals for Computer Driven CRT Display Systems in the Control of Generating Stations, in: 6th Man Computer Communications Conference. National Research Council of Canada, Ottawa Canada, 177.

Broadbent, D.E., 1971, Decision and Stress, New York: Academic Press, 1971.

Bunker Ramo, 1967, Simulator Program Summaries - All Weather Landing, Canoga Park, Calif. Progress Report to FAA.

Companion, M.A., and Corso, G.M., 1977, Task Taxonomy: Two Ignored Issues, in: Proceedings of the Human Factors Society - 21st Annual Meeting.

Duncan, K.D. and Shepard, A., 1975, A Simulator and Training Technique for Diagnosing Plant Failures From Control Panels, Economics, Vol. 18(6), 627.

Eggenberger, M.A., 1965, An Advanced Electrohydraulic Control System for Reheat Steam Turbines, American Power Conference, Chicago, Ill.

Fenwick, C.A., 1966, Evaluation of Multi Channel Navigation Displays, Navigation, (13)

Hecht, H., 1979, Fault Tolerant Software, in: IEEE Transactions on Reliability, Vol. R-28, No. 3

La Cava, J.L., 1978, Human Factors Test of SCC interface for No. 2ESS, Personal Communication

Ropelewski, R.R., 1978, Air Inter's A-300 Autolandings Routine, in: Aviation Week and Space Technology, 45-57

Spady, A.A., 1976, Preliminary Report on Airline Pilot Scan Patterns During ILS Approaches, in: Proceeding of the Aircraft Safety and Operating Problems Conference. NASA Report SP-416

Selye, H., 1956, The Stress of Life, New York; McGraw-Hill.

Teichner, W., 1974, The Detection of a Simple Visual Signal as a Function of Time of Watch, Human Factors, 16(4), 339.

Toy, W.N., 1978, Fault Tolerant Design of Local ESS Processors, Proceeding of the IEEE, Vol. 66(10)

APPENDIX

Electronic Switching System (ESS) Overview

The basic function of a switching system is to interconnect communication paths. An example is illustrative (see Figure 1). Customer A wishes to speak to customer B. Customer A dials customer B's seven digit phone number. Switcher 1 to which customer A's line is connected routes the call to switcher 2 to which customer B is connected. Switcher 2 and customer B both have the same first 3 digits (i.e. 555). Switcher 2 routes the call to customer B (i.e., line 2222). Switching machines are contained in buildings called central offices. An average local central office has 15,000 incoming lines. Local central offices range from 2,000 to 50,000 incoming lines.

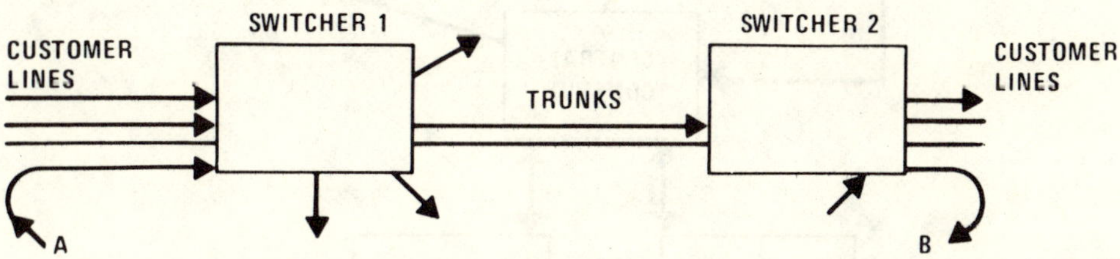

Figure 1.

Older switching systems were electromechanical and hardwired and as such were relatively inflexible to changes of incoming line size and features offered. Newer electronic switching systems have stored program control which allows flexibility - new programs and data can be easily added.

An ESS analyzes a call in discrete steps. These steps start when the customer lifts the receiver. Every few milliseconds, the system scans each customer's line and records its state: on hook, off hook, etc. If a change has occurred since the last scan then a program is consulted by the central control to see what action should be taken. If a phone has gone from on hook, then dial tone is given to the customer. While the customer dials, the program directs that all customer digits be recorded in the call store (Figure 2). After all digits are dialed, the central control consults the program for the next step. The next step is to route the call to the office involved. When the call is answered that state is registered in the call store. The whole process is characterized by continuous interplay between the programmed logic (program store) and the temporary memory (call store) which acts as an electronic slate. When the call is completed, the slate is wiped clean of information pertaining to that call.

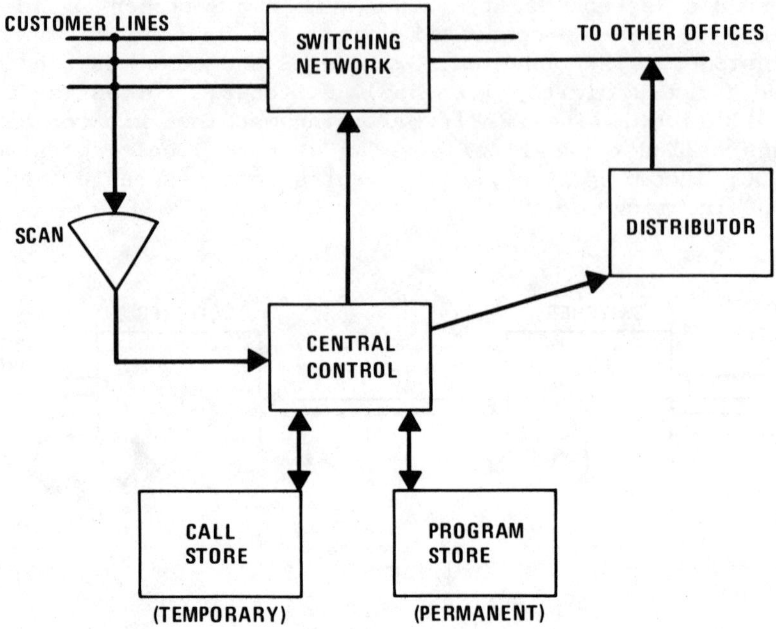

Figure 2.

The ESS design is strongly influenced by the fact that it must operate in real time. It must respond promptly to customers signals and data. It must also respond to trouble detection circuits designed into hardware to ensure dependable operation. In order to do this, the system has a hierarchy of programs. Some must be run on strict schedule while others may be run as required.

The central control incorporates an interrupt mechanism which momentarily seizes control of the system when a manual interrupt or trouble detector signal is received. The interrupt causes the system to stop its present program task, to store the place at which it was interrupted, and then to transfer control to an appropriate fault recognition program.

If a fault is found the system is reconfigured to bring in a backup unit. Then control is returned to the interrupted program. Diagnosis programs isolate the specific circuit pack(s) that caused the unit to be faulty.

An ESS has a design requirement that it will be out of service no more than a few minutes per year. To meet this objective of ultra reliability, redundancy of system components has been the approach taken to compensate for potential machine faults. Without this redundancy, a failed central control processor might cause a failure of the entire ESS. With duplication a standby processor takes over control and provides continuous telephone service. More than half of stored programs are for maintenance.

Duplicated in many ESS are the: central control, program stores, call store, data buses. Figure 3 shows redundant ESS components.

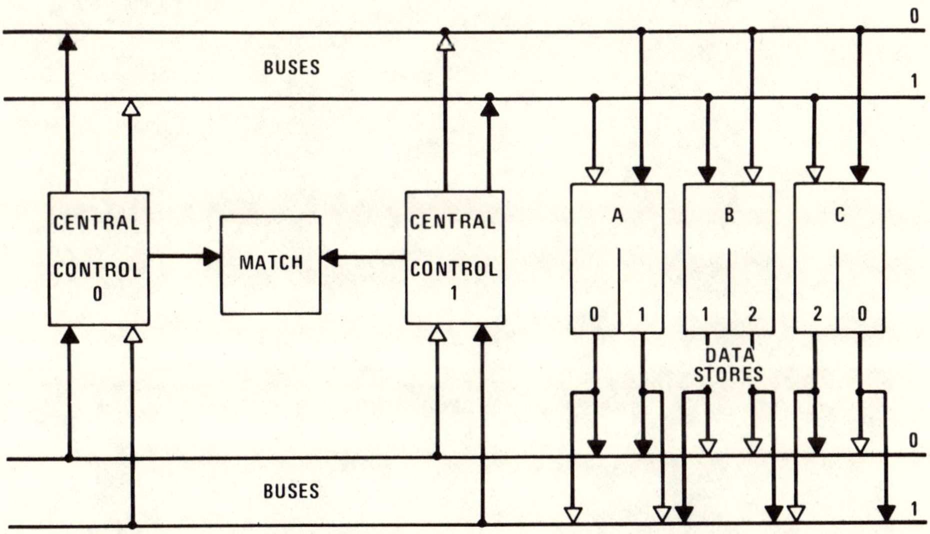

Figure 3.

Detection of errors is made in various ways for the different components. Central Controls have their various outputs compared. If a discrepancy occurs, then an interrupt occurs and a fault detection program is run to isolate which processor is bad. Then the system is reconfigured and tested for sanity. Then normal processing resumes. Resolution of a faulty central control causes one of the highest levels of interrupts. Other components (e.g., program store, call store, etc.) are also checked. For example, a program store is thought bad if logic or data checks are not met.

AIDING PROCESS PLANT OPERATORS IN FAULT FINDING AND CORRECTIVE ACTION

David A. Lihou

Chemical Engineering Department
Aston University
Birmingham B4 7ET, England

INTRODUCTION

An essential requirement for maintaining the efficient operation of process plants is effective fault finding, followed by taking the appropriate corrective action, soon enough to prevent loss. Partial deterioration and/or unrevealed failure of process equipments can combine to cause loss. Often, on being alerted to a faulty state, operators will find it difficult to distinguish between primary causes and their symptoms, which are secondary causes.

Fault Symptom Matrices

Fault finding relies upon the correct interpretation of a set of deviations in measured process variables, which relates to a single or multiple faulty state. The process variables and locations on the plant form a two dimensional matrix. The elements of the matrix are the symptoms (high, low, zero, normal, etc.) which process variables are expected to display, in response to a specific cause.

Fault symptom matrices have been used in fault-finding research for some time (Rigney, 1966); but there has been a recent upsurge of interest in their application to process plants. Shepherd (1976) and Morris (1976) have considered them as operator aids for ensuring optimal fault finding strategies. Similarly, Berenblut and Whitehouse (1977) have used decision tables to represent faulty states, which appears to be the basis of the Anticipator computer package described by Munday (1977).

Multiple faulty states can reinforce, reduce or eliminate certain symptoms. The following rules may be used in combining the corresponding elements of two fault symptom matrices:

Null AND null, No, Less or More gives null, No, Less or More, respectively;

Less AND Less gives Less;
More AND More gives More;
No AND Less gives Less;
No AND More gives Less;
Less AND More gives null.

Computer-Aided Fault Finding

Fault symptom matrices can be stored in a computer by means of a coded format which eliminates the need to store null elements. The method used here is in terms of Fault Symptom Equations, which correspond to the code used by Fussel (1973) to synthesise large fault trees from mini-fault trees. Similarly, Apostolakis et al. (1978) have simulated fault trees from event tables.

On-line, the computer monitors the state of the selected process variables and creates a State Matrix in terms of high (3), low (2) and zero (1) values. Variables which have normal values do not appear in the State Matrix. By comparing the State Matrix with its library of Fault Symptom Matrices, the computer can deduce the current faulty states on the plant.

Frequently, the State Matrix will be changing with time and, at any instant, a symptom may be reduced in magnitude, or eliminated, by a combination of faulty states. These dynamic changes in the State Matrix can bewilder operators, if they have not experienced a similar situation or been told what might happen. On interrogation, the computer would be able to advise the operator of the primary and secondary causes of the present faulty state. The computer may advise also, on the appropriate corrective action. Alternatively, operators can devise their own strategy of corrective action, using the computer-aided surveillance to determine whether their corrective actions are moving the plant back to a normal state, which is represented by a null State Matrix.

When State Matrices are diaplayed on a VDU, it helps operators to see zeros for the variables which are normal; although these elements are not stored in the Fault Symptom Matrices. Furthermore, it is more comprehensible if the number code of the stored matrix is displayed as letters: H for high corresponding to 3, L for low corresponding to 2, and Z for zero corresponding to 1. Z would only be used for flow or level measurements.

Fault Trees

The dynamic changes in the State Matrix, referred to above, represent the development of a route through faults trees to the deviant outcome(s). Thus, the facility of automatically generating fault trees for any named outcome is an essential aid to identifying the primary cause(s) of the current faulty states on the plant. Considerable research has been devoted to computer--aided generation of fault trees (Fussel, 1973, Apostolakis, Salem, and Wu, 1978, Lapp and Powers, 1977), including the handling of fault propogation upstream as well as downstream (Martin-Solis, Andow and Lees, 1977).

In the method presented here, fault trees are generated from Operability Study records. The cause and effect combinations for process lines are recorded as Cause Equations. Deviations in outlet lines, from major process equipment, are connected to deviations in inlet lines by means of the appropriate nodal responses in Symptom Equations.

Operator Response to an Alarm

When there is likely to be limited time available, for fault finding and corrective action, between an alarm sounding and a hazardous event, a logical sequence of fault finding actions, which minimises the probable duration of fault finding, must be worked out in advance. The purpose is to devise an optimal fault finding strategy, based on the Half Split approach, which minimises the probable time for fault finding, fault diagnosis and corrective action.

In an unpublished paper, Christer and Shields (1977) recommended that the time required to check each possible fault, should be divided by the probability that the fault has caused the alarm. The optimal order for checking faults is in ascending order of this quotient.

The operators must practice this sequence of actions, and be timed, in order to ensure that the strategy is optimal in reality. They must not deviate from this strategic sequence of fault finding in the event of a genuine alarm. They would follow flowcharts similar to Figures 9 and 10.

Case Study

This paper illustrates these types of aid by means of a practical example, based on the recovery of acetone, from an aqueous mixture, by distillation at reduced pressure. The cause and effect responses of the plant were deduced by an Operability Study (Chemical Industries Safety and Health Council, 1977). This information was written down as Cause Equations and Symptom Equations some of which are reproduced in the Appendix; more details of computer-aided Operability Studies are to be published shortly (Lihou, 1980).

OPERABILITY STUDIES

These are structured studies of the Process and Instrumentation (P & I) diagram of plants, aimed at discovering potential causes of hazardous consequences, or operating difficulties, which could lead to loss. In conducting the study, deviations in process lines are identified by Key Words. The first Key Word is a Property Word which may be used to define normal operating conditions. The second Key Word is one of seven Guide Words which describe the deviation. Possible causes of these devistions are sought and, if the consequence is hazardous, action is taken to minimise the chance of the cause or its consequences.

In the method presented here, for storing all the cause and effect combinations which are considered during an Operability Study, the Key Words are identified by number indices which are listed in brackets, following the line identification. This is followed by a string of causes, comprising the Cause Equation, separated by + to indicate OR and * to indicate AND. The indices used to define the Key Words are shown in Table 1. Also shown on this table are the components which occur in the acetone recovery process, which is used as a case study. When describing the deviation (Flow, As well as) it is convenient to identify the contaminating component by a third number. Similarly, concentration deviations should have a component number.

Causes due to faulty equipment are identified in the Cause Equation by the equipment identifying alpha-numeric code, followed by a single symbol in brackets. The meaning of this code is illustrated in Table 2.

L333(531) = L328(531)+ HE104(L)*{L333(32)+L303(33)}

This equation means Line 333 (Concentration, More, Acetone) is caused by L328 (Concentration, More, Acetone) OR HE104 tubes leaking AND {Line 333 (Pressure, Less) OR Line 303 (Pressure, More)}.

Table 1 - Meaning of index numbers in brackets following a line number or a node number

Index Number	Property Word	Guide Word	Component
1	Flow	No	Acetone
2	Temperature	Less	Water
3	Pressure	More	Steam
4	Level	As well as	Air
5	Concentration	Part of	Ammonia
6	Heat	Reverse	
7	Cool	Other than	

ACETONE RECOVERY PLANT

The P and I diagrams are shown in Figures 1 and 2. The latter shows the distillate receiver drum D101 in more detail than on Figure 1, because the flooding of D101 is a hazardous event which is used to illustrate the development of an optimal fault finding strategy, following the sounding of alarm LAH102.

In this plant, a liquid feed containing 98% acetone in water at 35°C and 3.5 bar absolute, is separated into a distillate product containing 0.3% water and a waste residue containing 0.01% acetone. The distillation column C101 contains 34 valve trays. The top of the column operates at 43°C and 457mm Hg absolute. The bottom of the column operates at 98°C and atmospheric pressure.

Most of the symbols on Figure 1 and 2 are common enough to be generally understood (Austin, 1979). Some less common instrument symbols are listed below.

LSH = High Level Switch
LAH = High Level Alarm
LXSL = Low Level Switch requiring resetting
LSL = Low Level Switch
LAL = Low Level Alarm
⋈ = Non-Return Valve

Nodes

The points where streams enter and leave major equipments are considered as nodes, and each node is given a unique number. If immiscible streams flow through the equipment and react or exchange heat or material, these streams are each given unique node numbers. For example, the liquid and vapour streams passing

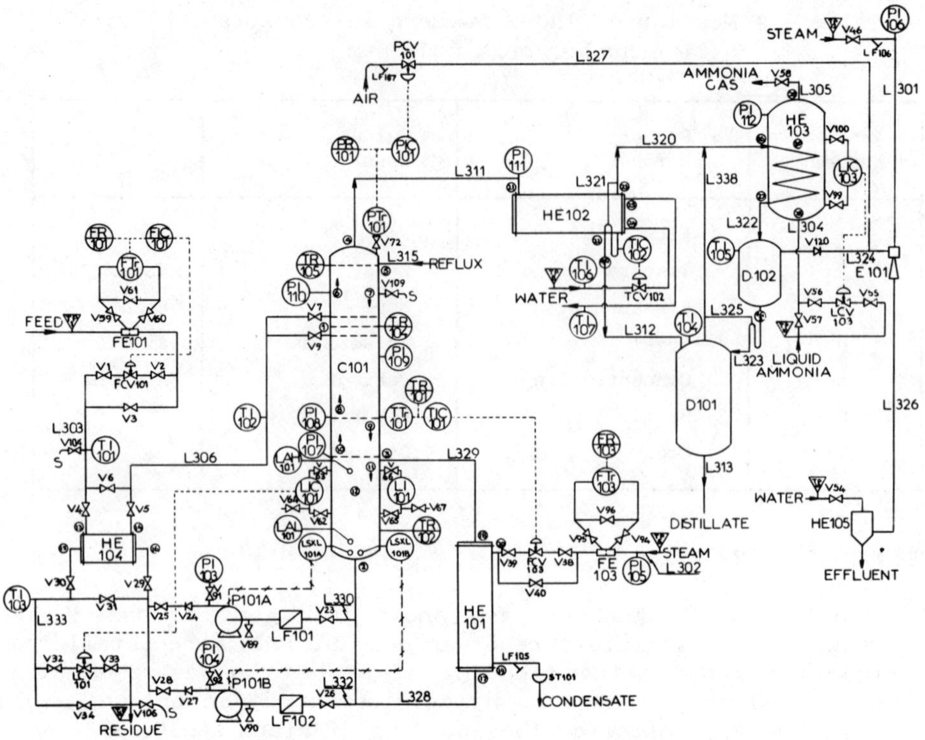

Figure 1. P&I Diagram for an acetone recovery plant

in a distillation column would be given separate node numbers for the enriching section above a side stream, for the enriching section below the side stream but above the feed plate and for the stripping section. The location of nodes is shown by their numbers in small circles on Figure 1. In the Symptom Equations, the nodes are identified by N and the node number; for example N1(13) means More Flow into C101 at node 1.

Examples of Cause Equations and Computer Storage

L303(13) = FE101(-1)+V61(1)+V59(0)+FIC101(1)+FCV101(1)
+V3(1)*V1(1)*V2(1).

This equation means Line 303 (Flow, More) is caused by FE101 indicating too low (eroded) OR V61 open OR V59 blocked OR FIC101 set too high OR FCV101 stuck open OR (V3 open AND V1 open AND V2 open).

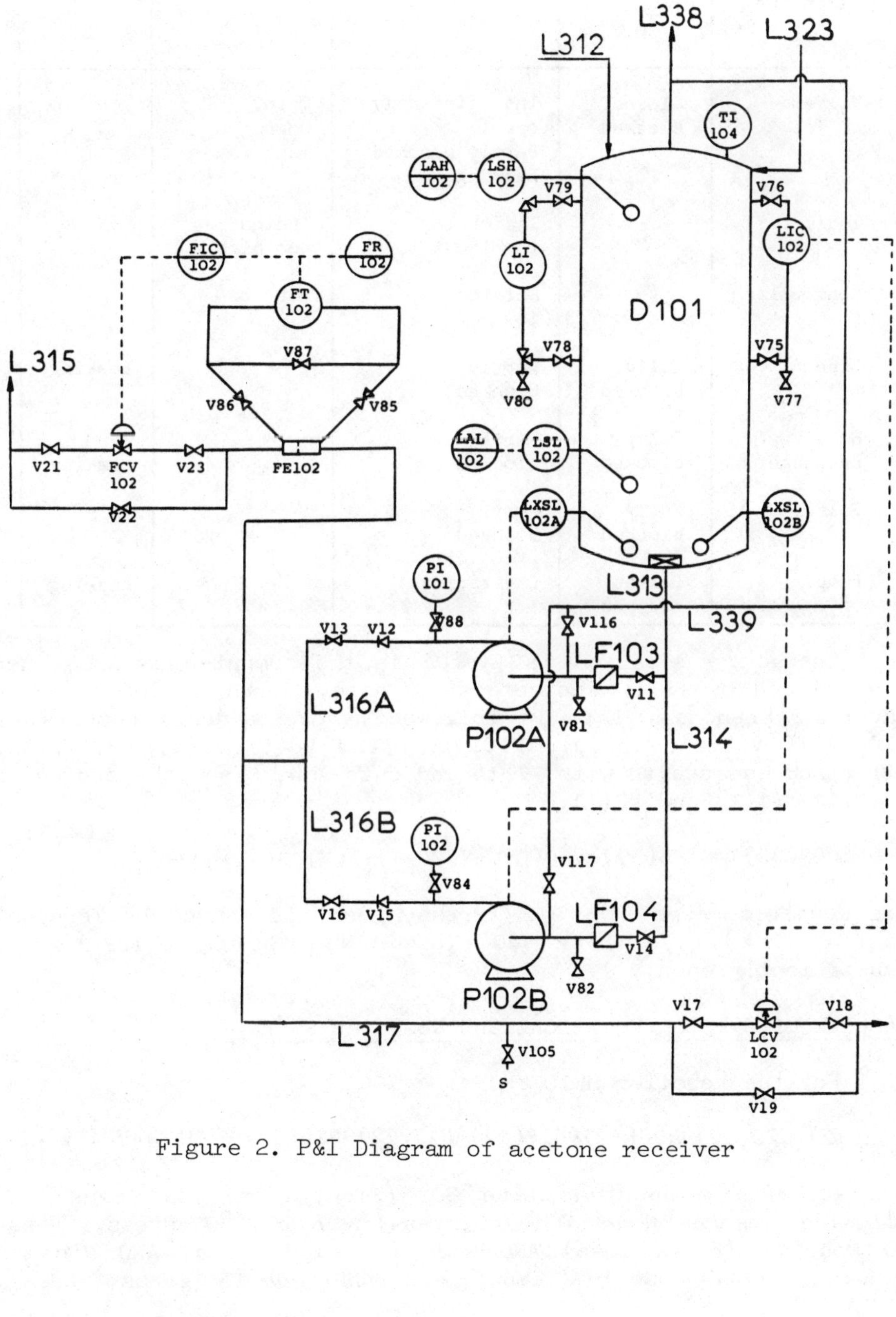

Figure 2. P&I Diagram of acetone receiver

Table 3 - Dynamic state matrices for events leading to acetone loss in the residue

Equipment	Index Numbers			
	0	-1	1	L
Valve	Closed. Blocked	Insufficiently open. Partly blocked. Leaking	Open. Open too much	-
Indicator	-	Indicating too low	Indicating too high	-
Controller	-	Set too low	Set too high	-
Line	Fully blocked	Partly blocked	-	Leaking
Heat Exchanger	Fully blocked	Partly blocked	-	Tube leaking
Filter	Fully blocked	Partly blocked	-	-
Pump	Failed stopped	Cavitating	-	Leaking

The above equation is stored in the computer as L303 then 13, followed by +1 to indicate an OR gate, then 6 to indicate the number of branches. This is followed by the address numbers of FE101(-1), V61(1), V59(0), FIC101(1) and V3(1)*V1(1)*V2(1). The latter address starts with -1 to indicate AND, 3 and the addresses of V3(1), V1(1) and V2(1).

$$L302(32) = TP6(32)+V40(0)*\{V38(-1)+FCV103(-1)+V39(-1)\}$$

This equation means Line 302 (Press, Less) is caused by Terminal Point 6 (Press, Less) OR V40 closed AND {V38 OR FCV103 OR V39 insufficiently open}.

Examples of Symptom Equations and Computer Storage

For the reboiler HE101:

$$L302(12)\rightarrow N20(12)*N20(32)*N18(12)*N17(12)*N19(12)*N19(22)*N19(32)$$

This equation means that Line 302 (Flow, Less) will cause the following symptoms Node 20 (Flow, Less) AND Node 20 (Press., Less) AND Node 18 (Flow, Less) AND Node 17 (Flow, Less) AND Node 19 (Flow, Less) AND Node 19 (Temp., Less) AND Node 19 (Press., Less).

HE101(L)→N18(13)*N20(13)*N19(12)

This equation means that HE101 (Leaking) will cause the following symptoms Node 18 (Flow, More) AND Node 20 (Flow, More) AND Node 19 (Flow, Less).

For the distillation column C101:

L329(12)→N3(12)*N10(12)*N10(32)*N11(12)*N11(22)*N12(22)*N2(12)*
N2(22)*N9(22)*N8(12)*N8(32)*N6(12)*N4(12)*N6(32)

This equation means that Line 329 (Flow, Less) will cause the following symptoms Node 3 (Flow, Less) AND Node 10 (Flow, Less) AND Node 10 (Press, Less) AND Node 11 (Flow, Less) AND Node 11 (Temp., Less) AND Node 12 (Temp., Less) AND Node 2 (Flow, Less) AND Node 2 (Temp., Less) AND Node 9 (Temp., Less) AND Node 8 (Flow, Less) AND Node 8 (Press., Less) AND Node 6 (Flow, Less) AND Node 4 (Flow, Less) AND Node 6 (Press., Less).

The above Symptom Equation is stored as L329 then 12 followed by 14 to indicate the number of nodal responses, followed by the sequence of numbers 312, 1012, 1032, 1112, 1122, 1222, 212, 222, 922, 812, 832, 612, 412, 632.

ACETONE LOSS IN RESIDUE

Imagine the situation that the operator takes a sample via V106 and finds that the acetone concentration exceeds the specified maximum of 0.01%. The Cause and Symptom Equations are stored in the computer as illustrated above; such that by calling L333(531), More Concentration of Acetone in Line 333, the fault tree shown as Figure 3 will be displayed on a VDU. The continuations would be obtained by calling the appropriate events; thus L329(11), No Flow in Line 329 would produce Figure 4 on the VDU and L329(12) would produce Figure 5.

Note from the Appendix that the Cause Equation for L330(531) is N2(22); i.e. Node 2 (Temp., Less). Node 2 is listed in the Symptom Equations for C101 and the only two faults producing N2(22) in this set of Fault Symptom Matrices is L329(11) and L329(12). Similarly, the Cause Equations for L329(11) and L329(12) contain only N18(11) and N18(12), respectively. Searching the Symptom Equations of HE101 for these nodal responses at N18, produces the first row of causes on Figures 4 and 5, respectively.

On Figure 5, the cause of L328(23) is N2(23) which is untenable with N2(22) higher up the fault tree, on Figure 3.

The complete fault tree, including the continuation of Figure 3 through L303(33) and L333(32) is very extensive; particularly the causes leading to a low pressure in Line 333.

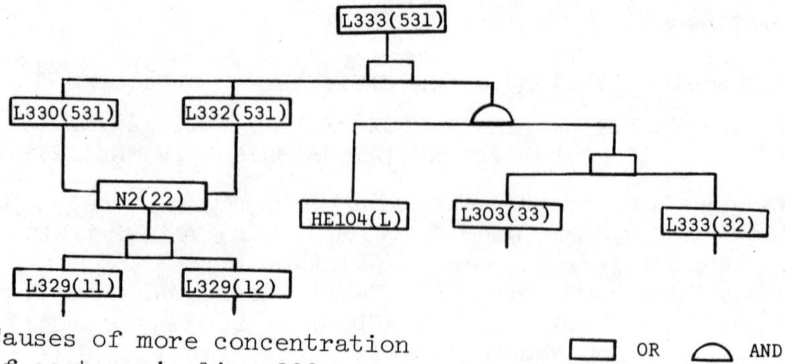

Figure 3. Causes of more concentration of acetone in line 333

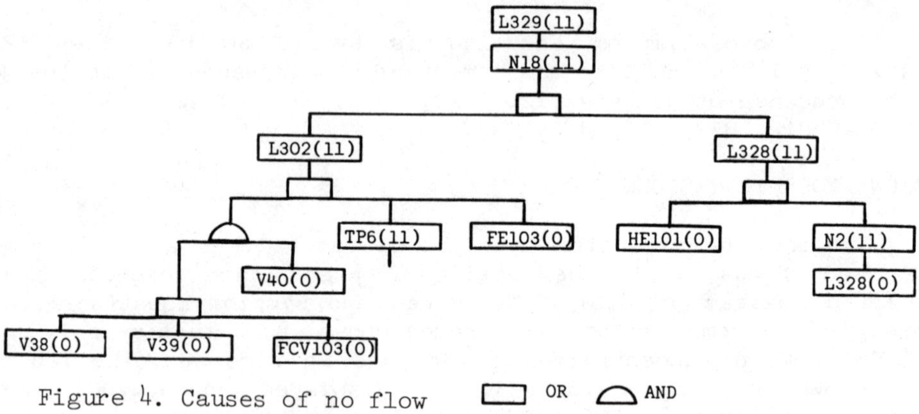

Figure 4. Causes of no flow in line 329

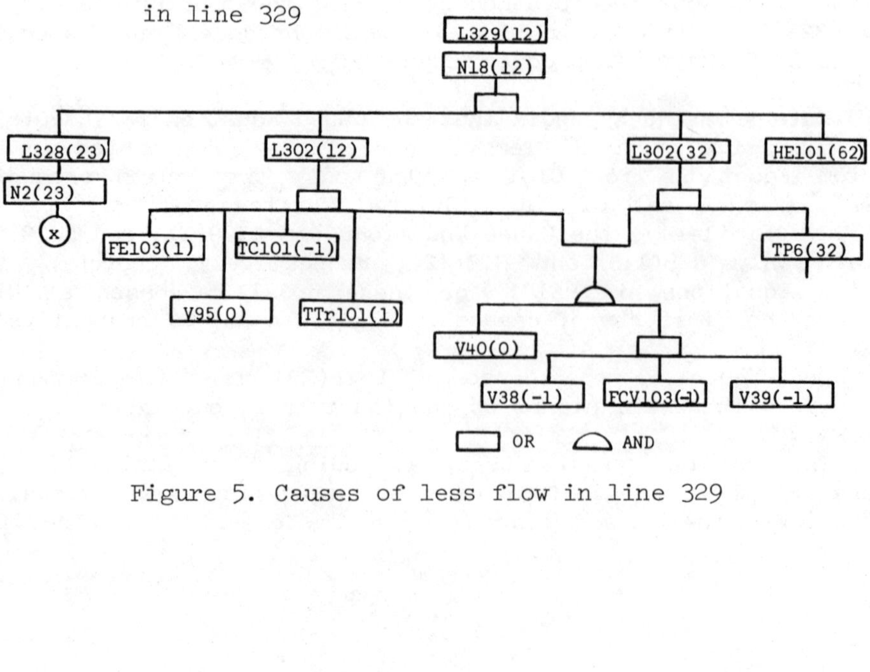

Figure 5. Causes of less flow in line 329

This is clearly a highly complex situation, requiring skillful fault finding to locate the primary cause of the faulty states. However, the evolution of the present State Matrix should provide clues for finding the primary cause. The Acetone Recovery Plant has numerous indicating instruments which could have been fed directly to an on-line computer. The advantage of such computer surveillance can be demonstrated by considering the evolution of the faulty states which started on Figure 5 with the temperature transmitter TTr101 reading too high; that is TTr101(1).

Events Leading to L333(531)

The evolution of the faulty State Matrix are described below and shown in Table 3 as nodal responses. The nodes are numbered in small circles on Figure 1. The appropriate Symptom Equations are listed in the Appendix.

Events 1. TTr101(1) causes L302(12) which results immediately in responses at the nodes around the reboiler HE101.

Events 2. From Figure 5 it can be seen that L302(12) causes L329(12) which results in further responses at nodes around the column C101 and the ultimate outcome L333(531) as shown on Figure 3.

In Table 3, the symptoms which remain normal are shown with 0; those which would be low are indicated by L. Measurable symptoms are underlined; but note that TIC101 and TR101 would both indicate normal although, in reality, the temperature at Node 9 is low. The error is due to TTr101 reading too high.

OPTIMAL RESPONSE TO LAH102

Figure 2 shows in more detail the instruments and equipment associated with the distillate receiver D101. It was calculated that after the alarm LAH102 sounds, the operator would have about 10 minutes to locate and correct the fault before D102 would fill and liquid acetone would be drawn into the ejector E101, shown on the right of Figure 1.

The events which could lead to a high level in D101 are shown as fault trees in Figures 6, 7 and 8. It is possible to obtain fault trees for L316B(11) and L316B(12) similar to Figures 7 and 8, respectively, by making the following exchanges: L316B replaces L316A, V14 replaces V11, LF104 replaces LF103, V82 replaces V81, P102B replaces P102A, LSXL102B replaces LSXL102A and V16 replaces V13.

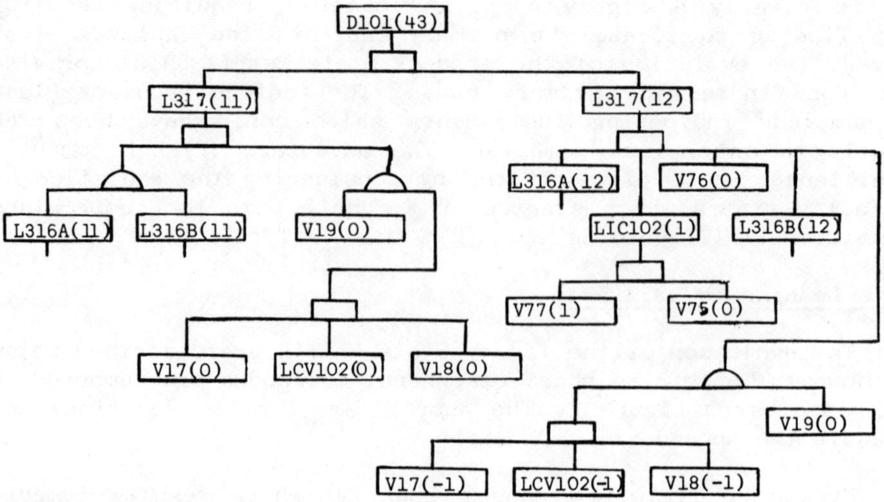

Figure 6. Causes of high level in Distillate receiver D101

OR AND

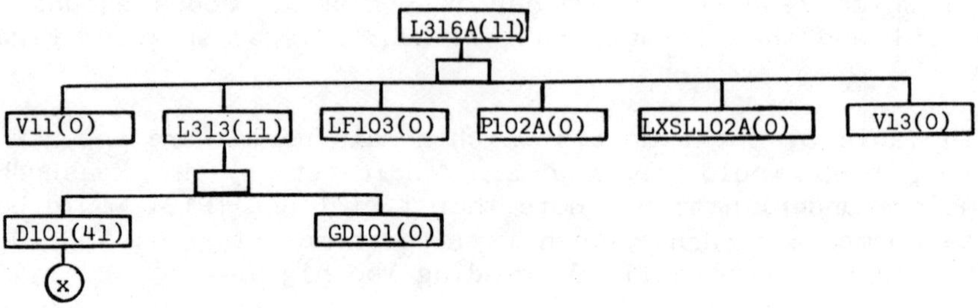

Figure 7. Causes of no flow in line 316A

OR AND

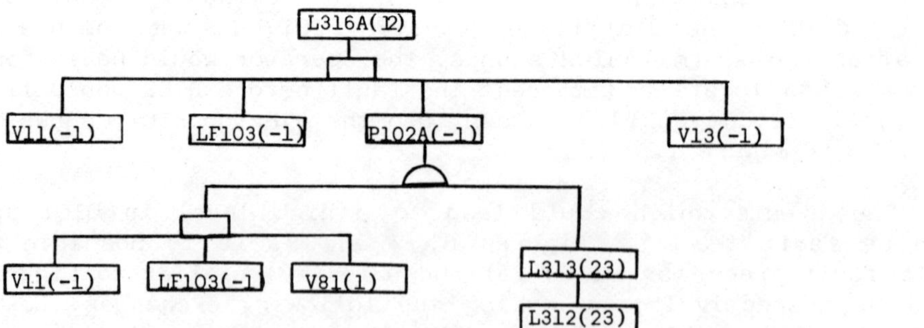

Figure 8. Causes of less flow in line 316A

OR AND

Event Probabilities

The probabilities were calculated using the Weibull distribution from equation 1 in which the time interval t was taken as 1 year.

$$F(t) = 1 - \exp\{-(t/\eta)^{\beta}\} \quad (1)$$

The characteristic age η is equal to the mean time between failures μ for those equipments which do not deteriorate with time; that is when $\beta = 1$. Where the shape parameter is not unity, η was calculated from μ by means of equation 2.

$$\eta = \mu/\Gamma(1+1/\beta) \quad (2)$$

The appropriate constants needed to evaluate equation 1 are listed in Table 4. These data are listed for the case where L316A is in use; the data for equipments in L316B are related to the corresponding equipments in L316A.

The low level switch LSXL102A is a safety device which is tested at intervals of 0.3 years, by deliberately causing a low level in D101. The probability that LSXL102A will stick when an unscheduled low level occurs in D101, can be estimated by its Fractional Dead Time.

$$FDT = (T/\eta)^{\beta}/(\beta+1) \quad (3)$$

For T = 0.3 and η and β from Table 4, FDT = 0.00025 per year. At least one genuine event D101(41) will occur in a year.

If LCV102 sticks, LCV102(0), V17 and V18 would be closed to carry out repairs. The probability that V19 will be closed in error was taken as 0.01 due to human error per maintenance action.

Prob.{V19(0)*{V17(0)+LCV102(0)+V18(0)}} = .01x.393 = .00393

Because V75 may become blocked by gradual deposition in the lower line to LIC102, it is purged periodically via V77. The optimal interval T between purging was calculated to minimise the probability that LIC102 will control high in one year. Taking the human error rate as 0.01 per action, and the Weibull parameters for V75 to define LIC102(1),

Prob.{V77(1)} = Prob.{V75(0)} = .01/T (4)

Prob.{LIC102(1)} = 1-exp{$-T^3$} + .02/T (5)

The minimum value of equation 5 was given by T equal to 0.2875 years or 15 weeks. With this purging interval Prob. {LIC102(1)} is 0.093 per year.

Table 3 - Dynamic state matrices for events leading to acetone loss in the residue

EQUIPMENT		C101									HE101			
Events	Prop. Words	Node Numbers												
		2	3	4	6	8	9	10	11	12	17	18	19	20
1) TTr101(1)	Flow	0	0	0	0	0	0	0	0	0	L	L	L	$\underline{L}$
causes	Temp.	0	0	0	0	0	0	0	0	0	0	0	L	0
L302(12)	Press	0	0	0	0	0	0	0	0	0	0	0	L	L
	Level	0	0	0	0	0	0	0	0	0	0	0	0	0
2) L302(12)	Flow	L	L	L	L	L	L	L	L	0	L	L	L	$\underline{L}$
causes L329(12)*	Temp.	$\underline{L}$	0	$\underline{0}$	0	0	($\underline{L}$)	0	0	$\underline{L}$	0	0	L	0
L330(531)*	Press.	0	0	$\underline{0}$	$\underline{L}$	$\underline{L}$	0	$\underline{L}$	0	0	0	0	L	L
L333(531)	Level	0	0	0	0	0	0	0	0	0	0	0	0	0

Table 4 - Failure data for equipments

Equipment	β	μ (year)	η (year)	Probability of fail to "danger" in 1 year
LF103	2	0.3	.34	.9998
GD101	1	5	5	.18
LCV102	1	2	2	.393
LXSL102A	3	2.6	3	.00025
V75	3	.87	1	.0235
TIC102	1	3	3	.2835
TCV102	2	2	2.3	.2212
HE102	3	4	4.6	.0102

The probabilities of TP8(23) and TP8(12) were taken as insignificant. TIC102(1), TCV102(-1) and HE102(72) were assumed to be independent events; so Prob.{L312(23)} = 1-.7165x.7788x.9898 = 0.448 per year.

Optimal Checking Sequence

The optimal strategy was determined by measuring the time taken to walk to different equipments and carry out prescribed observations. These times were divided by the probability that the equipment would be the cause of D101(43). The sequences shown on Figures 9, and 10 were devised so that equipments with the lowest ratio of Checking Time/Probability are checked first, and so on, in ascending order of this ratio.

Some instruments are in the control room, others are out on the plant. It was assumed that the operator would be in the control room when LAH102 sounds. There would be only one operator available to carry out the checks, under normal circumstances.

Check FIC102 and FR102. These instruments are in the control room and to observe them takes half a minute. The logical check procedure is shown in Figure 9 top. This strategy directs the operator to one of three further fault-finding actions. These are presented below. Note in relation to Figure 9 that the response of FIC102 to raising the set point may take up to one minute; but this time of 1.5 minutes is worth spending in that it will enable the operator to decide which of the following events need checking first:

1) LCV102 stuck and LIC102(1)
2) No Flow in L316A
3) Less Flow in L316A

Check LCV102 Stuck and LIC102(1). Checking times were measured as follows; probabilities were calculated as described in the previous section.

Go to LIC102	= 3min.
Check V75 open, open V77; is V75(0)?	= 1.5min.
Set LIC102(-1) and go to LCV102	= 2min.
Go directly to LCV102	= 2min.
Open V19	= 0.25min.

$$\frac{\text{Total Time to Check LCV102 Stuck}}{\text{Prob.\{LCV102 stuck\}}} = \frac{3+2}{.393} = \underline{12.7}$$

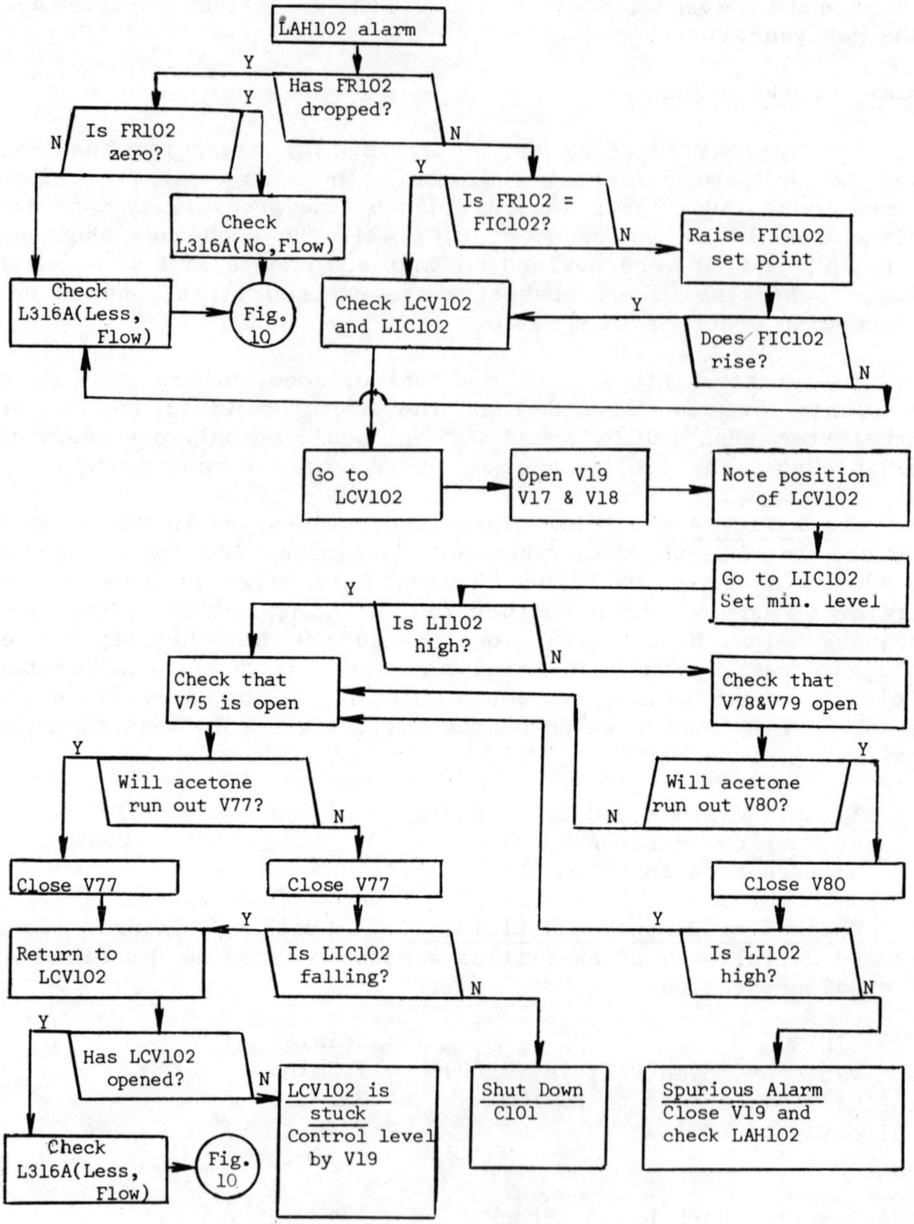

Figure 9. Sequence of control room checks and checking LCV102 and LIC102 on plant - following LAH102 alarm

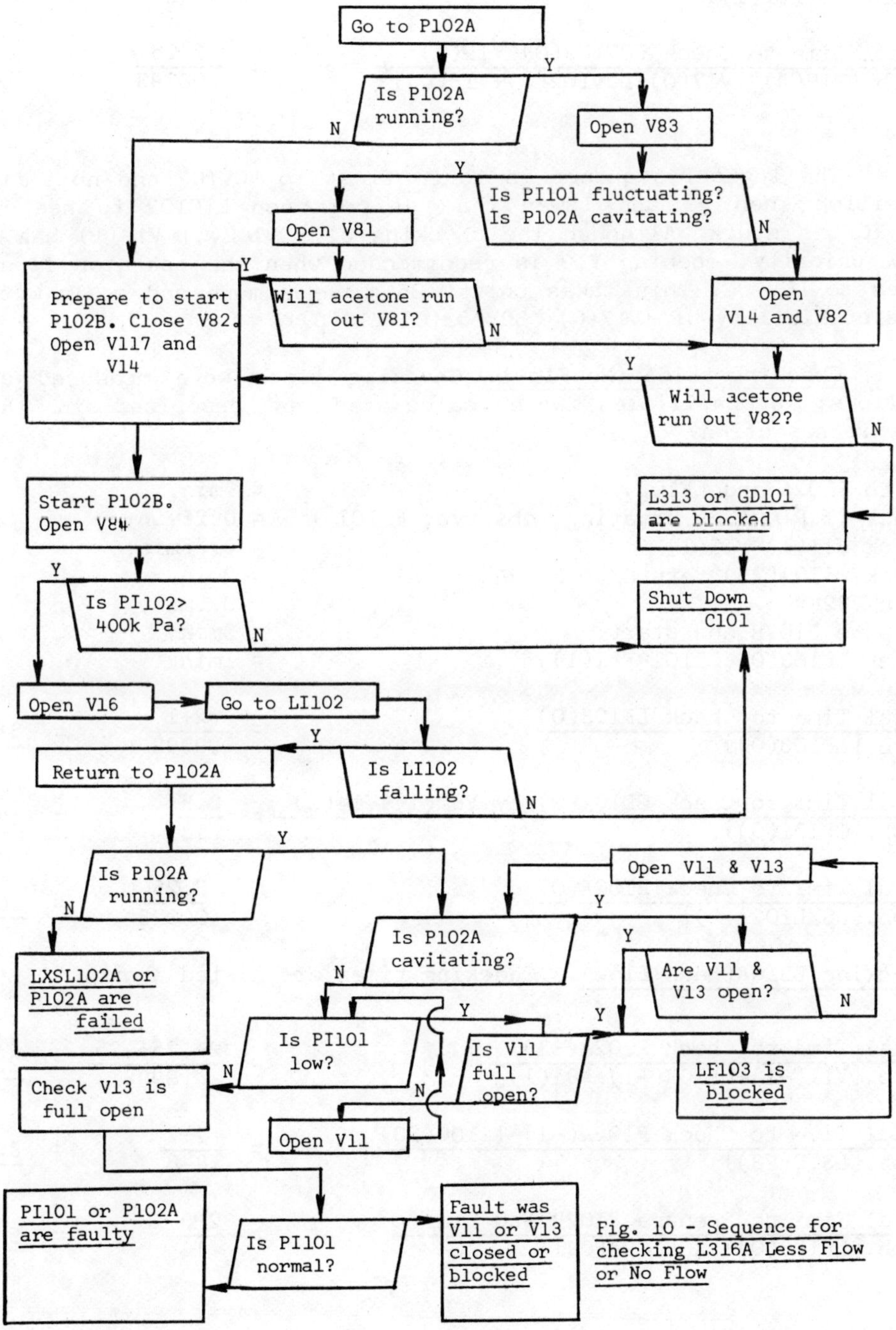

Fig. 10 - Sequence for checking L316A Less Flow or No Flow

Figure 10

$$\frac{\text{Total Time to Check LIC102(1)}}{\text{Prob.\{LIC102(1)\}}} = \frac{3+1.5}{.093} = \underline{48}$$

$$\frac{\text{Total Time to Check LCV102(0)*V19(1)}}{\text{Prob.\{V19(0)*\{V17(0)+LCV102(0)+V18(0)\}\}}} = \frac{2.25}{.00393} = \underline{573}$$

The logical sequence is to go first to LCV102 and note its position then go and lower the set point on LIC102 to see if LCV102 is stuck. Although the checking of LCV102(0)* V19(1) has a low priority, opening V19 is recommended when the operator first goes to LCV102. This takes only 0.25 minutes more and deals with the possibility of {V17(0)+ LCV102(0)+V18(0)}.

<u>Checking L316A (No Flow).</u> Checking times were measured as follows; probabilities were calculated as described in the previous section.

Go to P102A and LF103	= 2min.
Check if P102A is rotating, observe, PI101	= 0.25min.
Check V11(1)*V13(1)	= 0.25min.
Check TI104(23)*return to P102A	= 2min.
Open V82*V14	= 0.5min.
Prepare P102B and start	= 3min.
Check LF103(0),PI101*V81(1)	= 1min.

$$\frac{\text{Total Time to Check LF103(0)}}{\text{Prob.\{LF103(0)\}}} = \frac{2+1}{.9998} = \underline{3}$$

$$\frac{\text{Total Time to Check GD101(0) by V82(1)*V14(1)}}{\text{Prob.\{GD101(0)\}}} = \frac{2.5}{.18} = \underline{14}$$

$$\frac{\text{Total Time to Check P102A(0)}}{\text{Prob.\{LSXL102A(0)\}}} = \frac{2.25}{.00025} = \underline{9000}$$

<u>Checking L316A(Less Flow).</u> Checking times are listed above.

$$\frac{\text{Total Time to Check P102A(-1)}}{\text{Prob.\{V11(-1)+LF103(-1)+V81(1)\}}} = \frac{2+.25+.25}{.9998} = \underline{2.5}$$

$$\frac{\text{Total Time to Check P102A(-1)*TI104(23)}}{\text{Prob.\{L312(23)\}}} = \frac{2+2}{.448} = \underline{8.9}$$

$$\frac{\text{Total Time to Prepare P102B and Start}}{\text{Prob.\{LF103(0)+LXSL102A(0)\}}} = \frac{2+3}{1} = \underline{5}$$

The logical sequence which was specified for checking L316A(11) and L316A(12) is shown on Figure 10. Although checking that P102A is still running has a low priority (ratio = 9000), it is at the top of the tasks in Figure 10; because it can be done with a trivial additional time when checking the highest priority; namely LF103(0). A pressure gauge was installed, subsequently, in order to enable the operators to check for LF103(0) or LF103(-1) and to arrange for preventive maintenance of LF103.

ACKNOWLEDGEMENT

The assistance of Mr. R. Rahimi, a graduate of Aston University, in carrying out the Operability Study on this process is gratefully acknowledged.

REFERENCES

Apostolakis, G.E., Salem, S.L., and Wu, J.S., 1978, "CAT - A Computer Code for the Automated Construction of Fault Trees", Report EPRI-705.

Austin, D.G., 1979, "Chemical Engineering Drawing Symbols", George Godwin, London.

Berenblut, B.J. and Whitehouse, H.B., 1977, "A Method for Monitoring Process Plant Based on a Decision Table Analysis", Chem. Engr., London, 318:175.

Chemical Industries Safety and Health Council, 1977, "A Guide to Hazard and Operability Studies", Chemical Industries Assn., London.

Christer, A.H. and Shields, S., 1977, "Diagnostic Procedures in Fault Finding", paper submitted for Production Congress, I. Chem. E. Midlands Branch.

Fussell, J.B., 1973, "A Formal Methodology for Fault Tree Construction", Nucl. Sci. Engng., 52:421.

Lapp, S.A. and Powers, G.J., 1977, "Computer-Assisted Generation and Analysis of Fault Trees", in "Loss Prevention and Safety Promotion in the Process Industries", Vol. 2 DECHEMA, Frankfurt, p. 377.

Lihou, D.A., 1980, "Computer-Aided Operability Studies for Loss Control", paper to be presented at symp. on "Loss Prevention and Safety Promotion in the Process Industries", Basle, Switzerland.

Martin-Solis, G.A., Andow, P.K., and Lees, F.P., 1977, "An Approach to Fault Tree Synthesis for Process Plants", in "Loss Prevention and Safety Promotion in the Process Industries", Vol. 2 DECHEMA, Frankfurt, p. 367.

Morris, R., 1976, "Logical Approach to Fault Finding", Symp. on Diagnostic Techniques, Teesside Polytechnic, U.K.

Munday, G., 1977, "On-Line Monitoring and Analysis of the Hazards of Chemical Plant", in "Loss Prevention and Safety Promotion in the Process Industries", Vol. 2, DECHEMA, Frankfurt, p. 273.

Rigney, J.W., 1966, "Measurement and Prediction of Cognitive Loadings in Corrective Maintenance Tasks: A Bayesian Approach", Tech. Report No. 46, Dept. of Psychology, Univ. of South California.

Shepherd, A., 1976, "The Application of Fault Symptom Matrices to Improving Diagnostic Performance", Symp. on Diagnostic Techniques, Teesside Polytechnic, U.K. Sept.

APPENDIX

The complete Operability Study of the plant has been recorded as Cause Equations for lines and Symptom Equations for equipment. Only those equations which are needed to illustrate this paper are recorded below.

Cause Equations

```
L302(11) = FE103(0)+V40(0)*{V38(0)+FCV103(0)+V39(0)}+TP6(11)
L302(12) = FE103(1)+V95(0)+TIC101(-1)+TTr101(1)+V40(0)*{V38(-1)+FCV103(-1)+
           V39(-1)}
L302(32) = TP6(32)+V40(0)*{V38(-1)+FCV103(-1)+V39(-1)}

L303(33) = TP1(33)+HE104(-1)+L306(33)

L312(23)=TIC102(1)+TCV102(-1)+HE102(72)+TP8(23)+TP8(12)

L313(11) = D101(41)+GD101(0)
Note:  GD101 is the grille in the bottom of D101 at the inlet to L313.
L313(23) = L312(23)

L316A(11) = L313(11)+V11(0)+LF103(0)+P102A(0)+LSXL102A(0)+V13(0)
L316A(12) = V11(-1)+LF103(-1)+P102A(-1)+V13(-1)

L316B(11) = L313(11)+V14(0)+LF104(0)+P102B(0)+LSXL102B(0)+V16(0)
L316B(12) = V14(-1)+LF104(-1)+P102B(-1)+V16(-1)

L317(11) = L316A(11)*L316B(11)+V19(0)*{V17(0)+LCV102(0)+V18(0)}
L317(12) = L316A(12)+L316B(12)+LIC102(1)+V76(0)+V19(0)*{V17(-1)+LCV102(-1)+
           V18(-1)}

L328(11) = N2(11)+HE101(0)
L328(23) = N2(23)

L329(11) = N18(11)
L329(12) = N18(12)

L330(531) = N2(22)
L330(32) = P101A(-1)
P101A(-1) = N2(23)*{N2(32)+V23(-1)+LF101(-1)}+V89(1)

L332(531) = N2(22)
L332(32) = P101B(-1)
P101B(-1) = N2(23)*{N2(32)+V26(-1)+LF102(-1)}+V90(1)

L333(32) = {L330(32)+V25(-1)}*{L332(32)+V28(-1)}+V29(-1)*V30(1)+V29(1)*
           V30(1)*V31(1)
L333(531) = L330(531)+L332(531)+HE104(L)*{L303(33)+L333(32)}

D101(43) = L317(11) + L317(12)
```

Symptom Equations

L302(11)→N20(11)*N18(11)*N17(11)*N19(11)*N19(22)

L302(12)→N20(12)*N20(32)*N18(12)*N17(12)*N19(12)*N19(22)*N19(32)

L302(32)→N20(22)*N20(32)*N18(12)*N17(12)*N19(22)*N19(32)

L328(11)→N17(11)*N18(11)*N19(11)*N20(11)

L328(23)→N17(12)*N17(23)*N18(12)*N20(12)*N20(33)*N19(12)*N19(33)

L329(11)→N3(11)*N11(12)*N10(11)*N10(32)*N8(11)*N8(32)*N6(11)*N6(32)*N4(11)*
N9(12)*N9(22)*N12(22)*N2(12)*N2(22)

L329(12)→N3(12)*N11(12)*N10(12)*N10(32)*N8(12)*N8(32)*N6(12)*N6(32)*N4(12)*N11(22)*
N9(22)*N12(22)*N2(12)*N2(22)

A METHOD FOR OPTIMIZING HUMAN PERFORMANCE IN DETECTING AND DIAGNOSING MISSION AVIONICS FAULTS

W.B. Gaddes and L.R. Brady

IBM Federal Systems Division
Owego, NY 13827, U.S.A.

INTRODUCTION

Mission avionic systems have realized, and will continue to realize, significant advances in the application of automated maintenance test programs and on-line readiness testing. This is primarily due to the extensive use of digital computers, digital interfaces, and operational/diagnostic software. Dramatic advances have been made in the reliability of hardware in the quest for increased operational availability. Fault tolerance, parallel data busing, on-line diagnostics, and reconfiguration offer additional promises toward meeting the ideal "mission failure-free" avionics system.

However, current generations of mission avionics systems recently deployed and those in development could realize significant additional enhancements in operational availability through the application of a method developed by the authors. It ensures that human performance capabilities and limitations in fault detection and isolation are given due consideration during early design and development. As the developments of enhanced automated maintenance test programs and related diagnostic capabilities have proceeded, there has been a tendency to de-emphasize the role of the operator and maintainer in the detection and diagnosis of system faults, and to assume that the maintenance test programs will be the panacea. The maintenance test program capabilities are specified in terms of "percentage of faults isolated to one, two, or more replaceable assemblies", a statistical definition not particularly enlightening regarding the specific fault groups which might be transparent to or defaulted by the test programs.

The lack of emphasis on human performance by design and development groups has been due primarily to the lack of a clear method for defining the human role early in the development phase. The human role in the system, particularly with respect to human performance contributions in case of mission fault occurrence, is difficult to define on a timely basis. However, the consequence of this lack of definition is serious, often resulting in unacceptable meantimes-to-repair, technical manual deficiencies, and a large number of "no trouble found" field returns which place a burden on the logistics system. A dependence upon the statistical characterization of diagnostics in terms of "percentage of faults detected and isolated", is unacceptable. The human role must be defined early and provisions made to ensure that human performance is optimized in this critical area. The authors estimate that 8 to 10 percent of the fault cases in advanced avionics will require human operator or maintainer intervention and involvement using operational information or system functional behavior transparent to the maintenance test programs. It is this key area that the authors feel must be addressed early, to ensure that frantic efforts are not required during the first several years of deployment to correct deficiencies in technical manuals, training, and in data regarding system behavior under specified fault cases.

Human performance in the detection of and diagnosis of mission avionics faults must be given more than cursory treatment during early design and development, due to a still-significant number of residual faults that are either transparent to maintenance test programs or the test programs default, or are of a nature that require mission functional performance data to be assessed by the human.

The host of system performance standards and specifications and statistical characterization of maintenance test program capabilities is not adequate to handle the total population of expected faults; human performance must take us the last 5 or 10 percent of the way.

TYPICAL PROBLEMS ENCOUNTERED IN THE DESIGN, DEVELOPMENT AND INITIAL DEPLOYMENT OF MISSION AVIONICS SYSTEMS WITH HEAVY DEPENDENCE UPON AUTOMATED MAINTENANCE TEST PROGRAMS

A review of "lessons learned" during the early deployment of mission avionics systems with a heavy dependence upon automated maintenance test programs (MTPs) has shown that due consideration of human operator and maintainer contribution to fault detection and diagnosis early in design could have resulted in increased operational availability. Undue demands upon the logistics system caused by an excessive "no trouble found" return rate could have been avoided as well. The major areas of concern to a design team involved in advanced mission avionics development should be:

a. Reducing the number of "false positives" (incorrect error callouts) to an absolute minimum. When this error rate exceeds a threshold of approximately 5%, the maintainers lose confidence in the MTP. In those situations where the false callouts are recognized as a limitation, emphasis should be placed upon supplementing the MTP with avionics "functional string" performance characteristics to permit the human to carry the detection diagnosis process one step further in a manual process.

b. Minimizing the size of fault groups; in view of the demands that this characteristic has upon the logistics system. In those cases where the MTP cannot isolate to a single replaceable unit or subassembly, emphasis should be placed upon defining the human operator or maintainer contribution; again with respect to what operational or unique "avionics functional string" information might be uniquely available to the human.

c. Defaults by the MTP and those classes of failures that are transparent to (not detected by) the MTP must be assessed. These classes of failures generally are not covered in an adequate manner by training and technical manuals. Often these failure modes may be detected and diagnosed by a systematic approach to what avionics functional capabilities are affected and which are not; again a logical role for the human operator or maintainer.

d. Maintenance of and configuration management of the MTPs to reflect engineering changes in the avionics system. This is not an early design or development issue, but MTPs generally do not reflect current hardware configurations, lagging the hardware changes by six months to a year, adding to the "false positive" problem and further diminishing maintainer confidence in the test programs.

e. Technical manuals and supporting logistics and maintenance engineering technical data reflect an undue degree of dependence upon maintenance test program accuracy, completeness, and currency. Early in developments, there exists a decided lack of data regarding exactly what the human can or must do when a fault is not detected, the test program defaults, or an unacceptably large number of suspected assemblies are listed.

f. Technical manuals and related technical data, including training and maintenance trainer requirements reflect the "idealization" of the maintenance test programs as well.

g. System design, particularly with respect to the "windows" and controls provided for the human operator and maintainer, is usually deficient with respect to their availability to access key functional performance data, or to establish modes of particular interest to a given fault mechanism. This design area also reflects an undue confidence in the ability of the maintenance test program as characterized by the statistical parameters of "percentage of faults isolated to one or more replaceable assemblies". Very nominal improvements in data access or control/switching mode capabilities for the operator or maintainer have been shown to improve maintenance capabilities dramatically.

h. Design of maintenance test programs generally reflects the subsystem or unit interconnection in terms of how performance is measured or how faults are detected and isolated. The authors suggest that the MTPs be designed on the basis of "mission functional strings" of avionics units, to be more in consonance with the way that humans must detect and diagnose failures. If this structure is implemented, the MTP can, with human intervention, provide the necessary "windows" into the system for the data needed for human problem solving.

Maintenance test program design, development and test generally occurs on a priority and on a schedule which is secondary to the objective of getting the prime hardware designed, developed, integrated and tested. As a consequence:

a. Maintenance test program software development often suffers in quality of approach (engineers often write the diagnostic program), quality assurance suffers, and formal in-process reviews of diagnostic software are almost non-existent.

b. The MTP software, due to the lack of a structured high quality approach, becomes difficult to maintain.

c. On hardware MTP test and verification often lags prime equipment development, slips schedule and ends up incompletely verified or tested at time of system deployment.

d. Developmental problems in prime equipment hardware or in the operational software often mask the real performance characteristics of the MTP until operational deployment, with unhappy consequences.

e. Management attention is not focused upon maintenance test programming or upon prime hardware design for test point or test data access by the human, particularly during early design and development.

f. Mission avionics assets (hardware, lab test facilities, personnel) are often not available during the appropriate early stages of design and development.

g. The most expeditious manner of fixing MTP problems is often selected, resulting in "bandaids" which in the long term are unacceptable.

h. Early field experience is not captured in an acceptable format and in data which characterizes the human performance difficulties or positive enhancements. Field experimental data are generally reported in programming trouble reports (PTRs) and in other forms which mask the real human issues.

These characteristics found in several recently deployed systems are not new or unusual. They reflect, however, a lack of due consideration of human capabilities and limitations regarding detection and diagnosis of mission avionics faults, particularly that class of faults transparent to or ambiguously called-out by the MTPs. The method described by the authors in the next section has been applied to an advanced mission avionics system scheduled for deployment in the 1983 time frame, with good results. If applied, the human performance "lessons learned" as described herein will be alleviated to a significant degree, resulting in reduced dependence upon contractor technical services, decreased down time, improved maintainer confidence in MTPs, lessened demand upon the logistics system, and increased operational availability.

A method has been developed by the authors which ensures that the requirements upon human operator and maintenance performance is specified in quantitative terms early in the design and development phase. Conversely, the appropriate use of human performance capabilities in the detection and diagnosis of mission avionics faults is ensured. The method uses a technique of describing the mission avionics system in terms of "system functional strings" or characterization of which units contribute to particular mission-oriented functions. Once this functional contribution of the respective avionics units is defined, human performance in manual fault detection and diagnosis can be understood and steps taken to optimize the human role.

DESIGN OF AN ADVANCED MISSION AVIONICS SYSTEM WITH FULL CONSIDERATION OF HUMAN CAPABILITIES AND LIMITATIONS IN FAULT DETECTION AND DIAGNOSIS

Method Objectives

The method developed by the authors ensures that human performance requirements for fault detection and isolation are understood early in the development process, and that design considerations are implemented to enhance human performance in this critical area. The key elements of the method are as follows:

a. Definition of a maintenance philosophy for manual fault detection and fault isolation by operator and maintainer that is mission function oriented for those cases where faults are non-detected or transparent to maintenance test programs.

b. Definition of the manual information, procedures and decision logic required of the human to enable fault detection, isolation and repair in those cases where diagnostic programs, operational readiness tests, and built-in tests default.

c. Establishment of a common design engineering, diagnostic software development, and human factor engineering data base to ensure that all these disciplines are working to common criteria and data.

d. Definition of input data for technical manuals to ensure that the manuals reflect information required to effect the desired human performance in detection, isolation, and repair of mission avionics faults.

e. Characterization of the human performance in a form and to a level of detail adequate for training and training system definition.

f. Characterization of the manual detection and isolation performance and system conditions necessary for development test and evaluation.

g. Utilization of a technique which considers the avionics system as a compilation of "mission functional strings", and utilization of this functional definition as the basis for defining the human "window" or entry into the system and the human role in detection and diagnosis of faults.

The method described herein has been utilized to provide the necessary data characterization of, definition of, and implementation of those necessary system attributes, (hardware and diagnostic software) technical manuals, and training for human involvement when diagnostic programs default.

Description of Method - Systems Approach to Manual Maintenance (SAMM)

Figure 1 describes the major steps in the System Approach to Manual Maintenance (SAMM), in which the definition of requirements

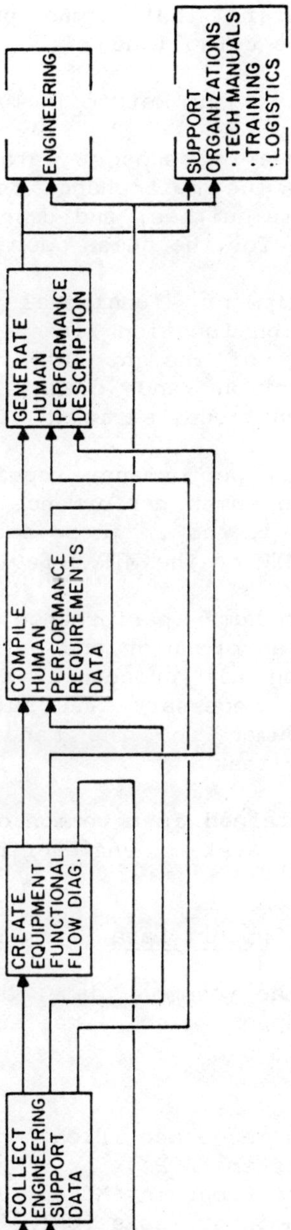

Fig. 1. Methodology for determining human performance requirements for manual maintenance.

for, and means for ensuring that human performance in fault detection and isolation are established.

The four basic steps in the method include:

a. Collection of engineering support data (1.0) provides the characterization of the maintenance test programs (MTPs), identification of ambiguities, and description of the means of access ("window") for the human operator and maintainer.

b. Creation of an equipment "functional flow diagram" (2.0) oriented along mission function lines. This is an important step; and the heart of insight into human performance in manual fault detection and diagnosis. The system is characterized by "functional strings" of units.

c. Compilation of human performance requirements data (3.0), both with respect to human performance in those cases where the MTPs are utilized, and in those cases where faults are transparent to the MTP or the MTP defaults.

d. Generation of system human performance descriptions (4.0) - The role of the human operator and maintainer is described in detail, including all procedural steps and a detailed description of the necessary and sufficient information required of the human for the manual fault detection, diagnosis and repair task.

These data are maintained in a common data base utilized by design engineering, systems engineering, and the support disciplines.

<u>Collection of Engineering Support Data</u> (Item 1.0)

In creation of the common data base, the following engineering and related support disciplines are contacted; and the data as described is obtained:

a. Systems Engineering
   1. Programming Performance Specifications (PPSs):
      . Operational Program (AOPs)
      . Maintenance Test Programs (MTPs)
      . Operational Readiness Tests (ORTs)
   2. Cable Interconnection Data
   3. Troubleshooting logic diagrams
   4. System mission functional requirements

b. Reliability Engineering
   1. Reliability block diagrams
   2. Failure modes effects criticality analyses (FMECAs)

c. Maintainability Engineering
   1. Maintainability analyses, predictions
   2. Maintenance test definitions

d. Logistics Engineering
   1. Logistics support analysis (LSA)
   2. Special tools, test equipments

e. Maintenance Engineering
   1. Maintenance task descriptions, task times, other maintainer characterizations

f. Human Factors Engineering
   1. Operator and maintainer task analysis
   2. Human performance capabilities, limitations
   3. Man-in-loop test definition

g. Software Engineering/Development
   1. Maintenance test program descriptions
   2. MTP logic diagrams

This data will form the basis for characterizing and describing the total system and operational conditions under which human operators and maintainers will function under design conditions when the MTPs do their job. A complete and accurate characterization of these conditions are essential to ensure that human performance is optimized here as well. But more importantly, this knowledge is critical to proceeding to define the requirements for human performance under manual conditions (MTP in error or defaults).

Analysis of the data provides the system analysis structure of equipment functional groups, function within the groups, and specific operational modes. Figure 2 shows the system analysis structure.

Mission functions are then identified and the units of equipment involved in the function are identified. The reliability flow diagrams; FMECAs, and system block/interconnection diagrams are used to develop a first-cut equipment/function matrix.

Creation of Missions Avionics Equipment Flow Diagram (Item 2.0)

The conceptual basis for the SAMM approach is the structuring of the mission avionics units into human task/equipment/mission functional groupings or "string" flow diagrams. These function flow diagrams serve as the basis for understanding what information is available through what mechanisms or "windows" to enable the human to perform manual fault detection and diagnosis.

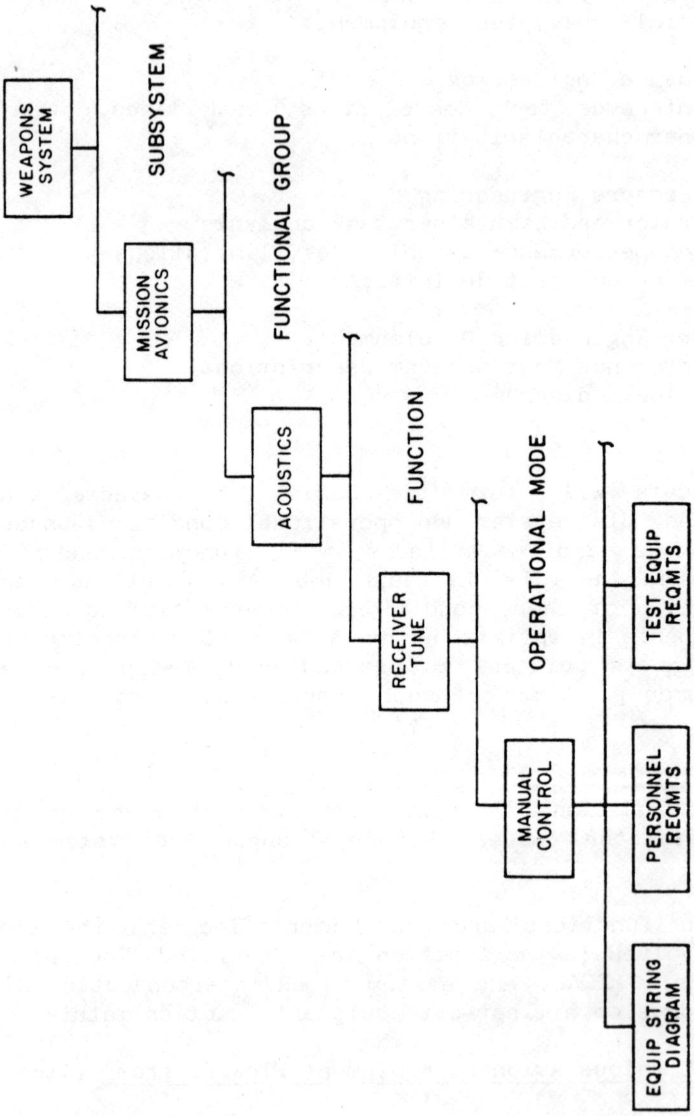

Fig. 2. Sample mission avionics system analysis structure.

The development of the diagrams utilize the following:

a. Mission avionics equipments by unit, grouped by mission function, defined by mission essentiality and criticality, and modes of operation.

b. Equipment interconnection and cabling data.

c. Operator and maintenance task/function analyses.

d. Reliability block diagrams.

e. Maintenance engineering data.

f. Failure modes and effects criticality analyses (FMECAs).

Each unit (and shop replaceable assembly if organization maintenance defines removal and replacement to that level) is analyzed using interconnection data to determine functional signal flow through the string. This identifies the serial and/or parallel relationships of units and in conjunction with an operator/maintainer task analysis, describes the "windows" or human interfaces available for manual fault detection and isolation. Once completed, each unit is reviewed with respect to that MTP function(s) allocated to that unit, and for accessability of function by the human operator/maintainer through some "window". These data are compiled in a common data base.

The completed equipment flow diagram provides the following data for analysis of human performance requirements for and potential contribution to manual fault detection and diagnosis. The data include:

a. Human task/equipment/mission function signal flow, forming the basis for manual fault detection and diagnosis.

b. Cable and connector identifiers listed by functional mode for manual troubleshooting of this class of fault.

c. Identification of individual units/subassembly tests by functional mode; includes those covered by the MTP as well as those requiring human involvement.

d. Identification of operator and maintainer "windows" (displays, controls, CRTs, printouts, etc.) for potential use by the operator/maintainer for detection and diagnosis. Figure 3 illustrates a typical top level equipment functional flow diagram. The initial identification of human involvement in troubleshooting will be derived from the equipment functional string diagrams. The intent is to develop methods for

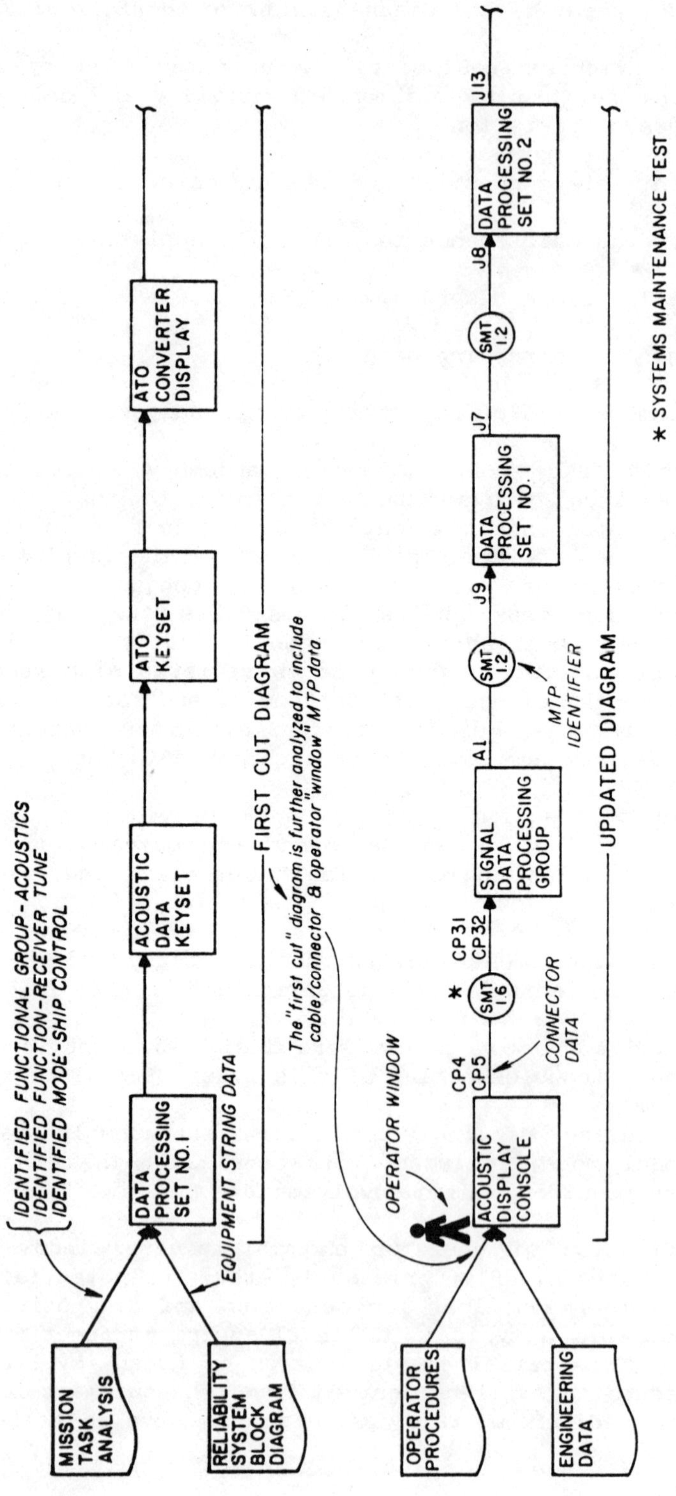

Fig. 3. SAMM flow diagram development.

obtaining information (both airborne and on the ground) regarding contribution of respective units, subassemblies, and their interconnections to failure modes and effects as seen as impacts in functional strings through available "windows".

Compile Human Performance Requirements Data (Item 3.0)

The Systems Approach to Manual Maintenance (SAMM) utilizes four forms to support the definition of human performance requirements for manual fault detection and diagnosis.

The data are developed in a manner permitting summarization or "roll-up" of requirements for the entire mission avionics system or if required the weapon system.

Data Sheet Number One - Mission Avionics System Maintenance Summary

This data sheet contains all related maintenance summary information at the system level; and generally correlated to a group of equipments or units. This is the final form to be completed in the process, and consists of summary information regarding skill or occupation specialty codes, general special purpose test equipment requirements, and other human performance descriptions pertinent at the subsystem level.

The items addressed are:

1. Mission avionics subsystem description consists of a functional description of each subsystem, its purpose and major interfaces, within the mission avionics system.

2. Major unit/functional group summary.
   Describes the functional relationships of units (and subassemblies if appropriate). Specific units and their failure modes and effects are described with respect to subsystem functional "strings". This is the level at which the logical relationship of units to functional strings is defined in terms of manual fault detection and diagnosis.

3. Subsystem common and peculiar support equipment and human performance requirements by functional groups.
   Describes the common and peculiar support and test equipment as well as the human performance requirements for detection and diagnosis of faults, by functional groups.

Data Sheet Number Two - Functional Group Maintenance Summary

This data sheet identifies the mission function and units

involved in the functional group or "string" necessary for completion of a mission function. The major elements addressed the data sheet two include:

1. Functional group description. A brief description of the functional group of units (and/or subassemblies) is included here.

2. Functional groups of units involved in the accomplishment of the particular mission function under analysis. The system Logistics Support Analysis (LSA) control number and mission equipments required to execute a function are correlated on this form. The development of a "system level" approach Logistics Support analysis is described in a separate paper (IBM, 1980).

<u>Data Sheet Number Three - Function Maintenance Summary</u>

The respective units comprising the mission avionics system are generally configured or utilized in a particular manner (mode) to perform a given mission function. Data sheet number three is completed for each function within a given functional group.

The types of information entered upon this data sheet include:

1. Function description. A description of each function is given and the operational model (M-1 through M_n) of each function are identified.

2. Function equipment by mode. These data provide a maintainer with another aid for fault isolation. The table in Step 2 correlated the Weapon System equipments to each operational mode.

3. Special consideration summary. All support equipment, personnel requirements, estimated repair times and predicted task frequency data for a function is contained within Step 3.

<u>Data Sheet Number Four - Human Performance Requirements for Manual Fault Detection and Diagnosis</u> (As a Function of Operational Mode)

This is the lowest level of detail developed in the Systems Approach to Manual Maintenance (SAMM), and reflects the heart of the analytic process. Each specific mode of operation is addressed for each mission function identified. This data sheet contains equipment flow diagrams by mission function, special support requirements, and human performance requirements as well as task descriptions.

For each operational mode of a given mission function there will be a complete set of number four data sheets compiled. The nature of the data compiled on these sheets is as follows:

1. Description. A definition of the particular mode of operation under analysis.

2. Major components. A cross reference of units, subassemblies to SAMM reference numbers and to the Logistics Support Analysis (LSA) control numbers.

   This permits the bridge to be made to the source documentation describing human performance/tasks.

3. System power. An area of importance is the power source, distribution, and control (e.g., switches and circuitbreakers) for each subsystem.

4. Equipment oriented block diagram. The SAMM function flow diagram representing the conceptual basis for the manual approach to detection and diagnosis is described. All connector, interconnection data, and troubleshooting rationale stem from this source.

5. Control and display requirements. Those controls and displays, the "windows" to the system, are identified both in tabular as well as illustrative format. Cross references are made to other sources of human performance task descriptions.

6. Principles of operation. A block by functional block description of the SAMM functional flow diagram; including inputs and outputs as well as equipment function. This information is utilized by the human in working with the functional diagrams and fault "truth tables".

7. Interconnecting cable list. The interconnection cabling and connectors are identified with respect to the functional flow diagram, to identify those fault events where the human must trace continuity and perform other functional checks at the interconnection level.

8. Avionics system operational checkout. The procedures for operating the system to determine functional response are identified here. The checkout procedure is standardized as much as possible, with tailoring.

9. Manual fault detection and isolation procedures. The suggested sequences for human tasks in fault detection and isolation are included here.

   Utilizing this data, the tasks and human performance is assessed. These data serve as source data for the technical manuals.

10. Mission debriefing forms. The data contained in Step 9 and supplementary information is used to develop a mission debriefing form; as necessary to capture those mission operational functions and events of direct correlation to the required human performance for manual fault detection and isolation under post-mission conditions.

CONCLUSIONS

The implementation of automated maintenance test program (MTPs) has been a significant factor in reducing down time and in ensuring that increasingly sophisticated mission avionics systems are maintainable by personnel with lower skill levels and length of training. The accuracy and completeness of MTPs is increasing, and significant improvements in operational availability (Ao) are anticipated.

However, recently deployed mission avionics systems as well as the current generation of systems in advanced stages of development and test still require due consideration of the requirements placed by the system upon the human operator and maintainer. The extra "5 to 10 percent" of the way toward detecting, diagnosing, and repairing <u>all</u> mission avionics faults still depends upon the human element. This is particularly true of that class of "non-detects" or those faults transparent to diagnostics detected only in operational difficulties.

The authors have developed and applied a method for ensuring that the requirements imposed upon human operators and maintainers are defined early in the design and testing phase of mission avionics developments. The problem of playing "catch-up" in technical manuals and training, as well as engineering change activity on the hardware, will be greatly alleviated.

REFERENCES

A System Level Approach to Logistics Support Analysis (System LSA) IBM Federal Systems Division, Owego, NY Internal Report Number 80-535-1.

TRAINING

CHAIRMAN: A. SHEPHERD

SECRETARY: R. M. HUNT

TRAINING

Andrew Shepherd

Chemical and Allied Products I.T.B.
Hollombe Brook, Bury, Lancs.
England

CHAIRMAN'S REVIEW

INTRODUCTION

Most of the papers concerned with training presented at this symposium have discussed specific proposals for dealing with specific training problems or classes of problem. This has meant that contributors' views on general training issues are not always prominent. It has also meant that contributors have rarely considered whether and in which ways the tasks they are concerned with are comparable with other sorts of diagnostic task from other industrial, commercial or military contexts. This may be crucial since there are some marked differences between, for example, the two main types of operation discussed, namely process control tasks and electronic maintenance tasks. Some of these difference reflect the frequency of fault situations, the type and availability of information and the nature of the relationship between faults and their symptoms.

In starting this review, I have two concerns about mis-representing what contributors have said. Firstly, many of the ideas discussed are integral features of complete approaches to training problems - this is how it should be - but taking ideas out of these contexts may well distort them. Secondly, by trying to focus on a set of main issues, I am conscious of failing to represent many of the other things that different contributors have said. Most of these omissions are concerned with specific details of hardware, software and experience with specific approaches and are obviously best left for discussion in the individual papers.

A BIAS TOWARDS DIAGNOSIS RESEARCH

The contributors discussing training have tended to focus on the issue of diagnosis training rather than detection training, that is, the problem of identifying a failure knowing that some symptom has occurred, rather than monitoring system performance in order to identify whether processing conditions are deviating from specification. An exception is Wickens and Kessel's paper which argues that operators required to monitor automated systems benefit from a period of manual operation of that system, a view largely supported by Bainbridge.

This bias towards diagnosis research is arguably legitimate in that process plant and electronics systems are often instrumented to signal clearly when off-specification conditions arise - Sheridan points out, for example, that in nuclear power plants, detection of system failures and alerting of the operator is done by machine. The bias may also reflect the need, when conducting detection training research, to use costly and time-consuming real time simulation, whereas diagnosis training research permits acceleration of real time by only concentrating on failure situations. But in view of the fact that systems managers often desire this monitoring and detection skill in their operators, usually so that they can detect and rectify those situations that develop insidiously with time, detection training research deserves greater attention. There has been enough discussion of detection behaviour at this conference to suggest that such training research would not be at a loss for a theoretical basis.

Another question concerning the relationship between detection and diagnosis is that in those systems where both activities are feasible, is it legitimate to treat them separately? Curry observes that detection, diagnosis and remedial action are generally assumed to be three separate tasks, an assumption obviously shared by those contributors discussing training problems. But the situation is not so straightforward. Bainbridge describes the process operator as continually updating an internal model of the process, a model which first of all can be used to compare the state of a plant with desired operating goals, i.e. failure detection, and then can be used to select appropriate control actions, i.e. diagnosis. But conceding that detection strategy can influence diagnostic strategy does not undermine the separate treatment of diagnosis in training research. Rasmussen points out that we are not constrained to base system design on the mental activities that are used by experienced operators, but rather on the mental processes that the operator could use.

Accepting this notion and treating detection and diagnosis separately, expedites training design considerably. And it is doubtful whether anything is lost by not attempting to replicate the behaviour of experienced operators. Many operators of complex systems, whose competence rests with their experience are not particularly good at their jobs anyway.

THE CRITERION PROBLEM

Trainers must choose a criterion which they feel will confidently predict performance in the real situation. Duncan describes task analysis as the means of identifying required operator goals, with Leplat emphasising the need for further examination of the ways in which operators go about realising goals. In addition, we should give some thought to the conditions prevailing in the real situation when performance is required. Training designed to enable operators to diagnose infrequently or under conditions of stress should be assessed by an appropriate criterion. But the issue of actually measuring transfer of training has received little attention at this symposium, although Duncan refers to the difficulty of collecting data from the real situation to satisfy transfer measures, a point reiterated by Dellner.

Most attention concerning training criteria has centred on the different kinds of training measures adopted in research. Any diagnosis training criterion has several facets, including versatility, accuracy and economy. Different real situations demand different combinations of different values of these facets and training criteria adopted in research try to reflect these. Freedy and Lucaccini, for example, discuss the economic gain of information and accuracy in maintenance tasks, but appear to place less emphasis on versatility. Whereas Marshall and Shepherd, considering process operator performance, measure versatility and accuracy, but place less emphasis on economic gain of information on the grounds that most information is readily available on the control panel, i.e. it does not "cost" anything. Also, the process operator rarely needs to consider sheer speed in diagnosis, but is more concerned to diagnose within sufficient time to avert a shutdown or a serious situation developing. In different types of organisation, different levels of maintenance require different levels of skill, as Nawrocki points out. Thus we may seek rapid, non-versatile and perhaps less accurate performance from first levels, provided suitable versatility and accuracy are exhibited at the support levels. In certain production inspection situations, too much accuracy is taken as an indication that the inspector in possibly spending too much time.

The state of the art of diagnosis training is probably not sufficiently developed to make worthwhile much speculation of different subtle variations in performance criteria, but the versatility question is worth considering further. Duncan considers situations where:

(i) the set of faults to be distinguished are known and can all be included in the training programme;

(ii) a complete set of faults cannot be specified for training purposes, yet fault-finding is still required only within a specified system;

(iii) transfer to different contexts is required.

This third type should be qualified further. Whereas Rouse reports evidence of a context-free diagnostic skill, it is doubtful whether this skill would transfer to some of the probabilistic situations that Brehmer describes, which presumably would benefit from some kind of Bayesian strategy. There are other characteristics of fault-symptom relationships that limit transferability. The more clearly defined flow of information in an electronics system may permit a range of diagnostic strategies which could not be used to diagnose process plant, where referal of symptoms is considerably confused by control loops, heat exchange systems and pressure changes. The opposite is certainly true. To try to cope with the problem of complex symptom referal in process plant, Marshall and Shepherd describe how operators can be taught a set of diagnostic rules-of-thumb. Such rules can deal with faults that have not been anticipated and they can be used in other plants, i.e. other contexts. But there is clearly a limit to their freedom from context since they are irrelevant with respect to, say, electronics or mechanical systems.

A further aspect in choosing performance criteria which was discussed is whether one measures the product of diagnosis, such as accuracy, speed or versatility, or its process, that is, the way in which the trainee tackles a problem. Process measures involve somehow probing what information the trainee selects and the inferences he draws from it. I shall consider these process measures in greater detail later when I look at how they are used to give knowledge of results in shaping diagnosis performance.

Duncan implies that paying attention to processes in diagnosis is a key to establishing versatile and even context-free performance. This is an interesting notion, but one which must be considered against Rouse's experiments which demonstrate context--free transfer without any apparent effort to shape the trainee's strategy. And Nawrocki gives us an account of the possibility of improving diagnostic skill by playing Mastermind.

INSTRUCTIONAL ISSUES

The main themes that are to be included under this general heading are:

- Pre-instructional issues;
- Task simulation; and
- Knowledge of results, which also includes some discussion of the way in which training programmes are adapted as a consequence of a trainee's progress.

Pre-Instructional Issues

Duncan outlines three general approaches to diagnosis training.

(1) The "educational hypothesis", which assumes that trainees need an understanding of the laws governing systems functioning from which they can develop appropriate diagnostic strategies.

(2) Algorithmic procedures, which involve detailed analysis of the system to be diagnosed and the emphasising of the relationship between gross elements such as symptoms, indications and functional entities etc. Nawrocki gives a good outline of this approach to diagnosis training.

(3) Task analysis, which Duncan argues is important for establishing managements real goals and showing how diagnosis integrates with the overall task of the operator, but does not in itself suggest training conditions.

The educational hypothesis in its purest form is rarely adopted for training, most serious trainers and training researchers accepting the need for some form of practice at the task. It is interesting to observe, however, the general reluctance by most people to ditch completely any form of "educational" input. Towne, for example, in describing a computer system for generating fault-finding problems and providing appropriate support material explains the need to "assure that problem presentation does not outpace lecture schedules". This may simply reflect his clients' preferences or prejudices, but at the same time, many of us have a nagging feeling that this kind of input helps in some way. Duncan reports process plant trainees trained in some sort of plant theory as performing better than trainees without such training, although not much better. Far better was to train them with diagnostic rules-of-thumb, a kind of plant theory that had been "processed" to assist in diagnosis. But even these rules were accompanied by a simple account of the normal flows of the plant. It seemed justified on the grounds that this kind of technical story served to teach names of parts and relationships so that the rules could be used and, in some respects, an understanding of how plant functioned helped the trainee rationalise and remember the rules. My reason for dwelling on this point, which has not really occupied many of the contributors, is that is appears that a great deal of time, cost and effort is still devoted to teaching theory of normal operation in diagnosis courses in military and industrial contexts, without much justfication and without much idea of the form or detail it should adopt.

Algorithmic procedures, as Bond emphasises, have been and continue to be very successful, but are limited with respect to their capability in coping with faults that have not been anticipated.

Nobody spoke of the problem of selection, but some interesting comments were made concerning the preparation of trainees for diagnosis training. Rouse comments on qualities of maturity and motivation helping trainees make best use of computer aiding in fault finding training. Bond comments on the apparent electronics knowledge of computer technicians, even though their training was problem oriented. Whether this merely signifies a general interest in electronics or whether theory and diagnostic practice reinforce one another in any way is a matter of conjecture. Perhaps an important characteristic of theory preceding diagnosis practice is that it motivates and disciplines the trainee in a rather general way, rather than specifically aiding diagnostic strategy.

Against this background of uncertainty concerning the role of some kind of theory or technical story, it is interesting to assess the value of research such as de Kleer and Brown's into reasoning about mechanisms. It would appear that this kind of skill would be principally appropriate to the type of diagnostic strategy that Rasmussen describes as <u>search by hypothesis and test</u>, which he points out imposes an excessive memory load on the operator and which, it would appear, none of the contributors proposing training methods have encouraged. On the other hand, these reasoning skills may be very appropriate to <u>detection</u> training in view of the conclusions drawn by Wickens and Kessel and Bainbridge. It must also be added that we might expect search by hypothesis and test at a higher level of diagnosis where, for example, the first level of operators have failed to sort out the problem and the problem is deemed to be one of a more technical nature to be solved by a group of more qualified system experts with more time and assistance to hand.

Before leaving the question of pre-instructional issues we should also consider the broader context of the task. Nawrocki discusses the organisational factors governing the structure of the maintenance system, hardware alternatives to diagnosis and job performance aids. Each of these affects the task to be trained. The influence of job performance aids is particularly pertinent to this symposium. The task to be trained changes considerably if the operator is given computer aiding, for instance, which: in Rouse's paper helps the operator to remember the information and hypotheses he has already considered; in Lihou's paper enables the operator to compare a current state matrix against a library of fault-symptom matrices; or in Lees' paper aids the operator in ordering and interpreting alarms. Perhaps it would be better to consider complete operating regimes, rather than fragmenting the problem into its training, interface and performance aiding components.

Patrick and Stammers also emphasise the problem of handling organisational change when a new training system is being introduced.

Task Simulation

For a number of reasons, it is accepted by contributors that practice must take place using some kind of simulation:

- training in the real situation may be hazardous (Patrick and Stammers)
- faults in the real situation may be infrequent or unrepresentative (e.g. Patrick and Stammers; Dellner; Marshall and Shepherd)
- simulation training provides greater opportunity for the trainee to experiment with strategy (Svanes and Delaney)
- the sheer cost of recent technologically sophisticated military hardware precludes its use for training purposes (Nawrocki)
- simulation training permits greater control of the training process (Duncan; Patrick and Stammers)
- simulation training can more easily enable the accessing of a variety of training aids (Towne; Svanes and Delaney)
- simulation training may be the only means of preparing operators for infrequent yet crucial operations (Svanes and Delaney)

In particular, Patrick and Stammers review the benefits of using computers both to simulate tasks and to drive simulators. Among other things, they note the versatility of the computer as a training device, a point clearly demonstrated by Towne; Freedy and Lucaccini; and Svanes and Delaney. Towne, for example shows how the computer can simulate physical and functional characteristics of prime equipment in functional and malfunctional modes, as well as exploiting its capability in helping trainees during training and recording progress. A prominent feature of Freedy and Lucaccini's approach is the computer's capability of computing from raw performance data rather more sophistosticated measures of a trainee's progress which can then be used to select appropriate K.R. and generally determine the course of instruction. Svanes and Delaney demonstrate how the modelling capability of the computer can be exploited within training programmes to give trainees insight into the consequences of different planning and problem diagnosis activities. I should like to emphasise at this point that each of these three papers contain considerable detail concerning hardware and software used, which in no way can form part of this review.

With the question of simulation comes the question of fidelity. The fidelity issue it not one that has been explored to any great extent at this symposium. Patrick and Stammers note the legitimacy of a variety of different ways of representing tasks for trouble shooting training, including circuit diagrams, pictures of apparatus, verbal descriptions etc. as well as more sophistocated devices. Duncan makes the important distinction between a simulation required for criterion testing and one required for training - a simulation required for criterion testing may be concerned with aspects of perceptual fidelity, whereas that for training must have conceptual fidelity, an argument to be offered also by Leplat and Patrick and Stammers. Thus, for example, for criterion testing diagnosis from a control panel, a simulator should look something like the control panel in question and preserve the essential feature of presenting all information simultaneously, whereas training control panel diagnosis is best achieved by withholding information until items are specifically requested, since greater control is thereby exercised over the learning process. As well as gaining training facility by withholding information, the serial processing of information which is imposed is consistent with the type of fault finding strategy being sought. In many of the papers presented, however, the real task involves withheld information anyway, in which case this line of argument is less relevant.

The other important issue is what data base should be used for diagnosis training. Here there is some consensus, since most contributors exploit some form of fault-symptom matrix. In addition, Towne stores a variety of photographs of equipment in different operational modes to provide feedback to the trainee when some change to the system is being effected.

Knowledge of Results

The K.R. issue reflects several of the criterion issues. Some forms of K.R. are product oriented. For example, Towne informs trainees whether they are right or wrong in a diagnosis of a particular problem, giving them another attempt at the problem if they are wrong and the opportunity to use the "HELP" facility on the computer should they require it. Marshall and Shepherd also employ a right/wrong feedback during training exercises, but their remedial procedures lay in the hands of the instructor running the training course.

Two papers provide detail on process K.R. Duncan discusses three measures to assess the quality of any of the trainee's moves, be that a request for further information or an attempted diagnosis. These measures are based on the idea of "consistent fault set", that is, those faults that are still logically consistent with information so far gained in the diagnosis. These measures show whether the trainee makes a diagnosis before he has logically exhausted all alternatives, whether he seeks information which can in no way help to reduce his uncertainty and whether he realises when he has sufficient information to make a diagnosis. In short, these measures determine whether the trainee progresses logically and whether he knows what the diagnostic process is all about. it is by using this form of K.R. that the context free diagnosis that Duncan reports was achieved. The appropriateness of the consistent fault set measures ties in with Rouse's "fuzzy set model" of diagnosis. Freedy and Lucaccini adopt a more complex, though in some respects similar process measure to adapt instruction programmes to the needs of individual students and to provide appropriate K.R. In addition to formalising information gain from a particular test, by considering set size reduction and taking note of the probabilities of different outcomes of tests, they consider the cost of making each test. Therefore they are concerned with the economy of strategy, not just its quality. The similarities between Duncan's and Freedy and Lucaccini's measures are worth noting. The differences almost certainly reflect the contexts in which they conduct their research. Information is less costly in the process control situation, whether it is presented on a panel or must be called up on a V.D.U.. Whereas strategies in maintenance situations may need to take much more account of the economy of information gain.

CONCLUDING REMARKS

It is clear that there are a variety of issues and themes which can be identified in training research examining markedly different industrial, commercial and military tasks. But less obvious is the extent to which we can transfer research findings from one of these contexts to another. Differences arise with respect to so many factors, including: operator goals; frequency of performance; training budgets and other organisational factors. It is not clear therefore, whether we are all contributing to a common body of knowledge and can genuinely share methodologies, theories and research results, or whether we just share some similar ways of construing our own problems such that we can look to other contexts for ideas that we might try out ourselves, or discard if we see them to be irrelevant.

DISCUSSIONS OVERVIEW

DIAGNOSIS OF RARE EVENTS

The relative infrequency of fault situations in process plant was noted by one questioner who asked what sort of training conditions were most appropriate in these circumstances. The reply placed emphasis on the importance of task simulation to enable these rare events to be represented during training and to provide opportunities for refresher training at intervals during normal operation. The benefits of trainees learning rules as a basis for performance rather than learning rote was also stressed. Research has demonstrated the superior long-term retention and transfer of rule-based diagnosis over learning diagnostic strategies by rote.

The distinction was drawn between those rare events that are merely <u>infrequent</u> and those that are unforeseen. This prompted discussion of the distinction between the operator, who would try to cope with problems as they arise, and the specialist, who would provide a more "expert" back-up. It was argued that training for such specialists, e.g. plant superintendents or engineers, should be much more within the context of the task or plant than the rather general theoretical training such people typically receive, e.g. university or college education.

The question was raised concerning how operators recognise when to call in the specialist. This reflects the problem of when the operator changes his goal from trying to bring the plant back to target operating conditions, to trying to move to an appropriate safe, yet suboptimal state. The response to this question was not entirely satisfactory, although it was suggested that proper diagnostic training should reduce phenomena such as "cognitive lock-up", which obviously lead to protracted and unsuccessful diagnosis.

TRAINING FOR "GROUP" DIAGNOSIS

Speakers were asked whether and how training in diagnosis by groups should take place. The point was made that while group decision making may arrive at a consensus acceptable to the group, it was unlikely to arrive at a better decision than could be reached by individuals. It was also emphasised that it is rarely effective to teach the individual skills of team members within the group situation and that individual adaptive programmes should be used for this. Group "training" might be thought of more as group "practice", once individual skills had been established.

The observation was made that with new computer based instrumentation philosophies, operators would increasingly be working in greater isolation from one another. This would reduce the occasions for overseeing colleagues' work and might cause unwarranted assumptions by operators that colleagues had successfully completed the tasks allocated to them. Computers, it was suggested in response, could be used to make explicit the relationships between tasks or even, in principle, change the allocation of functions between operators during periods of crisis.

MOTIVATING TRAINEES

It was observed that all training proposals offered by speakers were rational and that nobody had spoken of motivation of the trainee. It was also suggested that trainees were likely to be bored by the tasks and the training methods offered. This was refuted by speakers who claimed that rather than boring them, trainees found the training programmes stimulating and enjoyable. But it was conceded that this had not been reported and rarely featured as a topic for discussion in this kind of training research. The point was made that, in any event, motivation of the trainee should always be a standard consideration in the design of any training programme.

THE DATA BASE FOR DIAGNOSIS TRAINING

The question of the data base required for diagnosis training was raised on different occasions. An important general point, made in relation to computer-assisted diagnosis training, was that data in the form of a fault-symptom matrix required far less computer memory capacity than that required for a broader mathematical model of a process and was often perfectly adequate for generating fault conditions suitable for diagnosis training. The choice between a fault-symptom matrix and a mathematical model should be made against the background of the overall task, since several other sub-tasks may also require the generation of considerable process data - a general model might serve several ends. This was just one of several occasions when the need to consider diagnosis training within the context of overall task training or overall human factors considerations was emphasised.

A different issue was whether it is better to generate the entries in a fault-symptom matrix according to some kind of mathematical procedure or whether the trainer should choose to rely on the experience of operating personnel to generate likely symptom patterns. The apparent benefits of the mathematical procedure must be weighed against the time and effort required for its generator and whether it will cope with the ideosyncratic behaviour of different units in different parts of the system. Another point emphasised that generating fault-symptom matrices was often in itself an excellent training exercise for senior operating personnel.

TRAINING FOR FAULT DIAGNOSIS IN INDUSTRIAL PROCESS PLANT

K.D. Duncan

University of Wales
Institute of Science and Technology
Cardiff, South Glamorgan, CF3 7UX

INTRODUCTION

The classical formula for training is simple enough. To train someone to do anything requires only: (1) opportunities to practise; (2) tests to check performance after practice; and, if practice and testing do not of themselves suffice, (3) hints, explanations or other information not intrinsic to performing the task. Industrial fault diagnosis training can present serious difficulties on all three counts.

In contexts where fault diagnosis can be crucial, e.g. in the control rooms of chemical factories or nuclear power stations, practice and testing in the operating situation will often be out of the question, and various problems of simulating the task arise. Moreover the criteria of fault diagnostic skill may turn out to be complex. Often the operator is expected to diagnose faults which no one has foreseen nor, perhaps, understood, even after the fact. What should the content of any simulated practice or testing be? Finally, when it comes to extrinsic information, the waters are even muddier. Just what sort of explanation will help a trainee to learn fault diagnosis or, in operational terms, will enable performance superior to that which is achieved by practice alone? These three problems of simulation, criteria and extrinsic information are obviously interrelated. I will turn first to the problem of extrinsic information and to what I have, perhaps unfairly, called the "educational hypothesis".

APPROACHES TO THE PROBLEM

The "Educational Hypothesis"

Broadly speaking the "educational hypothesis" is that an understanding of the physics, chemistry and engineering of the system will be sufficient to support diagnosis of the system's malfunctions. Courses designed to foster such understanding are well established, often leading to nationally recognised qualifications, e.g. the Chemical Technician's Certificate of the London City and Guilds Institute (City and Guilds Institute, 1974). The effectiveness of this approach has been perhaps most frequently questioned in the field of electronics fault-finding. Reviews of this research are to be found in Standlee et al. (1956), Wallis et al. (1967), and Duncan and Gray (1975a).

Reservations about the classical educational approach to fault diagnosis in electronic systems are well illustrated in the study reported by Williams and Whitmore (1959). These investigators carefully constructed two measures, one of the degree of understanding of how a complex electronic system worked, and the other of ability to fault find in that system down to component replacement level. The first of these measures was at its highest after leaving the training school at which point in time the second measure was at its lowest. The two measures crossed over in the course of a three year follow-up study, at the end of which the performance measure had reached its high point and the measure of understanding of how the system worked had reached its low. Clearly findings like these raised doubts as to the extent to which mastery of "theory" is readily applied in practice.

Now although one cannot rule out the possibility that theory, in some form, may constitute effective extrinsic information, a simpler interpretation would be that learning fault-finding only takes place with practice in the field, where theory is forgotten. Not surprisingly, alternatives to the educational approach were, and continue to be, sought especially for fault diagnosis in high capital cost military and industrial systems. One alternative line of attack sets out to reduce fault diagnostic problem solving to simple algorithmic procedures. Another reaction is to appeal to task analysis techniques.

Algorithmic Procedures

In the nineteen sixties and seventies, training research which explicitly concerned itself with reducing fault-finding problems to fault-finding procedures became well established on both sides of the Atlantic. In North America the work of Shriver and his associates is a well known example of this approach. The solution proposed by these investigators for the electronic fault finding tasks of the day was, essentially, a set of procedural guides consisting, first, of a symptom list which enabled the operator to look up the blocks in which the fault was to be found. Secondly a guide was supplied which specified "good" signals for the boundaries of the various blocks in the system, providing the fault-finder with unambiguous cues as to the block within which he should search. A third aid consisted of a list of resistances for within-block search. For each block were provided the resistances of small sets of components, generally chains of five or six condensers, resistors, etc. between an accessible test point and earth, (Shriver, 1960; Shriver, Fink and Trexler, 1964; Shriver and Trexler, 1965).

The extrinsic information necessary to enable fault diagnosis is thus very well specified. It was also pointed out by these researchers that the technology then current often produced tasks which operators cannot in principle be expected to perform efficiently on the basis of an understanding of electronic theory. They argued that not even designers of electronic equipment can do what the maintenance man was sometimes expected to do in fault location, namely determine, from his understanding of electronic theory, what a wave-form or voltage at a check point should be, or work out the set of components which, if malfunctioning, could produce a given abnormal reading. Engineers, when designing systems, recognise that theory alone is not sufficient, that "bread boarding" is also needed - because there is usually some capacitance in resistors and inductors, some inductance in resistors and wire, and some resistance in inductors. The point is further underlined by Shriver's policy of resolving doubt about symptoms by deliberately inducing the failure, (Shriver, Fink and Trexler, 1964).

The work of Shriver and others in the United States which effectively reduced fault diagnostic problem solving to procedure following was paralleled in Britain by techniques of ordering various problem solving tasks into branching sequences of (usually binary) decisions known as "logical trees" (Jones, 1964) or "algorithms" (Horabin et al., 1967). Algorithmic procedures for fault diagnosis of a complex industrial acid purification process - Figure 1 - were reported by Duncan (1971 and 1974) and for the task of diagnosing failures in the crude distillation unit of oil refineries, a decision tree - Figure 2 - was developed by M.J. Gray (Duncan and Gray, 1975b).

Algorithms, or more precisely diagrammatic representations of algorithms, continue to enjoy popularity in British industrial training circles. Performance aids of this sort illustrate rather clearly both the strength and the weakness of approaches which reduce fault-finding problems to algorithmic procedures. Manifestly these performance aids provide all the extrinsic information necessary to distinguish between a set of possible faults. But what about faults not in the set?

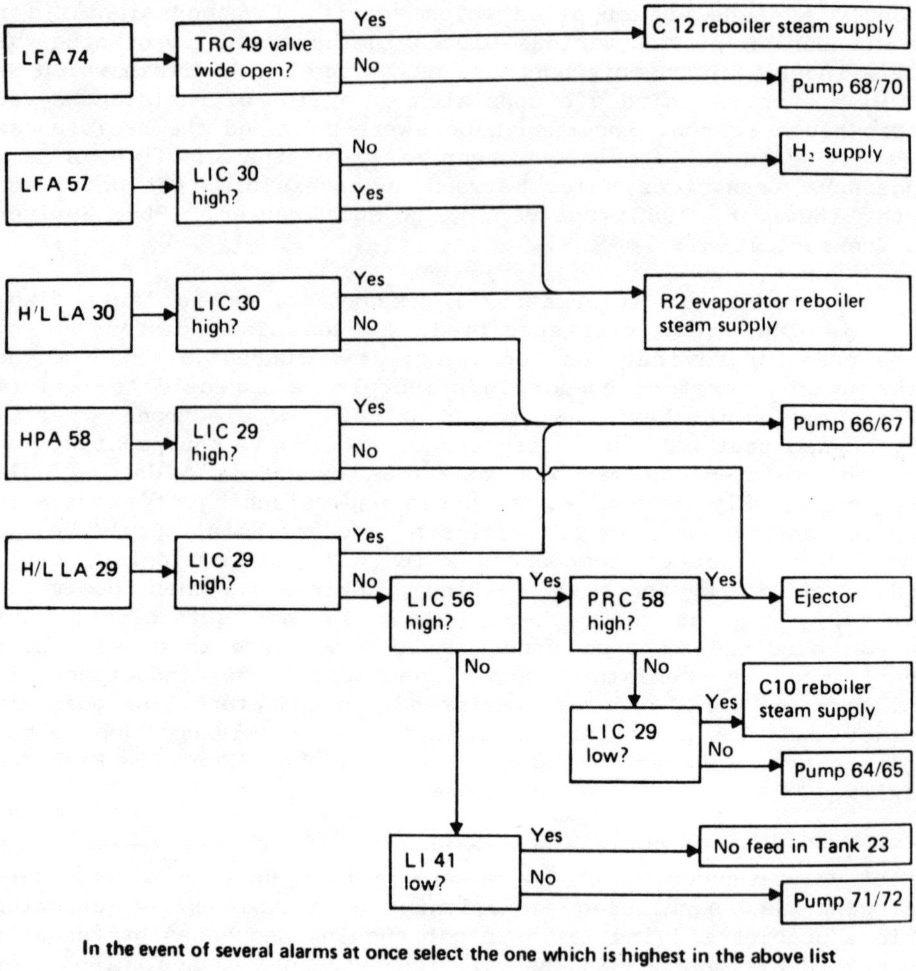

Fig. 1. Decision tree for acid purification process fault diagnosis task

Chemical factory and oil refinery operators were expected to diagnose rapidly and efficiently any of the list of faults specified by management, and the algorithms in Figures 1 and 2 enabled them to do this. However, operators were also expected to diagnose faults in general, not just those distinguished by the algorithm. It was admitted that the set of faults envisaged by management was not exhaustive. So the criterion of human performance, at least in these two industrial situations, was versatility or capability to diagnose faults not previously experienced. The second ingredient in the simple formula for training with which I opened this paper - tests to check performance following practice - must somehow recognise this criterion problem.

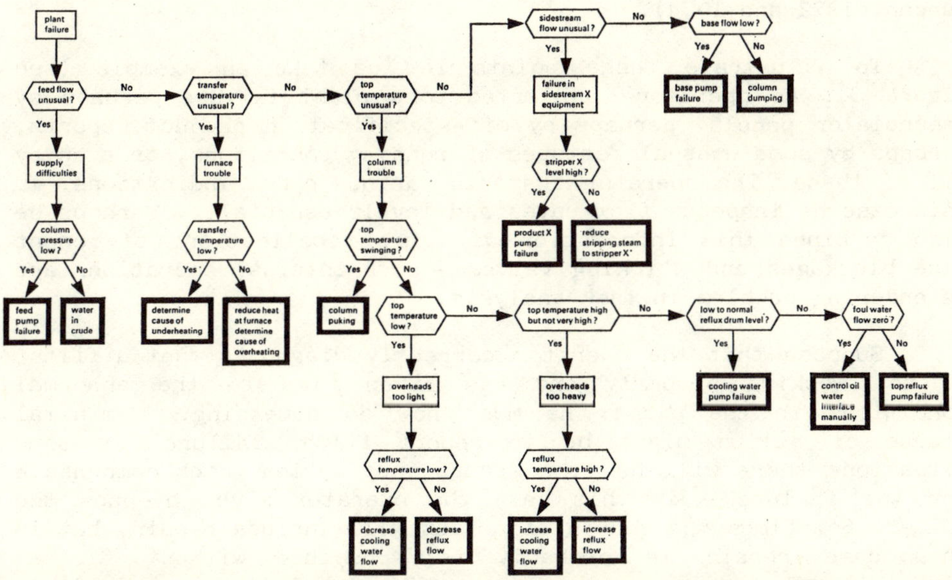

Fig. 2. Decision tree for diagnosing faults in crude distillation units

Task Analysis

The 1960's disenchantment with the approach of education also revived interest in methods of task analysis to meet the needs of industrial training. "Analyse the task and the training problem is solved" was a widely held belief and task analysis became a policy issue for many of the newly established Industry Training Boards. Task analysis, predictably, does not resolve the difficulties of industrial fault-finding training, but it may serve to pinpoint where these difficulties lie. Specifically, task analysis recognises that fault diagnosis is usually only part of an overall task and that fault diagnosis may emerge at various levels of description. It also provides a rationale for placing the fault diagnostic components of tasks into part-task or whole task training schemes.

The method of analysis which I will briefly describe has been applied in a number of contexts, including capital-intensive industry. The unit of analysis or "operation" is any activity, no matter how long or how short its duration and no matter how simple or complex its structure, which can be defined in terms of its objective. To cope with the several levels of description which may be appropriate when dealing with human behaviour, any operation may be conceived of as a plan which controls subordinate operations (each of which may be conceived of as controlling other subordinate operations and so forth, Annett and Duncan, 1967; Duncan, 1972 and 1974).

To illustrate these points let us take an example (see Figure 3). An operator is alerted to a plant failure perhaps by annunciator panels, perhaps by off-specification product reports, perhaps by some unusual features of running conditions, or even by all of these. The operator inspects various panel indications, in this case he inspects flow rates and levels especially. Perhaps he also combines this information with his recollections of recent line blockages and sticking valves - such internal operations are an enduring problem in task analysis.

Suppose that the operator correctly diagnoses that a filter is not working properly and has given rise to the abnormal conditions in the plant. He must now do something. A minimal course of action might be to report filter failure. In some situations there will be other courses of action which compensate for the failure - in this case the operator might by-pass the filter. Sometimes the course of action will include repair, but in this case repair is probably not possible without further diagnosis. What is wrong with the filter? If it is part of his task the operator will undertake subordinate operations to diagnose the fault in the filter, e.g. he will replace filter leaves with "bungs", observe the effect of replacements on the filtrate, and inspect filter leaves for holes. Replacing leaves with bungs will reduce the rate of filtration, but inspection of the filtrate will give a clue as to whether the filter leaves removed include faulty ones - see Figure 4.

Whether at the level of panel indications, or at the level of filter leaf inspection, fault diagnosis is conceived of as a plan and subordinate operations which are selected and arranged in sequences specified by the plan. The subordinate operations may include information seeking; actions such as reporting procedures, compensating procedures, repair procedures; and further diagnosis at a "lower level" in the task. Often subordinate operations such as inspecting instruments, injecting signals, sampling product and the like do not pose ssrious difficulties for training research. The central problem in fault diagnosis training is rather how to enable operators to execute plans.

A statement of a plan must incorporate at least an objective, e.g. identification of failure, and the operations diagnose plant failure

PLAN 1 (diagnose plant failure)

note instrument readings

note recent faults

diagnose pump failure

diagnose flow control failure

diagnose feed failure

.
.
.
.

diagnose filter failure

PLAN 2 (diagnose filter failure)

remove leaves

replace leaves

inspect filtrate sight glass

clean leaves

inspect leaves

reject faulty leaves

- any operation may be restated as a PLAN and subordinate operations
- on different occasions, subordinate operations may be performed in different sequences to achieve the PLAN
- operations may be restated until "effective" training or plant operating instructions can be specified for both PLANS and operations
- "effective" is defined by $p \times c$, where p is the probability of inadequate performance and c is the cost of inadequate performance, e.g. injuries, lost production, explosions etc; most of the operations in the example would probably need redescription to meet this criterion
- note that the analysis specifies all subordinate operations for PLAN 2 but fails to do so for PLAN 1

Fig. 3. Illustration of how one component of a task can be analysed - see text

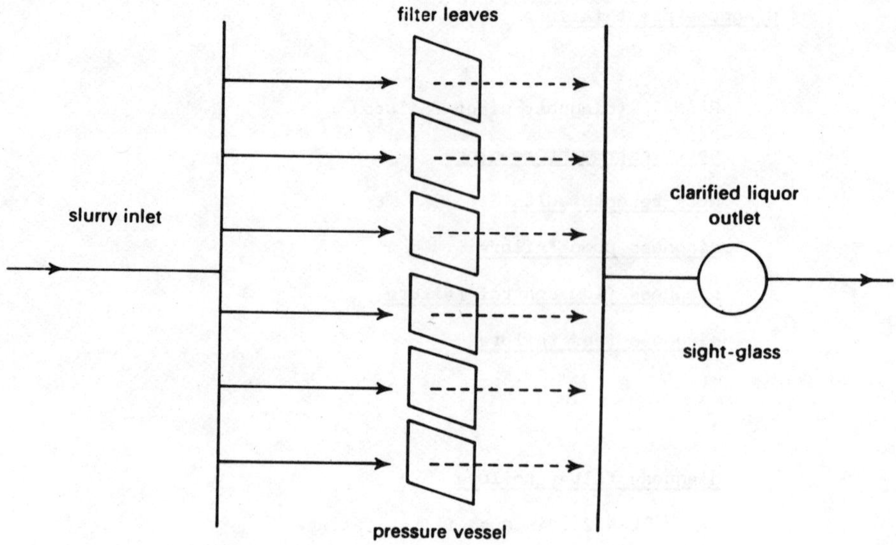

Fig. 4. Niagara Filter (after Shepherd, 1975). Slurry is pumped into the pressure vessel where it passes through one of several filter leaves into a common manifold then through the outlet where it may be observed through a sight glass. Leaves removed for inspection are replaced by bungs resulting in a slower filtration rate. Slurry characteristics determine time taken to clean the leaves and inspect for faults.

(reading instruments, removing components for inspection etc.) from which subsets must be selected and arranged in such sequences as will achieve the objective. Any promising hypotheses as to how operators may be enabled to execute the plans entailed in fault diagnosis should be explored and if possible tested. But plans themselves should as far as possible be stated in psychologically neutral terms. Plans are seen as the product of hypothetical psychological processes. Given time, resources and a task of major importance, successive revisions of the task hierarchy may eventually leave only plans entailing mastery of simple algorithms (Shepherd and Duncan, 1980).

To return to our example of an operator's task, the plan to fault-find the filter may specify removing half of the leaves then, depending on filtrate appearance, removing another quarter of the leaves and so on (Figure 4). In other words, a "half-split" strategy might enable the operator to carry out this plan. Alternatively, the number of leaves which may be removed at a time may be restricted to permit some minimum level of filtration dictated by plant priorities. One of the difficulties that may be encountered in search tasks of this kind is that plant management have not clearly specified what their priorities and trade-offs are, so that the operator is in no position to work out and execute an optimal plan (Shepherd, 1975). But in principle a strategy can be worked out for such tasks. If the operator has difficulty in understanding the strategy, i.e. if he cannot, in this example, plan the optimal removal and inspection sequence to identify the faulty leaf, then a job aid expressing strategy in algorithmic form will usually suffice.

Now, changes in filtrate appearance can for all practical purposes be attributed to a hole in one or more filter leaves. Put another way, the set of possible faults can be exhaustively specified, or as nearly as makes no difference. In contrast, there will often be no obvious strategy for interpreting control room panel indications since the failures to be distinguished by the operator cannot be exhaustively specified. Algorithms such as those represented in Figures 1 and 2 have limited usefulness and might even discourage operators from learning more general strategies. Yet the case for retaining the operator in the system may be that he will somehow generate plans which select appropriate courses of action for failures which have not been foreseen.

This brings us back to the criterion problem. By what test shall we assess versatility? Also, if algorithms are suspect when versatility is demanded, what extrinsic information will be effective? I have argued that task analysis is, by definition, neutral as to how the operator is enabled to execute plans. On the other hand, given training in subordinate operations and a clear statement of the objective, operators may discover strategies which effectively support diagnostic plans in novel as well as in more familiar situations. Indeed something of this sort seems to occur; in the absence of instruction in appropriate strategies, a minority of experienced process operators may develop useful rules-of-thumb for rather difficult control panel tasks (Shepherd et al., 1977).

A COMPARISON OF TRAINING METHODS

In the investigation just referred to (Shepherd et al., 1977) two questions are addressed which were raised at the start of this paper and which have been central to the subsequent discussion, namely, the effectiveness of different sorts of extrinsic information in enabling fault diagnostic plans, and the question of what criteria to apply in estimating fault diagnostic versatility. The empirical study of these questions also necessarily entailed addressing the third question raised at the start of this paper, namely the question of simulation.

Three groups practised identifying the fault corresponding to static instrument panel arrays showing various normal and abnormal indications. Instrument panel indications corresponding to sixteen faults were back projected to life size from pre-prepared slides. Prior to practice, two groups (the "theory" and the "rules" groups) were given a simplified account of how the plant worked in terms of inputs, outputs, product flow and control loops. Trainee subjects were instructed individually and questioned to ensure that they had mastered this rudimentary "theory". The "rules" group were in addition exercised in applying the diagnostic rules shown in Figure 5. The third "no story" group had no prior instruction of either sort.

(a) Scan the panel to locate the general area of failure, i.e. Feed, Reactor/Heat-Exchange complex, Column A or Column B.

(b) Check all control loops in the affected area. Are there any anomalous valve positions?

(c) High level in a vessel and low flow in associated take-off line indicates either a pump failure or valve failed 'closed'. If valves OK (see b), then pump failure is probable diagnosis.

(d) High temperature and pressure in column head associated with low level in reflux drum indicates overhead condenser failure - provided all pumps and valves are working correctly (rules b and c).

(e) If the failure is in the Reactor/Heat-Exchange complex, determine whether it is in the reactor or the heat-exchange system. A failure in the heat-exchange will produce symptoms in Column A but not in B. A failure in the reactor will produce symptoms in both columns.

(f) If the failure is in the feed system, check whether it is in stream X or stream Y. Because of the nature of the control system, a failure in the Y stream will produce associated symptoms in both the X and Y streams. A failure in the X stream will show symptoms in the X stream only.

The diagnostic rules differ in generality. The first four rules (a-d) are rather general in that they might apply to many diagnostic tasks of this kind. The last two rules (e and f) deal with plant-specific difficulties. Both are examples of a general problem in diagnosis, symptom referral, enabling diagnosis despite symptoms referred by a feed-forward loop (rule e) and by a feed-back control loop (rule f).

Fig. 5. Diagnostic rules

The results of a test which followed (Figure 6) suggest that extrinsic information in the form of instruction prior to practice is effective for "new" faults, i.e. arrays not previously seen by subjects during practice. Explanation of how plant works has some effect, offering perhaps some support for the "educational hypothesis". But the most encouraging performance was achieved with the addition of extrinsic information in the form of the diagnostic rules shown in Figure 5. A propos the criterion problem, it is argued by these investigators that generalisation to novel failures within the same system, i.e. the same plant and instrument control panel, is a useful performance criterion.

SIMULATION FIDELITY

In trying further to improve training, in particular training of fault diagnosis by applying rules-of-thumb, we faced a difficulty inherent in the training sequence I have just described. We could never be sure to what extent the diagnostic rules, or indeed anything else learned prior to practice, were subsequently applied or abandoned by trainee subjects during the simulated fault diagnosis task. It seemed that any attempt to ensure that the rules were applied would necessarily involve intervention during the course of solving the fault diagnostic problem. In what sense would practice then be a simulation of the task?

The rationale for the simulation used in these studies was the classical one, namely fidelity to the psychological demands of the real control room situation. To the extent that such fidelity was achieved, a substantial correlation would be expected between fault diagnosis with the simulator and the diagnosis of real plant faults - although the possibility of collecting such data was recognised to be remote (Duncan and Shepherd, 1975). Simulation fidelity in this sense in clearly appropriate for performance testing or for assessing the effectiveness of a training scheme. It is not necessarily appropriate for any simulation in the course of training. For training purposes it is surely necessary to simulate only those features of the task, if any, which in some way help a novice subsequently to perform it. I realise that simulation fidelity in this second sense may be even more elusive, but the distinction seems to me an important one. It is a central issue in the later work which I now describe.

In the original simulation argument, a distinction was made between (1) presented information, e.g. the readings of standard, fixed function, level, pressure, temperature, or flow indicators or recorders, and (2) withheld information, e.g. readings which must be retrieved by the operator via multipoint instruments, or computer terminals. It was argued that simulation fidelity entailed preserving this distinction. But in a subsequent training

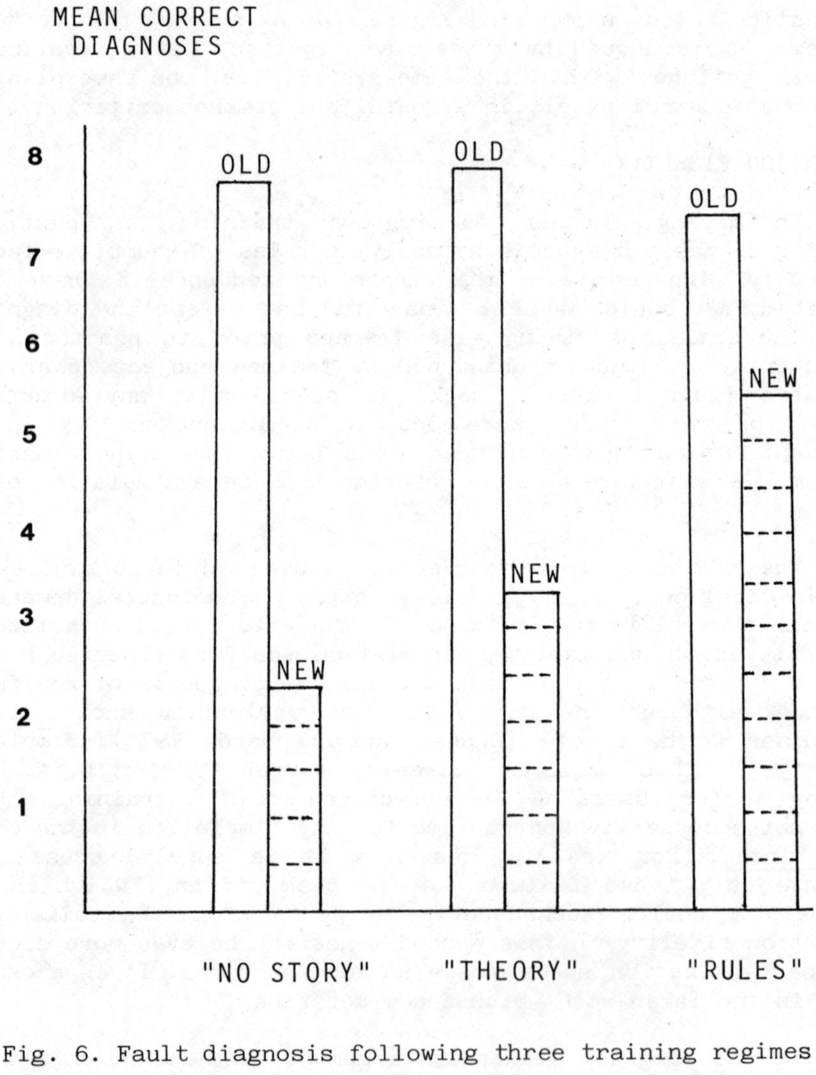

Fig. 6. Fault diagnosis following three training regimes

experiment we decided to violate this distinction by withholding all information during training until it was sought by the trainee. There are several advantages of withholding information. At any stage of the fault-finding problem, the instructor knows which indications the trainee has so far obtained and which not. The instructor can intervene when he is sure that a trainee is not applying diagnostic rules. And he can ensure that trainees retrieve information in a sequence which is at least consistent with diagnostic rules.

Instruction in the application of diagnostic rules is thus integrated into the fault diagnostic process, instead of first occurring en bloc before leaving the trainee to practise fault diagnostic problems in ways which remain for the most part hidden - as in our previous experiment. The sharp separation of strategic instruction and problem solving practice is of course a feature of many training schemes and is most starkly manifest in further or higher education courses.

This greater degree of control over how trainees solve fault-finding problems entails a departure from simulation fidelity in the first sense - that is fidelity to the psychological demands of the control room task. But we were simulating features of the task in a way which would improve fault diagnostic strategy and, arguably, fault diagnostic performance - simulation fidelity in the second sense. The question is therefore whether training by withholding information subsequently transfers to the original task in which indications are all presented on the control panel.

As things turned out in this investigation, subjects trained to diagnose faults by retrieving withheld information, later performed without difficulty when they were transferred to the original control panel task where all indications were presented. Indeed, the results suggested that comparable performance on the control panel fault diagnostic tests could be achieved with a drastically reduced training session, i.e. practice of far fewer fault finding problems (Marshall and Duncan, 1980). The technique of withholding information to enhance the application of rules-of--thumb in fault diagnostic problem solving has been incorporated in a training package so that it may more readily be understood and applied in industrial settings. Our first experiences with the use of this package are encouraging and are described in the paper by Shepherd and Marshall (this volume).

It seems to me that in this and in many fault diagnostic tasks, the conceptual features of the task, as against the perceptual features, are the ones which demand priority in the design of training. I recognise that there are important, long debated, psychological issues involved, such as the distinction between serial and parallel processing, between sequential identification of isolated features and pattern perception. I also acknowledge that in the last resort empirical investigations, possibly in the form of comparison studies, will probably be inevitable for particular industrial applications. I do not underestimate the perceptual factors in fault diagnosis. Rather I would suggest that a more useful practical remedy for the perceptual difficulties in fault diagnosis may lie in ergonomics rather than training. My colleague Dr. John Brooke has shown that ability to perceive relevant test points and thus minimise redundant testing, may be improved by choice of diagrammatic representations of the problem. But conceptual difficulty as indicated by premature diagnoses, or diagnoses based on insufficient information remains much the same across different problem representations. This distinction rests on using indices of the problem solving process such as redundant tests and premature diagnoses (Brooke and Duncan, 1980). The need to use such "process" measures, as distinct from overall measures such as accuracy and time, is an argument to which I will return.

CRITERIA

In the task analysis illustration, two cases of fault diagnostic problem were distinghished: (1) the case where faults can for practical purposes be exhaustively specified, e.g. fault-finding the filter and (2) the case where faults cannot be exhaustively specified of which control panel diagnosis is a widely encountered example.

The Criterion Problem: Case 1

In the case where faults can be exhaustively specified, the major criterion, finding the fault, is well defined and will usually be attainable either by plant experience or by systematic training. However a second criterion, the efficiency of fault location, is often important and may present training problems. Dale (1958) noted that inefficient searching strategies may sometimes be preferred because they impose less load on working memory. He also noted a tendency to gamble in subjects tackling his electronic fault-finding tasks. Duncan (1972) observed that operators searching for a water leak in a chain of heat exchangers may construe sucess solely in terms of finding the fault. This is an insidious feature of search tasks wherever operators have only plant experience rather than systematic training to guide them. Operators may simply not recognise the efficiency criterion without extrinsic information in some form.

In the filter fault-finding problem, a plan for filter leaf removal based on the half-split strategy will minimise the number of filter leaves which have to be inspected for holes. But instruction to emphasise that the criterion of sucessful performance is minimal checking will probably be needed, and such instruction, e.g. in applying half-split strategies, as Goldbeck et al. (1957) discovered, may not be a simple matter of explanation and demonstration of the half-split technique. To reinforce a point made earlier, effective training may depend on intervening during the fault location problem, by indicating to the trainee who makes sub-optimal checks what the optimal check would be, or alternatively preventing him from making sub-optimal checks at all. Even then, with careful interventions during problem solving practice, the operator may "revert" to inefficient fault location when left to his own devices on the plant, for the reasons noted by Dale. Thus in the filter fault-finding problem, the appeal of simpler sub-optimal strategies or of gambling may be increased or decreased by the properties of the liquor being filtered. In some instances cleaning a filter leaf and inspecting it for holes may be extremely tedious, on other occasions less so.

In the first case, then, where faults can be exhaustively specified, precise criteria can usually be worked out and accurate assessment of training is feasible. But to bring trainees up to criterion performance during training is one thing; to ensure that they will subsequently continue to perform to the same standards may require rather careful consideration of the payoffs and sanctions present in the operating situation.

The Criterion Problem: Case 2

Turning now to the second case where faults cannot be exhaustively specified, we encounter the difficulty that generalisation must somehow be assessed. One measure of generalisation which has already been described, requires the operator to diagnose failures he has not previously encountered, on the basis of training in failures in the same system, i.e. the same plant and the same control panel (Shepherd et al., 1977). Another important possibility is described in the paper by Rouse, namely transfer from a context free or minimal context task to an industrial fault location task. To the extent that such transfer is possible, instruction can be designed which should facilitate fault diagnostic performance in a variety of industrial settings. Perhaps this approach is the way to rehabilitate more general, more "educational" courses, as distinct from the plant specific courses which have been adopted because of perceived shortcomings of the teaching in technical colleges and other centres of higher education.

A third possibility, though not much work has been reported, is transfer from one context to another context. To the extent that such transfer takes place, the claims which are sometimes made for long experience in industry could be defended and the nature of experience which should ideally be planned for industrial process operators could be more closely specified. In our laboratories we have studied transfer from one minimal context to another minimal context showing small but significant improvements in diagnostic skill. These were enhanced by interventions during the problem solving process. The nature of these interventions will now be described in relation to the general issue of augmenting information available to the subject during practice of fault diagnostic problems.

Product v. Process Measures

Fault diagnosis can be measured in terms of correct solutions or solution time, but "product" measures such as these, although clearly important operational criteria, may offer little insight into the nature and quality of the fault diagnostic process. There is little prospect of appropriate training influencing the fault diagnostic process unless some measure of the process exists. Indeed, in this sense, the issues of what extrinsic information to provide, and what criteria to apply, are the same.

Possible measures have been proposed for the fault diagnostic process when training is based on a fault-symptom matrix (Shepherd, 1974; Duncan and Gray, 1975). At any stage during fault diagnosis practice, (provided that information is withheld and must be sought by the trainee), there will be a set of faults which is consistent with the information so far acquired - the "consistent fault set" (CFS). Three measures based on CFS are:

(1) premature diagnoses or diagnoses attempted when CFS is greater than one;

(2) redundant questions or questions which do not reduce CFS when CFS is greater than one;

(3) extra questions or questions asked when the size of CFS is one.

Duncan and Gray (1975) report comparisons, between the scores of experienced and inexperienced fault finders, and comparisons between trainee performance before and after training, which lend some empirical support to the use of these CFS measures.

If CFS measures are valid indices of fault diagnostic skill, then training should be improved if the trainee is provided with CFS measures of his performance during the solution of fault diagnostic problems. We now have some evidence that such provision of information based on CFS does indeed facilitate learning to solve fault diagnostic problems (Brooke et al., 1978) and that subsequent learning to solve problems of the same form but in a different context is also improved (Brooke et al., 1980). In other words, it seems that orienting the trainee towards reducing CFS enables him to acquire a measure of context free problem solving skill.

What we have tried to do, in old fashioned language, is to improve knowledge of results. Knowledge of results in this or any other learning situation will only be as effective as the information which it provides. The information provided will be effective only to the extent it is intelligible to the trainee. We could for instance give statements about "bits" of information which would be entirely adequate, qua information, but their intelligibility to trainees is another matter. CFS statements have the advantage that they use only whole numbers in the range from one to the number of possible faults at the start of the problem.

Besides informing trainees when they ask extra questions, redundant questions or make premature diagnoses, we have used a computer to display on a screen the efficiency of a test or check before the trainee makes it. Thus a trainee is always in a position to choose the most efficient check or test point. If for example the trainee proposed to test the unit at the end of an 8 unit chain, he would be informed that the outcome of such a test would be a CFS of either 1 or 7. I hasten to add that we have not worked with problems which are as simple as the classical serial search problem (Miller, Folley and Smith, 1953; Dale, 1958). First, there are many systems from electronic circuits to continuous process plant in which ideal checks or moves which dichotomise the possibilities do not exist. Second, systems are common in which the tests or checks may have more than two outcomes. Third, there are fault-finding problems which, unlike serial search, lack an obvious system structure to indicate unambiguously which move or check is efficient. The problems which we employed all had these three features.

CONCLUSION

The classical educational approach tends to ignore practice in actually solving fault diagnostic problems and all the difficulties of simulation fidelity which that entails. Criteria may not be explicit, and tests and examinations usually deal with knowledge of a technology in the terms in which it is taught, rather than with ability to diagnose faults. The appeal of this approach is the assumption that humans can develop strategies which enable them to solve practical problems if they are first provided with some explanation or account of how the system works. The possibility of rather general, context free, skills developing on this basis is attractive. The empirical evidence however is not encouraging. The case can perhaps be argued more optimistically for one variant of the educational approach which sets out carefully to select theoretical principles, "functional fundamentals", and insists on teaching these theoretical principles in a "functional context" (Duncan and Gray, 1975a). But historically these ideas tended to lose out in the face of fault diagnosis by procedures and algorithms.

Algorithmic procedures certainly have the advantage of emphasising practice of the fault diagnostic task, and powerful extrinsic information in the form of diagrams, job aids, or check lists is typically provided. Such extrinsic information is powerful for the further reason that it is usually integrated into the fault diagnostic process. The criterion is, by implication, all the failures which the procedures or the algorithm distinguish which is at once a precise criterion but a limited one, since the problem increasingly encountered in large industrial installations is that possible failures cannot be exhaustively specified.

Task analysis has the merit of emphasising performance, and may be particularly valuable if the method copes with the level of description problem as the method briefly outlined in this paper sets out to. Task analysis should identify those parts of the task where difficult fault diagnosis is involved and also indicate where any fault diagnostic instruction should figure in part-task or whole task training schemes. As to criteria, both plans, and operations governed by plans are emphasised. This is important when one considers that many practically oriented industrial training schemes provide an impressive repertoire of repair operations but, as anyone who owns a motorcar may appreciate, tend to neglect the plans which govern appropriate sequences of operations when, for example, a motorcar will not start. Task analysis is neutral as to what extrinsic information is provided but may usefully draw attention to components of the task where extrinsic information is essential.

Finally how can existing methods and techniques best be used and what further work needs to be undertaken to make better fault diagnosis training possible? Task analysis can probably be relied on to design plant specific training schemes in particular. The important distinction between the case where faults can be exhaustively specified and the case where they cannot, will emerge in the course of analysis. Where faults can be exhaustively specified, procedures or algorithms can usually be worked out and will often be the training solution of choice. However, when generalisation to novel or little understood fault diagnostic problems is required, we move into areas of uncertainty where further research is essential.

Two sorts of extrinsic information seem to offer promise, namely the diagnostic rules which successful experienced operators develop, and systematic information which goes beyond correct or incorrect diagnoses and diagnosis time. Indices of the problem solving process such as CFS measures need to be further researched. Extrinsic information is probably most powerful when it is integrated into practice of the fault diagnostic task. The necessary integration may be achieved by withholding intrinsic task information. Finally, variants of the traditional educational approach such as the functional context or functional fundamentals approaches should not be neglected.

The sophisticated human factors specialist may find my conclusions a mixture of rather prosaic advice and inponderables. I apologise for the inponderables. But I make no excuse for the prosaic advice since I am continually impressed by the number of hazardous industrial situations where such advice has either never been given or is simply ignored.

ACKNOWLEDGEMENTS

This work was supported by the Chemical and Allied Products Industry Training Board and the Social Science Research Council.

REFERENCES

Annett, J. and Duncan, K.D., 1967, "Task Analysis and Training Design", Occup. Psychol., 41, 211-221.

Brooke, J.B. and Duncan, K.D., 1980, "Effects of System Display Format on Performance in a Fault Location Task", Department of Applied Psychology, UWIST, UK.

Brooke, J.B., Duncan, K.D. and Cooper, Carolyne, 1980, "Interactive Instruction in Solving Fault-Finding Problems", International Journal of Man-Machine Studies, 12.

Brooke, J.B., Duncan, K.D. and Marshall, E.C., 1978, "Interactive Instruction in Solving Fault Finding Problems", International Journal of Man-Machine Studies, 10, 603-611.

City and Guilds Institute, 1974, "Chemical Technicians Certificate: Syllabus (Extract)", in E. Edwards and F.P. Lees (Eds), The Human Operator in Process Control, Taylor and Francis.

Dale, H.C.A., 1958, "Fault Finding in Electronic Equipment", Ergonomics, 1, 356-385.

Duncan, K.D., 1971, "Long-Term Retention and Transfer of an Industrial Search Skill", The British Journal of Psychology, 62, 439-448.

Duncan, K.D., 1972, "Strategies for Analysis of the Task", in J. Hartley (ed), Strategies for Programmed Instruction: An Educational Technology, Butterworths, London.

Duncan, K.D., 1974, "Analytical Techniques in Training Design", in E. Edwards and F.P. Lees (Eds), The Human Operator in Process Control, Taylor and Francis.

Duncan, K.D. and Gray, M.J., 1975a, "Functional Context Training: A Review and an Application to a Refinery Control Task", Le Travail Humain, 38, 81-96.

Duncan, K.D. and Gray, M.J., 1975b, "An Evaluation of a Fault Finding Training Course for Refinery Process Operators", Journal of Occupational Psychology, 48, 199-218.

Duncan, K.D. and Shepherd, A., 1975, "A Simulator and Training Technique for Diagnosing Plant Failures from Control Panels", Ergonomics, 18, 627-641.

Goldbeck, R.A., Bernstein, B.B., Hillix, W.A. and Marx, M.H., 1957, "Application of the Half-Split Technique to Problem--Solving Tasks", J. Exp. Psychol., 53, 330-338.

Horabin, I.S., Gane, C.P. and Lewis, B.N., 1967, "Algorithms and the Prevention of Instruction", Cambridge Consultants (Training) Limited, Cambridge.

Jones, S., 1964, "Why Can't Leaflets be Logical?", New Society, 102, No. 16.

Marshall, E.C. and Duncan, K.D., 1980, "Information Display and Process Control", in R. Easterby and H. Zwaga (Eds), Visual Presentation of Information (in Press).

Miller, R.B., Folley, J.D. Jr. and Smith, P.R., 1953, "Systematic Trouble Shooting and the Half-Split Technique", Technical Report 53-21(b), Lackland AFB, Human Resources Research Centre, July.

Shepherd, A., 1974, Personal Communication.

Shepherd, A., 1975, "A Classification Scheme for Decision Making Tasks in the Chemical Industry", in T. Singleton and P. Spurgeon (Eds), Measurement of Human Resources, Taylor and Francis.

Shepherd, A. and Duncan, K.D., 1980, "The Analysis of a Complex Planning Task", in K.D. Duncan, M. Gruneberg and D. Wallis (Eds), Changes in Working Life, Wiley.

Shepherd, A., Marshall, E.C., Turner, Ann and Duncan, K.D., 1977, "Diagnosis of Plant Failures from a Control Panel: A Comparison of Three Training Methods", Ergonomics, 20, 347-361.

Shriver, E.L., 1960, "Determining Training Requirements for Electronic System Maintenance: Development and Test of a New Method of Skill and Knowledge Analysis", Tech. Rep. 63, Human Resources Research Office, Alexandria, Virginia.

Shriver, E.L., Fink, C.D. and Trexler, R.C., 1964, "FORECAST Systems Analysis and Training Methods for Electronics Maintenance Training", Res. Rep. 13, Human Resources Research Office, Alexandria, Virginia.

Shriver, E.L. and Trexler, R.C., 1965, "Application and Test of the FORECAST Concept of Electronics Maintenance on Navy LORAN Equipment", Tech. Rep. 65-3, Human Resources Research Office, Alexandria, Virginia.

Standlee, L.S., Popham, W.J. and Fattu, N.A., 1956, "A Review of Trouble Shooting Research", Indiana University, Institute of Educational Research, Bloomington, Ind. (Res. Rep. 3).

Wallis, D., Duncan, K.D., and Knight, M.A.G., 1967, "A Review of Electronic Training Research in the United States Armed Forces", SP(N) Report No. 5/67.

Williams, W.L. Jr., Whitmore, P.G. Jr., 1959, "The Development and Use of a Performance Test as a Basis for Comparing Technicians with and without Field Experience: The Nike Ajax Maintenance Technician", Human Resources Research Office, The George Washington University, Washington, D.C. (Tech. Rep. 52).

A FAULT-FINDING TRAINING PROGRAMME FOR CONTINUOUS PLANT OPERATORS

E.C. Marshall and A. Shepherd

UWIST, Dept of Applied Psychology
Llwyn-y-Grant
Penylan, Cardiff, UK

Chemical Allied Products Industry Training Board
UK

INTRODUCTION

This symposium with its title, Human Detection and Diagnosis of System Failures, clearly implies that, at least in the immediate future, complex systems may have to resort to the skills of the human operator when problems arise during operation. However, the human attributes particularly appropriate to fault-finding are not inherent in the organism; operators of complex systems must be trained if they are to be efficient diagnosticians. This paper describes the development of a training programme specifically designed to help trainee process operators learn to recognise process plant breakdowns from an array of control room instruments. Although developed originally to train fault-finding in the context of continuous process chemical plant, it is probable that the techniques we are going to describe may prove to be equally effective in other industries. For example, power plants, crude oil refineries and oil production platforms all involve continuous processes which are operated from a central control room.

THE TRAINING PROBLEM

To develop any diagnostic skill, trainee operators must be given systematic and well prepared practice in fault-finding procedures. However, in order to improve plant efficiency and reliability, modern industrial processes have come increasingly to rely on complex control equipment. This leads to a paradox - these technological improvements reduce the frequency of process breakdowns, hence trainee operators have little opportunity for practising fault-finding in the control room. In addition, modern plant with its stress on efficiency and energy conservation, incorporates sophisticated process techniques which can further complicate diagnostic procedures in the event of a failure. In any case, training in the control room will be less than satisfactory not only because breakdowns are infrequent and irregular, but when they do occur, experienced operators will be far too preoccupied to take any interest in their trainees. Carefully designed simulated fault-finding exercises which can be practised outside the control room may well be the only practicable solution to the pressing problem of training diagnostic skills.

An essential aspect of any fault-finding training programme for chemical plant operators, we would argue, is that it should enable the trainee to diagnose faults that he has not previously experienced on the plant or practised during any training regime. For this reason, a major part of the programme is devoted to the teaching of diagnostic rules-of-thumb which we have shown, are powerful aids in the development of versatile fault-finding skill. In order that new trainees can become familar with fault-finding procedures relevant to their own control panel, the training package includes a facility to enable instructors to construct mock-ups of their own particular instrument arrays.

THE RESEARCH AND DEVELOPMENT OF THE FAULT-FINDING TRAINING PROGRAMME

Much of the research, the results of which have led to the production of this training programme, has been reported in detail elsewhere, for example Duncan and Shepherd (1975), Shepherd et. al. (1977), Marshall and Shepherd (1977), and Marshall and Duncan (1980). Therefore, in this paper we shall restrict description of this research and development to a brief summary of the major findings.

Simulation

An adequate simulation of control panels for the purpose of training these diagnostic tasks is achieved by presenting trainees with static pictures of the control panels on which are displayed instrument indications corresponding to the fault conditions which must be distinguished. Such a simulation preserves three arguably essential features: (i) panel layout; (ii) instrument design; (iii) approximate display size. The simulated control panel is constructed by mounting printed magnetic tiles, which represent individual plant instruments, on a large magnetic board. Pointers and labels, similarly made from magnetic tiles, are superimposed on the instruments and these can be adjusted to represent different indications. Symptom arrays for each fault are photographed using a good quality camera and slides thus produced are then back-projected on to a screen enlarged to life-size. This then provides a powerful training aid since it enables the trainee to make rapid comparisons between confusable faults and also allows him a great deal of practice in a relatively short time.

The process plant simulated in our experiments included unit operations, controls and features associated with thermal economies typically found in chemical plant design. The plant is instrumented with 33 instruments and 15 alarms as shown in Figure 1.

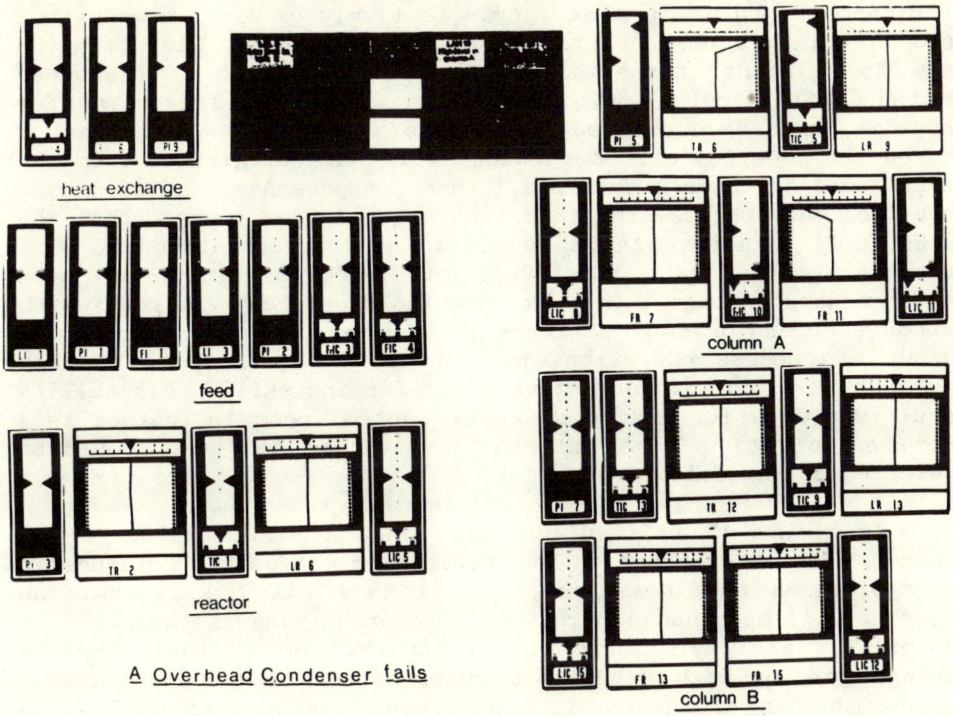

Figure 1. The simulated control panel used in our training experiments

Diagnostic Rules-of-Thumb

At the outset we would like to give a definition of just what we mean by a diagnostic rule-of-thumb. A rule-of-thumb is any principle or heuristic that can assist an operator in inferring the cause of a plant failure from an array of disturbed panel indications. From our experience it seems probable that skilled operators can achieve high standards of diagnostic accuracy with quite a small number of such rules.

Various commentators on the Three Mile Island incident have criticised operator training on the grounds that there was "a lack of emphasis on the comprehensive knowledge of theory, principles of operation, kinetics, thermo-dynamics, and so on, which would enable operators to correctly interpret information available to them in the control room", Greenberg, (1980). However, in the past we have repeatedly stated, on the basis of our research findings, that a weakness of conventional theory of this type is that it teaches trainees to reason from the fault to expected symptoms, whereas in the diagnostic task he is presented with an array of symptoms and must infer the fault that causes them. The advantage of diagnostic rules is that they effectively reduce the amount of information that the operator is using by directing his attention only to relevant indications among the whole mass of disturbed readings. Such rules may have diverse origins. They may, for instance, have been derived on the basis of reaction kinetics or indeed thermo-dynamics. Some rules will have originated from the experienced operator's detailed internal representation, or mental model, of the process. The important point is, of course, that the rules, once derived, can be taught to trainee operators who will subsequently use them effectively in diagnosis without any need for an understanding of the way in which the rules were generated. Of course operators may still be interested in learning about the theory of process operation, but we would stress that this may have little to do with effective fault-finding skill. This ability to derive rules for making sense of control room indicators is a prime example of a human attribute which, we believe, should be exploited to the full.

It should be pointed out that only a small proportion of trainees may be capable of generating effective diagnostic rules-of-thumb for themselves, but it is highly likely that the majority will be cabable of using these principles if they were so taught. Certainly we found in all our experiments that practice during training with diagnostic rules-of -thumb gave a dramatic improvement in trainees' accuracy when they had to fault-find novel plant failures.

Presented vs. Withheld Information

In the simulated fault-finding exercises we have so far described, trainee operators must use the array of information to make the appropriate diagnosis. In this case as in most control rooms, all the information necessary for diagnosis is presented. Since there is no way of knowing which information the trainee is using, it is impossible to know whether the trainee operator is using any rules learnt during training. To ensure that trainees do, in fact, apply rules when practising the solution of fault-finding problems we have developed a training regime which has yielded encouraging results. This technique entails withholding information which in the real situation is presented on the control panel. All the indications formerly presented on a panel are now retained by the instructor. To retrieve withheld information the trainee must request from the instructor any piece of information that he feels he requires in order to solve the diagnostic problem.

Withholding information in this fashion can ensure that trainees at least attempt to apply rules. Effective instructor intervention is also made possible. For example, frequently a trainee will move towards a diagnosis of pump failure, without checking the position of a valve in a control loop. If this is the case the instructor can make the trainee check the valve position. It should also be noted that in subsequent test sessions instructors can also monitor how consistently the trainee uses any instruction received during training.

Using this technique means that an important feature of the real task, presented information, has been abandoned. Simulation fidelity may have suffered because now the trainee is constrained to acquiring information in a serial fashion. A chief concern of our research was to assess how well subjects trained in the withheld mode transfer to the presented mode. We found that subjects had no difficulty in making this transition. In the training programme we have retained this important final test as a crucial indication of the trainees skill. Although the majority of practice during training is of the withheld type the trainee must be required to show that he can cope with a presented display.

Understanding Control Loops

One of the most valuable diagnostic rules which experienced operators use is always to check that control loops in the affected area of the plant are taking appropriate action. During our investigations we repeatedly observed that trainee operators, volunteer subjects and experienced operators had difficulty in applying this apparently straight forward rule. We suspect that this may be due to problems experienced when relating process behaviour to instrument readings on the panel. For example, a level controller, regulating level in a vessel by controlling liquid output, opens the valve if the level becomes too high and restricts the valve if the level falls. We have used the graphics terminal on our mini-computer to produce a simple dynamic simulation of such a level control loop. This dynamic model illustrates the relationship between level and valve position and in addition it demonstrates the way in which these parameters are presented on a typical control room instrument. When the trainee is familiar with normal operation of the loop we can introduce common failure situations; for example loss of feed, or pump failure. Figure 2 shows the valve failed closed. At a later stage we can remove the "plant" and show only the panel instrumentation. Using video films or photographs of this simulation, in conjunction with withheld practice trainees can, in a very short time, achieve 100% diagnosis scores on the level control loop. We believe that this constitutes a useful introduction to the mysteries of plant instrumentation and will help to prepare trainees for dealing with large arrays of instruments.

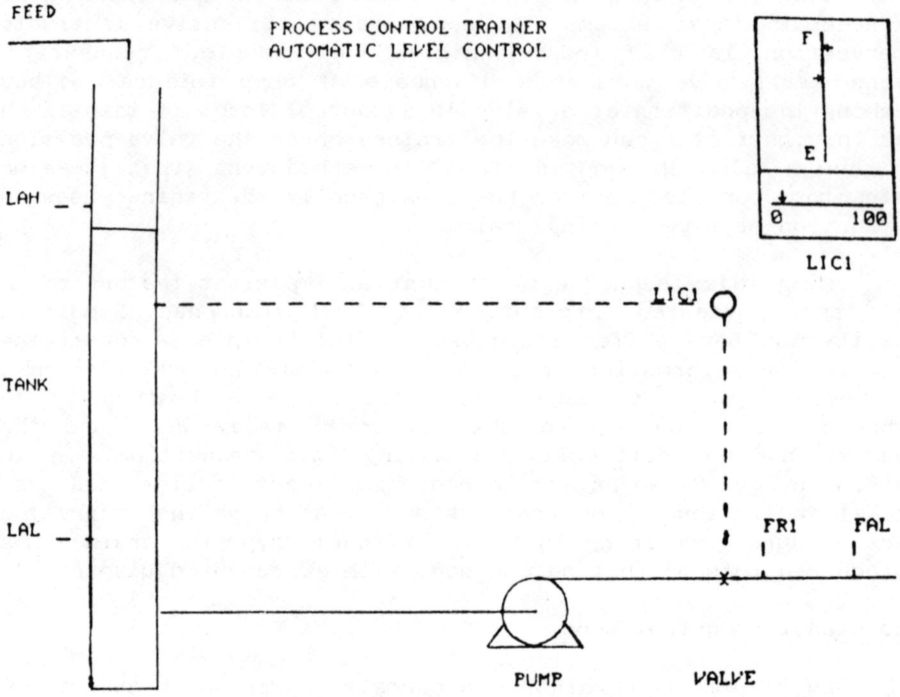

Figure 2 Computer generated schematic diagram of an automatic level controller and its associated panel instrument. In this example the valve has failed in the closed position.

THE TRAINING PROGRAMME

The training programme then, uses the previously described components, i.e. diagnostic rules-of-thumb, withheld and presented information, and the introduction to control loops in an integrated course for trainee operators. As far as possible the kit seeks to exploit hardware usually found in well equipped training departments, such as slide projectors, video cassette recorders, etc.

The programme is divided into three main sections. Each section is itself composed of a number of training modules. After working through each module trainees are tested to ensure that they can understand and apply the material they have just learnt.

Section 1 - Familiarisation Training

This section introduce the trainee to instrumentation in general and elementary fault diagnosis. More experienced trainees may not require to work through all the modules in this section, but it is essential the trainee should be able to pass the final tests provided before attempting Section 2.

The section starts with a tape/slide programme in which trainees are taught the name and function of instruments commonly found on control panels. They practice reading instruments, to note whether values are higher or lower than their set points, and to interpret valve output positions. The subsequent modules in this section use either video-tape or a tape/slide programme of the computer graphic simulation to introduce the trainee to the notion of automatic control, the way in which instruments monitor and control processes, and elementary fault diagnosis.

Section 2 - General Diagnosis Training

In this section trainees learn general principles of diagnosis in the context of the simulated plant that we have described earlier. The process incorporates many fault-finding problems common to a process plant. Trainees learn to apply diagnostic rules-of-thumb and they practice fault-finding in two modes, where instrument indications are either (a) presented or (b) withheld. Both training modes are used because as we have pointed out it is important to ensure that operators can easily transfer from one mode to the other.

Section 3 - Designing Plant Specific Training

This section provides instructors with the information they require to enable them to use the magnetic board and instruments (which are provided in the kit) to construct a control panel mock-up, and how to use a fault-symptom matrix to generate fault-finding exercises. Later in this paper we will describe in detail an exercise which exploited this third section of the training programme.

As a supplement to the three sections we felt it was essential to offer a short course for the training staff who will be responsible for utilisation of the programme to show them how to use the programme to its full advantage. This is in sharp contrast with elaborate general-purpose plant simulators which are not only expensive, but rarely provide the trainer with any help in the development or administration of effective training exercises. Perhaps we should emphasise that although Sections 1 and 2 of the programme in themselves provide a useful introductory course for operators they also are intended to provide models which trainers can use and adapt to generate exercises and training aids relating to their own particular fault-finding difficulties.

FIELD TRIALS

The Chemical and Allied Products Industry Training Board provided research funds to enable us to organise production of six prototype training programmes. These have been supplied to Chemical Manufacturers and Technical Colleges in the U.K. Representatives from these organisations have also attended the short course for instructors in the use of the programme.

Technical Colleges

The technical colleges who have purchased the programme are using it to supplement their existing process operator training courses. In this way apprentices are given opportunity for practice with panel operation and fault-finding procedures before they have had any real process experience. We have shown that the programme has advantages in the training of overseas students as the structuring of the exercises helps to overcome language difficulties.

Industrial Manufacturers

Several major chemical manufacturers are using the programme to give fault-finding training to operators and apprentices. In one plant we are collaborating with plant staff in the development of pre-commissioning training for operators of a new plant that will be going on-stream later this year.

We have also carried out a major exercise with another large chemical producer to design a training programme to improve fault-finding on an existing plant. This is described in detail below.

PLANT SPECIFIC TRAINING

An application of the plant specific training section of the programme, was developed for a large continuous process plant.

The Process and its Instrumentation

The process is a central component within a network of integrated continuous plants producing a textile polymer. It involves partial oxidation of a volatile and highly inflammable hydrocarbon in the presence of a catalyst. The operator is required to control two identical units each being operated from a similar control panel. Each control panel is approximately 3.5 metres long by 1.5 metres deep. There are some 50 automatic recorder controllers, most sharing two traces, a bank of ammeters relating to the various agitators and two panels of annunciators each composed of some 30 individual alarms - approximately 200 indicators in all. Because of the hazard inherent in the process the plant is equipped with a sophisticated interlock system, which automatically shuts down sections of the process in the event of out-of-tolerance indications.

The Diagnosis Task

Efficient fault-finding skill is important for two reasons. Firstly, in some cases, the skilled operator can forestall an unnecessary shut-down by recognising symptoms on the panel before they have become sufficiently serious to trip alarms or cause an interlock. Secondly, after an interlock, the operator must diagnose which remedial action should be taken before any attempts are made to put the plant back on stream.

For a number of reasons fault-finding on the oxidation panel is complicated.

(i) The whole process is pressurized. Thus in the event of a pressure control failure symptoms will be referred throughout the train within a relatively short time.

(ii) The process involves a number of elaborate recycle loops and contrary flows of vapour and liquid. This again means that symptoms will be referred across the plant.

(iii) Within the reactors themselves the physical chemistry responsible for the consistency and make-up of substances is quite elaborate.

(iv) Although the instruments are well laid out and clearly labelled, most of them are of the same type as shown in Figure 3, and hence may be a source of confusion.

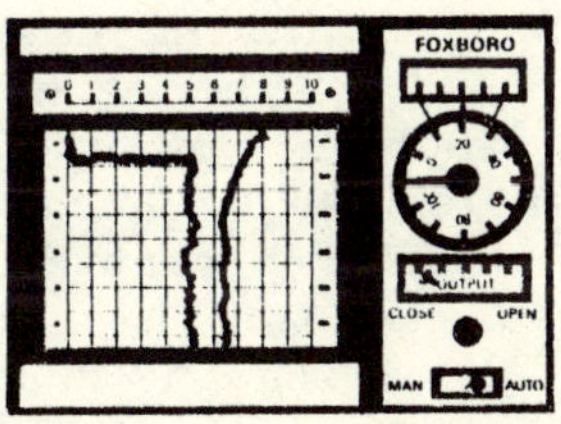

Figure 3 A typical panel mounted instrument. This controller is showing a rapid step change. This indicates a pump failure or a valve failed closed as explained in rule (i).

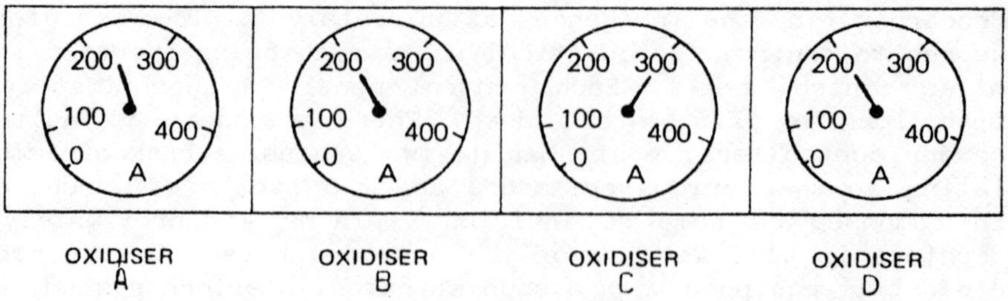

Figure 4 Panel mounted ammeters. These refer to agitator power consumption. Oxidiser C is showing a higher than normal reading. This indicates a problem in reactor C, rule (iii).

Preparing the Plant Specific Programme

Although the operator could possibly be required to deal with several hundred possible plant failures, only 20 faults were strictly necessary during training, to give operators sufficient practice with various fault-finding procedures. For example, when considering the various control loop problems, there was no need to include each possible controller failure for every one of the 50 controllers. A senior plant supervisor took several weeks to complete a fault-symptom matrix comprising the 20 faults and the appropriate 200 odd indications.

Guided by the models in the general diagnosis section of the programme, the supervisor also prepared three tape/ slide sequences to introduce trainee operators to the process: a) instrumentation and control of the plant; b) fault-finding the control loops; c) an explanation of the physical chemistry of the reaction.

Diagnostic rules-of-thumb, appropriate to the plant, were prepared during a workshop involving assistant foremen from the plant. During the first part of the workshop, participants worked through the first two sections of the fault-finding training programme - "Familiarisation" and "General diagnostic training". In the second part of the workshop participants worked through the material specifically designed for this process, corresponding to section three of the training programme - "Specific plant training". After participants had practised fault-finding from simulated panel arrays presented by back-projection, they were asked to develop and, hopefully agree on, a set of diagnostic rules-of-thumb to cope with a failure on the panel. After two or three hours of argument and discussion between the assistant foremen and supervisors, six diagnostic rules were accepted as suitable for versatile plant diagnosis. Each assistant foreman was asked to predict symptoms for a process failure which he had experienced. These additional failures then provided a further opportunity to test the efficacy of the rules. The six rules are shown in Table 1.

TABLE 1 THE DIAGNOSTIC RULES DEVISED FOR FAULT-FINDING THE OXIDATION PROCESS

(i) Check for step changes on each recorder controller. This rule allows operators to locate pump failures or valves which have stuck closed because only these failures cause the trace to show a rapid fall-off to zero. This is illustrated in Figure 3.

(ii) Check all pressure controllers to make sure that they are functioning correctly.

(iii) If there are no step changes or any pressure control faults, then the operator should check that the ammeters recording agitator current consumption in the reactors are reading normally, see Figure 4. High amps suggest a high level or high density of mixture in the reactor. Low amps indicate low density or excessive boil-up.

(iv) Check individual oxidisers to see whether the heat load or boil-up is normal. Any deviation here points to a feed problem in a reactor.

(v) and (vi) These rules require the operator to check flows and levels in other equipment further upstream in the process.

The Operator Training Programme

Armed with the failures to be trained and the set of diagnostic rules, a one week's course for training process operators was organised. Training was carried out by a training supervisor with wide experience of the process. Four operators were trained at a time, and to date seven such courses have taken place.

To measure how well operators could fault-find the process and to check on how much they improved throughout the training programme we carried out three validation tests. Each validation test consisted of a series of process failures presented on the back-projection screen. The process failures chosen for these tests had not been seen by trainees during any training practice. A strict time limit of two minutes was allowed for diagnosis of each failure. The test was thus extremely rigorous as the trainees had little time to contemplate this large array of presented information.

(i) Validation Test A - Trainees were given this test on the first day of the course as soon as they reported to the training room. They attempted six failures. The average score obtained by the 24 operators attempting this test was 2 correct out of 6, that is 33%. The highest score achieved by any individual was 4 out of 6.

(ii) Validation Test B - This test again consisted of 6 process failures but was carried out after trainees had been given some fault-finding training. They had worked through Section 1, Familiarisation training, and Section 2, General Diagnosis training, of the programme, and they had attempted successfully the tests given at the end of Section 2. In addition they had seen the three tape/slide sequences that had been prepared to explain various aspects of the oxidation panel. We were interested to see whether this introduction, coupled with fault-finding practice on the simulated plant, gave any improvement in fault-finding skill on the real panel. The average score for the 24 operators improved to 2.4 out of 6, i.e. about 40% accuracy, so there was a slight, but not statistically significant, improvement in fault-finding ability.

(iii) Validation Test C - This test was given on the last day of the course. Trainees by now practised the diagnostic rules especially developed for the oxidation panel and they had practised applying these rules both in the withheld and presented modes. They had all attempted a test where they had been given 20 presented failures to identify. The 20 failures comprised a mixture of 10 failures they had practised and 10 that they had only practised previously in the withheld mode. All operators achieved high scores on this test; the average for the 24 operators was over 90% accuracy. In the final validation test C operators were presented with 10 failures they had never seen before. They achieved an average accuracy of 63% which represents a considerable and significant improvement over performance in the first two validation tests.

We are now in the process of inspecting the data in greater detail to see whether there is any pattern apparent in the diagnostic errors because we expect this may well suggest remedial exercises to further improve accuracy in identifying novel process failures. Production and training staff at the plant are pleased with the results obtained to date and we are especially encouraged by the enthusiasm and cooperation shown by process operators attending the fault-finding courses. We hope to extend this exercise to include other sections of the process later in the year, when we will also include any remedial work suggested by analysis of test data.

THE BENEFITS OF THE TRAINING PROGRAMME

In conclusion we would like to summarise the advantages to be gained by using this kind of training programme.

(i) The technique gives systematic practice in fault-finding which is in no way possible "on-the-job" because of the irregular and unpredictable nature of process faults.

(ii) The training regime concentrates on teaching operators how to seek and evaluate <u>relevant</u> information from a large array of disturbed panel indications.

(iii) Training using diagnostic rules provides plant operators with strategies to help them cope with process failures they have not experienced previously.

(iv) It provides a way in which experienced operators can make explicit the kind of rules-of-thumb and heuristics that they use. Otherwise, we suggest, operators may well devise useful rules but they will keep them to themselves.

(v) The training regime provides a structure in which rules can be understood and practised by new operators and apprentices. In our experience there is never any formal way in which new operators are made aware of any useful diagnostic strategies.

(vi) Plant specific training programmes can be continually updated to include any novel process failures as they occur.

(vii) Because they are encouraged to practice predicting symptoms of given process breakdowns by manipulating the magnetic instruments on the simulated panel, trainees learn how to relate various states of the process to the array of panel indications. In this way trainees gain a thorough understanding of the process, in addition to acquiring fault-finding skill.

ACKNOWLEDGEMENTS

We wish to acknowledge the help and advice given by colleagues during the development of the training programme. We are also grateful for the assistance received from colleges and industrial organisations. In particular we are indebted to Imperial Chemical Industries Ltd., for their cooperation in the development of plant specific training in fault diagnosis.

The work described in this paper was supported by funds from the Chemical and Allied Products Industry Training Board and by the Social Science Research Council.

REFERENCES

Duncan, K.D. and Shepherd, A. (1975), A Simulator and Training Technique for Diagnosing Plant Failures from Control Panels. Ergonomics, 18, 627-641.

Greenberg, J., (1980) Human Error the Stakes are Raised. Science News, 117, 122-125.

Marshall, E.C. and Duncan, K.D. (1980) Information Display in Process Control Training. In Eds. Easterly R., and Zwaga, H., Visual Presentation of Information, In Press.

Marshall, E.C. and Shepherd, A. (1977) Strategies adopted by Operators When Diagnosting Plant Failures from a Simulated Control Panel. In Human Operators and Simulation. Institute of Measurement and Control, London.

Shepherd, A., Marshall, E.C., Turner A. and Duncan, K.D. (1977) Control Panel Diagnosis: a comparison of three training methods. Ergonomics, 20, 347-361.

THE ROLE OF COMPUTERS IN TRAINING FOR PROBLEM DIAGNOSIS

J. Patrick and R. B. Stammers

Applied Psychology Department
University of Aston in Birmingham
Birmingham B4 7ET, UK

INTRODUCTION

A theme in human factors that emerged in the early sixties was the application of the systems approach to occupational training. The main impact of this approach has been on the large scale training problems of military organisations although the industrial area has also been influenced (e.g. Butler, 1972). A variety of systems approaches have been proposed although recently there has been an attempt at rationalisations as for example in the Interservice Procedures for Instruction Systems Development (IPISD) in the USA. Systems thinking has developed over the same period in which computers have become increasingly available. Consequently, their role in training systems has been the subject of a continuing debate. This paper will review these issues in relation to training for problem diagnosis.

Training systems can be viewed at both the macro and micro levels. The macro level is concerned with the control of a flow of trainees through a process which converts them into an output of individuals with the necessary level of competence. Therefore, the system at this level is involved in the overall management of training. The training system's scale and structure will depend on, for example, such factors as the number of trainees, the number and location of training facilities and the financial and manpower resources available. These factors will determine the macro features of the system, such as whether it is large or small, permanent or temporary, centralised or distributed and will therefore begin to delimit the possible training solutions including any computer involvement.

Issues at the micro level are concerned with the interaction of the learner with some instructional environment. The ultimate design will be influenced by the efficiency of alternative methods of, for example, presenting information and providing remedial support.

Computers can therefore become components of the training system because of decisions either at the macro or micro level. For example the use of computer managed instruction would be the result of a decision at the macro level whereas the use of a touch sensitive screen would involve a decision at the micro level. The main focus of this paper will be the justification for computers at the micro level of the training system and how they can improve the instructional environment for problem diagnosis. Although many of these issues have been discussed during the last two decades (e.g. Rigney, 1962) a reappraisal is timely given the potential of recent hardware and software developments.

It is important to realise that there are many configurations of an instructional system ranging from total, partial to no computer involvement. Decisions about the nature and extent of the use of a computer are not as straightforward as they might appear. The training manager has to make a value judgement concerning when to use the computer in the training system. Seltzer (1971) attempts to externalise these value judgements and suggests that a computer should be used either when it provides a unique solution to an instructional problem (regardless of cost) or when it is the most cost effective option. However, the judgement of uniqueness of a training solution is not always easy given the difficulty of assessing instructional benefit. A tutorial in theoretical principles might be equally effectively accomplished by conventional or computer based techniques. In the area of problem diagnosis whilst the instructional environment could be provided by an instructor, the speed and storage capacity of a computer may be justifiably claimed to provide a unique solution (Patrick and Stammers, 1977). Uncertainty concerning the efficiency of various instructional solutions will gradually be resolved by future research and development.

PROBLEM DIAGNOSIS AND COMPUTER BASED TRAINING

In this section the potential role of computer based training in problem diagnosis will be more closely examined. In doing this, there are essentiallly four areas which need to be considered.

(a) What are the elements of problem diagnosis and what should form the content of the training programme?

(b) What are the contextual factors which may constrain the training solution?

(c) What instructional features are required in the training programme and to what extent does computer based training offer a unique or cost effective solution?
(d) Finally, what are the problems of implementation and evaluation of such training?

The effects of these issues are interrelated although they will be considered separately below.

Problem Diagnosis Performance

A variety of different situations can be subsumed under the heading of problem diagnosis. The most commonly cited example is that of a technician troubleshooting a piece of electronic equipment. Similarly a doctor may be said to diagnose a patient's illness. A less obvious example can be taken from the educational context where a teacher is tutoring a student. A feature of this situation is the teacher's ability to determine any misunderstandings of the student which necessarily involves some subtle diagnostic skills (Stevens et al, 1979).

In order to identify the training content of any of these situations or to determine the commonality of psychological demands between them, the application of some task analysis technique appears warranted. Some recent studies in troubleshooting (Frederickson & Freer, 1978; Malone, 1975) have used a conventional task analysis approach. One of the difficulties with techniques available for task analysis is the description of the sometimes complex cognitive processes underlying problem diagnosis. Essentially the task analyst is confronted with a problem of how to represent the state of knowledge of an expert and novice troubleshooter. Some initial inroads into this area have been made by Greeno's (1978) work on procedural representations and propositional networks although further work is required before such ideas can be embodied and applied in analytical devices.

An ideal approach would be the construction of a taxonomy of problem diagnosis tasks or subtasks with each category being mutually exclusive and exhaustive and having different training requirements. Initially consideration needs to be given to the elements embedded in problem diagnosis tasks.

(a) Manifestations of a problem state have to be identified. The extent to which these are given to the diagnostician will vary along a continuum. For example the activities of a process controller or technician may range from monitoring a 'normal' system for any symptoms of malfunction to acknowledging an obvious alarm state.

(b) After identification of a problem state the second step involves gathering and/or interpreting relevant information from the system in order to diagnose the problem. Information may be either recalled from human memory or some external store. A search may be initiated within the system to obtain information which was not present in the initial problem display. In electronics troubleshooting, a search is likely to involve using test equipment to find certain circuit values, whilst in a tutorial conversation the probe is a carefully constructed question presented to the student.

(c) Thirdly hypotheses must be proposed and revised in the light of information derived.

Consequently, problem diagnosis involves an iterative process between elements (b) and (c). This is central to the diagnostic process as any search should involve an efficient elimination of possible problem states.

The psychological demands of the above elements of problem diagnosis need to be determined. Malone (1975) elaborates the requirements which troubleshooting imposes in the context of maintenance training although these may be considered common to a variety of other systems.

> "An understanding of system states and interfaces.
> An understanding of the relationships existing among system elements and components, under normal and failure conditions.
> A knowledge of sequences, procedures and techniques.
> The ability to make decisions including discrimination among cues and selection of alternative sequences."
>
> Malone (1975, p.34)

The psychological demands which these requirements impose on the performer will depend on both the nature of the system in terms of the quantity and complexity of the inter-relationships within it and also the level of training of the performer. At this point it is worthwhile considering the problem of designing training if all the elements of the problem diagnosis task have to be learned. One approach is that of Gagné who in a variety of publications has proposed a series of principles concerned with the sequencing of instruction in the domain of what he refers to as intellectual skill (Gagné, 1975). A hierarchy of eight types of learning is identified such that the simple types of learning from the prerequisites for the more complex forms. The application of this scheme to the training of problem diagnosis skills would suggest that the trainee should learn to recognise and discriminate system components before proceeding to learn various system concepts (e.g. electronic or functional) and that complex rule learning and problem solving should be encountered last.

It is the training of these higher level intellectual skills in problem diagnosis which provides the greatest instructional difficulty for the trainer. Simpler forms of learning in problem diagnosis can be taught by a variety of conventional means (e.g. pencil and paper) and consequently computer based training is not likely to offer a unique training solution. However, the critical aspect of problem diagnosis involves the ability to interpret a display and "interrogate" a system in an efficient manner. It is in this area that computer based training is claimed to offer a unique training solution. Such claims will be examined in a later section which details the instructional environment necessary to facilitate the learning of such skills.

Contextual Factors in Problem Diagnosis Training

In this section some general contextual factors will be briefly considered which may affect the training of problem diagnosis. It is important that possible training solutions are considered not only in terms of instructional merit but also in terms of the external factors affecting the training system. Essentially for the training manager these factors are constraints or resource limitations which may be as diverse as finance, availability of equipment and attitudes of instructors.

There are a number of reasons why training in the real situation may not be viable. These include:

(a) Errors, which characterise the early stages of learning, may have unacceptable "costs" in the real situation. For example, this may apply to problem diagnosis with equipment which is expensive and/or hazardous and to medical diagnosis with patients in hospital.
(b) The real situation may be either unavailable or inaccessible for a variety of reasons.
(c) It may be difficult to present a representative set of faults and it may only be possible to examine those which occur fortuitously in the training period.

These factors suggest that problem diagnosis training should use some simulation which attempts to reproduce the major psychological demands of problem diagnosis behaviour as detailed in the previous section. This has been the trend for a number of decades in the training of maintenance troubleshooting skills. Fink and Shriver (1978) in a useful review suggest that a variety of types of simulation have been found to be cost effective compared to actual equipment trainers when teaching such skills. At this point it is important to note that a computer itself does not have to produce the simulation of the task. This might be achieved by a book, diagram or equipment mockup. On the other hand if the instructional features afforded by a computer in some aspect of the training of problem diagnosis skills constitute a unique training solution, then a computer generated simulation does not have to be justified in its own right. If, however, the simulation is necessarily computer-driven, then other instructional roles for the computer can be explored.

A frequent requirement for a problem diagnosis training solution is that it needs to be versatile. This is particularly necessary in the context of training troubleshooting for a variety of pieces of equipment or for equipment which is constantly being revised or replaced. In this situation the simulation could be of a general purpose nature such that it can be used to adequately represent a variety of pieces of equipment or it could be of a specific nature but capable of being conveniently updated or changed.

An often overlooked contextual factor which is important when a new training solution is being considered, as in the case of computer based training, is the need to successfully effect organisational change (Lawler, 1979). It is easy to underestimate the ramifications throughout an organisation which the establishment of computer based training might produce. One approach to this problem is the use of transactional evaluation (Seidel, 1978) which is a procedure enabling people to perceive the consequences of the projected change. Often a novel training solution or device is intended to be used alongside a conventional instructional system. Consequently, the favourable reactions to the training innovation by instructors and students is critical. This may well be affected by the degree of face validity which the solution possesses. Bryan and Regan (1972) review a variety of features which promote a more favourable reaction by instructors.

A final issue concerns the requirements or objectives of the training system for problem diagnosis. This is related to the use of different evaluation measures. The period of time over which the trainee is expected to retain often unpractised skills is important. The level of performance which is required or can be tolerated in the transfer to the real situation is also an important determinant of the training solution (Hammerton, 1967). Since the criticality of human problem diagnosis is likely to increase in the future with more routine diagnosis being accomplished by automated test systems, then the characteristics of retention and transfer of such skills need to be better understood than they are at present. One of the subsidiary benefits of computer-based facilities in scattered locations is that it is relatively easy to check levels of proficiency after training and when necessary implement refresher training programmes.

Instructional Features of Computer Based Training for Problem Diagnosis

This section will explore the main features of an instructional system which are required for the training of the higher level cognitive processes involved in problem diagnosis. It will be argued that the computer can provide many of the requisite instructional features although it is limited in some respects. The storage capacity and speed of operation which a computer provides coupled with decreasing hardware costs offer the basis for a promising training solution. On the other hand whilst new micro-technology may offer a second opportunity for computer based training (Sugarman, 1978) any training solution should be primarily examined on its instructional merits and costs. It is to be hoped that the widespread promulgation of computer based training for its own sake, which has occurred to some extent in the past, will not be repeated.

The task representation. Some representation of the task has to be presented to the trainee. The literature on troubleshooting in electronics maintenance reveals that a variety of representations of the task have been used with varying degrees of success. These include circuit diagrams, pictures of the apparatus, verbal descriptions and functional diagrams. In terms of instructional effectiveness it may be possible to extrapolate some recommendations from the evidence concerning the importance of fidelity of simulation. Fidelity has typically been interpreted as referring to equipment or physical similarity although as Caro (1977) notes this aspect is not always the determinant of training effectiveness. The simulation should represent the psychological demands of the task in order to promote both learning and transfer to the real situation. Consequently the use of some form of functional representation, despite low physical fidelity, has been effective in a number of troubleshooting training studies (summarised by Garland and Stainer, 1970; Fink and Shriver, 1978). Whilst computer involvement in the representation of the task is an expensive solution, there are advantages which it affords over other training solutions; flexibility and the opportunity to drive complex representations of the task. It is important however to retain a broad view even in these contexts of the possible training configurations in which a computer may be involved. It may provide the total task representation or it may be combined with other pieces of equipment.

A computer can offer flexibility with respect to the nature of the task representation. Some functional diagram may help develop an efficient search strategy and interrogation of the system by representing the inputs and outputs of components of the system. On the other hand the concept of functional context training (Duncan & Gray, 1975b) suggests that such skills should not be learnt in a vacuum. The relationships between functional elements, technical diagrams and actual components have to be learned to some degree. Consequently the fast generation of alternative representations of the same task with a computer may be an important instructional feature for training and particularly for transfer to the real situation. At present there is no evidence to support this popular notion although these principles are embodied in for example the AIDE system (Towne & Rigney, 1979). The ECII simulation which has been evaluated in a number of maintenance contexts (eg McGuirk et al, 1975) can provide various combinations of task representation under computer control. Simulations for problem diagnosis training may have varying degrees of physical and psychological fidelity. At one extreme full scale realistic simulations have been developed for complex systems such as nuclear power plants. At the other extreme the trainee can interact with a teleprinter to practice fault finding. In between can be found examples where the trainee interacts with schematic representations of the task, by pressing buttons on a panel, eg early versions of the ECII device (Finch, 1971) or where the interaction is with a display of system components, eg a circuit diagram (May et al, 1977), questions being input via a keyboard.

Other approaches to problem diagnosis training have involved using a computer terminal in connection with other representations of the task. An early feasibility study by Rigney et al (1966) paired a computer terminal with a piece of actual equipment, whilst a later evaluation by the same group (Huggett et al 1968) used a terminal with the real equipment on-line to the computer. Computer terminal plus simulated equipment has also been suggested (Daniels et al, 1975) and working systems have been produced using a terminal in conjunction with printed manuals and circuit diagrams (Parker and Knight, 1976).

It can be seen that the range of possibilities are large and that the training system designer's problem remains one of choosing an appropriate representation within given resource limitations. The choice is difficult given the paucity of data comparing the transfer of training from these various representations to on-the-job problem diagnosis.

A final issue is that a computer based task representations can provide a "dynamic" visually presented task with which the trainee can interact. This idea stems from the work of Paivio (1969) which details the facilitative effects of imagery on the learning and retention of verbal material. There is some evidence that this notion can be beneficially applied to computer generated graphics when complex concepts are being trained. Supporting evidence comes from a study by Hansen and Dick (1969) concerned with the learning of kinematic relationships and a series of studies into the understanding of concepts in science (eg Rigney & Lutz, 1976). Hammond (1971) adopts a slightly different perspective and suggests that the difference between student and computer generated displays could provide useful feedback in the training of medical diagnostic skills. Some further research is required to investigate the instructional potential of interactive and animated graphics before clear recommendations can be made. Nevertheless, if future findings are also favourable then computer involvement may well constitute a unique and effective training solution in this area.

Feedback and feedforward. Information concerning the appropriateness of certain problem diagnosis behaviours and strategies can be presented to the learner in two ways during problem solution activities. Information may be presented after the trainee has made a response or choice in the system (feedback) or it may be presented before any behaviour is required (feedforward). A distinction can also be made between different types of feedforward information which may either be general in nature (perhaps involving broadly described strategies or instructions) or specific (involving projected payoffs and consequences of certain search actions). In the general context of research on learning it is interesting that feedback information has received greater attention than feedforward information which may be a reflection of the behaviourist tradition.

Before either type of information can be presented to the trainee, it is necessary to be able to evaluate performance in some clearcut manner. In problem diagnosis training there are a variety of criteria which may be used for evaluation and which are likely to vary with the nature of the task. Performance can be evaluated against some "ideal" solution which might be based on such criteria as probability, criticality, cost or some amalgamation of these. In the context of training refinery process operators, Duncan and Gray (1975a) propose that performance can usefully be evaluated by reference to the "consistent fault set" and consequently inadequate performance can be characterised in terms of premature diagnoses, redundant questions and extra questions. Hartley and Sleeman (1973) describe how Bayes therorem can be used to find the best "attribute" of a patient to investigate at each stage of the medical diagnosis process. Clearly the generation of any information pertaining to the diagnosis process depends in both situations upon reliable task information gathered from subject matter experts. As soon as this can be specified, together with an "ideal" path based on some relevant criteria, then a computer can be used to evaluate diagnostic performance. As the data base of the task becomes larger and more complex, then it is likely that the computer will begin to afford a unique instructional solution. This is the case in the context of some electronics troubleshooting training as, for example, the use of the "expected utility" model to evaluate performance in the ACTS system (Knerr and Nawrocki, 1978).

Let us now consider how information concerning diagnostic performance should be conveyed to the trainee during problem solution. The importance of receiving extrinsic feedback during training has been extensively reviewed. However, even in studies which are primarily concerned with simple perceptual-motor skills it is evident that extrinsic feedback has to possess a variety of features in order to promote learning:

(a) It should occur before the student's next response
(b) It should be relevant to the task and useable by the student
(c) It should be clear and undistorted
(d) It should be related to the cues or feedback inherent in the task which will be available after training.

Whilst these principles are straightforward in many simple learning situations, their application to the area of problem diagnosis is not so clearcut. The reason for this is that having decided the criteria of "ideal" performance, feedback to the student on any deviation from this has to be presented in a manner which is useful. But simply the student has to understand why his interrogation of the system was suboptimal. In turn the tutor in order to accomplish this, must be able to represent the state of knowledge of the student together with possible gaps and misconceptions. These latter features need to be diagnosed by the tutoring system and only then is it possible to provide feedback which will improve the student's cognitive structure associated with the problem diagnosis process.

The alternative means of providing information is by some form of feedforward. In a variety of simple learning situations, techniques such as cuing, prompting and guidance have been found as effective as feedback techniques. At one extreme, in the problem diagnosis area, feedforward may take the form of strategy advice presented as an "advance organiser" (Ausubel, 1968). This may help the absorption and comprehension of new conceptual relationships and could be provided not only at the beginning of the training session but also at strategic points during the problem diagnosis process. At the other extreme, feedforward may tell the student what to do next and why or it may provide the projected consequences of certain actions. However, the sole use of prescriptive feedforward information is unlikely to be effective for training problem diagnosis. Firstly, there is evidence to suggest that the student will benefit from being active in the learning process (eg Belbin, 1969). Secondly, the provision of useful information during diagnosis is dependent upon inferring weaknesses from the problem diagnosis behaviour. Alternatively, it is likely that the student will benefit from some feedforward in the early stages of learning when errors are more common. The optimal mixture of feedback and feedforward information at various stages of learning in problem diagnosis needs to be empirically determined. However, whatever mixture of information is provided it is probable that "fading" the amount, as learning progresses, will also be an effective teaching technique. This sort of facility was available in the TASKTEACH system (Rigney & Towne, 1974) which allowed the student to switch from an instructional mode to a self test mode of operation, and to decide when to take a criterion test. In addition extra information of a feedforward nature could be requested from the system.

A flexible computer program for providing different sorts of information to the problem diagnosis trainee is described by Brooke et al (1978). The data base of the program is a fault--symptom matrix, the fault finder in isolating a particular problem can be provided with either advance information about the efficiency of a choice (feedforward) or with feedback on a system test once made. Research programs of this kind should provide more information on the relative efficiencies of these modes of instruction.

The data manipulation power of a computer is likely to be vital in storing and evaluating student's responses with respect to a data base except in fairly simple problem diagnosis tasks. The artificial intelligence movement has suggested that in order to provide useful information during problem diagnosis it is necessary to develop both student and expert models (eg Brown et al, 1975). The hypothesised states of knowledge represented by these models can be compared and used to provide information to the learner concerning any misconceptions or weaknesses. Thus the ACTS system (Crooks et al, 1978) sets out to "aid" the fault finder's decision, by giving advance information in the form of instructions which convey to the learner the costs and efficiencies of projected tests on the equipment under study. The SOPHIE laboratory and its Articulate Experts described by Brown et al (1976) provide intelligent feedback and feedforward to the student during electronics troubleshooting. The natural language dialogue is a particularly appealing facility, although of course this has to be paid for in computer capacity.

Individualised and adaptive instruction. It is generally accepted that instruction should be tailored to the individual needs of a trainee. Such adaptivity can be accomplished in a variety of ways including self pacing, provision of feedback, learning about the learner and adjusting the level of difficulty of problems. In the previous section the problem of providing useful evaluative feedback was discussed in the light of the deviations between student and expert models. This type of information cannot only form the basis of feedback during solution but can also be aggregated and used to select the nature of the next problem presented to the learner. This is not so easy, as measures of subjective difficulty may not directly correspond to the criteria involved in any "ideal" solution. A truly adaptive instructional system would not only explore the efficacy of different "difficulty" levels for each student but would also evaluate alternative means of representing the task and different instructional styles. These relationships have not yet been fully explored. However, there is some evidence that matching the cognitive style of the trainee with the instructional system may facilitate learning (Pask and Scott, 1972). One possible advantage of using a computer where training technology is only gradually developing, is that its fast, flexible closed-loop features can be employed to learn about the learner. Speculatively the same computer-based research system would be used for research and subsequent training application.

Implementation and Related Issues

This paper has discussed the potential role of computer-based training for problem diagnosis tasks. It has focussed on the direct instructional uses of computers and has emphasised the factors of task representation, provision of feedforward and feedback information and the adaptivity of the instructional environment. In these areas computers can have an important and sometimes unique contribution. Even in this direct instructional usage the training manager is faced with some difficult decisions concerning the resources which would be required to support a computer's role. Decision on hardware will not be easy at a time when experts are still discussing the relative advantages of distributed time-sharing systems and stand-alone devices. Similarly software provision for a novel training applications can be expensive. The further development of sophisticated general purpose software, such as the philosophy behind such systems as TASKTEACH, is clearly a step in the right direction.

A computer may be indirectly involved in problem diagnosis performance. Computer support can be provided for the diagnostician as a job performance aid and the device can also function as a tutor. A different sort of computer aiding can be available for the instructor. Training materials (eg programmed texts) can be produced for off-line use by an instructor interacting with a computer-based system. General purpose programs can help the instructors or subject matter experts to produce well-structured materials according to established instruction principles.

The paper has suggested that it is important to maintain a wide view of roles (both direct and indirect) for a computer in the training of problem diagnosis. Its degree of involvement in the training configuration can vary considerably. Of course justification of the use of computers may not be necessary in organisations in which they are readily available. Examples would be computer controlled processes, power generation plants and data-handling systems in industry and commerce. In these situations the training can be embedded in the existing system and instructional features can be included in the system software. Thus a terminal that usually functions as the system/user interface can become a computer controlled training facility with, incidentally, a high degree of face validity for the trainee. Such applications would appear to be particularly relevant to some problem diagnosis situations which are characterised by a demand for maintaining high levels of troubleshooting competence over time periods of no practice.

The final issue to be briefly considered concerns evaluation. Typically computer based training systems have been evaluated in a variety of ways. A common concern has been with the impact of training on the trainee in terms of both attitudes and learning gains. The instructional benefit of computer based training is often gauged against more conventional methods. Nevertheless if the computer's instructional role can be varied, the evaluation should perhaps be more concerned with the efficiency of different configurations of computer based training. Measures of external validity, involving transfer to the job situation are critical in the area of problem diagnosis training although there is a lack of such evidence. This may partially be explained by the paradox that the reasons which often justify computer usage in this area (eg lack of opportunity for practice) are those which also make it difficult to evaluate training in the real situation. Also many of the computer based training systems for problem diagnosis are still in their research and development stages and it is to be hoped that more implementation will be achieved in the future.

REFERENCES

Ausubel, D.P., 1968, "Educational Psychology: A Cognitive View", Holt, Rinehart and Winston, New York.

Belbin, R.M., 1969, "The Discovery Method of Training", HMSO, London. Training Information Paper No. 5.

Brooke, J.B., Duncan K.D. and Marshall, E.C., 1978, Interactive instruction in solving fault finding problems, International Journal of Man-Machine Studies, 10: 603-611.

Brown, J.S., Burton, R., Miller, M., Dekleer. J., Purcell, S., Hausmann, C. and Bobrow, R., 1975, "Steps towards a theoretical foundation for complex knowledge-based CAI", Bolt, Beranek and Newman, Boston, Report No. 3135.

Brown, J.S., Rubinstein, R. and Burton, R., 1976, "Reactive learning enviroment for computer assisted electronics instruction", Bolt, Beranek and Newman, Boston, Report No. 3314 (AD-A035302).

Bryan, G.L. and Regan, J.J., 1972, Training system design, in:"Human Engineering Guide to Equipment Design", H.P. Van Cott and R. G. Kinkade, eds., US Government Printing Office, Washington, D.C. pp 633-666.

Butler, F.C., 1972, "Instructional System Development for Vocational and Technical Training", Educational Technology Publications, Englewood Cliffs, N.J.

Caro, P.W., 1977, "Some factors influencing Air Force simulator training effectiveness", Human Resources Research Organisation, Alexandria, Va., Tech. Report No. 77-2 (AD-A043239).

Crooks, W.H., Kuppin, M.A. and Freedy, A., 1978, "Application of adaptive decision aiding systems to computer-assisted instruction. Adaptive computer training system", US Army Research Institute for the Behavioral and Social Sciences, Alexandria, Va., Rep. No. TR-78-A6, (AD-A056900).

Daniels, R.W., Datta, J.R., Gardner, J.A. and Modrick, J.A., 1975, "Feasibility of automated electronics maintenance training (AEMT) Vol. I: Design development and evaluation of an AEMT/ALQ-100 demonstration facility", Warminster, Pa., Naval Air Development Center, Rep. No. NADC-75146-40 (AD-A020873).

Duncan, K.D. and Gray, M.J., 1975a, Scoring methods for verification and diagnostic performance in industrial fault finding problems, Journal of Occupational Psychology, 48: 93-106.

Duncan, K.D. and Gray, M.J., 1975b, Functional context training: A review and application to a refinery control task, Travail Humain, 38: 81-96.

Finch, C.R., 1971, "Troubleshooting instruction in vocational technical instruction via dynamic simulation", Department of Vocational Education, Pennsylvania State University, University Park, Pa.

Fink, C.D. and Shriver, E.L., 1978, "Simulators for maintenance training: Some issues, problems and areas for future research", Air Force Human Resources Laboratory, Brooks Air Force Base, Tex. Rep. No. AFHRL-TR-78-27 (AD-A060088).

Frederickson, E.W. and Freer, D.R., 1978, "Basic electronics skills and knowledges", US Army Research Institute for the Behavioral and Social Sciences, Alexandria, Va., Res. Note 79-5 (AD-A068191).

Gagné, R.M., 1975, Taxonomic problems of educational systems, in: "Measurement of Human Resources", W.T. Singleton and P. Spurgeon, eds., Taylor and Francis, London, pp. 13-23.

Garland, D.J. and Stainer, F.W., 1970, "Modern Electronics Maintenance Principles", Pergamon, London.

Greeno, J.G., 1976, Cognitive objectives of instruction: Theory of knowledge for solving problems and answering questions, in: "Cognition and Instruction", D. Klahr, ed., Erlbaum, Hillsdale, N.J.

Hammerton, M., 1967, Measures of the efficiency of training devices Ergonomics, 10: 63-65.

Hammond, K.R., 1971, Computer graphics as an aid to learning, Science, 172: 903-908.

Hansen, D.N. and Dick, W., 1969 "Memory factors in computer-controlled maintenance training", Naval Training Devices Center, Orlando, Fl., Tech. Rep. -NAVTRADEVCEN - 68-c-0071--1 (AD697980).

Hartley, J.R. and Sleeman, D.H., 1973, Towards more intelligent teaching systems, International Journal of Man-Machine Studies, 5:215-236.

Huggett, G., Davis, J. J. .and Rigney, J. W., 1968, "Computer aided technical training using electronic equipment on-line with CAI system", Electronics Personnel Research Group, University of Southern California, Los Angeles, Tech. Rep. No. 59 (AD-672 189).

Knerr, B. W., and Nawrocki, L.H., 1978, "Development and evaluation of an adaptive computerized training system (ACTS)", US Army Research Institute for the Behavioral and Social Sciences, R and D Utilization Rep. No. 78-1 (ADAC65839).

Lawler, E.E., 1979, Applying motivation theory to work organisations, Paper presented at the NATO Conference, "Changes in the Quality and Nature of Working Life", Thessaloniki.

McGuirk, F.D., Pieper, W.J. and Miller, G.G., 1975, "Operational tryout of a general purpose simulator", Air Force Human Resources Laboratory, Brooks Air Force Base, Tecas, Rep. No. AFHRL-TR-75-13 (AD-A014794).

Malone, T.B., 1975, Requirements and concepts for fully generalized maintenance training systems, in: "New Concepts in Maintenance Trainers and Performance Aids", W.J. King and J.S. Duva, eds., Naval Training Equipment Center, Orlando, Fl., Rep. No. IH-255 (AD-A017 216). pp. 31-37.

May, D.M., Crooks, W.H., Purcell, D.D., Lucaccini, L.F., Freedy, A. and Wellman, G., 1977, "Application of adaptive decision aiding to computer assisted instruction", US Army Institute for the Behavioral and Social Sciences, Alexandria, Va., Rep. No. TR-77-A26 (AD-A055657).

Parker, G.R. and Knight, K.R., 1976, Can computer-assisted learning reduce the cost of equipment training, Royal Air Force Educational Bulletin, No. 13 pp. 7-15.

Pask, G. and Scott, B.C.E., 1972, Learning strategies and individual competence, International Journal of Man-Machine Studies, , 4: 217-253.

Patrick, J. and Stammers, R.B., 1977, Computer assisted learning and occupational training, British Journal of Education Technology, 8: 253-267.

Rigney, J.W., 1962, Potential uses of computers as teaching machines, in: "Programmed Learning and computer-Based Instruction", J.E. Coulson, ed., Wiley, New York, pp. 155-170.

Rigney, J.W., Bond, N.A., Mason, A.K. and Macaruso, R.B., 1966"Training corrective maintenance performance on electronic equipment with CAI terminals: I A feasibility study", Electronics Personel Research Group, University of Southern California, Los Angeles, Tech. Rep. No. 51 (AD-646651).

Rigney, J.W. and Towne, D.M., 1974, Computer aided performance training for diagnostic and procedural tasks. Journal of Educational Technology Systems, 2:279-304.

Seidel, R.J., 1978, "Transactional evaluation: Assessing human interactions during program development", Human Resources Research Organization, Alexandria, Va., Rep. No. HumRRO--PP-8-78.

Seltzer, R.A., 1971, Computer-assisted instruction -what it can and cannot do. American Psychologist, 26:373-377.

Stevens, A., Collins, A. and Goldin, S.E., 1979, Misconceptions in student's understanding, International J. Man-Machine Studies, 11:145-156.

Sugerman, R., 1978, A second chance for computer-aided instruction, IEEE Spectrum, August, pp 29-37.

Towne, D.M. and Rigney, J.W., 1979, "A developmental micro-processor-based system for OJT and JPA management in electronics maintenance". Behavioral Technology Laboratories, University of Southern California, Los Angeles, Rep. No. NAVTRAEQUIPCEN-76-C-0023.

COMPUTER-BASED MAINTENANCE TRAINING IN THE MILITARY[1]

Leon H. Nawrocki[2]

US Army Research Institute
P.O. Box 281
Ft. Monroe, VA 23651, USA

The purpose of this paper is to provide a context for the problem domain which the technical issues of this conference address. In the writing of this paper it became increasingly clear to the author that the issue of simulation for maintenance training could not be dealt with independent of the overall structure and history of the maintenance training arena. It is also worth pointing out that this paper is the forerunner of an anticipated effort to document the state-of-the art and recommended research directions in support of an overall effort to improve maintenance effectiveness in the US military. The result then is a paper which does not address explicit technical issues at the level of the majority of the other conference papers, but rather provides an overview within which it is hoped that there will become visible a linkage between scientific inquiry and requirements in the real world.

---

[1] The opinions expressed in this paper are solely those of the author and do not necessarily represent the opinions and policies of the United States Department of the Army.

[2] I would like to express my appreciation for the timely information and technical assistance provided by my colleagues at the US Army Research Institute, in particular Ms Helena Barsam whose efforts in searching for and evaluating often obscure documents formed the basis of this paper.

Before continuing, it is important to note the bias of the author, which should assist the reader in separating opinion from fact, should there be instances where this might occur. This bias is based on the author's personal belief that the ultimate solution to increasing the efficiency of maintenance performance rests with the "man" portion of the man-machine interface, and secondly that people can be taught the logical thought processes involved in detection and diagnosis within a reasonable amount of time. In the case of the latter, this belief has yet to be demonstrated beyond question, but the activities of this conference may very well prove to be a significant step in that direction.

The format for this paper is to first address the nature and scope of the problem of maintenance in the US military (hence forth references to military or military agencies should be assumed to refer to the United States military unless otherwise indicated). The unique characteristics of military maintenance, including constraints, will be reviewed followed by a discussion of alternative approaches to improving the current system. These approaches generally fall into three clusters, those dealing with the management procedures and processes within the maintenance establishment, those dealing with hardware solutions and those dealing with training solutions. In the last category, a case will be made for employing the confluence of technologies in computer science, instructional development and simulation technology. Finally, a brief review will be provided regarding the current efforts within computer based maintenance training simulation and thoughts concerning directions which appear fruitful to pursue.

PROBLEM AND SCOPE

While not a totally new problem, the military has become increasingly aware that equipment maintenance may be the most critical issue in terms of meeting the overall military mission. Within the last decade, the level of sophistication of major systems and individual equipment has increased dramatically. While many argue that complexity of equipment does not necessarily mean more complicated equipment, it seems clear that one must differentiate between the operation and the maintenance of the equipment. Thus, hardware can be designed for ease of use, but almost invariably this results in less reliable and more sensitive equipment, hence increased maintenance, an activity for which the costs are often less visible. Head (1978) demonstrated that the qualitative sophistication of weapon systems has dramatically increased the cost of maintenance manpower, training and support costs. Moreover, Pyatt (1972) reviewed several tactical airborne radar systems and found that the actual reliability of delivered equipment was consistently and significantly less than that promised in the original hardware specifications. Hence, the evidence, albeit limited suggests that hardware design solutions are unlikely to alleviate the maintenance requirement.

In addition to the equipment side of the issue is the people side. The military, for the most part, is faced with a relatively high turnover of personnel and new recruits are generally lacking in formal skills or experience simply as a function of the age market which forms the basis for input to the military. Hence, the military must devote a substantial portion of its resources to providing initial training, much of which is "lost" in later turnover. While such turnover may very well be exhibited in industry, industry tends to tap into the already experienced market and need provide less and often no specific training.

The magnitude of the maintenance requirement in the military is often surprising even to those who intuitively realize the probable scope of maintenance requirements given the size of this establishment. In 1976, it was estimated that there were .71 systems per person in the Army (which generally has a lesser requirement for sophisticated systems) and since that time over 40 major and 400 different minor hardware systems have been introduced or are in the development stage. Best estimated indicate that maintenance is 25-30% of the military budget. Moreover, the total maintenance costs of a piece of equipment throughout its life cycle are often expected to exceed its acquisition costs. In fact, the life cycle cost of repair labour alone on the A-7D aircraft exceeds all other costs combined, including operation (King, 1978). Over $1.8 billion is spent annually on specialized skill training, most of this on technical maintenance skills. The number of trainees can be staggering. For example, the Navy provides training in basic electronics to over 22,000 students annually, the Air Force trains nearly 6,000 students a year in aircraft maintenance and the Army provides training for over 7,000 students on track and wheel vehicle maintenance and repair on an annual basis (FY 80 Military Manpower Training Report, 1979). These represent only a small portion of the total student training in maintenance skills. In fact, there are about 240,000 maintenance technicians in each of the military services at any one time, of whom nearly 25% are relatively new with personnel costs of $5 billion a year not including training costs (Shriver, 1975).

With a system as massive as described, it is not surprising that difficulties arise, and these can only be exacerbated by the increase in quantity and sophistication of equipment. The US Government Accounting Office reviewed the military maintenance system in 1978 and concluded that equipment deficiencies were frequently not properly recognized or corrected, that maintenance was often done improperly and that on-the-job training of technicians was far from adequate. One indication of the type of problem occurring is that it is estimated that about 30% of components or units originally diagnosed as faulty are in fact operating correctly (Rowan, 1973) and some recent evidence suggests that for some equipment the rate may be as high as 60% incorrect fault diagnosis. All of this is within an environment in which training resources are strained and on-the-job training time to maintain and enhance skill proficiency must take second place to just keeping up with required daily work.

The situation described is clearly not so bleak as to be insoluble or there would be little reason for continuing this paper. Therefore, let us consider some of the actions which have already occurred or are ongoing to assist in managing the military maintenance system. Basically, these fall into two general areas: first, efforts to better assess the system itself in order to insure adequate procedures for managing and evaluating system performance, and second, activities which focus on increasing the effectiveness of the technician. The first although of less direct concern for this conference, will be addressed briefly to separate structure and organizational factors from the remaining discussion.

ORGANIZATIONAL FACTORS

The general structure of the military maintenance system is based on performance at three levels (the Army actually uses four levels, but two of these are increasingly melded together for practical purposes). The organizational rationale is less important for this purpose than the functional distinction. At the lowest and most immediate level is organizational or operational which involves minimal technical skill and is generally concerned with detection and diagnosis of major component faults. At a more intermediate level, (direct or general) components may actually be disassembled and sub-components replaced. Lastly, the fixed depot type maintenance includes fine component checks and repairs. In a kind of rough analogy, first level maintenance is about what the average home repair would involve, intermediate level is about the level expected at a local repair shop service outlet, and the third level is what one might expect at a major dealer center or factory. As might be expected, the level of maintenance reflects personnel experience, and thus time in military. Repair at the first level tends to be modular unit replacement and at the third level actual sub-component checkout and actual repair.

In a recent analysis of the military maintenance system organization, several differences were observed between the military and civilian maintenance environment (Drake et al., 1977). The military environment tends to require greater specialization, in part because of the structural separation of the maintenance levels, and in part because equipment readiness takes precedence over cost. Average technical experience is considerably less than in the civilian sector (the average civilian employee tends to have already received formal training and 1-3 years job related experience). It is also worth noting that both civilian and military technicians find formal on-the-job (OJT) training the most useful. But, the military emphasis on daily readiness of equipment, in conjunction with high volume, provides much less incentive for such training time. Moreover, the structural separation of maintenance levels is such that technicians at all levels in the military feel a lack of continuity in their input to the maintenance process and less exercising of multi-level skills. Thus, the analysis recommended that incentive conditions and actions to increase job enrichment are at least one means which could be addressed to improve performance of the total system.

On the more pragmatic side, it has been realized for some time that some difficulties in military maintenance may be due to incorrect or inefficient organizational procedures. The Army has initiated a substantial effort to provide a more accurate information and evaluation system for maintenance tasks in order to determine the extent to which problems arise from procedures or reflect training needs (Harper, Simpson and Fuller, 1979; Harper, 1980). This effort has resulted, to date, in more accurate measures at the first and second maintenance levels to separate these problems. The model provides measures to determine if problems are due to inefficient use of manpower, parts supply flow, system overload or ineffective technical work. The last is measured by job turnback rate, task completion time, inefficient technical actions, supervisor ratings, frequency of exposure to maintenance tasks and frequency of critical task exposure. This system is in the process of being field validated following preliminary evaluation indicating the general method is workable and functional. An unanticipated fallout from preliminary work has been that the information system can also assist in identifying basic equipment design problems leading to new tools or procedures in support of the maintenance technician. Until such a system is fully validated and implemented the actual portion of maintenance failure correctable by training will remain unclear, other than it appears substantial.

The point of the preceding overview is simply to make clear that maintenance problems must be approached from a systems standpoint if the extent of technical training needs, and hence the value of supporting research on detection and diagnosis models, is to be recognized. In addition, evaluation of proposed alternatives for the non-procedural aspects (to include training) is critical from a cost and effectiveness standpoint. This need has also been recognized by recent work on a Maintenance and Diagnostic Analysis Model (Mills and Wolf, 1978). Although the model emphasizes evaluation of the effectiveness of implementing automated diagnostic equipment, the same procedures appear to be relevant for evaluating the impact of training improvement.

HARDWARE ALTERNATIVE

Assume that the total maintenance system has been examined and that the problem is localized to inadequate personnel performance and furthermore, assume that the problem is one of fault discovery rather than actual repair or replacement, what then are the options? Basically, three general approaches have been proposed, each with advantages and disadvantages. The first of these is the use of Automated Test Equipment or ATE (a historical overview of the development of the ATE concept is contained in Lustig, 1973). The advantages are obvious in that the need for sophisticated training supposedly is reduced if not eliminated altogether (General Dynamics, 1970). On the other hand, Navy and Air Force ATE expenditures have risen to about $1.2 billion annually and this is expected to increase, particularly as the Army has recently initiated purchase of ATE for several electronic systems. The growth in military use of ATE is reflected by the development of data banks to provide logistic and technical information for ATE designers as well as a quarterly ATE newsletter under Department of Defense sponsorship.

Recently though King (1978) and King and Hemel (1978) have pointed out that ATE is far from a panacea. Not only is the current and expected cost likely to exceed the cost of more conventional approaches, but ironically, one of the biggest problems with ATE is the repair and maintenance requirement for ATE itself! For example, the largest use of ATE to date has been for the Navy's Versatile Avionics Shop Test (VAST). However, following the introduction of this ATE system, substantial new and additional training, training materials, and even a new skill area had to be added in order to optimize the use of the system. Similarly, another ATE system in the Air Force, a Converter/Flight Control Test Station, was found to require a high degree in training towards its use as well as maintaining the system itself (Baum et al., 1979). Moreover, the cost of the system, its low reliability and student safety factors resulted in the need for a training simulator to teach the operation and use of the ATE.

Beyond this problem, it is also the case that ATE is generally restricted in use to special purpose electronic equipment, and therefore, a limited range of applications. In short, ATE is one approach, but with sufficient drawbacks to suggest, it is far from the ultimate solution.

JOB PERFORMANCE AIDS

A second approach to improving diagnostic performance is through the use of job or performance aids. The use of such aids has received substantial military support, particularly since the late 1950's when several projects were initiated to analyze existing maintenance training procedures with regard to determining specific areas which could be improved through the use of technical manuals, flow charts, tables and other job aiding devices (Rowan, 1973). The most ambitious current program in the military is the Army's Skill Performance Aid System (Klesch, 1979). The program was originally conceived as a result of reviewing the utility of existing technical manuals. Such manuals were found to be generally written beyond the level of detail necessary to perform required maintenance, especially at the first and second levels of maintenance. In addition, the manuals were written primarily as engineering design documents with little, if any, attention to providing diagnostic techniques or strategies. Since that time, a coordinated effort between the training and equipment developing communities has resulted in trial versions of simplified and semiinstructional technical manuals. Preliminary samples are in the final development and evaluation stage. The major advantages are expected to be a lowered requirement for initial training and an alleviation of the problem of limited job experience.

As with ATE, however, certain disadvantages occur with job/performance aids. Manual version of these aids can become unacceptably bulky. For example, it is estimated that the manuals in support of the Army's XM-1 tank may exceed the physical volume of the tank itself. There is also the problem of updating as errors are discovered or equipment modifications are made (which occurs with astonishing frequency) to equipment over time. In part, the storage problem can be overcome through alternative media (e.g., microfiche, videodics) and the updating problem through the use of computer generated procedural aids. An excellent example of a system for combining these features is the Logic Model System which employs a portable display and embedded logic circuits based on maintenance dependency charts (Andre, 1978). The system provides the technician with check strategies based on the results of previously requested measurement inputs. The entire system weighs about 20 pounds and is being evaluated by the Air Force for electronics maintenance.

Of course, using alternative delivery systems for job aids requires at least an initial heavy cost investment for development and acquisition. In any case, there are additional disadvantages remaining regardless of the aiding media. Space limitations may be exceeded for first and second level maintenance, particularly in cramped quarters within ships and aircraft. Also, the Air Force discovered from experiences in a massive job aid development effort that complete reliance on aids failed to account for all possible problems, and if the aids failed there was no other solution source (Joyce, 1975). Moreover, while the use of aids reduced the difficulty of quickly training new technicians, their use also reduced the opportunity to increase the sophistication level of these technicians through OJT, resulting in fewer senior level technicians. Related directly to these two problems is the comment by King (1978) that as equipment sophistication increases even the designer cannot accurately predict all the potential malfunctions and causes (a situation which is increasingly the case for computer circuits designed by computers!), and the use of aids relies on having only a new highly qualified technician to monitor maintenance, which can be a liability in a critical tactical environment.

In terms of the training value of job/performance aids, there is little hard evidence that the combination of aids and experience will provide a certain degree of OJT. One particular problem is that practice is restricted to high frequency tasks, regardless of criticality, as tasks are not directed by training needs but by equipment failure. Expanding the domain of knowledge of a technician is also unlikely as aids are designed to fit the task requirements of the position rather than as general instructional aids. Neither are aids designed to make explicit to the technician the rationale behind the approach and thus, do not provide for generalized training. It is worth noting that at an early conference on military maintenance, attendees concluded that general technical knowledge would be preferable to specific job training but that such an approach was constrained by lack of effective methods for inducing such generalized knowledge (Shriver, 1957).

TRAINING AND SIMULATION

By now, the alert reader will have guessed that the third approach to improving diagnosis is by training the technician. The sheer magnitude of the training cost within the military as reviewed earlier in this paper is sufficient evidence that the need for training is recognized and that it will continue despite partial solutions such as ATE and aids. On the other hand, two issues have become of increasing concern in the training arena. First there is the problem of using actual equipment for training, and second there is the issue of task specific training versus general knowledge and procedures.

As operational equipment becomes more costly, there is increasing reticence to use such equipment in a strictly training mode. Not only is there the possibility of damaging expensive equipment and possible safety hazards, but using actual equipment for training reduces the number of items critically needed in the operating environment. Even when such equipment is available, the lack of flexibility reduced the range of maintenance experience to which a trainee can be exposed. An obvious solution, and one which is receiving increasing attention is the use of simulators.

A concerted effort to investigate the use of simulation for maintenance training began in 1976 with the formation of a military task group to plan for research and development in the area of maintenance training simulation. Since that time, several formal within and between service groups have been formed to coordinate and plan for increased simulation. Simulators have been frequently used for training of system operators, and a number of principles have been developed for the design of such simulators (Gerathewohl, 1969; Smode, 1971; Training Analysis and Evaluation Groups, 1972). Unfortunately, considerably less is known about the use of simulation for maintenance skills and most of the existing principles have not been validated in this context (Miller, 1975). As Kinkade (1979) has pointed out the situation is aggravated by current procedures to develop maintenance training simulators. Unlike operator skills, such as aircraft pilot, maintenance skills and requirements are often addressed late in the development cycle of new systems. The personnel who do address these issues are usually trainers interested in employing existing curricula and often lacking the expertise to perform the job/task analysis necessary to design cost effective simulators. Also, such personnel tend to operate in a more traditional manner regarding the use of training aids and devices.

In those few instances in which comparisons have been made between simulation for maintenance training versus other alternatives the simulation route looks promising. Kinkade (1979) notes that the Navy use of a Fault Isolation Simulator for automatic boiler control trainees reduced mean course throughout time by 50%, even though the simulator included greater training in diagnostics than previously. A preliminary evaluation of a part-task simulator for British radar display diagnosticians produced significant improvement in terms of fewer checks and less time taken by trainees given fault finding problems (Cunningham, 1977). In terms of cost, an interesting comparison by King (1978) indicated that the cost of actual equipment for training for six systems was $26.2 million whereas training simulators for the same systems were about $4.0 million.

Such evidence, plus a general dissatisfaction with current training methods has resulted in increased attempts to systematically explore the design of maintenance training simulators. Most of this effort has occurred in the military context. Little is available from civilian industrial efforts, at least in formal documentation. In a recent literature search, this author was able to discover about 30 articles on maintenance training in industry since 1970, and none of these addressed the use of simulation. This probably reflects both the tendency for industry to concern themselves directly with their own internal organizational needs and the availability of a trained, experienced technician pool. The lack of design criteria has resulted in attempts to determine the need for maintenance simulation via expert judgment (Herrick, Wright and Bromberger, 1977) and by application of known principles from early simulation fidelity research (Clark and Gardner, 1977). However, Purifoy and Benson (1970) have pointed out that the key to successful use of simulation will require attention to instructional strategies specific to maintenance training and early input to the system acquisition process.

Given the lack of empirically derived principles for the design of maintenance training simulators, to include analysis of instructional requirements, the military research community is currently emphasizing the development and evaluation of general purpose simulators. For example, two current efforts within the Navy are the Class A Electronic Equipment Maintenance Training System (EEMT) and the Rigney System. The EEMT is a general purpose simulator geared toward support of the Navy basic electronics technician courses. The design is based on the outcome of a task commonality and training requirements analysis as well as determination of fidelity requirements. The system is computer based and can provide a combination of two and three dimensional mockups of a wide range of Navy electronics equipment. The effort will include a cost and training effectiveness evaluation. The Rigney System is a more sophisticated (from the standpoint of software) system in the early development stages and is directed toward training of more advanced technical skills. This system provides an automated instructional capability based on adaptive controller functions. The adaptive capability permits the optimizing and scheduling of the instructional sequence and a means for automatically generating the instructional sequence based on individual trainee performance (Rigney, 1976; Rigney et al., 1978). Thus, the EEMT is a somewhat traditional simulation concept emphasizing hardware fidelity while the Rigney System is a more radical approach which takes into account an implicit model of the human learning process.

The Army is concentrating its current effort in the development of a general purpose computer-based simulator with logic modules to provide for mechanical maintenance training as well as electronic. The Army Maintenance Test and Evaluation Simulation System (AMTESS) is oriented toward developing a broad based family of computer-based simulators with tailoring to specific maintenance areas accomplished via flexible software and modular components. Like the EEMT, the AMTESS effort is being designed on the basis of an analysis of training requirements and fidelity options, albeit across a wider range of tasks (both mechanical and electronic systems). However, the goal is to provide sufficient computer software capability to permit the introduction of specific instructional strategies, such as the Rigney System, as these become more clear. It is assumed that the approach of a modular family of simulators will provide a flexible capability and low cost simulation suitable for current and long range needs. Preliminary broadboard design alternatives have been proposed and an evaluation of these alternatives in a field training environment is in the planning stage.

It is currently too early to be specific about the physical characteristics of the simulators just described. An interesting commonality though is the attention to relating the hardware design to systematic analysis of the instructional process and an implicit approach that there are common strategies in maintenance training. As indicated earlier in this paper, it has been recognized that general training is more advantageous than specific training, but less is known about providing the former. As a result, the military has increasingly leaned toward providing specific technical skill training. Some possibilities for reversing this trend and yet doing so cost effectively may now exist. Shriver (1966) points out that a better approach to electronic diagnostic training may be to stress the relationship between gross elements, such as symptoms, indications, functional entities, and so forth, rather than providing knowledge of electronic theory per se. In effect, Shriver suggests that instruction should be geared toward the presentation of relationships for direct use and not require the construction of these relationships. This idea has been further pursued by Brock (1977) who argues that in ridding maintenance training programs of theory nonessential for maintenance performance, we may have failed to distinguish between design engineer theory and those aspects of theory which provide the trainee with a conceptual framework for diagnosis of problems. As an illustration, one can readily imagine diagnosing automobile engine failure with little knowledge of venturi principles or electron flow. On the other hand, it is unlikely successful diagnosis could be accomplished without a grasp of the relationship between fuel and ignition, the major component functions and a general approach to narrowing down the alternatives (i.e., logic).

Fortunately, the confluence of technologies makes the concept of generalized knowledge instruction more practicable than in the past. Advances in the application of instructional theory (such as the Instructional Systems Development Model currently employed by the military) and the capability of computer driven simulators to rapidly process and analyze trainee inputs make the concept of individualized instruction possible and thus the implementation of instructional strategies heretofore too complex for massive training programs. What remains is to determine the most effective means for using these technologies. Three Army supported research efforts are underway to examine potential methods for training logic and diagnostics.

The first of these is referred to as the Adaptive Computerized Training System (ACTS). Not really a specific hardware system, ACTS is basically an instructional model and the supporting software to support the model. The concept of adaptive instruction has been used in several different ways. The most common has been as a procedure for adjusting learner criteria so as to gradually move the learner toward the desired instructional outcome without discouraging progress; i.e., an instructional incentive approach. The same phrase has also been used to refer to the branching and sequencing of instructional material as a function of learner progress, analogous to the Rigney System approach. The ACTS approach provides a method for modeling learner performance so as to track discrepancies between the learner and the "expert" diagnostician, such that analysis of these discrepancies can employ the adaptive techniques more commonly used (Knerr and Nawrocki, 1978). Another feature of ACTS is the assumption that utilities form the basis of the difference between experienced and inexperienced diagnosticians. Hence, the model tracks learner input in terms of the implied utility for an action. Both an expected utility and multi-variate mathematical model have been investigated as a basis for implementing ACTS. Preliminary evidence indicated the ACTS approach does in fact improve diagnostic behaviour and a field evaluation using existing hardware is planned in the near future (May, Crooks and Freedy, 1978). A more technical description of this work is provided elsewhere in these proceedings by Dr. Freedy.

Another approach being addressed is the investigation of techniques for teaching generalized troubleshooting logic. The Framework for Aiding the Understanding of Logical Troubleshooting (FAULT) assumes that there is a level of knowledge in diagnostics which can be abstracted from specific tasks and that this knowledge can be taught and will transfer to the specific task. Experimental evidence to date has indicated that practice on context-free (abstracted) troubleshooting tasks does indeed transfer to specific skills. However, the extent to which computer aiding in the learning process can systematically enhance this process remains somewhat ambiguous. This effort is discussed in detail elsewhere in the proceedings by Professor Rouse. Future directions are oriented toward clarifying the effects of computer aiding, determining the impact and measurement of problem complexity on skill transfer, and from this developing a cognitive model(s) of diagnostic performance.

A third, and relatively newer direction has been the examination of the use of logic games as a means for imparting diagnostic skills (Knerr, Simutis and Johnson, 1979). In the first of a series of planned experiments, subjects were provided with three types of training prior to transfer to computer presented logic circuit troubleshooting problems. One group received traditional instruction in the use of logic charts and logic diagrams, the second played a computerized version of a logical thinking game called MASTERMIND which requires identifying a four digit set of randomly generated numbers via a yes/no questioning sequence. The game control group was given a non-logic card game (Blackjack). Both the MASTERMIND and Blackjack groups received limited instruction similar to the traditional group following exposure to the games. While transfer data were not conclusive, it was clear that the MASTERMIND group produced significantly greater transfer than the Blackjack control group, although less than traditional instruction. The results were sufficiently encouraging to suggest that logic games can induce logical thinking procedures with a decrease in required training time. As motivation to play games is generally greater than that devoted to formal instruction, logic games may prove to be effective in that their use may motivate learners to provide themselves with greater practice on their own initiative. A follow on experiment is underway to compare a more pure training strategy of logic gaming versus traditional instruction.

In summary, several means are being pursued for improving maintenance performance in the military ranging from organizational changes to revisions in training procedures. One of the more innovative and promising directions is the combination of techniques for imparting generalized diagnostic knowledge with the capabilities of computer-based simulation. It is clear that what is needed is greater empirical evaluation of proposed models in terms of skill transfer and application requirements.

REFERENCES

Andre, W.L., 1978, "Helicopter Fault Isolation Equipment Evaluated by Army, Navy Air Force", Army Research Development & Acquisition Magazine, September-October 1978, 24-25.

Baum, D.R., Clark, C., Coleman, T.P., Lorence, S., Persons, W. and Miller, G., 1979, "Maintenance Training System: 6883 Converter/Flight Control Test Station", Air Force Human Resources Laboratory, AFHRL-TR-78-87.

Brock, J.R., 1977, "Simulation in Maintenance Training: Now That I've Thrown Out the Bath Water, Where is the Dear Baby?", Presentation at American Educational Research Association, New York, NY.

Clark, C. and Gardner, J.A., 1977, "Designing Simulators for Practicality", Proceedings, Naval Training Equipment Center Industry Conference.

Cunningham, D.J., 1977, "Electronic Maintenance - Training Analysis and Course Design", Admiralty Research Lab, ARL/APU/R10 (England).

Drake, K.R., Sanders, M.S., Cooks, W.H. and Weltman, G., 1977, "Comparative Studies of Organizational Factors in Military Maintenance", Perceptronics, PTR-1043-77-10.

General Dynamics, 1970, "CATE SACE Study. Systems Analysis/Cost Effectiveness of Computer Controlled Automatic Test Equipment", Contract DAAN07-67-C-0440.

Gerathewohl, S.H., 1969, "Fidelity of Simulation and Transfer of Training: A Review of the Problem", Federal Aviation Administration, FAA-AM-69-24.

Harper, W.R., Simpson, H.K. and Fuller, R.G., 1979, "Development of a Maintenance Performance System for Maintenance Jobs", Anacapa Sciences, Inc., Army Research Institute Contract MDA-903-78-C-2007.

Harper, W.R., 1980, "Development and Evaluation of New Performance System for Maintenance Jobs", Anacapa Sciences, Inc., Army Research Institute Contract MDA-903-78-C-2007 (Draft).

Head, R.G., 1978, "Technology and the Military Balance", Foreign Affairs, 56, 771-789.

Herrick, R.M., Wright, J.B. and Bromberger, R.A., 1977, "Simulators in Aviation Maintenance Training: A Delphi Study", Naval Air Development Center, NADC-78015-60.

Joyce, R.P., 1975, "Performance Aids and Technical Training: Complementary Sources", Applied Sciences Associates Inc.

Kinkade, R.G., 1979, "Maintenance Trainer Vs. Flight Trainer Procurement: Some Striking Dissimilarities", Third Biennial Maintenance Training and Aiding Conference, Naval Training Equipment Center.

King, W. J., 1978, "New Concepts in Maintenance Training", Aviation, October-November 1978, 24-26.

King, W.J. and Van Hemel, P.E., 1978, "Maintenance Training Technology and/or Our National Defense Posture", Presented at AFSC/NACMAT Science and Engineering Symposium, San Diego, CA.

Klesch, J., 1979, "On-the-Job Training and Aiding: The Army's SPA Program", Third Biennial Maintenance Training and Aiding Conference, Naval Training Equipment Center.

Knerr, B.W. and Nawrocki, L.H., 1978, "Development and Evaluation of an Adaptive Computerized Training System (ACTS)", Army Research Institute for the Behavioral and Social Sciences, Utilization Report 79-1.

Knerr, B.W., Simutis, Z.M. and Johnson, R.M., 1979, "Simulation Approaches to Maintenance Training", Presented at Military Operations Research Society, Vandenburg Air Force Base.

Lustig, J., 1973, "Trends in the Development of Automatic Test Equipment", New York U. Bronx School of Engineering and Science, Defense Advanced Research Projects Agency Contract N00039-68C-3579.

May, D.M., Crooks, W.H. and Freedy, A., 1978, "Application of Adaptive Decision Aiding Systems to Computer-Assisted Instruction: Experimental Studies", Army Research Institute TR-78-A4.

Military Manpower Training Report for FY 80, 1978, Department of Defense.

Miller, G.G., 1974, "Some Considerations in the Design and Utilization of Simulators for Technical Training", Air Force Human Research Laboratory, AFHRL 74-65.

Mills, G.F. and Wolf, K.A., 1978, "A Method for Evaluating Diagnostic Aid System in Army Land Vehicle Maintenance", Rand, Defense Advanced Research Projects Agency Contract R-2123-ARPA.

Purifoy, G.R. Jr. and Benson, E.W., 1979, "Maintenance Training Simulators Design and Acquisition: Summary of Current Procedures", Air Force Human Research Laboratory, AFHRL TR-79-23.

Pyatt, E.A., 1972, "Study: The Cost of Weapons", Presentation at Solid State Device Reliability Workshop, Warrenton, VA.

Rigney, J.W., 1976, "Principal Thrusts and Deficiencies in Research and Development on Simulation in Maintenance Training", Proceedings of the First International Learning Technology Congress & Exposition on Applied Learning Technology, Volume IV, 21-23.

Rigney, J.W., Town, D.M., King, C.A. and Moran, P.J., 1978, "Field Evaluation of the Generalized Maintenance Trainer-Simulator. I. Fleet Communications System", Defense Advanced Research Projects Agency Contract N00014-75-C-0838.

Rowan, T.C., 1973, "Improving DoD Maintenance Through Better Performance Aids", Defense Advanced Research Projects Agency, AS-758-713.

Shriver, E.L., 1975, "Maintenance Training Systems, from Technical Manuals to Automated Job-Performance Aids and Training Devices", Naval Training Equipment Center, NAVTRAEEQUIPCEN--IH255.

Shriver, E.L., 1957, "The Proceedings of the First Hum RRO Maintenance Conference", Human Resources Research Organization.

Shriver, E.L. and Trexler, R.C., 1966, "A Description and Analytic Discussion of Ten New Concepts for Electronic Maintenance", Human Resources Research Office, TR 66-23.

Smode, A.F., 1971, "Human Factors Inputs to the Training Device Design Process", NAVTRADEVCEN TR-69-C-0298-A.

Training Analysis and Evaluation Group, 1972, "Analysis of the Transfer of Training, Substitution, and Fidelity of Simulation of Training Equipment", Naval Training Equipment Center, NAVTRAEEQUIPCENTAEG-2.

A GENERAL-PURPOSE SYSTEM FOR SIMULATING AND TRAINING COMPLEX DIAGNOSIS AND TROUBLESHOOTING TASKS

Douglas M. Towne

Behavioral Technology Laboratories
University of Southern California
1845 S. Elena Ave. 4th Floor, Redondo Beach, CA 90277

INTRODUCTION

As military hardware becomes increasingly complex and costly, the suitability of using the prime equipment to meet the bulk of training requirements diminishes. At the present time, "general purpose" trainers can be developed which have the capability of providing realistic experience in operating or maintaining the prime equipment. Such trainers not only can simulate the functional and physical characteristics of the particular prime equipment, but they can deliver tutorial and special problem assistance when required by individual students; they can rapidly select, initiate, and administer problems based upon individual student performance; and they can record detailed performance data for instructor use. Furthermore, it is possible to provide "universal" applicability of both hardware and software, such that the trainer can be rapidly switched to a different prime equipment once the "courseware" is produced.

Behavioral Technology Laboratories, (BTL), a research laboratory at the University of Southern California, has developed and tested such a system. The design features which distinguish our approach from others are as follows:

1. Hardware consists entirely of commercial off-the-shelf equipment; it is unmodified, and retains manufacturers' warranties and maintenance services.

2. Software is based on UCSD (University of California at San Diego) Pascal, a high-level structured programming language which is standardized and widely used. The executive program

generates all student-computer interactions, based upon (1) the data base, which represents the content information, (2) the specific actions taken by the individual student, and (3) the specifications and constraints supplied by the instructor. This approach allows new prime equipments to be "installed"solely by content experts who analyze the normal operation of the prime equipment and its reactions to selected failures. These data (and graphic materials) along with a short data list representing instructor specifications (which "shape" the delivery of training), produce a simulation and instructional dialogue which appear to have been custom-written. Yet the foregoing is accomplished without re-programming or making hardware modifications.

3. A small number of programs are required which are specific to the particular hardware employed in the trainer/simulator. The bulk of the programs which handle communications with the student and manipulation of the various peripherals are written independently of specific equipment characteristics. This approach allows upgrading the trainer hardware as newer devices evolve, with minimum impact on software.

The specific input-output media employed are as follows:

- A standard CRT upon which is displayed computer-generated tutorial materials (based upon the student's actions), graphic representation of test equipment displays (waveforms, multimeter readings, etc.), and other administrative communication from the executive program.

- A random-access microfiche retrieval unit with back-projection screen, for presentation of color images of front panels, test points, modules, units, boards, wiring diagrams, block diagrams, etc. The resolution achieved by micrographics allows direct presentation of complex physical configurations and detailed diagrams.

- A touch pen with which the student makes all entries. Displayed switches are set by touching the new setting on the image. Test equipment is connected by touching the appropriate test point with the touch pen. Requests and other responses are made by touching the desired command on a "menu" placed on the front panel of the trainer/simulator.

IMPLEMENTING TRAINING AND SIMULATION

As noted above, a new prime equipment may be implemented on the trainer/simulator without making changes to the computer program. This is done by developing a detailed content data base which provides all necessary data to allow the executive program to display appropriate images to the student in response to his actions. The steps involved in producing this content base are 1) perform a functional analysis of the normally operating equipment, 2) analyze the effects of various selected malfunctions on the equipment, 3) construct the data base containing the relationships obtained in step 1 and 2, and 4) prepare color images (35 mm. color slides) of the equipment in all possible states, as defined in steps 1 and 2.

Each of these steps will be discussed below.

Functional Analysis

A functional analysis of the normally operating equipment to be simulated is performed, 1) to determine the normal responses of all front-panel indicators and all test equipment measurements in all modes of operation, and 2) to provide a physical map of the system to guide simulation.

A functional analysis consists of two primary sections:

1. A multi-leveled, structured set of diagrams, depicting the entire equipment in one gross representation, followed by a series of successively more detailed diagrams, each related to a unique section or block from the next higher level. The representations will include classical block diagrams and schematic diagrams, organized for convenient (and computer-managed) handling. Figure 1 illustrates the structure of this diagram set. Associated with each "block" or unit, (including the top level) is the following instructional information:

   a. A statement of the purpose of the unit, described in a level of detail that is appropriate to understanding the unit's functions at the current level of decomposition.

   b. A statement of the theory of operation of the unit, described in terms of the ways in which its next lower sub-elements combine to perform the unit's functions.

   c. A pictorial representation of the physical unit, if the subject unit exists as a physical entity. For example, the "frequency matching" function might be accomplished by a viewable module or board, whereas the "timing control" function might be distributed throughout the system. In this latter case, a list of physical elements comprising the subject function is produced.

2. Associated with the foregoing material (1a - 1c) is a digital index, stored on disc, by which the instructional program can retrieve the desired diagram or explanation.

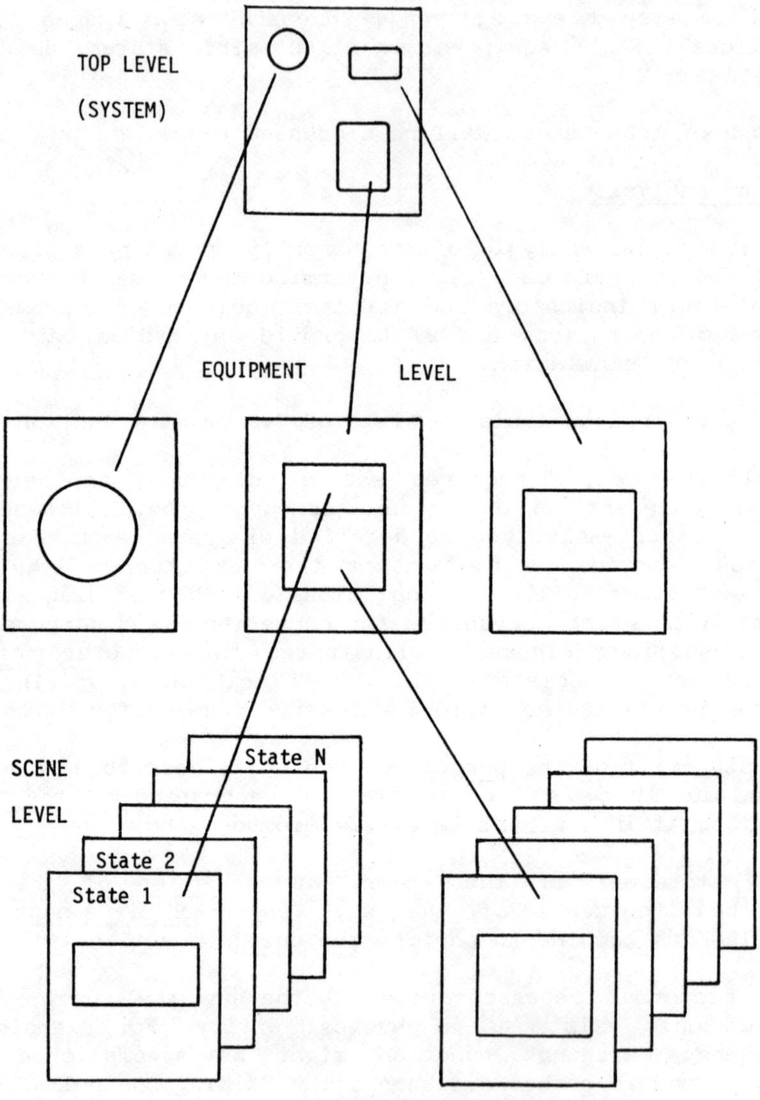

Figure 1. Hierarchical System Representation

Malfunction Analysis

The trainer/simulator simulates both normal operation and abnormal operation. We have learned that effects of malfunctions are most efficiently characterized (in the data base) as exceptions to specified normal operations. Thus data describing the effects of a malfunction can be limited to abnormals only. The result of the malfunction analysis, then, is a complete list of abnormal indications (front panel, test equipment, audio, etc.) produced in all possible configurations or modes.

The data base structure utilized provides for maximal grouping of equipment states, further reducing the volume of data required. For example, a particular malfunction may cause the power meter to read abnormally whenever the coupler unit is in "Forward Operation". This state (Forward Operation) can then be described in a short list of switch setting combinations. All references to that state, by malfunctions which evidence abnormalities in the state, are then single index references.

The tasks required to complete the malfunction analysis are as follows:

1. Compile a complete data base segment which specifies the normal indications or measurements at each sensible indicator or test point, in each configuration or mode. Techniques developed by BTL allow the simulator program to determine symptoms in all cases, yet the data base requires no complex mathematical models.

2. Develop a selection rule by which a large number of candidate malfunctions will be evaluated. This rule is formulated to apply higher scores to those malfunctions which:

   - are representative of a large class of frequent failure types;

   - are conceptually different from other malfunctions already selected;

   - produce symptoms which are especially instructive in highlighting critical function-symptom relationships or relationships which have been troublesome in past courses;

   - involve test equipment in either very commonly required ways or especially troublesome ways;

   - are either catastrophic to system operation or unusually difficult to detect.

3. Compile a substantial set of candidate malfunctions. This set may include malfunctions recommended by current instructors, malfunctions occurring with unusually high frequency in the field, as well as malfunctions identified by BTL technical personnel.

4. Apply the evaluation procedure to the set of candidates, producing a set of feasible malfunctions, ordered by value assigned. This list is then examined to determine the number of malfunctions required to represent the critical functional, physical, and instructional aspects of the maintenance workload.

5. Collect detailed sympton information for each malfunction in the problem set. As with the normal data, multiple abnormals are accommodated at any indicator, reflecting the multiple responses of the failed system in various modes of operation.

Data Base

The data developed in the functional analysis and malfunction analysis are then entered into a data file, and stored on 8-inch floppy disc in a special format designed to minimize the volume of information stored. The key device for doing this is a structure of conditions which serve as entry points to particular sections of the data base. During training the executive program evaluates those conditions until one is found to be true. This condition is a necessary condition expressing switch settings required to justify further processing of nested sufficient conditions. Once a sufficient condition is located which evaluates to true,the executive program has a direct index to the exact way in which each front panel indication and test equipment measurement will appear. Volume of data is further reduced by allowing malfunctions to share portions of those symptom blocks which are common.

Visual Representations

Visual representations of each section of the equipment are produced in each possible state. A large number of 35 mm color slides are made of all sections of the front panel, all related test equipment, the sides, top, and rear panels of all units, and all internal elements, down to the replaceable units. Views of test equipments and prime equipment front panel sections are replicated to the degree necessary to reflect "all possible" states of the section. The number of color images normally ranges between 800 and 1,200. A section is defined to retain functional integrity and be as large as possible without exceeding fifty states for the section.

Each malfunction is then assigned a problem number and an associated initial failure report, such as "....heading data is no good in local display mode". If desired, alternate initial failure reports may be associated with the same malfunction, thereby producing multiple problems per malfunction.

HARDWARE

The BTL trainer/simulator consists of three off-the-shelf commercial units:

1. Terak 8510a computer, with graphics and alphanumeric CRT and dual floppy disc drives.

2. Bruning Model 95 microfiche projector with RS232-C computer interface for random-access under computer control; 1,800-image capacity.

3. Science Accessories Corporation Graf-Pen, with RS232-C computer interface; 1 mm resolution.

This configuration provides two displays, (1) a computer--driven CRT used primarily to interact with the student in instructional and tutorial ways, and (2) a rear-projected microfiche screen, representing the real equipment, which responds to student actions by displaying high-resolution color images reflecting each action.

Alternatively, the CRT screen can display test equipment (multimeter, oscilloscope) readings as the sonic pen is touched to the desired test point displayed on the microfiche screen.

All student actions are performed via touching appropriate portions of CRT displays, color graphics displays, and fixed menu, with the sonic pen. Table 1, below, provides some example student actions.

The dual disc drives are employed such that each student "owns" a disc for the duration of a course. At the completion of each problem, whether successfully solved or not, data are written to the student's disc summarizing the final status of the problem and effectiveness with which the problem was handled. A list of items recorded is presented in a later section.

Table 1. Student actions and systems responses

STUDENT ACTION	ACCOMPLISHED BY TOUCHING	RESPONSE
Changing a switch setting	The desired (new) setting of the switch	A new image, with the switch in the new position (and all other indicators displaying as they would in that equipment state
Checking an indicator or reading	The section of the equipment containing the indicator	A new image which includes the desired indicator
Using test equipment	The desired test equipment name (on the CRT), then the test point displayed on the graphics screen	An image of the test equipment scope, meter, or dial display
Replacing a module unit, board, or component	REPLACE on menu	CRT confirms that replacement has been made; future symptoms depend on correctness of repair
Requesting assistance in symptom asessment, strategy development, or deductive malfunction isolation	HELP on menu	CRT requests identification of type of help requested then provides tailored interaction based on individual student's past actions

SOFTWARE

The software which controls the delivery of instruction and simulation consists of three major sections:

1. simulation software which senses and interprets student actions, determines the effect of those actions in the real equipment, and responds by altering the presentation of the equipment on the graphics screen and/or CRT;

2. pedagogical software which administers selection of problems, presentation of instructional material, provision of "help", recording of performance data, etc.;

3. utility software for creation and checkout of the data base, initialization of student discs, summarization and printout of student performance data, entry of problem selection constraints, etc.

All the foregoing is written in UCSD Pascal, a high level, structured programming language developed for applications on small microprocessors. Particular advantages of this approach are (1) Pascal is highly standardized, maximizing ease of augmenting and maintaining programs, (2) Pascal was specifically developed for ease of "transportability", i.e., implementation on new processors, and (3) the structured nature of the language tends to promote well-organized programs which are easier to develop and document.

An overview of the software structure is shown in Figure 2.

The four basic components are as follows:

1. the executive program, which controls the simulation and performs all pedagogical functions. The major constituents of simulation control are:

   a. student action interpretation - sensing and interpreting the significance of an action by the student, as indicated by a sonic pen "strike";

   b. state evaluation - determining the state of an equipment (including test equipment) from data describing a malfunction and assessment of the current mode;

   c. image display - causing the display to accurately reflect the new state of the currently displayed section, unit, etc.

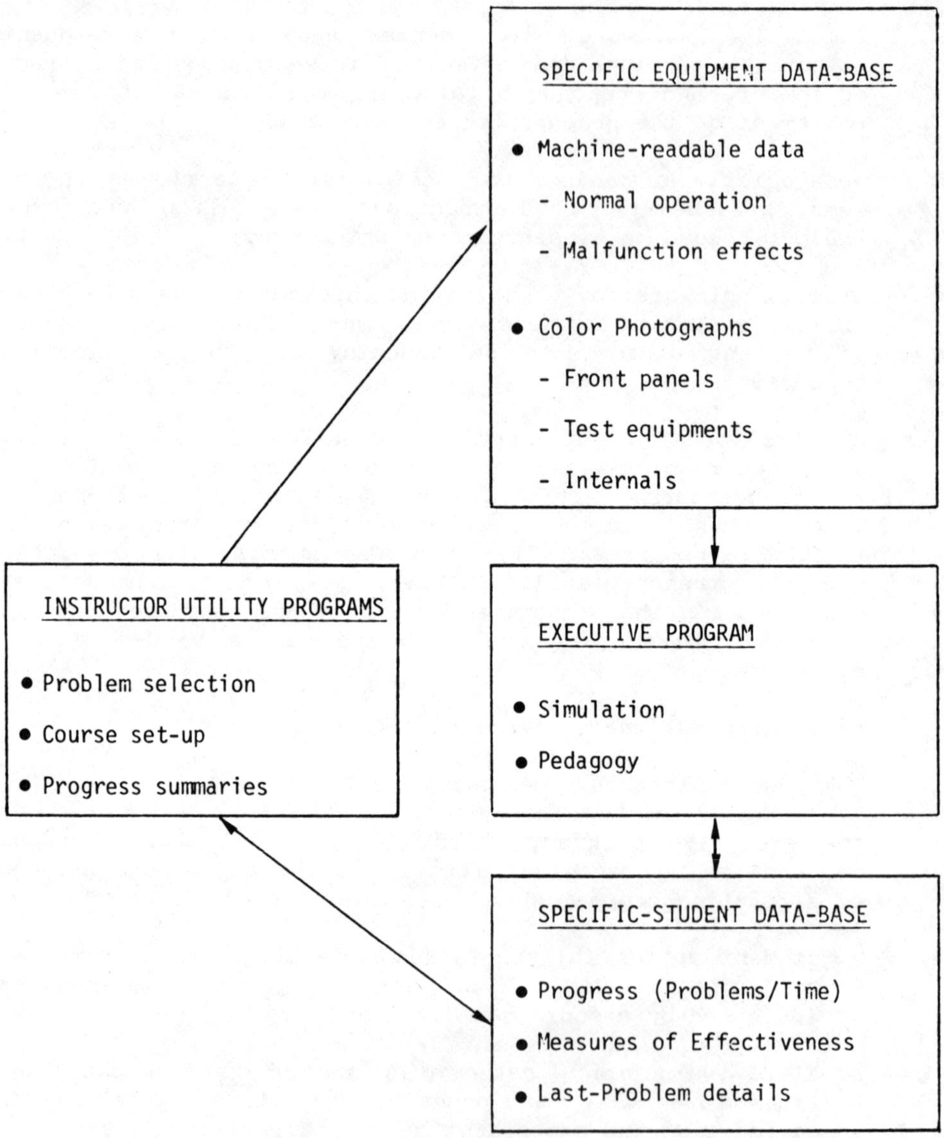

Figure 2. System software structure

The pedagogical functions consist of problem selection, administration of instruction and support, problem termination (requiring, allowing, or preventing same), and recording of performance data.

2. The specific equipment data base which contains machine--readable data characterizing the normal and abnormal responses of the equipment and images of the real equipment, its sub-elements, and associated test equipment. The contents of the specific equipment data base are shown in Figure 3.

Alphanumeric Data (on disc)

- Normal equipment states
- Abnormal equipment states
- Index of malfunction similarities
- Image Topographies (to interpret student actions)
- Image Index (state-to-image)

Micrographics (microfiche cassette)

- "Top"-level diagrams or photograph, of system
- Photographs of each equipment
- Photographs of each equipment-scene

Figure 3. Contents of Specific Equipment Data Base

3. The specific student data base which contains records of completed problems, measures of effectiveness for each problem, and microscopic detail for the last problem attempted. The latter facilitates re-runs of problems with instructor assistance.

4. Instructor utility programs concerned with problem selection constraints (to assure that problem presentation does not outpace lecture schedules), course set-up routines, and progress summaries. A number of special-purpose utilities are available for the use of instructors, to produce individual and class performance summaries.

FUNCTIONAL CHARACTERISTICS

This system has now been employed to simulate three complex electro-mechanical systems; a radar repeater, a shipboard communications system, and a satellite communications system.

In addition to simulating failed states, the trainer can simulate all possible normal states of an equipment. This capability is provided to allow students to operate the system at length, gaining a clear understanding of normal indications and measurements in all modes of interest. Obviously the system can also be used to train equipment operators.

Practice Mode

At the beginning of a troubleshooting problem the CRT screen presents a brief textual description of a malfunction complaint, representing the malfunction report placed by the equipment operator. On the micrographics display there will be a representation of the complete object equipment system. The student touches one of the pictured system-level equipments, and the total system image on the micrographics screen is replaced by an image of that equipment. The student touches the power control section of the equipment and it is replaced by a close-up image of that section; he touches the "on" position and a new image appears, with the power switch in the "on" position. Any meters or other displays in this image will reflect accurate readings for the object equipment in the currently simulated malfunction condition. The student might then touch "Higher Level" on the trainer's permanent menu, causing the current picture on the micrographics screen to be replaced by the next higher level image that includes that area (in the example, the new picture would be of the entire system component). The student could then choose to examine some other area of the current system component, or he could look at portions of a different system component. Whenever detailed scenes with visible indicators are shown, the values of those indicators reflect the actions thus far taken by the student (such as turning on the power on the aforementioned major system component).

In practice mode, if a student observes an indicator and is uncertain whether its value is normal, given all the settings he has made on the equipment, he can make use of the "HELP" function on the permanent menu. When this is touched, the CRT screen displays a message that tells the user whether there are any abnormal readings in the current micrographics display. This feature is not available to the student in the Test Mode (see below).

After having surveyed the system in a variety of switch setting states, the student user consults the Technical Orders and other system documentation, in order to develop preliminary hypotheses about the nature of the malfunction. At this point, it may be necessary to use test equipments, such as oscilloscopes or multimeters, to confirm a hypothesis. The student touches "Test Equipment" on the command menu, a list of available test equipments appears on the CRT screen, and a selection is made by touching the name of the desired unit on the CRT display. At this point the touch sensor device can be thought of by the student as the test equipment probe. When the student touches a displayed test point the system displays a photograph of the measurement or waveform which would be obtained. When the student is prepared to replace an element he uses the touch sensor to bring up a view of that element and touches "REPLACE" on the command menu. The CRT screen will confirm the pictured element has been replaced. The student can now take more test equipment readings and can move about the simulated equipment, checking for system malfunction symptoms. If none are detected, the student will touch "PROBLEM SOLVED" on the command menu. Only at this point will the student be told whether or not the problem has indeed been correctly "repaired". If it has been, the student will given the next problem scheduled for that point in the course. If a student has not correctly restored the equipment when he touches "PROBLEM SOLVED", a CRT message will inform him that the problem has not been solved and that he should continue to troubleshoot the problem.

Test Mode

Test mode is the same in many respects as practice mode. The major differences are that the "HELP" function is not available in the test mode, and a student's "Problem Solved" decision are irrevocable. The absence of the "Help" function will mean that students will have to make their own judgements, with the aid of technical orders and other documentation, as to whether a given instrument indicator or equipment reading is normal or reflects a malfunction state.

Data Recording

The BTL trainer records the following data for each problem attempted:

Problem number
Final solution state (solved, not solved, interrupted)
Number of incorrect claims of problem solution
Number of replacements made
Elements incorrectly replaced
Total time spent on problem
Number of usages of support functions (practice mode only)

APPLICATIONS AND EVALUATIONS

Over the past two years, three unique systems have been implemented on the trainer/simulator, as follows:

1. A Navy Fleet Communications System, consisting of six major equipments, twelve peripheral units, and three test equipments.

2. A Navy radar repeater (AN/SPA-66).

3. A Navy Fleet Satellite Communications System (AN/WSC-3), consisting of three major equipments, ten peripheral units, and three test equipments.

The last two of these implementations were accomplished entirely by technicians who were subject-matter experts but not programmers or specialists in CAI.

The first of these applications was tested in the U.S. Navy Advanced Electronics School, in San Diego, California. Each of twenty students worked 38 practice problems over 10 session-hours. The second application at the U.S. Navy Mobile Technical Unit-5 involved ten students each working 33 practice problems in 16 session-hours. Counting test problems worked on the simulator /trainer, we have data for more than 1,000 student-problems worked in nearly 400 student-hours.

The most significant findings (Rigney, Towne, King, and Moran, 1978; Rigney, Towne, Moran, and Mishler, 1978) were that (1) training transferred effectively to the real equipments (as evidenced by performance tests on the real equipment), (2) time to localize was reduced by a factor of two, and standard deviations by a factor of five, and (3) student and instructor acceptance was very high - according to questionnaires, they would divide their time approximately 50-50 between the BTL Trainer/Simulator and the real equipment.

REFERENCES

Rigney, J.W. and Towne, D.M., 1974, "Computer-Aided Performance Training for Diagnostic and Procedural Tasks", Journal of Educational Technology Systems.

Rigney, J.W., Towne, D.M., King, C.A. and Moran, P.J., 1978, "Field Evaluation of the Generalized Maintenance Trainer-Simulator: I. Fleet Communications System", Technical Report 89, Los Angeles: University of Southern California, Behavioral Technology Laboratories, October 1978.

Rigney, J.W., Towne, D.M., Moran, P.J. and Mishler, R.A., 1978, "Field Evaluation of the Generalized Maintenance Trainer-

Simulator: II. AN/SPA-66 Radar Repeater", Technical Report 90, Los Angeles: University of Southern California, Behavioral Technology Laboratories, November 1978.

Towne, D.M. and Rigney, J.W., 1979, "A Developmental Microprocessor-Based System for OJT and JPA Management in Electronics Maintenance", Los Angeles: University of Southern California, Behavioral Technology Laboratories, February 1979.

Towne, D.M., 1979, "The Automated Integration of Training and Aiding Information for the Operator/Technician", In: Third Biennial Maintenance Training and Aiding Conference, Orlando, Florida: Naval Training Equipment Center.

ADAPTIVE COMPUTER TRAINING SYSTEM (ACTS)

FOR FAULT DIAGNOSIS IN MAINTENANCE TASKS

Amos Freedy and Luigi F. Lucaccini

Perceptronics, Inc.
Woodland Hills, CA 91367

OVERVIEW

This paper describes the Adaptive Computer Training System (ACTS) which focuses on improving and sharpening higher-order cognitive skills in electronics troubleshooting. The application of decision models to training is reviewed prior to presentation of the features of the ACTS.

Although maintenance tasks rely heavily on a technician's knowledge and training about the maintained systems, such tasks can be viewed primarily as decision tasks. If the technician possesses sufficient knowledge of system parts and function, he applies it by making a series of decisions about which symptoms to look for, whether to repair or replace a malfunctioning part, and so on. ACTS is used in electronics maintenance training to address the quality of such decisions and the process of generating and choosing from among alternatives, rather than for the learning of specific procedural sequences.

ACKNOWLEDGEMENT

This research was supported by the U.S. Army Research Institute for the Behavioral and Social Sciences under Contract Number MDA 903-78-C-2039. The views and conclusions contained in this document are those of the authors and should not be interpreted as representing the official policies, either expressed or implied, of any office of the United States Government.

ACTS incorporates an adaptive computer program which learns the student's diagnostic and decision value structure, compares this to that of an expert, and adapts the instructional sequence so as to eliminate discrepancies. An expected utility (EU) or a multi-attribute utility (MAU) model is the basis of the student and instructor models which, together with a task simulator, form the core of ACTS. Earlier versions of the system used an expected value model (Freedy and Crooks, 1975; Crooks, Kuppin and Freedy, 1977). The student model is dynamically adjusted using a trainable network technique of pattern classification. The training content (instructions) and problem presentation sequence are generated with heuristic algorithms. ACTS is implemented on an Interdata Model 70 minicomputer and uses interactive graphics terminals for man/machine communication.

The present training system focuses on electronic troubleshooting. The student's task is to troubleshoot a complex circuit by making various test measurements, replacing the malfunctioning part, and making final verification measurements. The model of the student evaluates the student's selection of measurements and replacement of circuit modules. Troubleshooting provides an excellent application for the ACTS methodology because it is heavily dependent on judgment and probabilistic inference. In addition, troubleshooting is of great practical importance in numerous commercial and military systems, and it lends itself to economical implementation for training purposes.

Work to date has produced an operational system which demonstrates the feasibility of applying artificial intelligence techniques to computer-assisted instruction in a minicomputer--based training system. Experimental evaluations of ACTS have demonstrated that the adaptive decision model accurately learns the utilities of an expert technician and that students can effectively use the simulated troubleshooting task.

Additionally, instructions based on the utilities can further improve the decision performance of students; however, feedback of optimum choices immediately following the student's choice also seems necessary.

BACKGROUND: CAI AND DECISION MAKING

Individualized Instruction

A central theme in the field of educational technology is the creation of methods which allow individualized instruction. Training specialists and educational theorists recognize the importance of focusing on the individual student if significant advances in the efficiency and effectiveness of instruction are to be made (Crawford and Ragsdale, 1969); Glaser, 1965). Bloom (1968) has advocated the concept of mastery learning, in which instruction is designed and managed so that all students reach a given level of achievement, albeit at different rates.

The principles now included under the rubric of programmed instruction (PI), which grew out of pioneering work by Pressey, Skinner, and others, have facilitated the practical implementation of mastery learning techniques. Such principles, also claimed as advantages of PI, include: student-paced progression, immediate knowledge-of-results, individualized instructional sequencing, use of explicit performance objectives, diagnostic assessment, and the division of instruction into small discrete steps. These principles formed the basis for the multiplicity of programmed textbooks, teaching machines, and early CAI systems seen in the 1960's.

Adaptive Instruction

It has been recognized for more than a decade that true individualized instruction must include some form of adaptation to the individual student (Smallwood, 1962). However, while most researchers recognize the need to adapt instruction to individual differences, adaptation is usually made on the basis of response history. That is, the great majority of adaptive programs are made adaptive by the logic branching structure of the programs.

Central to the problem of adaptive CAI is the utilization of suitable criteria for optimizing learning effectiveness and the construction of decision rules for selecting instructional options. The development of adequate decision rules is very difficult in conventional adaptive CAI systems because a student's knowledge and skill level appear to be structured and fallible when viewed in the context of CAI.

Sophisticated optimization techniques for maximizing learning effectiveness have been used in several very elegant and highly adaptive CAI programs (Atkinson, 1972; Smallwood, 1971). However, these techniques have only been used for simple learning situations, which usually involve lower order cognitive skills such as memorizing lists of vocabulary words. This is because the optimization methods (developed from control theory) require a precisely stated learning model which predicts student response to alternate instructional options. As skills become more complex, it is less likely that simple mathematical learning models can be found.

A promising approach to adaptive CAI is the application of Artificial Intelligence (AI) techniques. AI techniques and theory, traditionally, have been concerned with the intellectually demanding tasks of problem solving and goal-directed decision making. These techniques are uniquely suitable for applications where unstructured environments are involved (Nilsson, 1965; Slagle, 1971). Natural language understanding and the heuristic programming approach to pattern recognition have been used in CAI systems which are based on information structure representations of the subject matter (Carbonell, 1970; Hartley and Sleeman, 1973; Koffman and Blount, 1974; Brown, Burton, and Bell, 1974). These systems utilize network analysis of the structures to generate instructional sequences, thus, the term "generative CAI".

Techniques of adaptive pattern classification can also be used to provide individualized instruction. Given a model of the student's behaviour, the pattern classifier adaptively adjusts parameters of the model until the model accurately predicts the student's performance. The model parameters then provide the basis for generating instructions and feedback. For the present decision training system, the parameters of an adaptive decision model are used as the basis for training the student in a decision task.

Adaptive Decision Modelling

Adaptive models of decision making attempt to learn the decision process of the human operator by (1) successive observation of his actions, and (2) establishing an interim relationship between the input data set and the output decisions (the model). Learning in this context refers to a training process for adjusting model parameters according to a criterion function. The object is to improve model performance as a function of experience or to match the model's characteristics to those of the operator.

There are two areas of research which attempt to establish useful adaptive decision models. The first, derived from behavioral decision research, is termed bootstrapping (Dawes, 1970; Goldberg, 1970). This procedure uses a statistical regression process to fit the parameters of the decision model to the decision maker's previous judgments. However, the bootstrapping technique is applied off-line to decisions which have been observed earlier.

A second approach to adaptive decision modelling involves trainable decision and classification networks. This technique is used as the basis of the ACTS system since it provides the capability to adjust model parameters on-line and to change model performance accordingly. The technique centers around adjustment of an expected value (EU) model of decision making. The decision network follows the decisions of the decision maker and adjusts its parameters to make it behave like the operator. In the EU model, the operator's subjective values for decision outcomes are the adaptively adjusted parameters.

The dynamic value estimation technique, developed by Perceptronics in the context of a decision aiding task (Crooks, Kuppin and Freedy, 1977), is based on the principle of a trainable multi-category pattern classifier. The value estimator observes the operator's choices among R possible decision options available to him, viewing his decision making as a process of classifying patterns of event probabilities. The value estimator then attempts to classify the event probability patterns by means of an expected utility evaluation, or discriminant function. These classifications are compared with the operator's decisions and an adaptive error-correction training algorithm is used to adjust pattern weights, which correspond to utilities, whenever the classifications are incorrect. Thus, the utility estimator "tracks" the operator's decision making and "learns" his values.

Decision Models in Maintenance Training

A maintenance technician makes a number of decisions while servicing the systems under his responsibility. He must decide whether the system is performing within tolerable limits, what symptoms of trouble to consider, what information to gather in troubleshooting, what test equipment to use, and so on. For these types of decisions, the technician must be trained to know the alternatives available to him, to estimate the odds on the outcomes of these alternatives, and to assign a value to each alternative. For example, in auto maintenance, the mechanic is trained to adjust the distributor with a "feeler" gauge or a dwell tachometer. He learns how accurately he is able to set the dwell angle with either instrument. The decision to choose one instrument or the other is influenced not only by the odds of setting the angle correctly, but also by the technician's stakes or values for each alternative. The feeler gauge may be preferred if it is right next to the mechanic in his tool box.

Decision training in maintenance should thus focus the student's attention on (1) listing the alternatives that he must consider, (2) estimating the odds of the various outcomes, and (3) evaluating the desirability of the outcomes. The adaptive EU decision model in the ACTS provides a method for instructing the student in these activities. The student is not trained to make a specific sequence of decisions. Rather, the parameters of the EU model are used to generate instructions about how to evaluate the decision alternatives. In the ACTS, adaptive sequential decision training is implemented within the context of electronic circuit troubleshooting. The student's task is to find a circuit fault by making final measurements to be able to declare that the device is repaired. However, the same principles can be applied to many other types of decision making tasks.

The training given in the circuit fault diagnosis and repair task is based on the assumption that the student has a good basic background in electronics but that his experience with trouble-shooting is limited. This might be the case with a student who has recently completed advanced military electronics training but has not yet performed troubleshooting tasks in his first permanent duty assignment. This skill level can be assessed either in terms of previous training received or in terms of performance on an entering test of electronics and troubleshooting knowlegde. It is assumed that the prerequisite laws of electricity, circuit component behaviour, circuit sub-systems, circuit diagrams, use of test equipment, and the like, have already been learned.

ACTS SYSTEM DESCRIPTION

The ACTS is an interactive computer program that models and simulates the four functional units of training: (1) the task being trained, (2) the student, (3) the instructor, and (4) the instructional logic. The organization of these four units in ACTS is illustrated in Figure 1.

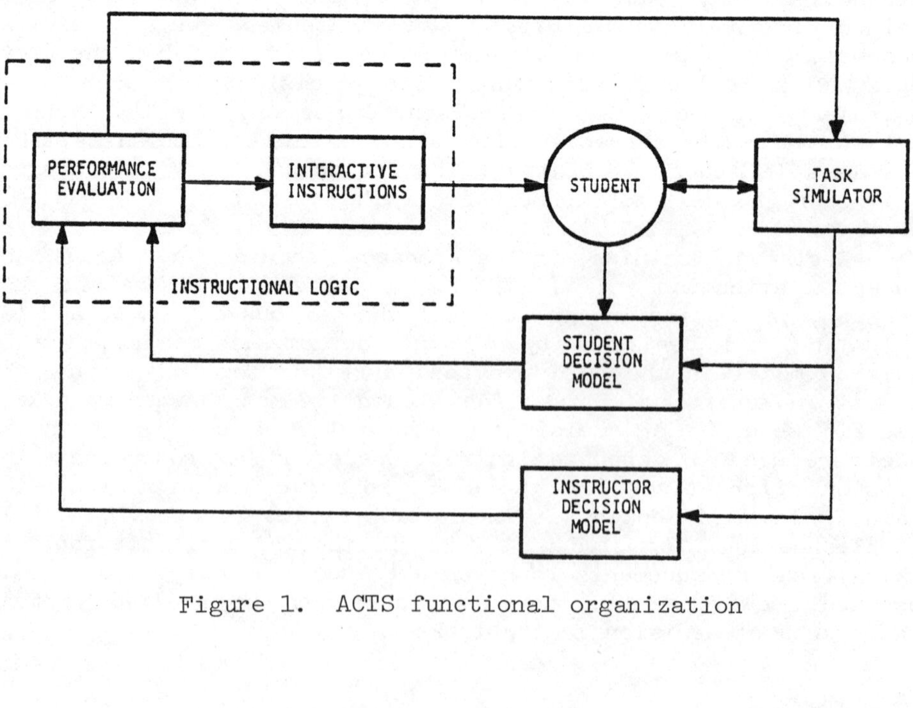

Figure 1. ACTS functional organization

Task Simulator. In ACTS, the student's decision task involves troubleshooting an electronic device. The troubleshooting task centers on a model of an electronic circuit in which faults can be simulated. The circuits currently used are a modular version of the Heathkit IP-28 regulated power supply and the U. S. Army A9000 power supply. The simulated circuits have 10 and 11 functional modules, respectively, which can be replaced and 32 and 23 measurements, respectively, which can be used to isolate faults. The operation of each power supply is simulated by the computer program, using a table-driven simulation of the fault system. The program simulates the results of checking symptoms, taking measurements, and replacing modules.

Training in the present system occurs with certain restrictions on the extent of circuit simulation. The student interacts with a terminal which contains a display of the simulated circuit; thus he cannot make such troubleshooting observations as smelling faulty capacitors, looking for burned resistors, or touching overheated semiconductors. In addition, the measurement results are presented in a semi-interpreted form (high, normal, low), rather than as absolute readings, (e.g., 3.6 volts, 1.25 mA), so that the student need not refer to a table of normal circuit levels. Although these modifications involve an abstraction of the troubleshooting task, it is assumed that they do not affect the critical decision making aspects of the troubleshooting task.

The circuit simulation was designed to meet several objectives. In addition to providing an environment for observing troubleshooting behaviour, the simulator gives the results of the student's choice of alternatives by displaying the results of measurements. Finally, the circuit model is designed to simulate the essential characteristics of decision making under uncertainty. Thus, the outcomes of the measurements are probabilistic, reflecting the fact that, in practice, fault locations are uncertain for the troubleshooter.

Student Decision Model. The student decision model is a mathematical decision model used in the ACTS to model the decision behaviour of the trainee and his instructor. The student decision model provides a method of describing or defining the student's behaviour. The ACTS then uses the model to infer the current state of the student's knowledge.

The decision model not only describes the initial state of the student's knowledge but it also tracks changes in the student's performance, adapting the model parameters to describe the student's improvements and errors. From this model of the student's behaviour, the ACT system gives instructions to improve the student's decision making.

A multi-attribute utility (MAU) decision model is used to represent the student. The MAU model is both a descriptive and normative model of decision masking which assumes that a "rational" decision maker selects the alternative with the greatest expected value. According to the model, decision making within the context of electronic troubleshooting involves three basic factors: (1) fault information gain, (2) commercial information gain, and (3) cost. The expected value of an action is then the sum of these factors weighted by specific utilities. The attributes and model are presented below:

Fault Information Gain:

$$A_{i1} = \sum Pij\ (F-Fij)/F$$

Commercial Information Gain:

$$A_{i2} = \sum_{j} Pij\ (M-Mij)/M$$

Cost:

$$A_{i3} = C_i$$

MAU:

$$MAU_i = \sum_{k} U_K A_{iK}$$

where

Pij = Probability that the j'th outcome will occur if the i'th alternative is chosen.

F = Current number of possible faults.

M = Current number of possible faulty modules.

Fij = Number of possible faults given current possible faults and the j'th outcome for action i.

Mij = Number of possible faulty modules given current possible faults and the j'th outcome for action i.

Ci = Cost of i'th action.

A_{iK} = K'th attribute for action i.

U_K = Utility for k'th attribute.

MAU_i = Expected utility of action i.

Given the available alternatives, attribute levels and utilities, the optimum choice is determined according to the maximum expected utility principle by calculating the expected utility for each possible alternative and then selecting that alternative with the greatest MAU.

ACTS uses the MAU model not only as the description of the student's decision making but also as the basis for estimating changes in his knowledge as inferred from his decision behaviour. A technique of artificial intelligence, known as the learning network approach to pattern classification, is used to estimate the student's utilities in the EU model (Crooks, Kuppin and Freedy, 1977). The utility estimator observes the student's choices among the possible decision alternatives, viewing his decision making as a process of classifying patterns of event probabilities. The utility estimator then attempts to classify the event probability patterns by means of a multi-attribute discriminant function. These classifications are compared with the student's choices and an adaptive error-correction training algorithm is used to adjust pattern weights, which correspond to utilities, whenever the classifications are incorrect. This utility estimator operates concurrently in real time as the student performs troubleshooting operations; thus, the MAU model continuously tracks the student's decision performance as it changes during the course of training.

Instructor Decision Model. The second decision model in ACTS is an MAU model of an expert decision maker's performance. This model is used (1) as a standard against which the utilities of the student model are compared, and (2) as a source of help in directing the student's activities and in suggesting alternatives. The instructor model has the same mathematical form as the student model, except that the utilities are preset and remain constant throughout a session. The utilities of this model are adaptively estimated prior to the training session by tracking the performance of an expert technician as he locates simulated faults or are set based on a priori expectations of expert troubleshooting behaviour.

The ACTS includes an algorithm for calculating the conditional probabilities of action outcomes. Conditional probabilities are of the form:

The probability of obtaining a particular measurement

outcome, given the previous measurement outcome history and the measurement.

These conditional probabilities are obtained by the ACTS algorithm from the apriori fault probabilities, PK, by the following formula:

$$P_{ij} = \left[\sum_{K \varepsilon Oij} PK \right] / \left[\sum_{K \varepsilon S} PK \right]$$

Where S is the current set of faults, Oij is the subset of S for which the outcome of action i is the j'th outcome. The apriori probabilities are obtained from an expert technician during the development of the task fault model.

Instructional Logic. The fourth major functional unit of the ACTS computer program is the instructional logic which selects the instruction and aiding information for the student. The instructional logic checks for convergence of the students utilities, compares the student's utilities with those of the expert, and compares the student's expended cost with that of the expert for the same problem. These three condition checks are then used to select or modify the following messages:

> Your choices indicate that you are inconsistent in your troubleshooting strategy. Before making a choice, consider carefully the uncertainty reduction, fault isolation, and costs associated with each choice.
>
> Congratulations. Your choices show that you are consistent in your strategy for troubleshooting. However, there may still be differences between your strategy and the expert's. If so, the next page will describe these differences.
>
> You appear to overemphasize: uncertainty reduction and underemphasize: cost.
>
> Congratulations. Your performance is identical to that of the expert. You are now a qualified troubleshooter on the IP28 circuit.
>
> Congratulations on repairing the circuit. Your total cost to debug the circuit was 190. The instructor's total cost would have been 120.

Prior to the troubleshooting session, the student is assumed to have completed the preliminary lessons on the power supply involved. Consequently, instructions in the troubleshooting unit are not focused on the type of measurements to make or the functions of specific components of subcircuits. Rather, ACTS instruction is directed toward training an inexperienced technician to evaluate the utilities of the alternative measurements he can make and to select those alternatives that are most effective, given their relative costs.

In addition to the instructions that are displayed on the basis of the student's decision performance, the ACT system also includes a HELP option which the student can select as desired. The HELP option uses the expert decision model to suggest which measurement to make, their tradeoffs, and their relative overall values.

MAN-MACHINE SYSTEM CONFIGURATION

The ACTS consists of four major hardware components: an Interdata 70 minicomputer with a 64K memory, a Tektronix 4024 CRT for each subject, a console communication device, and an experimenter communication device, which may be a teletype or a CRT. During training sessions, ACTS can operate automatically with no intervention required by an experimenter. However, a number of options for experimenter control are available. The experimenter can start or terminate a student session, change the choice of options available to the student, save or print the current state of the system, terminate ACTS operation, or change the starting weights of the model. Each student has exclusive access to his own terminal through which all instructional interaction occurs.

INSTRUCTIONAL APPROACH

Training Procedure. Training on the ACTS system is provided through a system of phased instructional presentations. A series of units on the given power supply is presented to the student. The material begins with the most basic information about power supplies and terminates with the troubleshooting unit, which consists of a number of circuit fault problems. For all instructions prior to the troubleshooting unit, the procedure is to present text material to the student, allow him to ask questions and receive answers, and then give the student a test. If he passes the test he is advanced to the next unit; otherwise, he repeats the current unit. When the student has completed all the preliminary units, he begins the troubleshooting phase of instruction.

Each troubleshooting problem consists of a single circuit fault which the student must locate and replace. On the display is shown a schematic diagram of an electronic circuit, plus printed messages which indicate possible actions and give information. The student selects his responses and types them in on a keyboard. The student can select from among a number of activities to isolate the fault in the displayed power supply circuit. The student can choose to take a voltage or current measurement, replace any circuit module, or request help. Following a student's command to perform these activities, the ACTS program displays the results of the simulated actitivy and then indicates the next allowable activities.

Interspersed among the fault problems, the ACTS presents the instructions which describe recommended circuit measurements and the conditions during which they should be chosen. After the instructions have been displayed, the fault problems are resumed. However, the student can request to see these instructions at the appropriate time by selecting the appropriate command on the display screen.

Consideration and Help. When the circuit is displayed, a malfunction is signaled by displaying overt symptoms in a table of symptoms and measurement outcomes. The student is then told that he will next be expected to input some action candidates for consideration. He may also ask for help at this point. Provided that the 'HELP' option is allowed, a help request will provide the student with the expert's considerations, as shown in Figure 2. After looking at these, the student may request help again - in which case certain tradeoff information for the expert's considerations will be displayed. This information includes the cost of each action, all outcomes and their probabilities for each action, and the fraction of faults to be eliminated by each outcome of each action.

The student next chooses his candidates for consideration. These candidates may be measurements and/or module replacements. The system then displays for him the value of each attribute for each of his considerations. At this point help may again be requested if the 'HELP' option is set to aid the student in choosing an action from amongst the considerations. The student may also choose immediately without help.

Action Selection and Help. If help is requested, the tradeoffs of the final considerations are then displayed. The message is the same as that used earlier to display the tradeoffs for the expert's considerations. Help may then be requested again, in which case, an expert ranking of the final considerations is presented. The attribute levels of the considerations are then re-displayed.

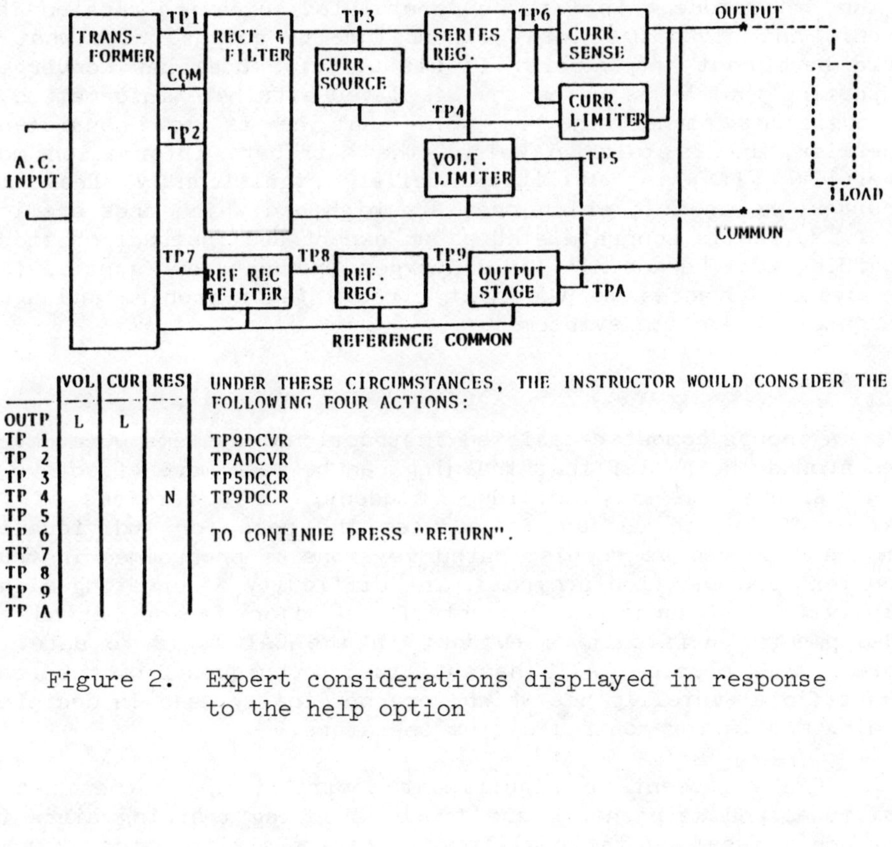

Figure 2. Expert considerations displayed in response to the help option

The student may then choose 'none of the above' in which case he will be asked for new considerations or he may type a choice from the list of considerations. If his choice is a measurement, its outcome is displayed in the symptom/outcome table. If his choice is to replace a module, the part of the display depicting the module is enhanced on the screen. If the chosen action did not repair the circuit, the cycle repeats with a request for new considerations. If the chosen action correctly replaces the faulty module, the overt malfunction symptoms are corrected on the screen and the system enters an evaluation phase.

Evaluation (Feedback) Phase. When the evaluation phase begins, the student is first congratulated on having repaired the circuit and given his total expenditure to compare with what it would have cost the expert. If his utility model has converged, indicating that he is using the displayed attribute information in a consistent manner, he is told that he is now consistent; otherwise, he is told to weight the attribute information more carefully. If his utilities differ significantly from the expert's, he is told which ones are high and which ones are low; otherwise, he is congratulated as an expert and instruction stops. Providing that he has not yet converged to the expert's utilities, the system advances to the next circuit fault problem and again presents malfunction symptoms.

EVALUATION

Although computer-assisted instructional systems were developed around the belief that training can be made more effective if it is adapted to the individual student, early versions of CAI systems failed to adequately address the need for individualization and served as sophisticated versions of programmed instruction text presentation programs. The difficulty of creating highly adaptive CAI lessons is undoubtedly a major factor behind the developmental difficulties evident in the CAI field to date. To address this problem, ACTS has utilized techniques from the realm of artificial intelligence which were originally used in decision--aiding models for control system operators.

ACTS represents a significant contribution to the area of individualized maintenance and troubleshooting training since the approach focuses on the quality of decision making rather than on the learning of specific responses and procedural sequences. Emphasis is placed on the identification of the student's decision value structure and the adaptive generation of instruction to resolve discrepancies from the desired (expert) value structure.

Results of initial evaluations of the ACTS approach (Freedy and Crooks, 1975) were concerned mainly with the behaviour or the adaptive decision model and the range of student behaviour to be expected. It was found that the expected value (EV) model quickly converged on the decision behaviour of students who exhibited consistent decision strategies. In these initial studies, students varied widely in rate of decision making and consistency of approach. When aiding (provision of the alternatives an expert would consider) and feedback (identification of the alternative an expert would choose) were given, students solved circuit fault problems at lower cost than without such assistance (Crooks, Kuppin and Freedy, 1977).

On the basis of these initial studies, several areas of improvement to ACTS were identified, including improved human factors design of the man-machine interaction and the need to reduce the number of utilities considered by the student. As a result the adaptive models of the original ACTS were replaced by more efficient multi-attribute utility models, and the ACTS was supplemented with production rules. The system was also modified to provide for simultaneous use by multiple students and for simultaneous use of multiple circuits.

More recent laboratory studies of the modified ACTS approach support and extend earlier findings (Hopf-Weichel, Freedy, Lucaccini and Purcell, 1979). As before, it was found that the student model was able to capture student values rapidly. With practice on troubleshooting problems and feedback regarding differences between the student's and the expert's value structures, student values gradually or, in some cases, rapidly approached those of the expert model. Figure 3 shows how student values for two utilities, cost of troubleshooting actions and commercial information gain (isolation of faulty modules), approached those of the expert model as training progressed. These results are consistent with earlier studies and provide some evidence that the group of six students on which these data are based had, in fact, acquired a value structure similar to that represented by the expert model.

Troubleshooting performance during training and on subsequent test problems is shown in Figures 4, 5, and 6. These figures present data for two groups of six subjects each, one trained on a series of 20 ACTS troubleshooting problems of graduated increasing difficulty (the experimental or ACTS group), and the other trained on a similar series of 20 troubleshooting problems of graduated increasing difficulty, using the actual circuit itself (the control or actual equipment group). Figure 4 demonstrates the superiority of the ACTS group during training, both in terms of the number of troubleshooting actions made and the estimated cost of these actions in time and equipment. The instructional features of ACTS, which are operative during training, clearly facilitate the performance of the ACTS group as

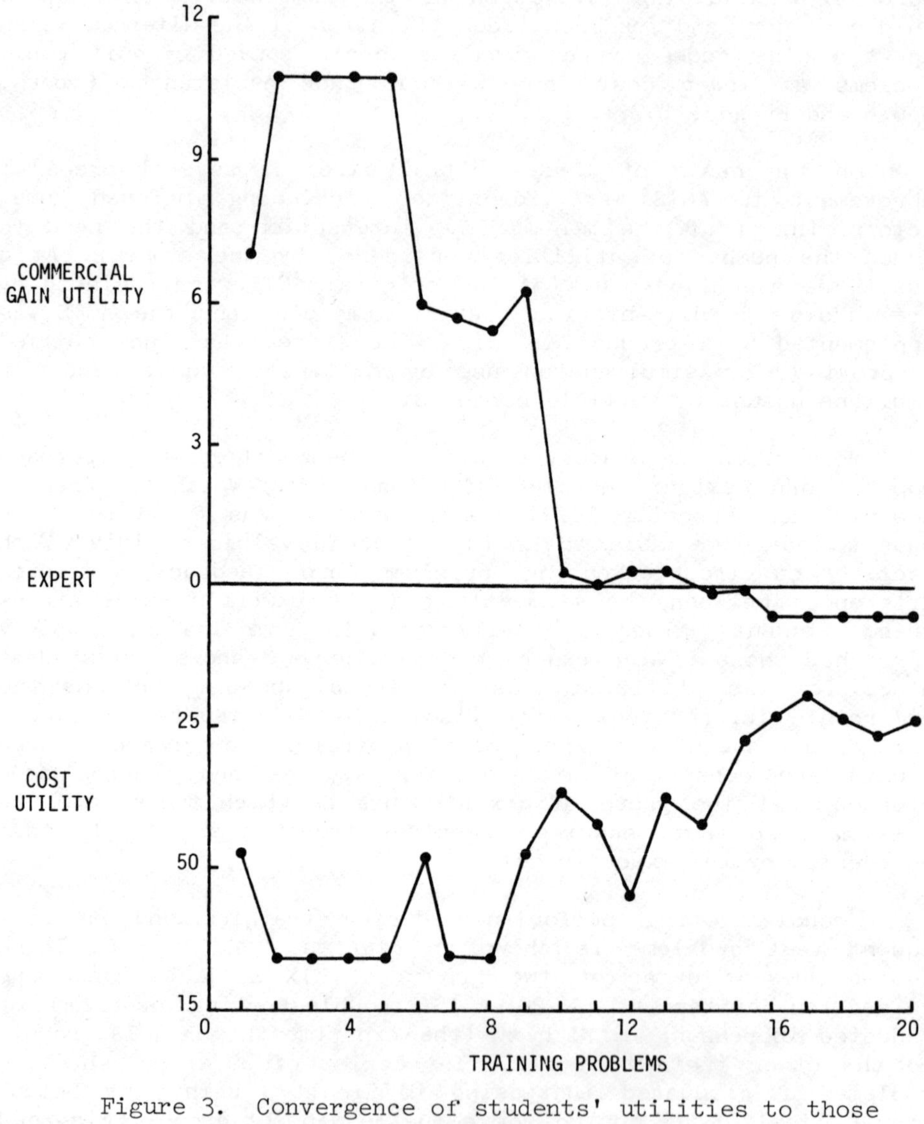

Figure 3. Convergence of students' utilities to those of expert model during acts training

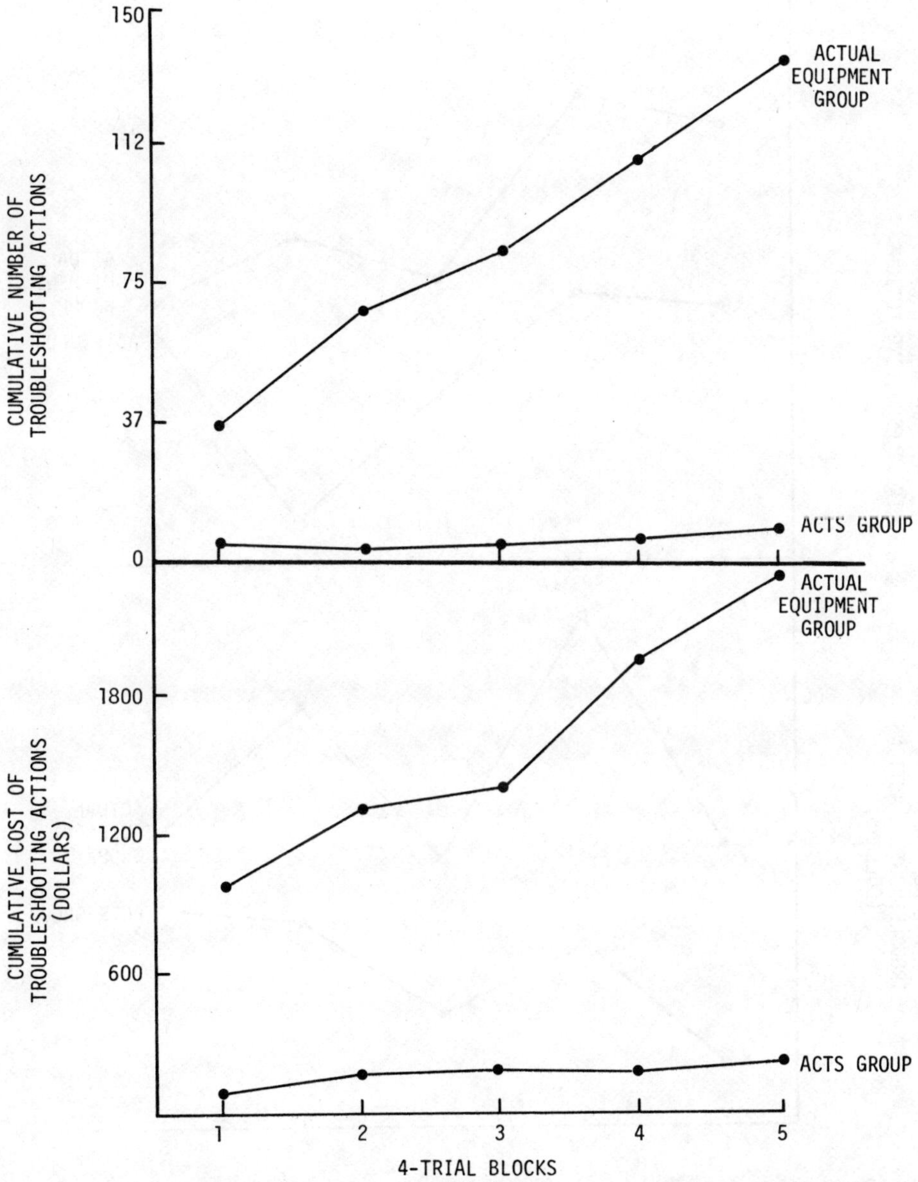

Figure 4. TRAINING: Troubleshooting of students during training on acts or actual equipment, compared to expert model

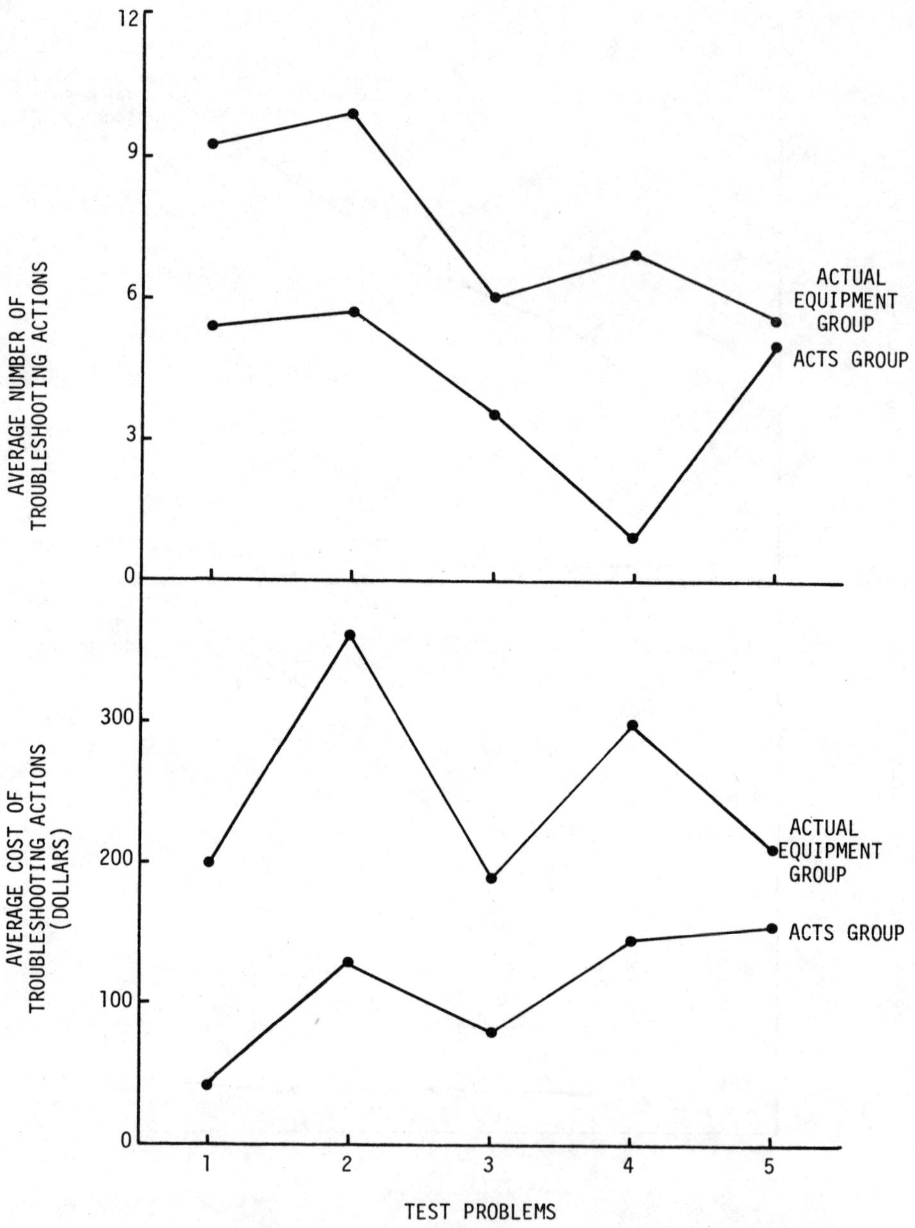

Figure 5. POST-TRAINING: Troubleshooting performance of students on text problems after training on acts or actual equipment, compared to expert model

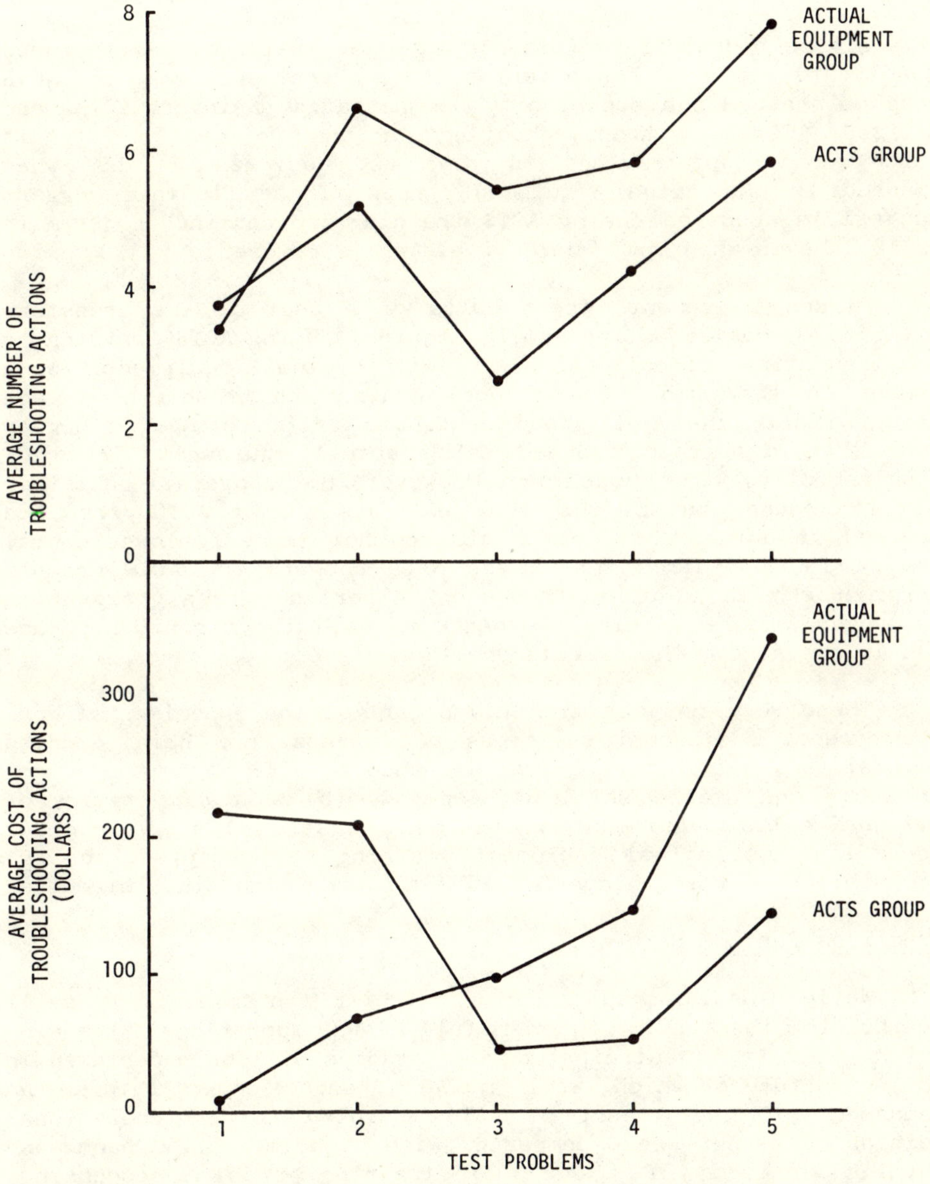

Figure 6. TRANSFER: Troubleshooting performance of students on test problems using actual equipment after training, compared to expert model

evidenced by the dramatically reduced costs and reduced numbers of troubleshooting actions taken as compared to those of the actual equipment group.

Figure 5 depicts performance when the instructional features of ACTS are removed. For a series of five test problems, the ACTS group maintained the superiority demonstrated during training and utilized 45% less troubleshooting actions to resolve circuit faults at a cost savings in time and equipment of 55% when compared to the actual equipment group. These findings suggest that skills acquired during ACTS training are retained and remain available even when instructional aiding is removed.

Figure 6 presents the results of a test of the transferability of troubleshooting skills acquired during ACTS training to troubleshooting circuit problems with actual equipment. With respect to the number of actions required to troubleshoot each circuit fault, the ACTS group continued to perform at a higher level than the group trained using actual equipment. Costs of troubleshooting were about equal, however, on an overall basis for the two groups, but by the last test problem the ACTS group had reversed an unfavorable cost ratio to show markedly lower costs. When it is considered that these data compare the ACTS group's performance without prior "hands-on" experience to a group which had several hours of direct experience with the circuit hardware used in the study, the results are impressive.

These results represent only a part of the experimental work in progress. Additional questions of interest now being studied include: (a) What is the optimal sequence of ACTS circuit problems for rapid and stable skill development, (b) what mix (ratio) of ACTS and actual equipment training is best, (c) in what order should ACTS and actual equipment training be presented, and (d) does similar learning occur for ACTS training with other circuits?

Directions for the Future

While initial results are particularly promising, it still remains to test ACTS in the full-blown operational training environment. It is anticipated that studies will be undertaken in the near future which will include, among other things, an assessment of the impact of ACTS training in the operational training setting when compared with traditional, "hands-on" training in terms of transfer of training, problem sequencing, order and ratio of ACTS and "hands-on" training experiences, and long-term retention of skill. It is hoped that field studies will bear out our conviction that ACTS can make a significant practical contribution to the training of electronics maintenance personnel. It should not be overlooked that the basic approach outlined above may have implications for improving the quality of human decision performance on related tasks, although the applicability of ACTS outside the electronics maintenance training area remains to be explored.

REFERENCES

Atkinson, R. C., October, 1972, "Ingredients for a Theory of Instruction, American Psychologist, 27(10): 921-931.

Bloom, B. S., 1968, "Learning for Mastery", Evaluation Comment, (UCLA Center for the Study of Evaluation), 1(2).

Brown, J. S., Burton, R.R. and Bell, A.G., March 1974, SOPHIE: "A Sophisticated Instructional Environment for Teaching Electronic Troubleshooting". Cambridge, MA: Bolt Beranek and Newman, Inc., Technical Report BBN No. 2790.

Carbonell, J. R., 1970, AI in CAI: "An Artificial-Intelligence Approach to Computer-Assisted Instruction, IEEE Transaction on Man-Machine Systems, MMS-11(4), 190-202.

Crawford, D. G. and Ragsdale, R. G., July 1969, "Individualized Quasi-instructional Systems for the 70's, Working paper, Ontario Institute for Studies in Education, University of Toronto.

Crooks, W. H., Kuppin, M.A. and Freedy, A., January, 1977, "Application of Adaptive Decision Aiding Systems to Computer-Assisted Instruction: Adaptive Computerized Training System (ACTS)", Woodland Hills, CA: Perceptronics, Inc., Technical Report PATR-1028-77-1.

Dawes, R. M., 1970, Graduate Admission: A Case Study, Oregon Research Institute, Technical Report 10(1).

Freedy, A. and Crooks, W. H., April 7-10, 1975, "Use of an Adaptive Decision Model in Computer-Assisted Training of Electronic Troubleshooting". Proceedings of the Conference on New Concepts in Maintenance Training Reserach, Orlando, Florida.

Glaser, R., 1965, "Psychology and Instructional Technology", In R. Glaser (ed.) Training Research in Education, New York: Wiley.

Goldberg, L. R., 1970, Man vs. Model of Man: "A Rationale Plus Some Evidence for a Method of Improving Upon Clinical Inferences, Psychological Bulletin, 73: 422-432.

Hartley, J. R. and Sleeman, D.H., 1973, "Towards More Intelligent Teaching Systems, International Journal of Man-Machine Studies", 5, 215-236.

Hopf-Weichel, R., Freedy, A., Lucaccini, L. F., and Purcell, D., November, 1979, Adaptive Decision Aiding in Computer--Assisted Instruction: Adaptive Computerized Training System (ACTS), Woodland Hills, CA: Perceptronics, Inc., Draft Annual Technical Report PDATR-1076-79-11.

Koffman, E.B. and Blount, S.E., 1974, A Modular System for Generative CAI in Machine-Language Programming, IEEE Transactions on Systems, Man and Cybernetics, SMC-4(4), 335-343.
Nilsson, N.J., 1965, Learning Machines, New York: McGraw Hill.
Slagle, J. R., 1971, Artificial Intelligence: "The Heuristic Programming Approach, New York: McGraw Hill.
Smallwood, R. D., 1962, "A Decision Structure for Teaching Machines", Cambridge, MA: MIT Press.
Smallwood, R.D., 1971, "The Analysis of Economic Teaching Strategies for a Simple Learning Model", J of Math. Psychol., 8:285-301.

SCAT: SYSTEM CONTROL ANALYSIS AND TRAINING SIMULATOR

Dr. Torgny Svanes & James R. Delaney

The MITRE Corporation
Bedford, Massachusetts 01730
U. S. A.

INTRODUCTION

Current and proposed military communications networks are required to provide effective service in stressed environments involving unusual traffic demands and losses of network assets. To maintain service under these conditions, a near real time flow of network status information, sufficient to support quick diagnosis, is necessary. In most modern networks, a human controller has the ultimate responsibility to interpret status information for the entire network or for some subset of it. The network controller must be experienced in the network problems and in the selection of appropriate corrective actions. In most cases, the network controller is given little opportunity to gain experience through experimentation and his or her only training medium is the operational network itself.

As part of a continuing effort in the area of system control of communications networks, The MITRE Corporation has supported the development of SCAT, a System Control Analysis and Training tool. SCAT is a communications network simulator designed to be an aid in network evaluation and problem diagnosis, the training of network controllers, and the analysis of network response to system control actions. SCAT accomplishes the simulation of a specific network through representation of topology and call processing in a series of descriptive tables.

The feature which makes SCAT especially appropriate as a network controller training tool and as a diagnostic support is its comprehensive interactive capability. The controller/trainee or diagnostician interacts with an ongoing network simulation via a repertoire of keyboard-selectable functions from a computer terminal. Many of the selectable functions generate displays which aid the simulation user in evaluating the effects of a given scenario setting and of the control interactions on network status and/or performance. The interactive capability enables the user to perform on-line experiments through which he or she can alter the flow of the simulation in response to network developments.

Some of the functions selectable via the terminal keyboard are:

Scenarios

A number of scenario-setting functions are incorporated into SCAT's design. These functions include satellite outages, node outages or degradation, link outages or degradation, and change in traffic patterns. Via the interactive program, these scenarios may be activated at any time during a simulation run. Running the simulation with these scenarios will familiarize the controller/trainee with network response, and offer experience in diagnosing network problems.

System Control

SCAT's design also incorporates a number of System Control functions, including system control options of present and planned military networks.

Save/Restore Feature

This feature allows the user to save the network status at any given point in time for later restoral. The feature may be used, for example, to compare the relative effectiveness of different control actions on the same scenario.

Network Status and Performance Reports

SCAT features an extensive statistics package which allows the operator to focus his attention on the status and performance of selectable portions of the network.

Key Event Monitoring

This function enables the operator to monitor a multitude of individual call, node-to-node, and network-wide conditions. The monitor interrupts the simulation when a preset condition is fulfilled, displays an explanatory message, displays a complete history of the call that has fulfilled the conditions, and allows the operator access to the other selectable functions before continuing the simulation.

SCAT is currently operational in an IBM time-sharing environment and is programmed in FORTRAN. It can easily be installed on a mini-computer system. In this configuration, SCAT can be deployed for use in the field both for training and direct operational support. The model has been applied to the European AUTOVON and the NATO Initial Voice Switched Network (IVSN).

DESCRIPTION OF A COMMUNICATIONS NETWORK

Communications networks are designed to transmit information in many forms, using various combinations of switching techniques. Information can be in the form of voice conversations, preformatted messages, data from a computer, or any other intelligent communication. Two of the techniques used to control the information flow through the network, from the sender to the receiver, are store-and-forward switching and circuit switching. Store-and-forward switching advances the information through the network on a step-by-step basis. At each step, the information is temporarily stored until the next routing step has been successfully completed. Circuit-switched networks construct a complete communication path through the network from sender to receiver before the information flow ensues. In the case of circuit switching, the constructed path which traverses a number of switching points is retained until the communication is completed. The current version of SCAT concentrates on circuit-switched communications networks.

The basic functions of a circuit-switched communications network are to recognize a request for service by a subscriber, construct a path through the network to the called subscriber, and recognize the end of call, so the network assets utilized can be made available to other calls. Figure 1 is an example of a communications network. Physically, a network has a given topology and connectivity. The network components are switching nodes and internodal links. In most networks, there is not full connectivity; i.e., not every node has a direct link to every other node. In determing a path for a call, network logic proceeds from one switch to the next via the connecting links, until a connection is established between the two parties. The procedures employed by the network in setting up a call comprise the call-processing policy. Functions performed during call processing include route determination, interswitch, signaling, and, in military networks, preemption policy. The switching nodes each have a pool of common equipments such as receivers and transmitters which are used during the setup phase of a call. The setup phase is usually only a short period of time (a second or two, or less) compared to the talking stage of a call (a few minutes on average). Consequently, only a small number of common equipments required for call setup are allocated to each switch relative to the number of call typically active at the switch.

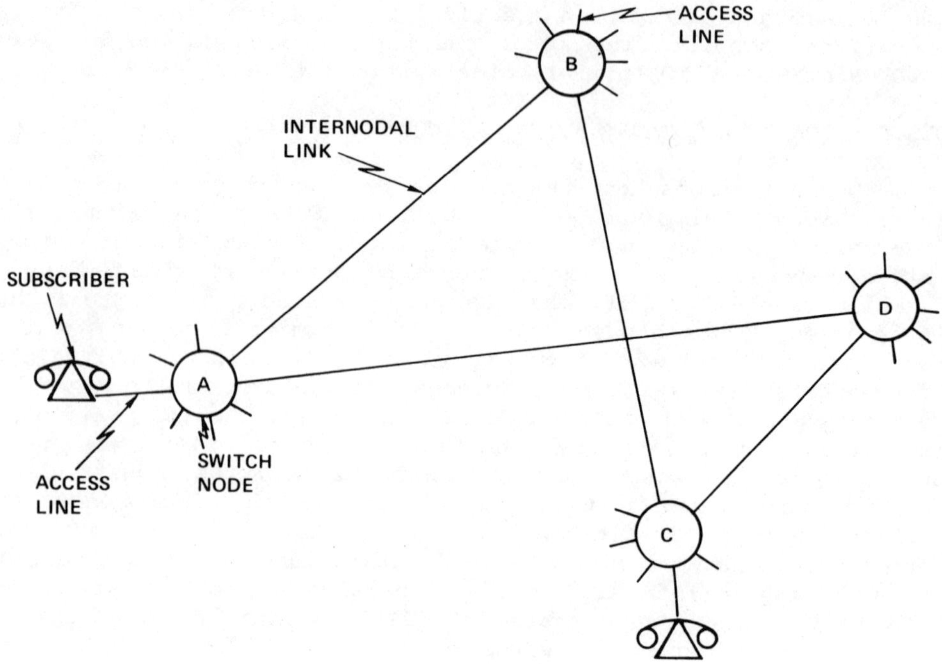

Figure 1. Communications network

CONTROLLING A NETWORK

Modern communications networks have manual and/or automatic facilities for controlling the efficient use of network assets, and particularly in military networks, the volume and composition of traffic carried by the network. An example of a control which affects network assets is the directionalization of traffic through a link. The implementation of this control allows seizure of a trunk in the link during call setup only by the switch at one end of the link. This control could be implemented to ease the call processing burden from a congested switch to a less congested, connected switch. An example of a control which affects traffic is line load control which controls the type of calls allowed to enter the network at a given switch. Call attempts originating on certain access lines are not acknowledged by the local switch node. This control could be implemented for congested switches to reduce call processing loads and to reduce common equipment seizure attempts.

The inclusion of viable controls in a network design, and the judicious implementation of the controls in the operational network, require a basic understanding of all components (transmission, switching, call processing, traffic, etc.) interacting in a network, both by the network architects and by the operational network controllers. It is the enhancement of the understanding of these complex interactions which the SCAT system addresses.

The design and implementation of network controls is further complicated in military networks because of the need to quickly reconfigure the network connectivity and/or subscriber composition in a tactical network, or to assess and respond to equipment degradation or changes in traffic volume and distribution in a stressed strategic environment. Since the forms and effects of externally generated stress conditions can never be fully anticipated, the question of visibility into the operational status of the network, to recognize problem areas and to evaluate the effects of control actions, must be addressed.

In summary, military communications networks are complex systems which must be designed to function effectively under stressed conditions. To maintain service under stress, controls are designed into the network architecture. Judicious selection and implementation of these controls require a high level of visibility into the network, an understanding of the visible parameters, and an awareness of the relationships among status and performance indicators and the available controls. There is a growing trend to incorporate dynamic controls into the switching logic itself, thus creating potential network "sanity" questions. Some degree of human involvement is always necessary; it is a question of how much.

A SIMULATION APPROACH TO CONTROLLER TRAINING

To provide a medium to aid network architects in the development of an effective network control system, and to train network controllers in the use of such a system, an interactive, discrete event simulator of the communications network is a natural approach. A simulator which models switch nodes, communications links, and call setup logic, and which introduces a traffic mix representative of the operational network, provides a realistic basis for the superimposition of additional functional software modules. These modules provide the logic which emulates the network visibility, status and performance measurements, information compilation, display generation, and the network controller station.

SCAT takes these desirable features of a simulation system one step further by enveloping the entire program design in an interactive mode of operation. The user of the simulation interacts with the network model, status and control logic, and additional scenario-definition logic via a computer terminal during the actual execution of the program. The user could be a network architect evaluating candidate designs of the network itself or the monitoring and control subsystems of the network. The user might otherwise be a network controller/trainee who becomes familiar with the network components and their performance during the simulation of realistic scenarios.

In the design of SCAT, we have attempted to provide a comprehensive set of visibility, control, status and performance reporting and scenario-defining mechanisms. Fully two-thirds of the program logic pertains to these and the operator-interface functions.

SCAT is designed an evolutionary system. The features which will be discussed are initial designs that are not intended to represent the ultimate in network visibility or control, neither from an analytical nor from a training point of view. However, the designers believe that the current SCAT system does present a comprehensive starting point from which expansion and refinement of existing features and the addition of new features will greatly aid communications network architects and controllers.

SCAT AS A TRAINING VEHICLE

For purposes of exposition, and to relate the present SCAT design to the topic of training for diagnosis, the following discussion assumes an application of SCAT as a training device for network controllers. In this context, some of the features to be discussed will be used chiefly by instructors, while others will be used by both instructors and trainees. As the trainee becomes more familiar with the visibility, control, and scenarios of the network and with SCAT itself, many of the features used initially by the instructors can be made available to the trainee. In this mode of operation, SCAT will provide a training vehicle which would eventually reduce the need for a full-time instructor.

Several of the interactive functions support training applications. The Hard Copy feature allows the transfer of SCAT reports to the line printer for session documentation, comparison to other reports, or for post-session analysis. These hard copy duplicates of CRT-terminal-displayed reports are useful in assembling training documents. Another SCAT function with training applications is the Statistics Network Partitioning. This feature allows the division of the network into any number of regions for performance statistics reporting. Networks are frequently segmented into regions for control purposes. Displays can be tailored to focus on the particular region of interest and visibility for which a controller trainee will be responsible. Since multiple partitionings are possible, training instructors can change partitioning to focus on regional groupings which best represent a particular training topic.

Another function that can be used for training is SCAT's ability to begin a session with a data base which reflects the network status after some amount of simulated time from a previous session. This feature allows an instructor to "pre-program" a particular scenario using the interactive scenario setting and control functions. When the scenario has been established, the instructor can save the network status data base on a disk file. By selecting the RESTORE function during initialization, the SCAT data base is initialized to the desired scenario rather than to a simulation time of zero. Any number of training scenarios can be preset in this manner.

INTERACTING WITH SCAT

The features previously described aid in the use of SCAT as a training tool. However, the most important training aspects of SCAT is that it allows the trainee to obtain status and performance information from realistic scenarios which parallel those seen in the network. Moreover, the trainee is able to implement control actions which duplicate those possible at the control station and to see the impact of those actions on the network status and performance. Either the trainee or the instructor can construct operational scenarios which emphasize a particular training subject or which reflect critical real-life scenarios in the operational network. This flexibility of SCAT's application in a training environment is possible because of the simulator's interactive, operator-selectable, functions. In an actual training session, there is complete flexibility in the combining and sequencing of the operator-selectable functions. Consequently, an instructor has a powerful, evolving training program; and the trainee is able to obtain "hands-on" experience and invaluable real-time experience before he or she ever sees an actual operational network control station.

The Interactive Features of SCAT

The operator-interface executive program has a flexible design structure. The operator, during an operator-interface period, may select a particular function by entering the unique code for that function via the terminal keyboard. Alternatively, the operator may select the main interactive function menu which displays the primary selectable function categories (see Figure 5). The operator is then led to the particular function of interest in the hierarchy via prompter messages and appropriate keyboard-entered responses. At each level in the functional hierarchy, the submenu for the level is displayed. The operator, at any level, has the option to return to the main menu at the top of the hierarchy. The more structured function-selection alternative allows a trainee to become familiar with SCAT quickly, while direct function selection allows the bypassing of the prompter message hierarchy once familiarity is achieved. Additionally, extensive editing of responses is done to eliminate inappropriate or illegal entries. This highly structured approach allows the inclusion of functions which require extensive definition of parameters by the operator without the operator having to memorize a lengthy list of responses. Each response in this approach helps to select the next prompter message, and the complexity of individual messages and responses is kept to a minimum. As shown in Figure 2, the interactive functions are categorized in functional groups. In the remainder of this section the functions within each group will be explained, highlighting their applications in a training environment.

TRAINEE'S VIEW INTO THE NETWORK

At the conclusion of the program initialization and at the beginning of each operator-interface period the network status map is displayed on the terminal screen. Figures 3A and 3B show the format and an example of the network status map. The map displays a composite of network resource status, traffic composition and distribution throughout the network, controls implemented, and scenario events invoked. The display presents a "snapshot" of key network parameters at the simulation time noted and a history of events and controls which have contributed to that status.

Since the updated network status map standardly introduces an operator-interface period, and because it consists of a comprehensive set of status parameters, the map is the keystone of the operator interface with SCAT. Further refinement and expansion of the status map as a network control station display is envisioned and is discussed in section 4.

As a controller-training device, the present status map display is not a realistic representation of the network status information which a controller has available on a global network basis. However, the purpose of SCAT is to aid the controller/-trainee in understanding the interrelationships among network components, especially within the contexts of externally generated stress and of the impact of control actions or inactions. This understanding is best fostered through a visibility into the

PRIMARY FUNCTIONS
1 SESSION CONTROL 2 NETWORK STATUS AND PERFORMANCE MONITOR 3 SCENARIO SETTING 4 NETWORK CONTROL 5 NETWORK MONITORING

1	2	3	4	5
● DISPLAY FUNCTION MENU ● SAVE/RESTORE DATA BASE ● RESTART FROM TIME ZERO ● DEFINE SIMULATION PERIOD ● HARD COPY CONTROL ● TERMINATE SESSION	● DISPLAY NETWORK STATUS MAP ● STATUS REPORTS • NODE ACCESS LINES • LINK CIRCUITS • COMMON EQUIPMENT • CALL HISTORY • NODE PAIR ROUTES AND ROUTE STATUS ● PERFORMANCE REPORTS • BY NODE • BY LINK • RESET ● REGIONALIZED STATISTICS • CALL SETUP TIMES • TRAFFIC LOAD • GRADE OF SERVICE • RESET	● TRAFFIC COMPOSITION • VOLUME BY NODE • DESTINATION DISTRIBUTION • PRECEDENCE DISTRIBUTION ● NETWORK DEGRADATION • NODE DEGRADATION OR OUTAGE • LINK DEGRADATION OR OUTAGE • SATELLITE OUTAGE ● CALL GENERATION ● CALL STATE TABLE DISPLAY & UPDATE	● NETWORK WIDE CONTROLS • CODE OR DESTINATION BLOCK • ORIGINATING CALL ACCESS CONTROL • ROUTING PROTOCOL SELECTION ● NODE CONTROLS • ORIGINATING TRAFFIC • CANCEL ALTERNATE ROUTING ● LINK CONTROLS • SKIP INROUTE SELECTION • BUSY OUT TRUNKS • DIRECTIONALIZE TRAFFIC	● CALL IDENTITY ● CALLS BETWEEN NODE PAIRS ● PRECEDENCE ● CALL SETUP TIME ● EXCESSIVE ALTERNATE ROUTING ● LOOPCHECK

Figure 2. SCAT Menu selectable functions

network which transcends that available in today's military communications networks. Of equal importance, the network visibility features of SCAT are also meant to aid in the advancement of the development of status reporting and presentation.

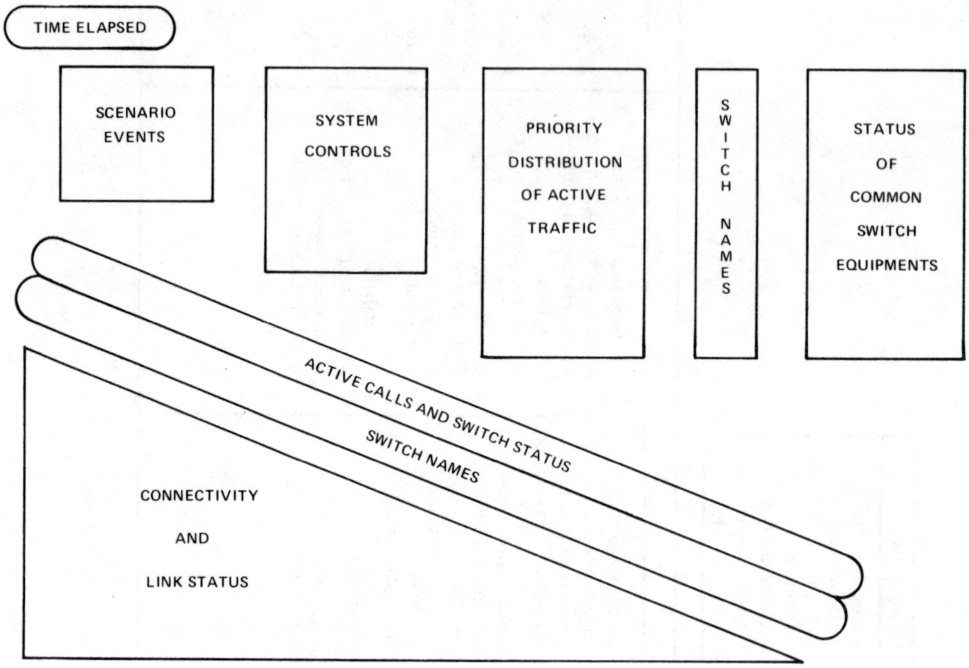

Figure 3A. Network status map format

TAILORING THE TRAINING SESSION

The session control functions (see Figure 5) which can be invoked during an operator-interface period serve to perform functions ancillary to the network model itself. The DISPLAY FUNCTION MENU function activates the primary function menu routine. The RESTART function clears the present data base and reads in the original initialization data file, if desired. This function is invoked primarily when a session has become muddled, and the operator simply wishes to start afresh. However, it can be used to run with different networks during the same session. The SAVE/RESTORE function was explained during the discussion of the initialization logic. SAVE/RESTORE is also a valuable intra-session tool. The SAVE part of the function saves the network status at the current simulation time. Not only the status, but the total SCAT data base including intermediate statistics arrays, are saved. Once a particular status has been saved, it can be RESTORED during any further operator-interface period or even during a later session.

SIM TIME		90:0		NETWORK STATUS MAP			PRIORITY					EI AVAIL	
ATB		EVENTS		CONTROLS		P	I	F	FO	ND	CC	RC	CO
0		60: 0 TRAF CHG		60: N0 4 CBK		25	18	12	2	N01	254	15	9
0		55: 0 TRAF CHG		0:0		2	8	6	2	N02	255	11	6
0		55: 0 N04 O/S		0:0		0	3	1	0	N03	254	3	3
		50: 0 TRAF CHG		0:0		0	0	0	0	N04	256	19	10
		45: 0 TRAF CHG		0:0		10	10	8	5	N05	254	10	6
				0:0		4	3	0	0	N06	256	5	3
(101)				0:0		11	20	7	0	N07	256	9	5
N01	(40)			0:0		8	4	2	3	N08	250	1	4
-1	N02	(14)				20	81	23	1	N09	724	32	24
-1	-1	N03	* 0 *			73	77	43	2	N10	811	77	49
: 0	: 0	: 0	N04	(76)									
: 1	: 0	-1	: 0	N05	(17)								
-1	: 0	-1	: 0	-1	N06	(51)							
: 0	-1	-1	-1	-1	-1	N07			(42)				
-1	3	-1	: 0	: 3	: 7	-1			N08	(232)			
2	-1	1	0	5	-1	1			-1	N09	(450)		
5	8	-1	-1	7	-1	1			-1	10	N10		

THE DIAGONAL OF THE TRIANGULAR MATRIX REPRESENTS THE NETWORK SWITCH NODES, AND THE MATRIX ELEMENTS REPRESENT THE INTERNODAL LINKS. AN ENTRY OF -1 IN THE MATRIX INDICATES NON-CONNECTIVITY BETWEEN THE NODES INTERSECTING AT THAT POINT. FOR EACH NODE ALONG THE DIAGONAL, THE NODE ID IS LISTED. ABOVE EACH ID, IN PARENTHESES USUALLY, ARE THE NUMBER OF CALLS CURRENTLY ACTIVE AT THAT NODE. THE PARENTHESES ARE REPLACED BY OTHER STATUS INDICATORS IF THE NODE IS NOT IN A NORMAL MODE OF OPERATION. FOR EACH INTERNODAL LINK IN THE MATRIX, THE NUMBER OF CHANNELS CURRENTLY OPERATIONAL AND NOT BEING USED BY A CALL IS LISTED.

AT THE TOP OF THE DISPLAY, THE CURRENT SIMULATED TIME, IN MINUTES, APPEARS. SCENARIO EVENTS AND NETWORK CONTROLS ARE LISTED ALONG WITH THEIR TIMES OF ACTIVATION. THE TRAFFIC CURRENTLY ACTIVE AT EACH NODE IS DECOMPOSED INTO THE FIVE PRECEDENCE LEVELS, AND THE NUMBERS OF EACH OF THE HIGHEST FOUR LEVELS AT EACH NODE ARE PRINTED TO THE LEFT OF THE VERTICAL COLUMN OF NODE ID'S. TO THE RIGHT OF THE VERTICAL COLUMN OF NODE ID'S THE CURRENTLY INACTIVE, BUT OPERATIONAL, COMMON EQUIPMENTS AT EACH NODE ARE ENUMERATED.

THE ENTIRE DISPLAY IS UPDATED AND RECREATED AT EACH OPERATOR INTERFACE PERIOD. THE OPERATOR SELECTS THE NODES DISPLAYED ON THE NETWORK STATUS MAP INTERACTIVELY.

Figure 3B. Network status map display

An important application of the SAVE/RESTORE function is to SAVE the network status after the network reaches a point of stability of operation and to RESTORE the data base at that point in future sessions instead of starting from time zero. For training applications, the instructor or trainee can SAVE a particular initial scenario and then try several candidate control strategies starting from the same point by RESTORING the same status prior to invoking each candidate control strategy. Each strategy could then be allowed to run for some designated simulation time, and one or more of the Network Monitoring functions could be used to compare the effect of each strategy on network performance. An aid to this type of session is the OFFLINE DATA COLLECTION function which allows the operator to save selected status and performance reports in a file for hard copy printing later. It would be difficult for a trainee to compare the results of multiple control strategy candidates without hard copy reports from each.

SETTING THE SCENE

Military communications networks must be designed to be flexible enough to accommodate high-priority traffic in a stressed environment. The call-processing logic and the control ensemble of military communications networks have as one of their design goals the successful completion of high-priority calls in times of stress. Consequently, it is imperative that the network controller be capable of recognizing the existence of stress in the forms of degraded network assets and/or the imposition of non-typical traffic patterns on the network.

SCAT has incorporated a set of scenario-setting functions in its operator interface. These functions, when invoked, alter the data base used by the network model to simulate the degradation or loss of network assets or to change the normal traffic profile to simulate subscriber response to the stress situation. Each of these functions also permits the revocation of a previously invoked scenario via a keyboard entry. The functions can be combined in any manner to reflect the impact of a particular scenario both on the network assets and on the network subscribers. These scenario-setting functions form the link between the idealized network model simulation and real-world situations in which the network controller must perform. In a training environment, these functions provide the testing ground on which the controller/trainee learns the subtle interactions of the network, the carried traffic, the network controls, and his or her visibility into the network status under a variety of stress scenarios.

The TRAFFIC function allows the operator to change the currently active traffic profile. SCAT allows the changes to be made for the entire network or for traffic being introduced at individual nodes. Parameters which can be altered include total volume, precedence distribution, destination distribution, or call-type distribution. The NETWORK DEGRADATION function imposes either the degradation of switch node assets including total switch outages or the degradation of internodal link assets including total link outages. Switch node degradation involves the reduction of the common equipment pools to simulate partial outage. A switch node total outage results in the termination of all calls currently active at the switch. Also, all links terminating on the switch are marked "out of service". Link degradation is accomplished by reducing the number of trunks on the link by one or more.

Link outage is accomplished by reducing the number of operational trunks to zero and simultaneously terminating all calls using a trunk on that link. SATELLITE OUTAGE is an expansion of the link outage, and of course, applies only to networks utilizing satellite trunks. During initialization, each trunk on every link is typed as terrestrial or satellite. When a SATELLITE OUTAGE is invoked, all satellite-typed trunks are put out of service, and calls using satellite trunks are terminated. If a particular link is comprised of a mixture of satellite and terrestrial trunks, the terrestrial trunks are not affected by this function. The CALL GENERATION function allows the operator to superimpose a call of his or her specification onto the network-model generated traffic. The call is defined by time of origination, node of origination, destination, type of call, and precedence of call. This function is usually used in conjunction with the call-monitoring function, which traces call setup through the network. This tandem of functions is extremely educational to a trainee; he or she can define calls which are meaningful to a particular stress scenario and follow the call setup through the network.

SIMULATING CONTROLLER ACTIONS

SCAT embraces an extensive repertoire of controls which are representative of controls currently available in military communications networks. The activation of a particular control is determined by responses to prompter messages. Controls interact with the network model and influence the network operations and polices until the operator deactivates them. AUTOMATIC ACCESS CONTROL is the only simulated control which monitors network operation and which is automatically activated and deactivated once the function is invoked. The function is invoked on a network-wide basis, but the control is activated on a node-by-node basis, dependent on the pool of available common equipments. The particular common equipment which is monitored is operator selectable, and the activation/deactivation of the control depends on defined pool threshold levels. The control results in certain categories of calls not being allowed access to the network at the controlled switch.

Other functions which control the admission of originating traffic at a switch node are LINE LOAD CONTROL and CODE BLOCK. The former is similar to AUTOMATIC ACCESS CONTROL without the automatic monitoring feature. The latter function does not allow network access to call destined for a particular controlled switch node. The controls can be limited to certain categories of originating traffic.

Functions which control internodal link access are DIRECTIONALIZATION and MAKE BUSY. The former allows seizure of a trunk on a controlled link for traffic in one direction only. The latter makes a designated number of trunks on the controlled link unavailable to any calls in either direction.

A control function which is related to the network call--processing policy is ROUTING PROTOCOL SELECTION. This control applies only to networks which have multiple protocols of which only one is invoked at a switch at any moment. In SCAT, currently, the operator selects a protocol via keyboard responses on a node-by-node basis. The protocol determined the degree of freedom a node has to route calls.

PERFORMANCE EVALUATION REPORTS

These functions provide visibility into the current network status during the operator-interface period, accumulate statistical data, and generate performance reports. Since network visibility is of primary importance to both the analysis and training aspects of SCAT, we have incorporated a comprehensive and flexible array of status and performance functions into it. The functions are comprehensive in that status indicators and performance counters are embedded into all facets of the model structure. The functions are flexible in that the operator can define report periods, reset counters, select sub-reports of interest, and even redefine the network regions for which reports are generated.

The status reports are formed from the SCAT data base at the time they are requested. Consequently, they reflect the current status of the parameters being displayed. The Network Status Map, Figure 3, has been discussed previously. Other status reports list the current status of access lines by node, internodal trunks by link, common equipment queueing by node and type of equipment, and scheduled simulation event queueing by time and call identity. These last status reports have been included primarily for communication network analysis since they presume an understanding of discrete event simulation and the SCAT data base structure. However, there is nothing to preclude a trainee, especially in the later stages of instruction, from using these reports.

The status reports provide an instanteneous "snapshot" of various network parameters. They supply insight into the network to a controller trainee because they freeze in time the complex interactions of the resource and traffic components. The trainee is then given the opportunity to view the network from many different structural and operational angles, to reflect on his observations, and to crystallize his or her understanding of some subtle point.

As illuminating as status reports can be, a true measurement of network performance requires the accumulation and compilation of information over time. SCAT provides statistical performance reports for both network resource groups (nodes and links) and network point-to-point subset pairs. Flexibility has been designed into the performance reports by permitting the operator to define the simulated time over which statistics are to be collected. Any of the reports can be generated during the operator-interface period. The statistics arrays can be reset via interactive functions at any time, or the operator can continue to build on the prior performance data for longer-range statistics.

The network resource performance reports are shorter-term statistics which summarize node or link activity. If the node performance report is selected, statistics are displayed, by call precedence, for all nodes in the network. The statistics for each node summarize the failure of switch nodes to advance calls to their destination during call setup. Some of these calls are dropped, others are rerouted, and higher-priority calls may preempt other calls of lower priority. In toto, these statistics reflect congestion at the switch nodes or on the internodal links during the period covered by the report. The operator may also select the performance report for a particular internodel link. For the report period, these statistics reflect the ability of the link to handle the offered traffic.

Longer-term statistics, which measure the network's ability to meet traditional communications network performance requirements, are generated on a regional basis. The operator selects the partitioning of the network into regions of interest. It should be noted that a region may range from a single node to the entire network. The statistics that can be selected are call setup times, traffic loads, and grades of service (GOS). The statistics represent averages between and within the selected network regions. Since the basic statistics are kept on a node-by-node basis, several reports on the same basic data with different regionalization may be generated during an operator-interface period.

MONITORING THE NETWORK SIMULATION

The network monitor enables the operator to monitor a multitude of individual call, node-to-node, and network-wide conditions. The monitor interrupts the simulation when a preset condition is fulfilled. It then displays a complete history of the call that has fulfilled the conditions, and allows the operator access to the other selectable functions before continuing the simulation. The various status reports available actually provide a global "snapshot" of the network at the time of the interrupts. The monitor may be invoked during an operator-interface period by selecting one or several of the functions in group 5 of Figure 2. The precise conditions under which the monitor is to be activated are specified by responding to prompter messages.

In the training context, the monitor may be used several ways. The most obvious application is in giving the trainee better insight into the network's behaviour under various conditions. He or she may, for example, monitor high-priority traffic in stressed network scenarios. When the monitor is activated, the network status information available gives a better understanding of the situation that caused the problem than a follow-up analysis of statistical data.

Another function particularly appropriate for training is the individual call monitor. This function allows the operator to follow the call setup to any degree of detail. He or she may use this to become familiar with network protocol and gain insight into the flow of traffic in the network.

An important application of the monitor is in verification of a network's data base. A Network's processing as well as its sizing and connectivity are described in SCAT's data base tables. The verification of the particular state tables and processing decision tables is greatly facilitated by the call monitoring function. Using the call generation function described previously, various types of calls may be fed to the network by the operator, and the calls may be traced, step by step, during setup.

A LOOK INTO THE FUTURE

A major goal in the design of the SCAT system has been a modular program structure to accommodate the expansion of existing features and the addition of new ones. Many of the current features were not in the original SCAT design, but have been added as the system developed. This modularity, in tandem with the table-driven network simulation design, provides a flexible yet sound platform on which to test other innovative ideas in the area of training system controllers of communications networks. Some of our current ideas are presented here; others will surely evolve.

MULTIPLE TERMINALS

As mentioned earlier, the concept of multiple computer terminals controlled from a centralized processor in which SCAT is the resident program is one we have considered. In this design, the SCAT executive program will have an initialization function that will define the operational configuration, regions of visibility, and regions of control for an application session.

The operational configuration defines the number of terminals, their logical unit assignments, and their SCAT function assignments. SCAT function assignments are the instructor station and/or one or more trainee stations. The instructor station controls the scenario functions and has access to all the operator interface functions, and full network visibility. The trainee stations have a specified visibility which can be a subset of the network, or the entire network itself, depending on the network control structure and the training topic. Each trainee station can invoke controls which influence its subset of the network. For a hierarchical control structure, the control repertoire and network visibility would be tailored for each level in the hierarchy.

This multiple-terminal version of SCAT would permit the training of several controllers simultaneously and would tailor the SCAT functions and the range of network status visibility to correspond to the actual situation. The coordination required in a distributed control design would be experienced by a controller during the training period.

The physical layout of the terminals could be designed to conform with actual controller station configurations. A hard copy on-line printer option could be included to relieve the trainee of remembering the immediate display history. This option would parallel the operational situation and aid the trainee in making control decisions.

In a multiple-terminal environment, additional software would be developed to accommodate the training function itself. Such logic could be a scoring mechanism whereby trainees could compare their progress with each other and through which friendly competitions could be developed. Grades of service, call setup times, or number of links to complete a call between two nodes could be used as scoring parameters.

TAILORING OF DISPLAYS AND MESSAGES

In either a multiple- or single-terminal implementation of SCAT, the status displays, performance reports, and system messages to the controller could be tailored to duplicate the actual network environment for content, format, and frequency of occurrence. Duplication of the displays, of course, would have to be done on an application-by-application basis.

COLOR GRAPHICS APPLICATIONS

The inclusion of color computer graphics in the SCAT design presents a wide range of possible applications. Many of the applications deal with the forms in which status and performance information are presented to a controller and in which the controller interacts with the network.

In the area of training, a sophisticated graphics terminal could replace or supplement the keyboard as the primary operator-input device with light-pen-sensitive touch panels. A menu of responses to prompter messages would be displayed on the terminal screen, and the operator would select a response by placing the light pen on the area of the screen containing the response. The chosen response would be blinked to reinforce the proper selection.

The present SCAT displays rely heavily on numerical representations of status and performance parameters. Color graphics supply a visual dimension whereby quantified parameters can be translated to the displayable color spectrum. Either a network map or a tabular display could be used. For example, dark, cool colors would reflect a lightly used link. As the link activity increases, the line representing the link on the screen could gradually be widened and displayed with a lighter color. If the link becomes congested, i.e., calls are not able to find a free circuit, then the widened, light-colored line could be periodically blinked. The period of the blink could be shortened to reflect preemption and precedence level of preempted calls. A rapidly blinking light-colored line would represent a congested link on which even high-precedence calls are being preempted. Consequently, the trainee would be relieved of the need to interpret numerical quantities. The color display would not only indicate possible problem areas, but a map would isolate the network area of concern. Other more sophisticated status and performancd measurement algorithms could also be used since they could be translated to the color spectrum.

Graphics displays would also accommodate the construction and meaningful placement of pictorial symbols to represent scenario and control actions. Out-of-service nodes or links could be represented by broken lines; a directionalization control on a link could be indicated by an arrow displayed over the affected link. Graphics also permit the telescoping of areas of interest. By telescoping a small region of the network, more detailed status and performance information could be added to the display for that region. Thus, the trainee would standardly be given very basic, simple structures to indicate a possible problem area in the network and a more richly informative display for the region when it is telescoped.

This discussion shows the applicability of a color graphics terminal for status display design and controller training. In combination with the existing SCAT system, a color graphics terminal would yield a comprehensive, instructional tool.

REFERENCES

Greene, T.V., et al., "No. 4 ESS: Network Management and Traffic Administration", Bell System Technical Journal, Vol. 56, No. 8 (Oct. 1977), pp. 1169-1202.

Lin, P. M., et al., "Analysis of Circuit- Switched Networks Employing Originating-Office Control and Spill-Forward", IEEE Transactions on Communications, Vol. 26, (June 1978), pp. 754-765.

Martin, James, Design of Man-Computer Dialogues, Prentice-Hall Inc., Englewood Cliffs, New Jersey, 1973.

Rosner, Roy, "Communications System Control for the Defense Communications System", Conference Record of 1978 International Conference on Communications, Vol. 2, pp. 31.1.1--31.1.5.

SUMMARY OF WORKSHOP DISCUSSIONS

SUMMARY OF WORKSHOP DISCUSSIONS

L.P. Goodstein

Risø National Laboratory
DK-4000 Roskilde, Denmark

ORGANISATION OF THE WORKSHOPS

The idea of providing opportunities for informal small group workshops was an important element in the planning of the conference in spite of the fact that the plenary sessions themselves were organised so as to enhance and stimulate discussion. At a relatively early point in time, potential participants were asked to rank order a set of eight topics which the conference organisers felt had direct relevance to the central theme of the meeting. These are listed below. On the basis of this (incomplete) pre-conference response, five of the original eight were chosen in a final selection round for all conference attendees and this resulted in a further reduction to the first three items on the list as the actual workshop topics. The chairmen of these were as indicated.

Models of Diagnostic Behaviour (Prof. N. Moray)

Usefulness of models; comparison of the different philosophical bases; limitations; evaluation and validation.

Models of System Structures (Dr. J. Wohl)

Commonalities of system structures across different physical domains; mechanical, chemical, information, etc.; and the use of these structures in diagnosis strategies and support (displays etc.).

Simulators (Prof. K. Duncan)

For training and evaluation (fidelity, etc.).

Design for Maintainability

Operability versus maintainability; use of theories and methods of diagnosis as a design basis.

Case Stories of Human Error

For the purpose of evaluating the various theories, models and support aids.

Management of Fault Diagnosis

Importance of organizational structure, incentives, etc. on performance.

Terminology

A search for common terms - one list was suggested; other preferences were requested.

Attention

The role of basic psychological theories of attention for describing monitoring performance in detection tasks.

The workshop groups were asked to devote the first two and a half hour session to identifying for each topic the important issues which thereafter were to be discussed and, if possible, resolved during a second two and a half hour session on the following day. This was of course done in an optimistic spirit since the main emphasis was on providing a catalyst for the debate and discussion process itself which could feed on each participant's interests, experience, enthusiasm and, perhaps, scepticism about alternate viewpoints.

INTRODUCTION TO THE SUMMARIES

Several of the workshops provided an opportunity for a more extensive discussion of topics covered in the plenary sessions. For example, the Workshop on Models of Diagnostic Behaviour was in fact a continuation of Session 2 on Theories and Models while the Simulation for Training Workshop served as a preface for the Training session on Friday. Therefore the interested reader is urged to refer back to the appropriate session summary in the course of perusing these workshop summaries. In addition the book index can be used in some cases to link concepts to specific authors.

By way of introducing the two workshops on modelling, the following representation can be useful as a framework for placing the two activities in perspective:

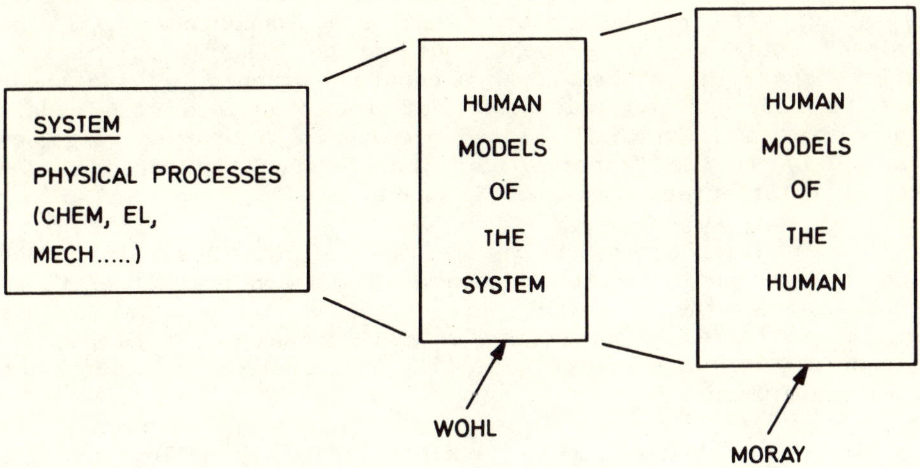

Thus Moray's group discussed human functioning in diagnosis based on a particular representation of the human while Wohl's group was interested in the modelling of systems by humans - designers, analysts, operators, trainers, etc. - as an aid in their thinking about the system.

EDITED SUMMARY OF CHAIRMEN'S REPORTS

The material presented in the following sections is an edited version of the summaries of the three workshops submitted by the respective chairmen.

The original intent of the Workshop on Models of Diagnostic Behaviour was to develop a taxonomy of diagnostic tasks which could be used to aid designers in deciding what difficulties were likely to be encountered by operators, and what kind of training and aiding might be optimal. That aim proved too ambitious. Instead, we have produced a table which relates the model of Rasmussen[1] to what we believe to be (1) the operations logically required to complete any diagnosis, and (2) the psychological processes used by the human operator when processing information with a view to completing diagnosis.

---

[1] For a brief description of the referenced model see Goodstein (this volume).

The classification scheme is shown in Figure 1.

The first column, "Receive information", "Comparison", ... "Action" is a list of the operations required for diagnosis. While the observed system is functioning normally only the first three operations occur: no discrepancy is detected between the current and recent state of the system of a kind indicating abnormality, so monitoring continues. Note that some departure from the "desired" state may be present, but not be judged or indicating that the system is abnormal. An aircraft disturbed by wind shear calls for corrective action but not for diagnosis of aircraft abnormality. All "normal" states include a tolerated range of error. Only if the "threshold of indifference" is exceeded the human operator enters a diagnostic phase.

In the diagnostic phase some kind of hypothesis is formed which will account for the presence of the abnormality. This is tested against the observable properties of the system; if the test is satisfied, appropriate action is taken. There may be subsequent iterations if the action is insufficient to satisfy the system requirements.

The three columns show the kinds of data handling involved when the task is at different levels of Rasmussen's classification.

Typical of a "skill-based" task is manually controlling a vehicle. We assume that after extensive practice, the skill has become automatised. Hence there is no conscious formulation of a hypothesis when, for example, a movement of the joystick no longer produces its customary affect, due to a change in plant dynamics. Rather, appropriate action occurs "reflexly". We note also that some diagnoses may result not in control actions but in judgements. "The system is now in State X" which, while not producing a control action, produces a message or at least a conclusion, after which someone else might intervene. (This would be particularly true of a medical diagnosis or a diagnosis in a monitoring, rather than a control task).

Rule-based situations eventually are those where a look-up procedure is available. The look-up may be in a printed manual, or as a learned list of options or may wholly or in part be invented by the operator. Normally the look-up is intended to be algorithmic and exhaustive. It goes beyond the data given and embodies what is formally known about system structure and function.

Knowledge-based situations are those which are typically (but not always) characterised by creative problem solving behaviour. This is often slow and shows the limited capacity of

				USUALLY RAPID → INCREASING TIME → USUALLY LENGTHY INCREASING USE OF CONSCIOUS INFORMATION PROCESSING AND INCREASING WORK-LOAD →		
		TYPE OF HUMAN OPERATION:		SKILL-BASED (DATA-DRIVEN)	RULE-BASED	KNOWLEDGE-BASED
		TYPE OF INTERNAL MODEL: AND EXAMPLE OF OPERATION:		TRANSFER FUNCTION INPUT/OUTPUT RELATIONS	DECISION TAELE STANDARD PROCEDURES IN MANUAL OR HEAD	FUNCTIONAL PROPERTIES STRUCTURES, RELATIONSHIPS
HUMAN OPERATOR FUNCTIONS	NORMAL SYSTEM FUNCTION	RECEIVE INFORMATION	1	SIGNALS E.G. PHYSICAL POSITION OF CURSOR	SIGNS E.G. VALUE OF DIGITAL GAUGE	SYMBOLS E.G. PATTERN OF INFORMATION STANDING FOR PHYSICAL CAUSALITY
		COMPARISON	2	SIGNAL VS TEMPLATE	SIGN VS ENTRY IN LIST	SYMBOL VS ITEMS IN MENTAL PICTURE OF STEPS TO GOAL
		DETECTION OF DISCREPENCY	3	ABSOLUTE MAGNITUDE OF RESULT OF COMPARISON	ABSOLUTE MAGNITUDE OF RESULT OF MULTIVARIATE COMPARISONS	
	DIAGNOSIS	FORMULATION OF EXPLANATORY HYPOTHESIS	4	X	LIST OF OPTIONS	PROBLEM SOLVING HYPOTHESIS GENERATION
		SEEK INFORMATION AND TEST HYPOTHESIS	5	X	SELECT SIGNS OR SIGNALS FOR TEST FROM SUGGESTIONS IN LIST	SELECT SYMBOLS, SIGNS, OR SIGNALS FOR TEST FROM HYPOTHESIS
	CURE	CHOOSE ACTION (COMPENSATION, CORRECTION)	6	NO CHOICE: REFLEX OR HABIT	SELECT ACTION	MODEL SELECTION PROCESS
		ACTION	7	MOVEMENT OR CLASSIFICATION	X	X

Figure 1. Classification table describing human diagnostic behaviour

the human as an information processor. It is almost always experienced by the operator as inducing high mental workload.

Note that neither the rule-based nor the knowledge-based behaviour results directly in action. Rather they modulate the choice of skills available to the operator, or (in the case of knowledge-based diagnosis) may modulate the choice of rules to follow both in diagnosis and action generation.

As we move from left to right across the table, several generalisations probably hold. Usually diagnosis takes longer. The system is likely to be more complex. The mental workload will be greater. The result of the diagnosis may be more like a reason than a cause. The principles of knowledge used will be more general. And probably system automation, both in control and recovery from abnormal states, will be increasingly required if the capacity of the human operator is not to be exceeded. Memory load will usually increase leading to errors.

We believe that one important requirement for successful diagnosis is for the operator to know which is the appropriate column for him to be in. Moreover, in a complex system, it may be necessary for him to move from one column to another at different stages of the diagnosis. If one considers the problem of aiding - computer, procedural, or human - the type of information data base and the type of display required will be quite different in the different columns. Both within a task and between tasks there is an interaction between the nature of the task and the cells in Figure 1. It is their joint properties which determine the display, information and aiding requirements. Moreover, these in turn may be further modulated by the costs of outcomes and the probability of abnormal states. If there is a frequent fault for which no procedure is known to lead to recovery, and which is rather subtle in its effect, it might seem to require a knowledge-based approach to diagnosis. But if it leads to a costly failure, it may be better to treat even a slight evidence for its existence as a reason to "scram" the system using a skill-based response.

The scheme in Figure 1 is therefore best regarded as a heuristic aid to thinking about the role of the human operator in abnormality diagnosis, in the light of some aspects of our knowledge of how the human processes information. As an exercise, we suggest that the reader consider system failures known to him. Where does each event fit in the table, and what does the table suggest he should do to improve the efficiency of the operator in diagnosing the cause of the trouble?

The primary objective of the Workshop on Models of System Structure was to discuss the "why's and how's" of modelling system structures; to review and compare significant methodologies; to discuss areas of application; and to develop the beginning of a taxonomy to aid in future modelling activities. Since the discussion was open and with few constraints, a variety of views were forthcoming without necessarily leading ultimately to a unanimity of opinion. Points discussed included:

- Definitions of terms.
- Purposes of modelling.
- Classes of systems.
- Classes of models.

In addition, the area of operator needs in the way of appropriate models of the system or process - especially in connection with control or diagnosis - was covered and thus, in a way, corroborated the acceptance by designers and others of the concept of mental models as a basic ingredient in the human "set of tools" for coping with the world about him/her[2].

Purpose of Models

Models of system structure are useful for three reasons:

(1) As aids in the analysis and understanding of systems;
(2) As aids in the synthesis and design of systems; and
(3) As on-line aids in the real-time operation of systems.

However, models which serve one purpose well may serve others ill. It has been said that a model should not be attempted until all of the questions to be asked of it have been identified. Models are only partly designed beforehand - then tend to "grow" in directions not necessarily relevant to the questions to be asked of them, due to programming constraints or opportunities, to the analyst's understanding of the system to be modelled (this may result in unnecessary representational detail), or to changes in stated purpose. Models for one purpose are often adapted for other uses, and while the class of questions which can be answered is thus increased, the number within a class may be reduced.

Classes of Models

Several examples were discussed. The main point seemed to be that the type of model depends mainly on the types of questions the model will be used to answer - however, no detailed classification of this relationshop was attempted.

---

2 See also the introduction to this volume.

Classes of Systems to be Modelled

Dimensions for classification include:

Structural complexity: Number of nodes, loops and variables; interdependency or interconnectivity; etc.

Functional complexity: Number of processes at a node; loop time constants; order of systems and subsystems.

Man-machine interaction: Level (operator, supervisor, manager, executive). Type: (procedure-following, tracking, monitoring, detecting, diagnosing, correcting, probing, anticipating, etc.).

Example

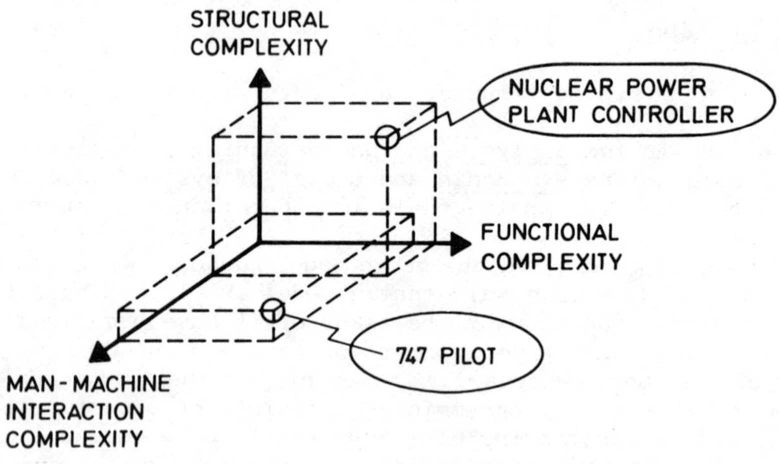

The discussion also indicated areas of concern about the applicability and limitations of modelling. To illustrate these, a short list of some of the key items is given here:

- How can modelling help to integrate people into systems?
- How can modelling help to integrate and represent the technical, physical, psychological, and verbal (e.g. protocol) aspects of system operation?
- What are the major problems and pitfalls in model validation?

- What models exist? How do they compare? What questions can be asked of them? How can operators, designers and others benefit from them?

- What is the most efficient combination or progression of models from verbal and/or graphic through mathematical and/or computer models to large realistic look-alike man-in-the-loop simulations?

- How can modelling improve the safety and availability of systems?

- How can models be used as aids to fault diagnosis; to design of man-machine systems (e.g. as an aid to function allocation between man and machine?

- Are there new improved modelling techniques? At what level of system or sub-system representation are they most appropriate? For example, can artificial intelligence methods such as production systems be used to model human cognitive activities in systems? What about graph theory for structural analysis?

- Can some of the other models .... heard about (e.g. Rasmussen, Rouse, Wohl)[3] be extended to other areas (e.g. software failure diagnosis)?

- What about practical modelling issues such as computer size constraints, run time requirements, measurability of real system parameters to check or calibrate model, etc.?

Artificial intelligence models can help understand what is going on in a system and why. But the expert knowledge base required (e.g. as in Stanford University Artificial Intelligence group's medical diagnosis program) cannot be developed for rare event diagnosis situations - there are no Three Mile Island diagnostic experts!

Operator Needs[4]

Operators, supervisors and controllers of complex systems require a model or several models of the system to help control and diagnose system operation, status and failure. However, it is

---

[3] See contributions by these authors in this volume.

[4] See also the previous workshop summary.

necessary to "match" the current or instantaneous representation of the system to the operator's current or instantaneous level of problem understanding in any given phase of diagnostic activity. For example, one could consider characterizing the level of system representation between the extremes of "aggregated broadscope" and "disaggregated narrowscope". In addition, the operations of truncation (eliminating processes from a functional representation) and/or simplification (simplifying the representation of a process) can serve to reduce the representational problem. From a control/display standpoint, this probably can be well supported by new computer graphics techniques and multilayer or "soft" control panels.

Two general comments were made:

Nuclear power plant design, control room design, and operator selection and training are done from completely different points of view, with the kind of result that the control room is designed primarily to support routine maintenance activities rather than normal and abnormal operations and activities. The total system context must be considered at all times.

There will always be major problems in acceptance of automation by highly-trained operational personnel. Changes must therefore be evolutionary in nature and backward-compatible with what they are already familiar with. Use of plant models for assisting operational decision-making can help speed this process.

The subject of training is fraught with difficulties - essentially because of the disparencies between our understanding of fundamental issues having to do with training, transfer and retention in connection with complex technical systems and the demands from industry, regulatory agencies and others for quick answers to urgent demands for "better training" as the cure for "inadequate" operator performance - especially in coping with abnormal situations. As an example of current trends, we see these days indications that full-scale control room simulators will be required in all American nuclear power stations.

While one can argue with this approach as representing an optimally balanced way to proceed, it is true that simulators offer considerable advantages and indeed, in some cases, may be a necessary component in a training program. Sometimes, however, the purposes of simulators seem to be confused since, in principle, they can be used for training, licensing, performance evaluation, demonstrations, as a design aid, etc. This Workshop on Simulation for Training attempted to assess the situation and, in particular, consider the areas of fidelity as well as simulation for fault diagnostic training (see also the training session summary).

As a general comment, the workshop agreed that to classical arguments for needing simulation, namely costs and hazards of practice with real equipment and plant, should be added the fact that simulation becomes necessary when learning is difficult or impossible with operational equipment. The present state of the art indicates a shift to simulation which changes or adds to features of the task to enable learning. This is distinct from the usual arguments for fidelity of simulation which are more appropriate for performance assessment.

Despite heroic efforts by simulator system salesmen to the contrary, fidelity is probably not best defined in terms of physical features of the operational equipment or plant. It is also doubtful whether fidelity is best approached from the logic or mathematics of the equipment or plant. In electronics, the nature of a signal at a test point may not be predictable because of changes with usage of both test gear and the equipment itself. In process plant "normal" running conditions drift over time. For the simulation of fault diagnosis in both technologies, a restricted matrix or look-up table is a more feasible proposition. At the other extreme, perceptual motor tracking, where the case is perhaps strongest for accurate simulation of the control loop, feasibility is questionable, e.g. manual control of the steam reboiler of a distillation column in open plant in all weather. Furthermore, there are grounds for predicting negative transfer when the simulation control function is very similar but not identical.

Expressed in another way, the concept of operator behaviour being categorizable as skill, rule or knowledge-based can also have significance for training and therefore for simulation fidelity - both process-wise and replica-wise. In addition to the more cognitive aspects of the operator's response to task demands, the workshop also considered the affective aspects. These were summarized as follows:

- Affective, the most serious of which is most probably stress.
- Non-contingent - in a hazardous environment fear for personal safety and safety of others.
- Contingent, e.g. fear of increasing personal and community hazards by mistakes; or loss of esteem.

At the present time the faithful simulation of affective task demands seems remote except perhaps for loss of esteem.

The foregoing fidelity arguments apply only insofar as they facilitate training. For example, a simulation is currently in use which provides more comprehensive representations of a communication network than in the operating station. This sort of help could be subsequently "faded out" until fidelity, in the task demand sense, is reached. Indeed a need was identified for more research into firstly, the provision of more information, or quite different information in a training simulation than exists in reality; and secondly the problem of transition from practice with a training simulation to testing with a simulation faithful to task demands.

With regard to simulation for fault diagnosis training, the different stages of the sequence of diagnostic activity (or activity closely associated with diagnosis) almost certainly have different simulation requirements. Again research is badly needed into the optimum simulation requirements for most of these. We identified five such stages which are common to electronic equipment and industrial process plant:

1. Scanning
2. Injecting signals
3. Symptom interpretation
4. Search (especially information-cost trade offs)
5. Component replacement/adjustment/repair

The sequence 1-4 or 1-5 can be iterative. Process plant probably imposes more constraints on iterations and offers fewer opportunities to inject signals - which would be reflected in the simulation concerned.

To the extent that diagnosis entails these five different sorts of activity in any process plant or any electronic equipment, the case exists for general purpose simulation, perhaps with the intention of teaching generalisable skills. Work reported in this volume supports this view in the case of stages 3 and 4. At the present time, process plants present many plant-specific features. Simulation of these would be a necessary supplement to any general training. The same can be said about electronic equipment although there is probably more variety of classes of electronic equipment, with considerable similarity within classes.

The relative emphasis given to simulation of general or specific features would probably therefore differ between process plant and electronic equipment training schemes.

INDEXES

PARTICIPANTS

Andersson, H.	Ergonområd AB Box 10032 S-650 10 Karlstad 10 Sweden
Aune, A.B.	SINTEF Div. of Automatic Control N-7034 Trondheim NTH Norway
Bainbridge, L.	University College London Department of Psychology London WC1E 6BT England
Barbet J.F.	Electricité de France Direction Études et Recherches 1 Ave du Général de Gaulle F-92141 Clamart France
Baron, S.	Bolt Beranek and Newman, Inc. 10 Moulton St. Cambridge, MA 03138 U.S.A.
Bergman, H.	Rijksuniversitet Psychologisch Lab. Varkenmarkt 2 Utrecht Holland
Bohr, E.	Institut für Unfallforschung TUV Rheinland e.V. Postbox 101750 D-5000 Cologne 1 West Germany

Bond, N.	California State University - Sacramento Dept. of Psychology 6000 "J" Street Sacramento, CA 95819 U.S.A.
Brady, L.R.	IBM Bodie Hill Road Owego, N.Y. 13827 U.S.A.
Brehmer, B.	University of Uppsala Institute of Psychology P.O. Box 207 S-75104 Uppsala Sweden
Brooke, J.B.	UWIST Dept. of Applied Psychology Llwyn-y-Grant Penylan, Cardiff CF3 7UX Wales
Curry, R.E.	NASA-Ames Research Center Mailstop 239-3 Moffet Field, CA 94035 U.S.A.
Delaney, J.R.	The MITRE Corporation P.O. Box 208 Mail Stop A412 Bedford, MA 01730 U.S.A.
Dellner, W.J.	Bell Laboratories 6 Corporate Place Piscataway, N.J. 08854 U.S.A.
Duncan, K.D.	UWIST Dept. of Applied Psychology Llwyn-y-Grant Penylan, Cardiff CF3 7UX Wales
Eekhout van, J.M.	Royal Dutch Navy Eksterstraat 115 1742 ER Schagen The Netherlands

Embrey, D.E. — University of Aston in Birmingham
Ergonomics Development Unit
Gosta Green
Birmingham B4 7ET
England

Ephrath, A.R. — University of Connecticut
Dept. of E.E. & C.S.
Box U-157
Storrs, Conn. 06268
U.S.A.

Felkel, L. — Gesellschaft für Reaktorsicherheit mbH
Forschungsgelände
D-8046 Garching
West Germany

Freedy, A. — Perceptronics Inc.
6271 Variel Ave.
Woodland Hills, CA 91367
U.S.A.

Gaddes, W.M. — IBM
Bodie Hill Road
Owego, N.Y. 13827
U.S.A.

Goodstein, L.P. — Risø National Laboratory
Electronics Department
DK-4000 Roskilde
Denmark

Hanes, L.F. — Westinghouse R&D Center
1310 Beulah Road
Pittsburgh, PA 15235
U.S.A.

Hol, J.O. — OECD Halden Reactor Project
Postbox 173
N-1751 Halden
Norway

Hollnagel, E. — Risø National Laboratory
Electronics Department
DK-4000 Roskilde
Denmark

Hunt, R.M.
University of Illinois
Coordinated Science Laboratory
Urbana, IL 61801
U.S.A.

Johannsen, G.
Forschungsinstitut für Anthropotechnik
Königstrasse 2
D-5307 Wachtberg-Werthhoven
West Germany

Jørgensen, F.V.
Carinavænget 5
DK-3460 Birkerød
Denmark

Kessel, C.
Hanasher 3
Nve Rom
Ramat Hasharon
Israel

Keyser de, V.
Université Libre de Bruxelles
84, Rue du Tabellion
B-1050 Bruxelles
Belgium

Laleuf, M.
Electricité de France
Direction Études et Recherches
1 Ave du Général de Gaulle
F-92 141 Clamart
France

Lees, F.P.
University of Technology
Dept. of Chemical Engineering
Loughborough Leicestershire LE11 3TU
England

Lihou, D.A.
Aston University
Chemical Engineering Department
Birmingham B4 7ET
England

Lind, M.
Risø National Laboratory
Electronics Department
DK-4000 Roskilde
Denmark

Lindskog, A.
Swedish Nuclear Safety Board
Box 5864
S-10248 Stockholm
Sweden

Mancini, G.	Euratom Joint Research Centre Ispra Establishment C.P. N. 1 I-21020 Ispra (Varese) Italy
Marshall, E.	UWIST Dept. of Applied Psychology Llwyn-y-Grant Penylan, Cardiff CF3 7UX Wales
Mehling, O.	Nukem GmbH Industriegelände D-6450 Hanan-Wolfgang West Germany
Michaels, A.	University of Stirling Dept. of Psychology Stirling FK9 4LA Scotland
Misenta, A.	Euratom Joint Research Centre Ispra Establishment Systems Analysis I-21020 Ispra (Varese) Italy
Montmayeul, R.	Electricité de France Direction des Études et Recherches 6, Quai Watier B.P. 49 F-78400 Chatou France
Moray, N.	University of Stirling Dept. of Psychology Stirling FK9 4LA Scotland
Nawrocki, L.	Army Research Institute 5001 Eisenhower Ave. Alexandria, VA 22233 U.S.A.
Parks, D.L.	Boeing Commercial Airplane Co. M.S. 47-08 P.O. Box 3707 Seattle, WA 98124 U.S.A.

Patrick, J. — University of Aston in Birmingham
Dept. of Applied Psychology
Gosta Green
Birmingham B4 7ET
England

Pau, L.F. — Ambassade de France
2129 Wyoming Ave. N.W.
Washington, D.C. 20008
U.S.A.

Petterson, E. — Statens Kärnkraftinspektion
Box 27106
S-10252 Stockholm 43
Sweden

Prendal, B. — Scandinavian Airlines System
Fack
S-16187 Bromma
Sweden

Rasmussen, J. — Risø National Laboratory
Electronics Department
DK-4000 Roskilde
Denmark

Reichert, G. — Gesellschaft für Reaktorsicherheit mbH
Bereich Systeme
Forschungsgelände
D-8046 Garching
West Germany

Reinecke, M. — Flugmedizinisches Institut d. Lw.
Abteilung IV
Ergonomie
D-8072 Manching
West Germany

Rouse, S. — University of Illinois at Urbana-Champaign
Coordinated Science Laboratory
Urbana, Illinois 61801
U.S.A.

Rouse, W.B. — University of Illinois at Urbana-Champaign
Coordinated Science Laboratory
Urbana, Illinois 61801
U.S.A.

Schumacher, W.	Frauenhofer Institut für Informations- und Datenverarbeitung Sebastian Kneipp Strasse 12-14 D-7500 Karlsruhe West Germany
Shepherd, A.	2, Kendal Road Holcombe Brook Bury, Lancastershire England
Sheridan, T.B.	Massachusetts Institute of Technology Dept. of Mechanical Engineering Room 1-110 Cambridge, Mass. 02139 U.S.A.
Sjölin, P.G.	Studsvik Energiteknik AB Fack S-61101 Nyköping Sweden
Skans, S.	LUTAB Snörmakarvägen 29 S-161 47 Bromma Sweden
Stammers, R.B.	University of Aston in Birmingham Dept. of Applied Psychology Birmingham B4 7ET England
Stassen, H.G.	Delft University of Technology Lab. for Measurement and Control Dept. of Mechanical Engineering Mekelweg 2 NL-2628 CD Delft The Netherlands
Stein, W.	Forschungsinstitut für Anthropotechnik Königstrasse 2 D-5307 Wachtberg-Werthhoven West Germany
Syrbe, M.	Frauenhofer-Institut für Informations- und Datenverarbeitung D-7500 Karlsruhe 1 Sebastian-Kneipp-Str. 12-14 West Germany

Svanes, T.
The MITRE Corp.
P.O. Box 208
Mail Stop A412
Bedford, MA 01730
U.S.A.

Talmon, H.
Delft University of Technology
Lab. for Measurement and Control
Dept. of Mechanical Engineering
Mekelweg 2
NL-2628 CD Delft
The Netherlands

Taylor, J.R.
Risø National Laboratory
Electronics Department
DK-4000 Roskilde
Denmark

Thermöhlen, J.
Militærpsykologisk Tjeneste
Christianshavns Voldgade 8
DK-1424 Copenhagen K
Denmark

Thijs, W.
Delft University of Technology
Lab. for Measurement and Control
Dept. of Mechanical Engineering
Mekelweg 2
NL-2628 CD Delft
The Netherlands

Thompson, D.
Stanford University
Dept. of Industrial Engineering
Stanford, CA 94305
U.S.A.

Towne, D.
University of Southern California
Behavioral Technology Laboratories
1845 South Elena Ave.
Redondo Beach, CA 90277
U.S.A.

Turksen, I.B.
University of Toronto
Faculty of Applied Science and Engineering
Toronto, M5S 1A4
Canada

Vakali, M.
University of Thessaloniki
Dept. of Psychology
Thessaloniki
Greece

Vees, C. TU Berlin
Inst. für Luft- und Raumfahrt
Sekr. F3
Marchstrasse 14
D-1000 Berlin 10
West Germany

Vulker, J.H.J.M. Royal / Shell Laboratory
Badhuisweg 3
NL-1031 CM Amsterdam
The Netherlands

Wahlström, B. VTT/SÄH
Vuorimiehentie 5
SF-02150 Espoo 15
Finland

Wewerinke, P. National Aerospace Laboratory
Anthony Fokkerweg 2
1059 CM Amsterdam
The Netherlands

Whitehouse, H.B. Insurance Technical Bureau
Albany House
Petty France
London
England

Whitten, W.B. Office of Naval Research
Dept. of the Navy
Arlington, VA 22217
U.S.A.

Williams, E.N.G. Research Branch, HQRFSC.
RAF Upwood
Huntingdon, Cambridgeshire PE17 1PJ
England

Winther, A. Danmarks Geologiske Undersøgelse
Kulbrinteafdelingen
Thoravej 31
DK-2400 Copenhagen NV
Denmark

Wohl, J. The MITRE Corp.
P.O. Box 208
Bedford, Mass. 01730
U.S.A.

Wreathall, J.
CEGB - Health and Safety Dept.
Courtenay House
18 Warwick Lane
London EC4P 4EB
England

Zwaga, J.
Univ. Utrecht
Psychol. Lab. R.U.
2 Varken Markt
Utrecht
The Netherlands

AUTHOR INDEX

SUBJECT INDEX[x)]

x) To allow easy referencing of the papers in this volume, subject terms are followed by the names of the authors of the papers where the subject is discussed.